AAPS Advances in the Pharmaceutical Sciences Series

Volume 20

The AAPS Advances in the Pharmaceutical Sciences Series, published in partnership with the American Association of Pharmaceutical Scientists, is designed to deliver well written volumes authored by opinion leaders and authorities from around the globe, addressing innovations in drug research and development, and best practice for scientists and industry professionals in the pharma and biotech industries.

For more details and to see a list of titles in the Series please visit http://www.springer.com/series/8825

Daan J.A. Crommelin • Jon S. B. de Vlieger
Editors

Non-Biological Complex Drugs

The Science and the Regulatory Landscape

Springer

Editors
Daan J.A. Crommelin
Utrecht University
Utrecht
The Netherlands

Jon S. B. de Vlieger
Top Institute Pharma
NBCD Working Group
Leiden
The Netherlands

ISSN 2210-7371 ISSN 2210-738X (Electronic)
AAPS Advances in the Pharmaceutical Sciences Series
ISBN 978-3-319-16240-9 ISBN 978-3-319-16241-6 (eBook)
DOI 10.1007/978-3-319-16241-6

Library of Congress Control Number: 2015940779

Springer Cham Heidelberg New York Dordrecht London

Printed on acid-free paper

Springer is part of Springer Science+Business Media (www.springer.com)

Contents

Introduction: Defining the Position of Non-Biological Complex Drugs 1
Daan J. A. Crommelin, Jon S. B. de Vlieger and Stefan Mühlebach

Part I Non Biological Complex Drugs

Polymeric Micelles 11
Ethlinn V.B. van Gaal and Daan J.A. Crommelin

Liposomes: The Science and the Regulatory Landscape 77
Daan J.A. Crommelin, Josbert M. Metselaar and Gert Storm

Glatiramoids 107
Vera Weinstein, Rivka Schwartz, Iris Grossman, Benjamin Zeskind and J. Michael Nicholas

Iron Carbohydrate Complexes: Characteristics and Regulatory Challenges 149
Stefan Mühlebach and Beat Flühmann

Drug Nanocrystals 171
Gerrit Borchard

Part II Characterization of NBCDs; Analytical Tools to Consider

Analytical Methods for Determining the Size (Distribution) in Parenteral Dispersions 193
David F. Driscoll and David F. Nicoli

NBCD Pharmacokinetics and Bioanalytical Methods to Measure Drug Release 261
Vishakha V. Ambardekar and Stephan T. Stern

Part III Closely related Complex Drugs

Low Molecular Weight Heparins, Biological Drugs close to Non-Biological Complex Drugs 291
Isabel Rodrigo, Sofía Caruncho, Concepción Alonso, Antonio Gómez-Outes and Barbara Mulloy

Nanoparticle Albumin-Bound Anticancer Agents 335
Neil Desai

Part IV Regulatory landscape and outlook

The EU Regulatory Landscape of Non-Biological Complex Drugs 357
Ruben Pita

Epilogue: What Did We Learn? What Can We Expect in the Future? Concluding Remarks and Outstanding Issues 381
Daan J. A. Crommelin and Jon S. B. de Vlieger

Index 389

Contributors

Concepción Alonso Biological Products and Biotechnology Division, Medicines for Human Use Department, Spanish Agency for Medicines and Medical Devices (AEMPS), Madrid, Spain

Vishakha V. Ambardekar Nanotechnology Characterization Laboratory, Cancer Research Technology Program, Leidos Biomedical Research, Inc., Frederick National Laboratory for Cancer Research, Frederick, MD, USA

Gerrit Borchard School of Pharmaceutical Sciences Geneva-Lausanne, University of Geneva, University of Lausanne, Geneva, Switzerland

Sofía Caruncho Biological Products and Biotechnology Division, Medicines for Human Use Department, Spanish Agency for Medicines and Medical Devices (AEMPS), Madrid, Spain

Daan J.A. Crommelin Department Pharmaceutical Sciences, Utrecht University, Utrecht, The Netherlands

Department Pharmaceutics, Utrecht Institute for Pharmaceutical Sciences, UIPS, Utrecht University, Utrecht, The Netherlands

Department of Pharmaceutics, Utrecht University, Utrecht, The Netherlands

Jon S. B. de Vlieger NBCD Working Group, Top Institute Pharma, Leiden, The Netherlands

Neil Desai Strategic Platforms, Celgene Corporation (Abraxis BioScience), Los Angeles, CA, USA

David F. Driscoll Stable Solutions LLC, Easton, MA, USA

UMASS Medical School, Worcester, MA, USA

Beat Flühmann Department of Global Regulatory Affairs, Vifor Pharma Ltd., Glattbrugg, Switzerland

Antonio Gómez-Outes Pharmacology and Clinical Evaluation Division, Medicines for Human Use Department, Spanish Agency for Medicines and Medical Devices (AEMPS), Madrid, Spain

Iris Grossman Teva Pharmaceutical Industries, Ltd., Petach Tikva, Israel

Stefan Mühlebach Department of Global Regulatory Affairs, Vifor Pharma Ltd., Glattbrugg, Switzerland

Department of Pharmaceutical Sciences Division of Clinical Pharmacy & Epidemiology, University of Basel, Basel, Switzerland

Josbert M. Metselaar Enceladus Pharmaceuticals BV, Naarden, The Netherlands

Barbara Mulloy Institute of Pharmaceutical Science, King's College London, London, UK

J. Michael Nicholas Teva Pharmaceuticals, Kansas City, MO, USA

David F. Nicoli Stable Solutions LLC, Goleta, CA, USA

Particle Sizing Systems LLC, Port Richey, FL, USA

Ruben Pita European Medicine Agency, EMA, London, UK

Isabel Rodrigo Biological Products and Biotechnology Division, Medicines for Human Use Department, Spanish Agency for Medicines and Medical Devices (AEMPS), Madrid, Spain

Rivka Schwartz Teva Pharmaceutical Industries, Ltd., Petach Tikva, Israel

Stephan T. Stern Nanotechnology Characterization Laboratory, Cancer Research Technology Program, Leidos Biomedical Research, Inc., Frederick National Laboratory for Cancer Research, Frederick, MD, USA

Gert Storm Department Pharmaceutics, Utrecht Institute for Pharmaceutical Sciences, UIPS, Utrecht University, Utrecht, The Netherlands

Ethlinn V.B. van Gaal OctoPlus B.V., Dr. Reddy's Laboratories Ltd., Leiden, The Netherlands

Vera Weinstein Teva Pharmaceutical Industries, Ltd., Petach Tikva, Israel

Benjamin Zeskind Immuneering Corporation, Cambridge, USA

About the Editors

Prof. Daan Crommelin is emeritus-professor at the Department of Pharmaceutics at Utrecht University. Until December 2011 he was scientific director of the Dutch Top Institute Pharma in Leiden. He is adjunct professor at the Department of Pharmaceutics and Pharmaceutical Chemistry at the University of Utah. Crommelin is co-founder of OctoPlus, a Leiden based company specialized in the development of pharmaceutical (mainly protein based) product formulations and advanced drug delivery systems. He published extensively and is on the editorial board of 10+ peer reviewed journals in the pharmaceutical sciences.

Dr. Jon S. B. de Vlieger obtained his doctoral degree in bio analytical chemistry from the VU University in Amsterdam. He is currently involved in the strategy department of the Dutch Top Institute Pharma, a not-for-profit organization catalyzing the development of medicines by establishing partnerships and actively managing research programs. For TI Pharma he coordinates several international public private partnerships, including the Non Biological Complex Drugs Working Group, an international network of scientific and clinical experts from academia, industry and regulatory bodies, with expertise in many aspects of the development and evaluation of NBCDs.

Introduction: Defining the Position of Non-Biological Complex Drugs

Daan J. A. Crommelin, Jon S. B. de Vlieger and Stefan Mühlebach

Contents

Introduction 2
Small, Low Molecular Weight Drugs Versus Biologicals and Non-Biological Complex Drugs 3
Small, Low Molecular Weight Drugs 3
Biologicals/Biological Medicinal Products 3
Non-Biological Complex Drugs (NBCDs) 5
  Definition of Non-Biological Complex Drugs: NBCDs 5
  The Regulatory Position of NBCDs 6
Content of the Book 6
Aim and Target Group 7
References 7

Abstract In the first chapter of this book the concept of non-biological complex drugs (NBCDs) is introduced. These are complex drug products but don't fall in the category of 'biologicals'. NBCDs were earlier defined as: medicinal products, not being a biological medicine, where the active substance is not a homo-molecular structure, but consists of different (closely related and often nanoparticulate) structures that

D. J. A. Crommelin (✉)
Department of Pharmaceutics, Utrecht University, Utrecht, The Netherlands
e-mail: d.j.a.crommelin@uu.nl

J. S. B. de Vlieger
NBCD Working Group, Top Institute Pharma, Leiden, The Netherlands

S. Mühlebach
Department of Global Regulatory Affairs, Vifor Pharma Ltd., Glattbrugg, Switzerland

Department of Pharmaceutical Sciences Division of Clinical Pharmacy & Epidemiology, University of Basel, Basel, Switzerland

D. J. A. Crommelin, J. S. B. de Vlieger (eds.), *Non-Biological Complex Drugs,* AAPS Advances in the Pharmaceutical Sciences Series 20, DOI 10.1007/978-3-319-16241-6_1

cannot be isolated and fully quantitated, characterized and/or described by physicochemical analytical means. The composition, quality and in vivo performance of NBCD are highly dependent on the manufacturing processes of the active ingredient as well as (in most cases) the formulation (Crommelin et al. 2014). Examples of NBCDs are iron-carbohydrate complexes, glatiramoids, liposomes, polymeric micelles, swelling polymers and many (other) nanomedicines. A number of these (and related) NBCD-families are dealt with in 8 chapters of this book. As complex drug products request sophisticated analytical means for characterization of their structure and in vivo performance, ample attention is paid to the analytical challenges in two separate chapters. Finally, a perspective regarding NBCDs is given from the regulatory side.

Introduction

In ancient times medicines consisted of complex mixtures, often from natural sources such as plants or animals. At the end of the nineteenth century the advent of modern organic chemical synthesis, and improved separation and analytical techniques provided the tools to produce pure bioactive substances. Modern medicine was born. Visionary scientists such as Emil Fisher and Paul Ehrlich developed the leading paradigms of modern medicine. Well characterized, pure, small molecules preferably synthesized through well designed and controlled organic synthetic chemistry routes became the leading group in our modern arsenal of medicines. Starting with drugs such as acetylsalicylic acid and salvarsan, followed by sulphonamides and penicillins, a myriad of small molecule bioactives was developed and used e.g. in infectious diseases, inflammatory diseases, cancer, pulmonary and gastro-intestinal diseases, mental diseases, or for pain killing and anaesthesia.

Over time we learned that the use of these pharmacologically active compounds could lead to serious side effect and even death. Those experiences were major drivers in the formation of regulatory bodies such as the Food and Drug Administration (FDA)) in the USA and similar institutions in other countries of the world. The last major development was the founding in 1995 of the European Medicines Agency (EMA)) located in London. The EMA 'is responsible for the scientific evaluation of medicines developed by pharmaceutical companies for use in the European Union'. Apart from the national agencies the World Health Organization (WHO) is also active in providing guidance documents and recommendations for national competent authorities.

Traditionally, the use of simple 'one-drug-is-best' treatment schedules has been discarded for a number of therapies where resistance formation may occur, such as in infectious diseases and cancer. But, recently, the idea that one drug, the more selective the better, would be the preferred approach to treat a disease is challenged by system pharmacologists, who propose to tackle complicated diseases such as 'metabolic syndrome' through different pathways. That means that combination therapy (prescription of different drugs to treat a disease state) is gaining popularity. But even in combination therapies the physician (and patient) will expect the use of pure, well characterized bioactives.

Small, Low Molecular Weight Drugs Versus Biologicals and Non-Biological Complex Drugs

The above may be true as the major paradigm in modern therapy, there are exceptions to these rules and these exceptions are growing in importance. Apart from the family of '*small, low molecular weight molecules*' the family of '*biologicals*' (bioactives derived from living material) is growing fast, in particular the group of monoclonal antibodies. These are considered as *biological complex* drugs. There are also complex drugs that aren't derived from living material. These drugs are the major focus of this book. They are called '*non-biological complex drugs*' (NBCDs). They have a number of characteristics that set them apart from small molecules and they don't fall under the definition of biologicals either.

These three categories will be briefly discussed in the following paragraphs.

Small, Low Molecular Weight Drugs

Small, low molecular weight drugs receive market authorization through a regulatory framework that evolved and is evaluated over many years by regulatory scientists. Generally speaking, regulatory bodies and the pharmaceutical industry are familiar with the process of assessing quality, efficacy and safety of innovative medicines and their generic, follow-on versions. But, there is still room for debate. The AAPS/Springer 'Advances in the Pharmaceutical Sciences' book series pays attention to these discussions (recent highlights: FDA Bioequivalence Standards, 2014; Global Approach in Safety Testing: ICH Guidelines Explained, 2013).

Biologicals/Biological Medicinal Products

Biological medicinal products are a different category. There are many definitions for biologicals in the literature. We use the one published last year in the AAPS J (2014). A biological product is a product derived from living material (such as cells or tissues) used to treat or cure disease. Biological products include a wide range of products such as vaccines, blood and blood components, allergenics, somatic cells, gene therapeutics, tissues, and recombinant therapeutic proteins. Biological products can be composed of sugars, proteins, or nucleic acids or complex combinations

Table 1 Characteristics of small molecule drugs compared to biologicals . (Reprinted with permission from GaBI Online—Generics and Biosimilars Initiative http://www.gabionline.net/Biosimilars/Research/Small-molecule-versus-biological-drugs, based on Declerck (2012) and Schellekens et al. (2010))

	Small molecule drugs	Biological drugs
Size	Small (single molecule)	Large (mixture of related molecules)
	Low molecular weight	High molecular weight
Structure	Simple, well defined, independent of manufacturing process	Complex (heterogeneous), defined by the exact manufacturing process
Modification	Well defined	Many options
Manufacturing	Produced by chemical synthesis	Produced in living cell culture
	Predictable chemical process	Difficult to control from starting material to final API
	Identical copy can be made	Impossible to ensure identical copy
Characterisation	Easy to characterise completely	Cannot be characterised completely the molecular composition and heterogenicity
Stability	Stable	Unstable, sensitive to external conditions
Immunogenicity	Mostly non-immunogenic	Immunogenic

API active pharmaceutical ingredient

of these substances, or may be living entities such as cells and tissues. Biological products are isolated from a variety of natural sources—human, animal, or microorganism based—and may be produced by biotechnology methods (Crommelin et al. 2014).

These biological products encompass a large array of different molecules and complex mixtures. Biotech derived proteins range in their molecular weight from 4 kDa up to 250+kDa, all showing complex secondary and tertiary structures. Blood and blood derived products are complex mixtures as well. A general feature is that these biological products contain (many) components that are difficult to fully characterize by physico-chemical means. Admittedly, this last statement needs some nuancing. For small proteins full physico-chemical characterization is now within reach of our modern technological analytical tools. But, for the larger and more complex mixtures, quality depends on strict source control/manufacturing control and robust downstream processes/filling and finishing. Table 1 lists communalities within the group of small molecules and of biologicals and the differences between those two families of medicines.

Non-Biological Complex Drugs (NBCDs)

Then, there is the category of complex medicines that don't fall under the above definition of biologicals: The category of *non-biological* complex drugs (NBCDs). Literature on this NBCD family is still rather limited and is mainly generated by the NBCD working group, an initiative hosted by the Dutch Top Institute Pharma, TI Pharma, in Leiden, The Netherlands). On their website (www.tipharma.com/NBCD) this NBCD working group presents itself as follows: 'The NBCD working group has been created to discuss appropriate and aligned science-based approval and post-approval standards to ensure patient safety and benefit with Non Biological Complex Drugs (NBCDs). The working group engages in activities to disseminate the corresponding scientific evidence to authorities, experts, health care providers and other relevant bodies with responsibility for treatment with NBCDs. Moreover the group is involved in scientific education and training on the above mentioned topics to relevant stakeholders'. Key publications on NBCDs are by Schellekens et al. 2014; Holloway et al. 2012; Schneider et al. 2012, and Schellekens et al. 2010. Some of these papers are meeting reports or write-ups of discussions with many stakeholders during international conferences over the last few years. The steering committee of this NBCD platform has initiated the publication of this book; the editors of this book are members of the steering committee.

Definition of Non-Biological Complex Drugs: NBCDs

The working group uses the following definition for NBCDs: they are medicinal products, not being a biological medicine, where the active substance is not a homomolecular structure, but consists of different (closely related and often nanoparticulate) structures that cannot be isolated and fully quantitated, characterized and/or described by physico-chemical analytical means. The composition, quality and in vivo performance of NBCDs are highly dependent on the manufacturing processes of the active ingredient as well as (in most cases) the formulation (Crommelin et al. 2014).

Examples of NBCDs are iron-carbohydrate complexes, glatiramoids, liposomes, polymeric micelles, swelling polymers and many (other) nanomedicines. Most of these NBCD-families are dealt with in the first 6 chapters of this book.

One of the major challenges when dealing with NBCDs, be it for novel or for follow on products is the characterization process. How to assess batch to batch similarity or the effect of changes in the manufacturing process (comparability assessment)? Our analytical toolbox has been growing fast over the last decennia, bringing the identification and determination of critical properties or components closer by. As an example, full characterization of key parameters for the assessment of similarity of a follow on liposome product to the originator is closer by than for glatiramoids or iron-carbohydrate complexes. In Chaps. 7 and 8 different aspects of recent developments in assessing the physico-chemical properties are discussed. But, even when we would be able to characterize these complex products to a large extent, we are still faced with the challenge of defining the relevant and meaningful product characteristics leading to changes in therapeutic performance.

The Regulatory Position of NBCDs

For developing generic versions of innovative medicines for small molecules with a well-described molecular structure an established set of protocols is in operation in most parts of the world. WHO, FDA and EMA were major players in the discussions and framing of these protocols, including the statistical procedures and acceptance levels.

Generic versions of small drug molecules receive marketing authorization when they are considered to be Pharmaceutically Equivalent (PE) and Bioequivalent (BE) to the reference (innovator) product. This paradigm does not hold for NBCDs. Differences in chemical composition, physical (nanoparticle) characteristics, (pre) clinical efficacy and safety experience were reported when NBCDs from different sources were compared. Therefore, it seems necessary to include preclinical and clinical assessments in a NBCD follow on product approval process. Because the generic approach is not more valid, exchangeability/substitutability between originator and follow on version has to be addressed specifically.

Both FDA and EMA see the challenges of developing follow on versions of NBCDs and have issued reflection papers or guiding documents for medicinal products from the NBCD family, recognizing the special position of these products in the regulatory landscape. We propose to follow for NBCDs a regulatory approach parallel to the one that was developed for biosimilars (Fig. 1).

There are families of complex pharmaceuticals that may not fully fall under the definition of NBCDs but share many of their characteristics such as Low Molecular Weight Heparins (LMWH) and albumin based nanomedicines such as abraxane. These closely related complex drugs are also dealt with in this book as they offer interesting insights in the issues under discussion (Chaps. 9 and 10). The communalities these type of products share with the NBCDs are complexity and heterogeneity, characterization issues, and the importance of strict control over the source material/manufacturing process to obtain reproducible products.

Content of the Book

This is the first book that pays comprehensive attention to the group of NBCDs. Chapters 2–6 deal with NBCDs, those that exist and those that we can expect in the (near) future (cf. the growing group of nanomedicines). Each chapter has a section on the regulatory experience specific to that NBCD group. These 'monograph' type of chapters are followed by two chapters on analytical techniques used to characterize NBCDs, and their pharmacokinetics, and then two chapters on closely related

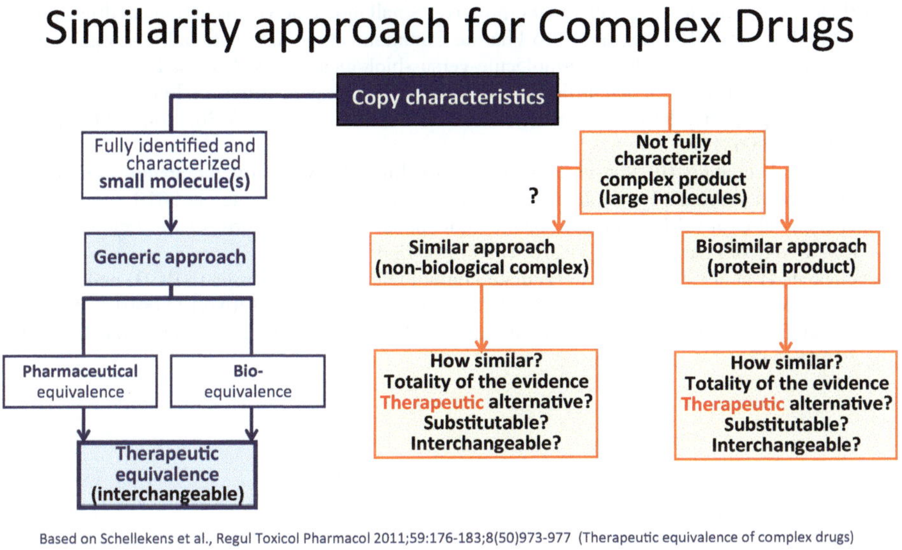

Fig. 1 Proposed similarity approach for complex drugs. (Based on Schellekens et al. 2011 (Therapeutic equivalence of complex drugs))

complex drugs, that aren't defined as NBCDs per se to discuss 'lessons learned'. Finally, the overall regulatory landscape is sketched by an author from the EMA.

Aim and Target Group

There is currently no book, nor comprehensive documentation on this topic other than some recent scientific papers. This book is meant to be used for years to come as a standard–reference work for NBCDs, and also to stimulate discussions on this topic and to further our thinking to ensure that decisions regarding the approval of complex drugs are made with the relevant scientific data on the table.

The target groups are firstly regulatory scientists in industry and regulatory bodies, secondly skilled health care professionals working in hospitals and, thirdly, those who work on the development of nanomedicines.

References

Crommelin DJA et al (2014) Different pharmaceutical products need similar terminology. AAPS J 16:11–14

Declerck PJ (2012) Biologicals and biosimilars: a review of the science and its implications. Generics Biosimilars Initiat J (GaBI Journal) 1(1):13–16. 10.5639/gabij.2012.0101.005

GaBI Online—Generics and Biosimilars Initiative. Small molecule versus biological drugs [www.gabionline.net]. Mol, Belgium: Pro Pharma Communications International. www.gabionline.net/Biosimilars/Research/Small-molecule-versus-biological-drugs. Accessed 14 Nov 2014

Holloway C et al (2012) Scientific considerations for complex drugs in light of established and emerging regulatory guidance. Ann N Y Acad Sci 1276:26–36. doi:10.1111/j.1749-6632.2012.06811.x

Schellekens H et al (2010) Non-biological complex drugs: How to show therapeutic equivalence. The Leiden proposal. Poster #R6341 FIP Pharmaceutical World Congress 2010

Schellekens H et al (2011) The therapeutic equivalence of complex drugs ('the Leiden Paper'). Regul Toxicol Pharmacol 59(1):176–183;8(50)973–977

Schellekens H et al (2014) How to regulate nonbiological complex drugs (NBCD) and their follow-on versions: points to consider. AAPS J 16(1):15–21. doi:10.1208/s12248-013-9533-z

Schneider C et al (2012) In support of the European Union biosimilar framework. Nat Biotechnol 30:745–747

Part I
Non Biological Complex Drugs

Polymeric Micelles

Ethlinn V.B. van Gaal and Daan J.A. Crommelin

Contents

Introduction, Overview Products in the Family 14
Design, Chemistry, Manufacturing and Control 20
  Design 20
  Manufacturing 28
  Control 34
Pharmacology, PK/PD 41
  Examples from the Clinic 41
Regulatory Status 55
  Quality Characterization 56
  Manufacturing 57
  Non-Clinical 58
  Clinical 59
  Generics 60
Prospects, Innovations 61
  Increased Understanding 61
  Increased Control 62
  First Products to Market 64
References 65

Abstract Polymeric micelles are nanoparticles formed upon self-assembly of amphiphilic (block co-)polymers in aqueous solutions. The resulting structure is a usually spherical nanoparticle with a hydrophobic core acting as a reservoir for poorly soluble active pharmaceutical ingredients (APIs) and a hydrophilic shell which provides colloidal stability and limits protein adsorption and opsonisation, resulting in long-circulation times. Since the physicochemical properties, and ultimately the in vivo distribution, safety and efficacy, of the final drug product are

E. V. B. van Gaal (✉)
OctoPlus B.V., Dr. Reddy's Laboratories Ltd., Leiden, The Netherlands
e-mail: ethlinnvangaal@gmail.com

D. J. A. Crommelin
Department Pharmaceutical Sciences, Utrecht University, Utrecht, The Netherlands
e-mail: d.j.a.crommelin@uu.nl

D. J. A. Crommelin, J. S. B. de Vlieger (eds.), *Non-Biological Complex Drugs*, AAPS Advances in the Pharmaceutical Sciences Series 20, DOI 10.1007/978-3-319-16241-6_2

highly dependent on the chosen polymer chemistry and manufacturing process, classification of polymeric micelles as nonbiological complex drugs is justified. This chapter provides an overview of the most important/common chemistry, manufacturing processes and control strategies used to manufacture polymeric micelles for medicinal products. Next, the pharmacology of polymeric micelles tested in the clinic is summarized and the relation between physicochemical characteristics and PK/PD as well as evaluation of choice and value of specific PK-parameters and required assay development are discussed. Regulatory aspects will be discussed based on currently available guidance of direct relevance for polymeric micelles as well as related guidance and suggestions for updates will be provided. The chapter will end with a preview of important developments and breakthroughs that can be anticipated in the (nearby) future and prospects for innovative and generic polymeric micelle drug products.

Keywords Polymeric micelles · Block copolymer · Nanoparticles · Chemistry · Manufacturing · Formulation · Characterization · CMC · Core · Shell · Solubilization · Amphiphilic · Drug-polymer conjugates · Generic · EMA · FDA · ICH · Drug product · Genexol · BIND-014 · NC-6004 · NK012 · NK105 · NK911 · PAXCEED · SP1049C

Abbreviations

DMF	Dimethylformamide
LAL	Limulus amoebocyte lysate
DMSO	Dimethyl sulfoxide
THF	Tetrahydrofuran
DMAc	Dimethylacetamide
AFM	Atomic force microscopy
API	Active pharmaceutical ingredient
ATRP	Atom-transfer radical-polymerization
AUC	Area under the curve
BSC	Best supportive care
CAC	Critical aggregation temperature
CDDP	Cis-dichlorodiammineplatinum (II)
cmc	Critical micelle concentration (NB compare CMC used for Chemistry, Manufacturing and Control)
CMT	Critical micelle temperature
CP	Cloud point
CPP	Critical process parameter
CQA	Critical quality attribute
DCM	Dichloromethane
DLS	Dynamic light scattering
DLT	Dose limiting toxicity
EMA	European medicine agency
EPR	Enhanced permeation and retention effect

EU	European union
FDA	Food and drug administration
GLP	Good laboratory practice
GMP	Good manufacturing practice
GPC	Gel permeation chromatography
HPMAm	*N*-(2-hydroxypropyl)methacrylamide
HPMAm-lactate	*N*-(2-hydroxypropyl)methacrylamide lactate
ICH	International Conference on Harmonization
IVR	In vitro release
LC	Liquid chromatography
LC-UV	Liquid chromatography:ultraviolet (detection)
MAA	Methacrylic acid
MBC	Metastatic breast cancer
M_n	Number average molecular weight
mPEG-b-p(HPMAm-Lac)	Poly(ethylene glycol):b-poly[N-(2-hydroxypropyl) methacrylamide-lactate]
MPS	Monocyte phagocytic system
MS	Mass spectrometry
MTD	Maximum tolerated dose
M_w	Weight average molecular weight
$M_{w(mic)}$	Weight average molecular weight of micelles
N_A	Aggregation number of micelles
NBCDs	Non-biological complex drugs
NCA	N-carboxyanhydrides
NCI	National cancer institute
NCL	Nanotechnology characterization lab
NIPAAm	N-isopropylacrylamide
NIST	National Institute of Standards and Technology
NMR	Nuclear magnetic resonance spectroscopy
NSCLC	Non-small cell lung carcinoma
NTA	Nanoparticle tracking analysis
P(Asp)	Poly (aspartic acid)
PDI	Polydispersity index
P(His)	Poly(Histidine)
P(Lys)	Poly(Lysine)
PAA	poly(amino acids)
PBLA	poly(beta-benzyl-L-aspartate)
PBLG	poly(gamma-benzyl-L-glutamate)
PCL	Poly(ε-caprolactone)
PCLLA	poly(caprolactone-co-lactide)
PDLA	Poly (D-lactic acid)
PDLLA	Poly (D,L lactic acid)
PDMAEMA	poly(2-dimethylaminoethyl methacrylate)

PEG	Poly(ethylene glycol). Also known as poly(ethylene oxide) (PEO) or poly(oxyethylene) (POE), depending on its molecular weight.
PEG-b-PDLLA	Poly(ethylene glycol)poly (D,L lactic acid)
PLA-PEG-ACUPA	PLA:PEG-S,S-2-[3-[5-amino-1-carboxypentyl]-ureido]-pentanedioic acid
PEG-P(Asp)	PEG-poly(aspartic acid)
PEG-P(Glu)	Poly(ethylene glycol)poly(glutamic acid)
PEG-PCL	Poly(ethylene glycol)poly(ε-caprolactone)
PEG-PPBA-P(Asp)	Poly(ethylene glycol)poly(butylene adipate)poly (aspartic acid)
PEO	Poly(ethylene oxide)
PEO-b-PPO-b-PEO	Poly(ethylene oxide)poly(propylene oxide)poly(ethylene oxide)
PGA	Poly(glycolic acid)
PICM	Polyionic complex micelles
PLA	Poly(lactic acid)
PLA-PEG	Poly(lactic acid)poly(ethylene glycol)
PLG	Poly(glycolic acid)
PLGA	Poly(lactic glycolic acid)
PLLA	Poly (L-lactic acid)
pNIPAAm	Poly(N-isopropylacrylamide)
PPBA	Poly(4-phenyl-1-butanoate) l-aspartamide
PPO	Poly(propylene oxide)
PSMA	Prostate-specific membrane antigen
PVA	Polyvinylacetate
PVP	Poly(N-vinyl-2-pyrrolidone)
QbD	Quality by design
RAFT	Reversible addition fragmentation chain transfer
RD	Recommended dose
ROP	Ring-opening polymerisation
RPT	Rabbit pyrogen test
SLS	Static light scattering
SPIO	Superparamagnetic iron oxide
TEM	Transmission electron microscopy
T_g	Glass transition temperature

Introduction, Overview Products in the Family

Nanocarriers are developed as a technology enabling the use of poorly water soluble new chemical entities and enhancing the efficacy and safety profile of existing and new drugs by altering their pharmacokinetic (PK) profile. Among the various nanocarriers under investigation, polymeric micelles are of interest because of their biocompatibility, biodegradability, ease of manufacturing and their extreme chemical versatility. Although their development commenced after that of liposomes and has progressed less far, several polymeric micelle products have reached clinical phase testing and one is marketed outside US and EU territories. Polymeric micelles represent approximately 7% of nanotechnology-related submissions at the Food and Drug Administration (FDA). A look at the overall pipeline of polymeric micelle drug products shows eight products progressing through phase I-III trials worldwide (Table 1) and a vast amount of preclinical research and development work described in the scientific literature. Based on the current status and ongoing efforts to further expand applicability and to optimize their characteristics, it is anticipated that in the upcoming 10 years 1–3 polymeric micelle products will enter the market and many others will enter the clinical evaluation path. A next phase will be reached within the next decade(s), when the first generic versions may appear.

Polymeric micelles are nano-sized particles formed upon self-assembly of amphiphilic (block co-) polymers in aqueous solutions above their critical micelle concentration (cmc) (see Fig. 1 for a schematic representation of various types of polymeric micelles). The active pharmaceutical ingredient (API) can either be physically encapsulated or chemically conjugated to the polymers inside the micellar core. The self-assembly of polymeric micelles can be driven by hydrophobic interactions, electrostatic interactions, or- less commonly- hydrogen-bonding or metal-ligand coordination reactions. The resulting structure is usually a spherical nanoparticle of typically 20–80 nm (Kwon 2003) with a hydrophobic core acting as a reservoir for poorly water soluble active pharmaceutical ingredients (APIs) and a hydrophilic shell which provides colloidal stability and limits protein adsorption resulting in long-circulation times. This circulation time is also a result of the size of the particles. Low molecular weight compounds up to 60 kDa, which translates into a size of 5–10 nm, are rapidly eliminated via the kidneys, whereas larger particles are removed via phagocytosis by the monocyte phagocytic system (MPS), or escape from the circulation via pores in sinusoidal endothelia in the liver (particles < 100 nm) (Kwon 2003; Moghimi et al. 1991, 2001; Ernsting et al. 2013). The upper size limit is dependent on splenic filtration and vascular fenestrations at sites of increased permeability (i.e. tumors, inflammation). Particles that exceed 300–400 nm are hampered in passing the splenic sinus and tend to get trapped (Ernsting et al. 2013). To allow extravasation of particles in tumors, their size must remain below the size limit of the vascular fenestrations, which is reported to range from 400 to 600 nm up to microns, depending on the tumor type (Ernsting et al. 2013).

Polymeric micelles are generally developed to meet a need for improved solubility, PK and/or biodistribution. The need for (non-toxic) solubilising agents stems from the fact that a major part of the existing therapeutic agents and the majority of new bioactive chemical entities are hampered in their use because of insufficient water-solubility, causing low and highly variable bioavailability (Rabinow 2004). For intravenous administration of poorly soluble APIs such as the potent anticancer agents paclitaxel and docetaxel, traditional solubilising agents such as Cremophor®EL and polysorbate 80 (Tween 80), respectively, have been used (ten Tije et al. 2003; van Zuylen 2001). However, quite high doses are needed and then

Table 1 Polymeric micelle products that have entered clinical testing (in alphabetic order)

Product (company)	Indication (phase)	API	Polymer	Micelle type	Average particle size (nm)	Drug loading (%)
BIND-014 DTXL-TNP (BIND Biosciences)	Hormone refractory metastatic prostate cancer (II) Non-small cell lung cancer (II) Solid tumors (I)	Docetaxel	PLA-PEG, PLA-PEG-ACUPA	Physically encapsulated/nanocontainer; targeted	100	10
Genexol®-PM (Samyang Corp.)	Metastatic breast cancer (M)[a] Metastatic non-small cell lung cancer (M) Ovarian cancer (M)[a] Lung cancer (III) Bladder cancer (II) Metastatic pancreatic cancer (II)	Paclitaxel	PEG-PDLLA	Physically encapsulated/nanocontainer	<50	16.7
NC-6004 (Nanocarrier)	Locally advanced or metastatic pancreatic cancer (III)	Cisplatin	PEG-P(Glu)-Cisplatin	Metal-ligand complexation micelle	30	39.0
NK012 (Nippon Kayaku Co.)	Colorectal cancer (II) Metastatic breast cancer (II) Multiple myeloma (II) Small cell lung carcinoma (II)	SN-38	PEG-P(Glu)-SN38	Physically encapsulated/nanocontainer	20	20.0
NK105 (Nanocarrier/Nippon Kayaky Co.)	Pancreatic cancer (II)	Paclitaxel	PEG-PPBA-P(Asp)	Physically encapsulated/nanocontainer	85	23.0
NK911 (Nippon Kayaku Co.)	Metastatic pancreatic cancer (II)	Doxorubicin	PEG-P(Asp)-Dox	Physically encapsulated/nanocontainer	40	n.a.

Table 1 (continued)

Product (company)	Indication (phase)	API	Polymer	Micelle type	Average particle size (nm)	Drug loading (%)
PAXCEED™ (Angiotech)	Psoriasis (II) Rheumatoid arthritis (II) Neurological disorders (PC) Secondary progressive multiple sclerosis (D)	Paclitaxel	PEO-*b*-PDLLA	Physically encapsulated/nanocontainer	n.a.	n.a.
SP1049C (Supratek Pharma, Inc.)	Gastrointestinal cancer (II) Colorectal cancer (I) Non-Small cell lung cancer (I)	Doxorubicin	PEO-*b*-PPO-*b*-PEO Pluronic L61, F127	Physically encapsulated/nanocontainer	30	8.2

[a] Marketed in Asia-Pacific region; I, II or III means: is in clinical phase I, II or III; for abbreviations: see list of abbreviations

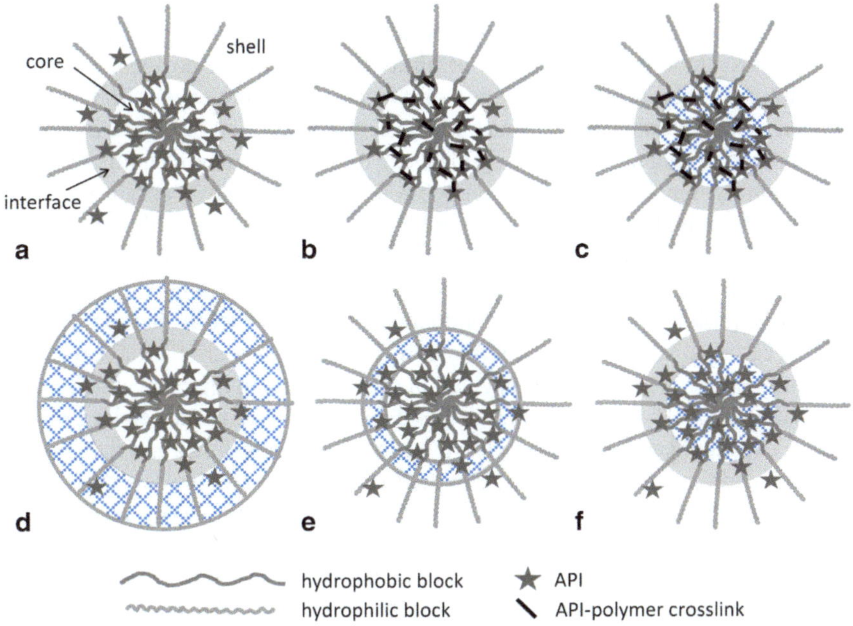

Fig. 1 Schematic representation of various types of polymeric micelles: **a** physically assembled micelles with physically encapsulated active pharmaceutical ingredient (API), **b** physically assembled micelles formed from polymer-drug conjugate, **c** polymeric micelles with a cross-linked core and chemically entrapped API, **d** polymeric micelles with cross-linked shell and physically encapsulated API, **e** polymeric micelles with cross-linked interface and physically encapsulated API, **f** polymeric micelles with cross-linked core and physically encapsulated API

such solubilisers have been shown to be toxic and to affect the PK of the drug to be delivered. Cremophor induces hypersensitivity reactions, hyperlipidemia, neurotoxicity and is thought to account for the non-linear PK of paclitaxel in Taxol® (Gelderblom et al. 2001; Sparreboom et al. 1999; van Tellingen et al. 1999). Polysorbate 80 is associated with hypersensitivity reactions and cumulative fluid retention (Engels et al. 2007) and was shown to be haemolytic. Administration of drugs formulated in such solubilising agents usually necessitates premedication with anti-allergic drugs. Polymeric micelles are explored as alternatives to the conventional solubilizers and have proven advantageous regarding safety, loading capacity, stability and control over drug release rate (Kwon 2003; Yu et al. 1998; Shuai et al. 2004a). Over 1000-fold increases in water solubility have been reached upon formulation in polymeric micelles, illustrating their solubilising efficiency (Zhang et al. 1996; Piskin et al. 1995). Polymeric micelles have a cmc in the order of 10^{-6}–10^{-7} M which is substantially lower than the typical cmc of 10^{-3}–10^{-4} M for traditional low molecular weight surfactants and explains their superior *in vivo* micellar stability (Adams et al. 2003).

Besides plain solubilisation, polymeric micelles can be used to optimise the pharmacokinetic profile and distribution pattern of existing or novel drugs. Traditional surfactants immediately dissociate upon dilution in the central compartment. The same applies to polymeric micelles with a low cmc. However, many polymeric micelles are developed with the aim to remain intact and release the drug in a controlled fashion. This release can be based on diffusion, ion exchange, chemical degradation or external triggers such as temperature, pH or ultrasound. In case of covalently entrapped drugs, the release can either occur via cleavage of drug molecules from polymers followed by diffusion out of the intact micelles, or via disassembly of the micelle into polymer-drug conjugates followed by cleavage of the drug from the unimers. Release or disassembly may also be obtained upon a hydrophobic to hydrophilic conversion of the core-forming block based on degradation or protonation (e.g. block copolymers containing L-histidine (Lee et al. 2003a, b), pyridine or tertiary amine groups (Tang et al. 2003). By altering the PK, toxic peak levels can be prevented and the exposure time can be prolonged resulting in continuous exposure which is thought to be beneficial in certain diseases. By preventing immediate release, particles have the chance to accumulate together with their payload in specific parts in the body and to increase local drug levels. The stealth properties and nanoscale size range render polymeric micelles excellent candidates to exploit the enhanced permeability and retention EPR effect pursued for many nanosized drug delivery systems. Passive accumulation into areas with increased permeability and retention, such as inflammatory sites or tumors, can be exploited for treatment of rheumatoid arthritis and cancer. Once accumulated in a tumor, smaller sizes are considered to be beneficial in terms of penetration into the dense tumor core (Popovic et al. 2010; Dreher et al. 2006). The relatively small size of micelles in comparison to many other nanomedicines therefore represents and important advantage. Finally, polymeric micelles can be decorated with ligands exposed on the surface to enable specific interaction with certain tissues/cells/receptors, a strategy generally referred to as active targeting. A prerequisite to exploiting the active targeting effect is that particles stay intact.

Besides classification based on intended use, polymeric micelles can be subtyped based on their chemistry, type of assembly, drug loading strategy, drug release mechanism, payload or administration route. Although alternatives to PEG are being explored, it remains the almost uniquely used hydrophilic block in copolymeric micelles. However, the hydrophobic block is extensively varied, with the most commonly used chemistries being poly(propylene oxide) (PPO, as extensively investigated by Kabanov et al. (2002; Kabanov and Alakhov 2002) and used in SP1049C), poly(amino acids) (PAA, the focus of Kataoka et al. (2000) and used in NK105, NK911 and NK012, as well as NC-6004), poly(esters) such as poly(lactic acid) (PLA; as used in BIND-014) and poly (D,L lactide) (PDDLA, used in Genexol-PM and Paxceed) (see Table 1). The most extensively studied and advanced types of assemblies are based on hydrophobic collapse of the core-forming polymer block. Alternatively, metal-ligand complexation or electrostatic interactions (forming polyionic complex micelles (PICM)) between the API and polymer are used as the driving force. Besides physical encapsulation, chemical entrapment is also explored as a strategy to obtain increased stability and control over release. The assembly can be further modified by chemical crosslinking of the shell, interface or core, in which case the micelle is transformed into a stabilized nanoparticle.

Although poorly water soluble small molecules are the most common subject of development for polymeric micelles, they can also be loaded with other molecular classes such as nucleotides, peptides and proteins. Lastly, the majority of polymeric micelles are developed for intravenous administration. Nevertheless, their applicability for non- or less invasive routes such as oral or subcutaneous delivery is also being investigated.

The flexibility in chemical structure of both the core and shell block offers the opportunity to tailor the micelles to carry specific drugs, to reach the desired target tissue, to modulate the release profile and to allow easy modification with targeting ligands. This versatility is a major advantage, but also requires careful and rational design.

Since the physicochemical properties, and ultimately the *in vivo* distribution, safety and efficacy of the final drug product are highly dependent on the chosen polymer chemistry and manufacturing process, classification of polymeric micelles as non-biological complex drugs (NBCDs) is justified.

As for other nanomedicines, the development of a polymeric micelle drug products starts with an initial design phase where polymer chemistry and micelle assembly are being selected based on previous knowledge from literature. The design is tested *in vitro* and *in vivo* and optimised if needed. The formulation considered optimal is then taken into further development up to clinical testing. This chapter will briefly summarise the most important design criteria, describe the synthesis, manufacturing and control for such selected formulations, and discuss their pharmacokinetics/pharmacodynamics (PK/PD) and regulatory status. Given the diversity of polymeric micelles, it is not possible to describe all. The body of this chapter therefore focusses on the most far advanced formulations, being polymeric micelle formulations of cytostatics for intravenous use that have reached the clinic. The basic principles described also largely apply to other systems, which may need fine-tuning and addition of extra conjugation steps or characterisations (e.g. ligand quantification).

Design, Chemistry, Manufacturing and Control

Being used for solubilisation, controlled release and/or improved disposition, several characteristics of micelles are of interest. For solubilisation, high loading efficiency and kinetic stability are desired. For controlled release, both stability of the micellar assembly as well as of drug encapsulation are essential. For improved disposition, stealth properties and size are the dominant parameters. In the research/early development phase of polymeric micelle drug development, the type of micelle, polymer blocks and assembly are selected based on *in vitro* screening studies. Once the building blocks have been defined, the actual synthesis and manufacturing route can be developed. This chapter will describe common grounds for design, chemistry and manufacturing and will highlight points of special interest for the various subtypes. The overall process for synthesis and manufacturing of drug-loaded polymeric micelles is depicted in Fig. 2.

Design

Micelle Stability

The chemical properties and the molecular weight of the hydrophobic block have a large impact on micelle stability and drug entrapment efficiency. In general, the cmc (and stability) of polymeric micelles increases with increasing length and hydrophobicity of the core-forming block. The length of the hydrophilic block has less impact, although a minimum core to shell ratio is required (Kabanov et al. 2002). Besides the chemical properties of the core, its physical state affects the kinetic stability. Micelles with a crystalline core (Kang et al. 2005; Slager and Domb 2003) or a frozen, glassy) core (T_g polymer $>37\,°C$) (Kwon and Okano 1996; Teng et al. 1998) were reported to have a greater kinetic stability than those with a liquid-like core (T_g polymer $<37\,°C$), resulting in slower disintegration and drug release. In addition, micelle stability was shown to be sensitive to stereochemistry. Studies with PEG-b-PLA based micelles either composed of a 1:1 blend of isotactic stereoisomers PDLA and PLLA (Poly (D-lactic acid) and Poly (L-lactic acid), respectively, or PDLA alone, or racemic PDLLA, revealed that the 1:1 blend exhibited the highest kinetic stability (Kang et al. 2005). This effect was ascribed to increased compactness and denser packing of the polymeric core due to strong van der Waals interactions. The highest level of control over micelle stability is obtained upon chemical crosslinking of the core or shell polymers (Rijcken et al. 2007a; Shuai et al. 2004b). In this case, a stable nanoparticle is formed which is no longer characterised by a cmc and is only dissociated upon chemical degradation. Through increased micelle stability and altered core structure, release kinetics of the API are also affected. However, drug release is still driven by diffusion. To further control release from stabilised micelles, the API can be cross-linked together with the

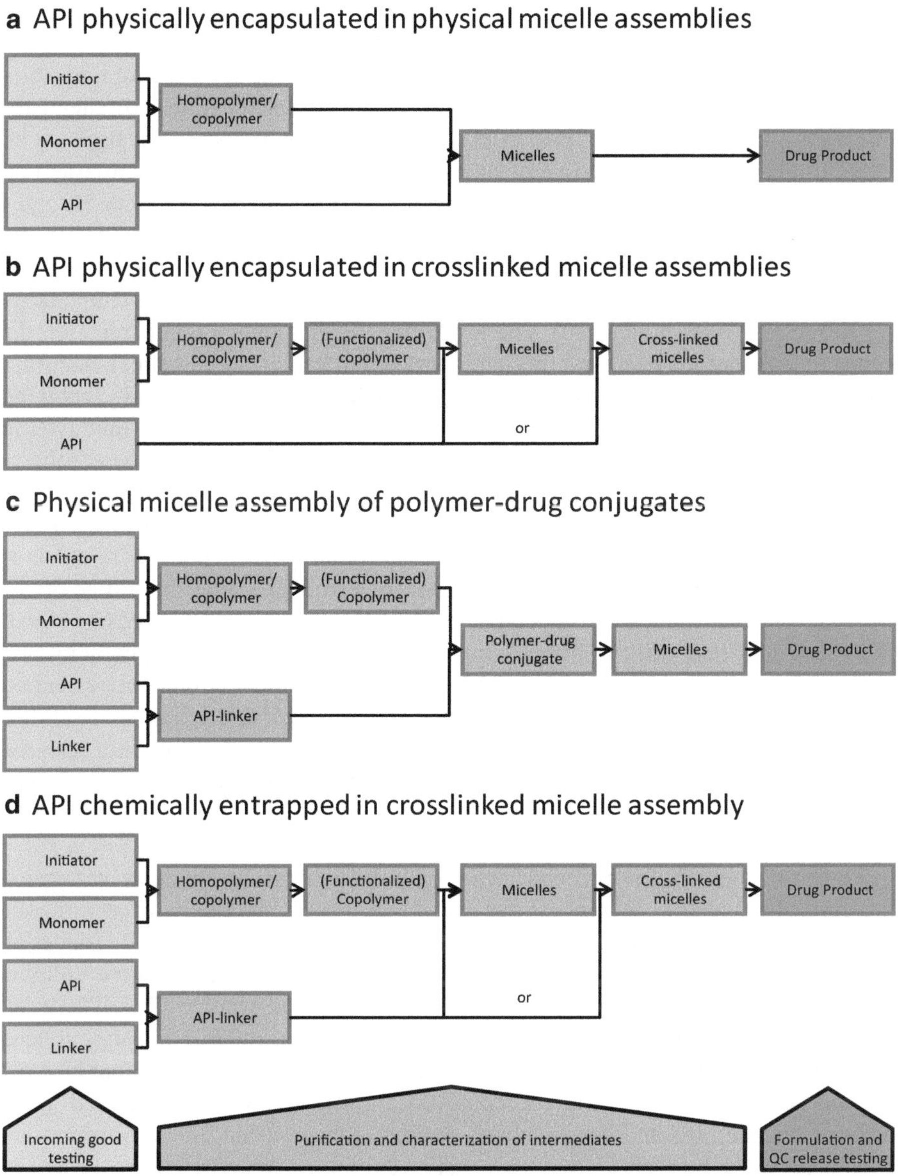

Fig. 2 Flow chart of synthesis, manufacturing and control of polymeric micelle drug products

micelle core (Coimbra et al. 2012; Crielaard et al. 2012; Quan et al. 2014; Talelli et al. 2010a, b, 2011). In this case, the drug-loaded micelle is converted into a stabilised nanoparticle with covalently entrapped drug, and can actually be regarded as a mono(macro)molecular assembly.

Drug Loading

The most important contributor to maximisation of drug entrapment is optimisation of the compatibility between the drug to be encapsulated and the (physico)-chemistry of the core-forming block (Tyrrell et al. 2010; Gadelle et al. 1995; Nagarajan et al. 1986; Liu et al. 2004). Predictive of such compatibility are structural similarity and comparable polarity. A commonly used equation in this respect is the Flory-Huggins interaction parameter χ_{d-p}, defined as: $\chi_{sp}=(\delta_d\delta_p)^2V_d/\kappa T$ with δ_d and δ_p being the solubility parameters for the drug and the core-forming polymer, respectively, V_d as the molar volume of the drug, κ being the Boltzmann constant and T the temperature (in Kelvin) (Tyrrell et al. 2010). The lower the Huggins interaction parameter, the better the (theoretic) comparability and entrapment efficiency. Practical examples include studies with poly(lactones) with different levels of hydrophobicity, showing increased encapsulation of indomethacin with increasing core hydrophobicity (Lin et al. 2003). Besides hydrophobic interactions, hydrogen bonding and crystallinity have also been identified to affect drug encapsulation (Shuai et al. 2004a).

In addition to initial matching of the core chemistry to the drug, modification and derivatization of the core block have proven to be effective strategies to enhance drug entrapment and micelle stability. Two of such strategies are exemplified by the work of Kataoka et al., who have developed PEG-polyaspartic acid (PEG-p(Asp)) polymers modified with 4-phenyl-1-butanol to increase core hydrophobicity and maximise encapsulation of paclitaxel (this formulation is under clinical evaluation under the name NK105) (Hamaguchi et al. 2005). A second, original approach for ultimate drug-core compatibility was to modify the core-forming polymer with the API to be encapsulated. PEG-p(Asp) was derivatized with doxorubicin and used for physical encapsulation of free doxorubicin (in clinical development as NK911) (Nakanishi et al. 2001).

Once the chemistry is chosen, further optimization can be obtained by varying the core block length. Increasing the core block length positively affects micelle stability, but also drug encapsulation efficiency (Xing and Mattice 1997; Tian et al. 1995; Elhasi et al. 2007). This effect is ascribed to the increase in core volume, hence increased capacity for drug incorporation (Xing and Mattice 1997; Tian et al. 1995; Shuai et al. 2004b; Elhasi et al. 2007; Aliabadi et al. 2007). For Poly(ethylene glycol)poly(ε-caprolactone) (PEG-PCL) micelles it was indeed shown that increasing the PCL block length while keeping the PEG block constant, resulted in increased micelle size and drug loading content (Elhasi et al. 2007). An optimum should however be identified, as increasing the molecular weight also results in increased crystallizability which in turn reduced the accessible volume for drug loading (Shuai et al. 2004a; Tyrrell et al. 2010; Kang et al. 2002).

Not only the selection of polymer composition and length, but also the procedure used for micelle preparation has an impact on the characteristics of the micelles formed. For example, 1.5-fold higher encapsulation efficiencies could be obtained when using an oil in water emulsion approach as compared to a dialysis-based method (see below) (Aliabadi et al. 2007). Similarly, increased encapsulation efficiencies were obtained for amphotericin encapsulated via solvent evaporation compared to the dialysis approach (Lavasanifar et al. 2001). In another study, rapid precipitation followed by dialysis proved superior regarding control over size (distribution) in comparison to the standard dialysis method (Kataoka et al. 2000). The relative concentrations of polymer, API and solvents all contribute to the final micelle characteristics. Drug loading increases with polymer concentration until a plateau is reached where the micelles are fully saturated (Xing and Mattice 1997; Hurter et al. 1993a, b). Drug loading capacity is also positively related to the concentration of the API (Shuai et al. 2004a, b; Tyrrell et al. 2010; Gadelle et al. 1995; Xing and Mattice 1997; Elhasi et al. 2007). The presence of drug inside the micellar core induces an increase in core volume via a direct effect, and also via an increase in aggregation number, which in turn leads to increased micelles and higher drug loading capacity.

Chemistry

Polymeric micelle drug products typically consist of an API (or multiple API's) and one or more polymers. The API can be encapsulated by physical loading into self-assembled micelles (with or without subsequent chemical conjugation steps) or by chemical conjugation to the amphiphilic polymer followed by self-assembly of the polymer-drug conjugates.

The amphiphilic block copolymer can be a diblock (AB-type), multiblock (ABA or ABC type) or graft copolymer (see Fig. 3 for a schematic representation). The hydrophilic block can be further modified with targeting ligands while the core-forming block can be modified with drug molecules or moieties that alter the hydrophobicity of the core and its affinity for the drug molecule of interest. The chemistry at the basis of each of these variants is the selection and synthesis of the core- and shell-forming blocks, which will be the focus of this chapter. Given that the possibilities for varying the identity as well as size of the shell-, but most importantly of the core-forming block are numerous, the resulting possible micelle types are limitless. Indeed, a wide variety of polymers and polymeric micelles have been exploited for drug delivery. This chapter does not aim to provide an overview of all possibilities, but focusses on those block copolymers that are most commonly used and/or have progressed furthest, reaching clinical evaluation. These are the copolymers of PEG and poly(esters), poly(amino acid)s (PAA), poly(propylene oxide) (PPO) or vinylic polymers.

Hydrophilic Block

PEG PEG is currently the most frequently used polymer to function as the micellar corona and counts as the gold standard in the field of (polymeric) drug delivery (Knop et al. 2010). Mostly, PEG with a molecular weight between 1–15 kDa is used as the hydrophilic block (Kwon 2003). Advantages of this polymer are its low cost of goods, low toxicity, low immunogenicity and its regulatory status (it is

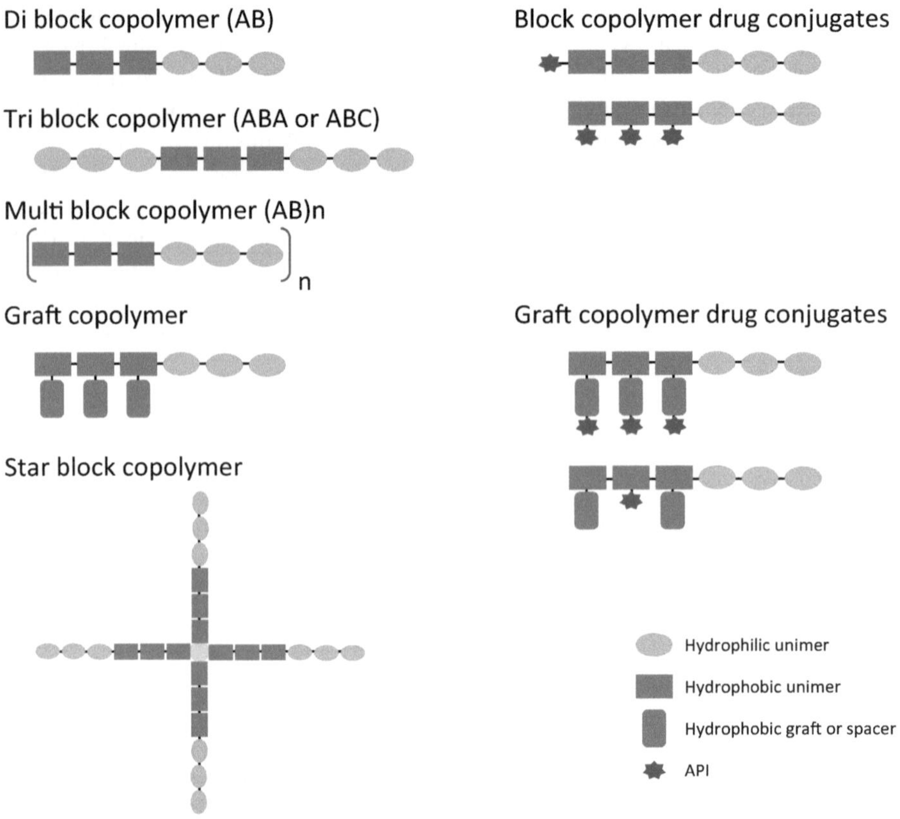

Fig. 3 Architecture of copolymers and polymer-drug conjugates

approved for internal use). PEG is known as an effective stealth agent. This effect is attributed to its high water solubility, high flexibility and large exclusion volume (Adams et al. 2003; Molineux 2002; Torchilin and Trubetskoy 1995; Woodle and Lasic 1992).

PEG is synthesised via ring-opening polymerisation (ROP) of ethylene oxide (EO) with a functionalised initiator, either a hydroxide or alkoxide (Gaucher et al. 2005; Tessmar et al. 2002; Nagasaki et al. 1995a, b; Zhang et al. 2004; Nakamura et al. 1998; Cammas et al. 1995; Xiong et al. 2012; Thompson et al. 2008). Alternatively, it can be prepared by derivatisation of commercially available HO-PEG-OH (or other homobifunctional PEGs) (Xiong et al. 2012; Thompson et al. 2008). The most convenient route is however to use commercially available functionalised PEGs. Several monofunctional, homobifunctional and heterobifunctional PEGs are commercially available.

Monofunctional or bifunctional PEGs can be exploited for initiation of polymerization. The most commonly used initiator is methoxypoly(ethylene glycol) (mPEG), functionalised with amino- (for synthesis of PAA block copolymers) or hydroxyl (for synthesis of polyester block copolymers) groups (Gaucher et al. 2005). These commercially available mPEGs can be converted into macroinitiators suitable for free radical polymerisation (Neradovic et al. 2001; Soga et al. 2004; Rijcken et al. 2005), atom-transfer radical-polymerization (ATRP) (Dufresne et al. 2004; Dufresne and Leroux 2004; Sant et al. 2004) or reversible addition fragmentation chain transfer (RAFT) polymerisation (Gaucher et al. 2005; Shi et al. 2003; Hong et al. 2004).

Alternatives to PEG Although PEG counts as the gold standard, it comes with certain drawbacks such as induction of hypersensitivity reactions, toxicity of side-products, degradation under stress, unexpected changes in the PK profile upon repeated administration, and non-biodegradability leading to accumulation (Knop et al. 2010). These disadvantageous properties have stimulated the development of alternative stealth coatings. Alternatively to PEG, functionalized poly(N-vinyl-2-pyrrolidone) (PVP) can be used for preparation of micelles with a PVP shell (Benahmed et al. 2001; Luo et al. 2004; Chung et al. 2004; Le Garrec et al. 2004; Lele and Leroux 2002). PVP is the next best explored option for the corona-forming block. PVP is non-ionic, biocompatible and is highly hydrophilic and flexible, similar to PEG. However, the availability of well-defined PVP macroinitiators is limited because its common synthesis route (free radical polymerisation) is characterised by poor control over molecular weight and chain end functionalization (Gaucher et al. 2005).

Other alternatives to PEG that are being explored are PVA (Luppi et al. 2002; Orienti et al. 2005; Zuccari et al. 2005, 2009), poly(ethylene imine) (Nam et al. 2003), poly(acrylic acid), poly (2-ethyl-2-oxazoline) (Kim et al. 2000a), polysaccharides (Rouzes et al. 2000), poly(asparagine), poly(N,N, dimethylamino-2-ethyl methacrylate), p(NIPAAM-co-N,N-dimethylacrylamide). Such coatings are still in the exploratory phase and their added value for stabilisation of polymeric micelle drug products in the clinic remains to be confirmed.

Hydrophobic Blocks

In contrast to the prominent position of PEG as the hydrophilic block, the hydrophobic block is extensively varied. Chemical categories of hydrophobic blocks include poly(amino acid)s (PAA), poly(esters), poly(propylene oxide)(PPO), vinylic polymers, poly(amine), poly(amine ester) and polycarbonates (as reviewed in (Tyrrell et al. 2010; Gaucher et al. 2005; Xiong et al. 2012; Torchilin 2007).

Block copolymers of PEG (or an alternative hydrophilic segment) and a hydrophobic block of choice can be synthesized by coupling separately synthesized polymers or by sequential synthesis. In the first route, the individual polymers must have 100 % end group functionality and must be present in amounts exactly matching the availability of the functional end groups (Kumar et al. 2001). This route is considered to have an increased risk of side products formation and is therefore not preferred (Gaucher et al. 2005). Sequential synthesis can be performed in single or multiple steps. If the monomers of the hydrophilic and hydrophobic block have a distinct reactivity, synthesis can be achieved by addition of all monomers at the same time (Kumar et al. 2001). Only after polymerisation of the most reactive monomer is completed, the second monomer will be polymerised. However, the best controlled and therefore preferred strategy is synthesis based on sequential monomer addition. This can be done by using a block segment functionalized with a macroinitiator.

PEG-PAA (PEG- polyamino acids) PAAs are a highly versatile class of polymers that have gained interest because of their biodegradability based on metabolization by proteases and their chemical flexibility. The multiple carboxyl and/or amine groups can be used for electrostatic complexation, but also for chemical modifications to form polymer-drug conjugates or to optimise the core to improve encapsulation efficiency and stability. Examples of PEG-PAA polymers used for polymeric micelles include poly(aspartic acid) (P(Asp)), poly(L-glutamic acid) (P(L-Glu)), poly(L-lysine) (P(L-Lys)), poly(L-histidine) (P(His)), and derivatives thereof such as poly(β-benzyl-L-aspartate) (PBLA), poly(γ-benzyl-L-glutamate) (PBLG).

The most common synthesis route for PEG-PAA is ROP of N-carboxyanhydride derivatives of various amino acids using bifunctional α-methoxy-ω-amino-PEG as the initiator (Gaucher et al. 2005; Nagasaki et al. 1995b; Zhang et al. 2004; Nakamura et al. 1998; Xiong et al. 2012; Schmeenk et al. 2005; Harada and Kataoka 1995). Other aprotic nucleophile or base initiators can also be used, but primary amines and alkoxide anions are most common. The reaction is performed in aprotic solvents such as DMF, toluene, dioxane and chlorinated alkanes (Xiong et al. 2012). The major challenge for this synthesis route is the presence of side reactions such as chain termination and chain transfer. These side reactions result in poorly predictable molecular weight and composition and formation of homopolymer contaminants which are difficult to remove from the desired copolymer (Xiong et al. 2012). Strategies for improved synthesis routes include living polymerization of N-carboxyanhydrides (NCA), forced polymerisation via the amine route, ROP at decreased temperatures (Xiong et al. 2012) and solid or liquid phase peptide synthesis (Van Domeselaar et al. 2003; Choi et al. 1999). Living polymerisation has the advantage of generating well-defined and predictable polymers with narrow weight distributions. A limitation of this method is the use of transition metal catalysts, necessitating rigorous purification. Strategies based on forced polymerisation via the amine route are based on the two competing processes that normally occur because initiators can be both nucleophilic and basic: "amine" and "activated monomer" reactions. The activated monomer reaction is induced by deprotonation of NCA by strong bases, producing an NCA anion which subsequently initiates chain growth via nucleophilic attack. By using strong nucleophiles such as primary amines (Aliferis et al. 2004) or primary amine hydrochlorides (Dimitrov and Schlaad 2003), the reaction can be forced to proceed via the amine route while suppressing the activated monomer route. ROP at decreased temperature was explored as an option to suppress the terminal reaction and enable a living polymerisation reaction (Vayaboury et al. 2004). Solid or liquid phase peptide synthesis is a highly controlled synthesis method that allows precise control over chain length and amino acid order. Despite this attractive feature, applicability is limited because the process is laborious and only suitable for synthesis of short chain polypeptides (Van Domeselaar et al. 2003).

PEG-poly(ester) The two most well-known examples of PEG-poly(esters) are poly(ε-caprolactone) (PCL) and poly(lactide) (PLA). Copolymers of glycolide (PGA), D,L-lactide and glycolic acid (PLGA), poly(caprolactone-co-lactide) (PCLLA) are other examples from this category. These polymers are synthesised by ROP of the monomer (lactide or ε-caprolactone) in the presence of stannous octoate (Yoo and Park 2001; Shuai et al. 2003; Kang and Leroux 2004), potassium naphtalene (Nagasaki et al. 1998) or calcium ammoniate (Piao et al. 2003) as the catalyst.

Drawbacks of this synthesis route are the use of toxic organometallic catalysts that need to be removed by purification, the harsh reaction conditions needed (elevated temperatures, organic solvents) and sensitivity to side reactions if conditions are not optimal (e.g. in case of the presence of water and impurities in starting materials) (Xiong et al. 2012). To circumvent the use of organometallic compounds, lipases are being investigated as an alternative catalyst (Xiong et al. 2012).

PEG-PPO PEG-PPO based copolymers are one of the most extensively investigated polymers for drug delivery. Polymers of the triblock architecture PEG-PPO-PEG are known as poloxamers and are commercially available as Pluronic® and Tetronic®. Several of these polymers have been approved by the FDA for use in pharmaceutical formulations (Chiappetta and Sosnik 2007). Pluronics have been widely studied by the group of Kabanov and are the basis of SP1049C, a polymeric micelle product which is currently in clinical trials. Pluronics are non-biodegradable polymers, but individual polymer chains with a size of 10–15 kDa range have been shown to be excreted through filtration by the kidneys. Besides functioning as a carrier, Pluronics have also been described to have intrinsic biological effects such as anti-infective and anti-inflammatory effects and inhibition of P-glycoprotein pumps.

PEG-PPO-PEG is synthesised by alkaline catalysed sequential monomer addition: polymerisation of PO to PPO, followed by anionic polymerisation of EO at both ends (Gaucher et al. 2005; Xiong et al. 2012). Using anionic polymerisation, polymers with a narrow molecular weight distribution can be obtained. On the other hand, unreacted PPO homopolymers, diblock co-polymers and triblock copolymers with lower degrees of polymerisations can be formed as contaminants that require (complex) purification procedures (Xiong et al. 2012).

Polymer-drug Conjugates Various designs for block-copolymer-drug conjugates can be used, including coupling of single or multiple drug molecules via a (degradable) linker directly to the hydrophobic block and coupling strategies using linkers and hydrophobic spacers. Several micelle-forming drug conjugates based on PEG-poly(esters) (Yoo and Park 2001, 2002), PEG-PAA block copolymers (Kwon et al. 1993; Yokoyama et al. 1990, 1992, 1998) and other polymers have been developed, as reviewed by Liu et al. (2012). In case of ester-based polymers, conjugation is usually obtained via formation of covalent bonds between the activated terminal hydroxyl group and reactive groups of the API, resulting in conjugates containing single drug molecules per polymer chain. For amino acid-based polymers, options are more diverse. The diversity and number of functional groups (amino acids may contain amino, hydroxyl and carboxyl groups) offers a wider range of synthetic routes as well as modification of single polymer chains with multiple drug molecules.

Manufacturing

Polymeric micelles are typically formed upon self-assembly in an aqueous environment. The process consists of several steps: (1) dissolving the polymers in a suitable solvent, (2) exposure to aqueous environment to induce micellation, (3) removal of organic solvent and (4) downstream processing to obtain a purified, stable, well-characterised drug product. Depending on characteristics of the polymer and API and on the desired process scale and quality, a micelle formation process can be selected. The choice for a manufacturing process has a direct impact on the final product characteristics. For example, it was shown that for the same API/polymer combination, 1.5-fold higher drug encapsulation could be obtained using the oil in water emulsion approach than for the dialysis method (see below) (Sant et al. 2004). Elsewhere, micelle size was seen to be affected by the manufacturing process, with rapid precipitation followed by dialysis allowing better control over size and polydispersity index (PDI) than the conventional dialysis method (Vangeyte et al. 2004). Figure 2 describes the principles of the most important manufacturing processes. The advantages and disadvantages of these techniques are briefly discussed in the following sections. The potential critical process parameters (CPP) for each process are summarised in Table 2. For the manufacturing of more complicated micelle products such as actively targeted polymeric micelles or those with covalently entrapped drug, cross-linked cores or shells, the same principles apply. In these cases, the copolymer used in the process may be modified to contain a targeting ligand, drug molecule or reactive groups that allow cross-linking or conjugation. In the latter cases, a synthetic conjugation/cross-linking step is performed after micelle formation which will affect the physicochemical properties of the micelle and the impurity profile and should be considered in the purification and characterisation of the final product.

1. Direct dissolution (Tyrrell et al. 2010; Gaucher et al. 2005; Kore et al. 2014)

Amphiphilic copolymers with a certain degree of water solubility can form micelles upon direct dissolution in water at a concentration above their cmc. Depending on the polymer, heating may be used to dehydrate the hydrophobic core to promote micelle formation (Gaucher et al. 2005). The API is loaded either by simultaneous addition with the polymer to the aqueous phase, or by addition to preformed polymeric micelles and solubilisation under stirring. This process is only suitable for

Table 2 Potential critical process parameters (CPPs) for the micelle manufacturing processes

Manufacturing process	Potential CPPs for each process	Product and process characteristics of special concern for each process
All	Polymer concentration	Micelle stability, encapsulation/entrapment efficiency, loading capacity, size
	API concentration	
	Ratio drug: polymer	
	Nature or organic solvent	
Direct dissolution	Order of addition of polymer and API	Encapsulation/entrapment efficiency, initial extra-micellar drug content, location of API inside micelles
	Mixing procedure (type and time)	
	Temperature	
Dialysis	Choice of solvent	Size, leachables and extractables, yield, selective loss of API or polymer, residual solvent, initial extra-micellar drug content, degradation of API (and/or polymer), endotoxin levels
	Dialysis membrane design	
	Cut-off	
	Material	
	Surface area	
	Stirring procedure	
	Temperature	
	Volume of water	
	Refreshment procedure	
	Time	
Co-solvent evaporation	Choice of solvents	Yield, selective loss of API or polymer, residual solvents, degradation of API (and/or polymer)
	Volume ratio of organic solvents	
	Volume ratio organic phase to aqueous phase	
	Mixing procedure	
	Temperature	
Oil-in-water (O/W) emulsion	Volume ratio organic phase to aqueous phase	Yield, selective loss of API or polymer, residual solvents, degradation of API (and/or polymer)
	Time	
	Temperature	
	Evaporation procedure (technique and time)	
Solution-casting/film hydration	Evaporation procedure (rotational speed, temperature, vacuum, time)	Yield, selective loss of API or polymer, residual solvent, degradation of API (and/or polymer)
	Surface area of round-bottom flask	
	Composition of reconstitution medium	
	Hydration volume	
	Hydration temperature	

Table 2 (continued)

Manufacturing process	Potential CPPs for each process	Product and process characteristics of special concern for each process
Freeze-drying/lyophilizing	Composition (amount of solvent)	Size, aggregation number, residual solvent, degradation of API (and/or polymer)
	Volume	
	Freeze-drying cycle	
	Time	
	Reconstitution medium	
	Reconstitution volume	
	Reconstitution time	
Heating of aqueous polymer solution from below to above CMT	Heating rate	Encapsulation/entrapment efficiency, extra-micellar drug content, residual solvent, degradation of API (and/or polymer)
	Order of addition of polymers and API	
	Volume (ratio) of organic solvent to aqueous phase	
Flash nanoprecipitation	Mixing time	Drug load, size (distribution)
	Polymer concentration	
	Organic phase flow rate	
	Aqueous phase flow rate	
	Volume ratio organic phase to aqueous phase	

water-soluble polymers (Pluronics®, charged polymers for formation of PICM) or moderately hydrophobic polymers. The cmc is the driving force for micelle formation: dissolving the polymer above its cmc results in spontaneous micelle formation and the API is encapsulated based on a preferred interaction with the core-forming segment and/or preferred uptake into the hydrophobic core of pre-formed micelles. Critical process parameters for this process include polymer concentration, API concentration, molar ratio of API to polymer, stirring/mixing procedure and time.

Advantages of this technique are its simplicity and absence of organic solvents. However, since this approach requires dissolution of the API and polymer, its applicability is limited to low to moderately hydrophobic polymers and drugs. Another disadvantage is the often low drug loading efficiency obtained through this procedure.

2. Dialysis (Fig. 2a; Gaucher et al. 2005; Taillefer et al. 2000; Kwon et al. 1995; Kim et al. 2001; Lavasanifar et al. 2000; Allen et al. 2000)

Polymer and API are dissolved in a water-miscible solvent (e.g. dimethylformamide (DMF), dimethyl sulfoxide (DMSO), tetrahydrofuran (THF), acetone, ethanol, dimethylacetamide (DMAc) (Tyrrell et al. 2010)) and subsequently dialysed against water. The slow exchange of organic solvent for water (non-solvent for core segment) is the driving force for micelle formation. Typical critical process parameters are the choice of solvent, polymer concentration, API concentration, molar ratio of API to polymer, ratio of organic solvent to aqueous phase. The manufacturing process can be optimised by varying the organic solvent and by crash-adding water to the organic phase prior to dialysis, which was reported to result in micelles with superior size distribution profiles (Vangeyte et al. 2004). This process is suitable for a wide range of polymers and APIs, but is less attractive for upscaling. Extensive dialysis over multiple days is generally required to remove the organic solvents. Such long processing time is undesired, not only for efficiency reasons, but also because of the potential impact on drug and/or polymer degradation and increased risk for microbial contamination. Even though polymeric micelles are in a size range that allows sterile end-filtration, endotoxins resulting from microbial contamination at early process steps should be avoided (Fig. 4).

3. Oil-in-water (O/W) emulsion (Fig. 2b; Kataoka et al. 2000; Jones and Leroux 1999; La et al. 1996; Kwon et al. 1997)

In the oil-in-water emulsion method (sometimes referred to as solvent-in-water emulsion), the API is dissolved in a water-immiscible solvent (e.g. chloroform, ethyl acetate, dichloromethane (DCM) (Lin et al. 2003). Upon exposure to water and emulsification by sonication or stirring, nanosized oil-in-water droplets are formed. The polymer may be dissolved in the solvent together with the API, in which case the exposure to water results in reorganization of the polymer chains into micelles with parallel encapsulation of the API. The polymer may also be dissolved in the aqueous phase, in which case preformed micelles swell upon exposure of the core to solvent and API is taken up into the micelles. The organic phase is subsequently removed upon stirring and evaporation.

As a 'green' and safer alternative to this procedure, supercritical fluid evaporation is explored (Kore et al. 2014). In this procedure, supercritical CO_2 is used as the dispersion phase instead of an organic solvent. Upon conversion of the CO_2 into a gas, the polymer reorganizes such that the hydrophobic block forms the core and the hydrophilic part protrudes into the water phase. An advantage of this technique, besides the circumvention of toxic organic solvents, is that high drug loading efficiencies can be achieved. However, the sizes of the obtained micelles are relatively big.

4. Film hydration (Fig. 2c; Zhang et al. 1996, 1997a; Kim et al. 2001; Zeng et al. 2004; Lee et al. 2005a; Lavasanifar et al. 2002).

This film hydration technique is also known under various names such as solution-casting, solvent hydration, solid dispersion or dry-down method. Polymer and API are dissolved in a volatile solvent (e.g. ethanol, chloroform, acetonitrile, DCM, DMF) in a round-bottom flask. Evaporation of the solvent by rotary evaporation or by nitrogen flushing results in formation of a thin film of polymer. An additional vacuum desiccation step can be applied to remove traces of organic solvents (Kore et al. 2014). Micelle formation is induced by addition of hot water (at the Tg of the block copolymer). During the film formation, the drug preferentially interacts with the hydrophobic polymer. The subsequent exposure to water results in hydration and dissolving of the hydrophilic segment while the interaction between hydro-

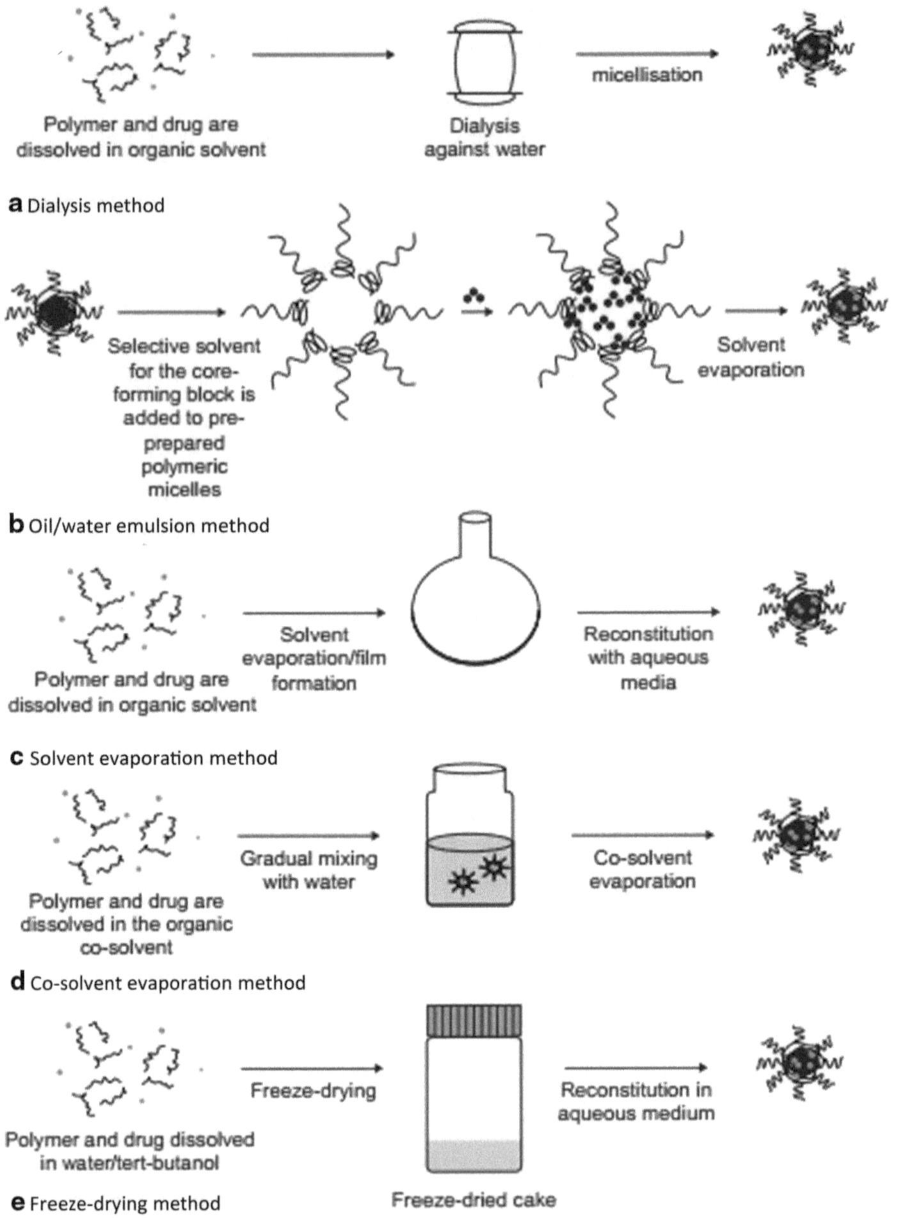

Fig. 4 Schematic picture of processes for polymeric micelle formation and physical drug encapsulation. (Reprinted with permission from (Aliabadi and Lavasanifar 2006))

phobic drug and the core-forming block is further increased, leading to micelle formation and encapsulation of the drug. Sonication can be applied to obtain uniform micelles. This process is only applicable to polymers that are relatively easy to reconstitute, which prerequisites a high hydrophilic/lipophilic balance (Aliabadi and Lavasanifar 2006).

5. Co-solvent evaporation (Fig. 2d; Shuai et al. 2004a, b; Aliabadi et al. 2005a, b; Liu et al. 2001; Jette et al. 2004)

Polymer and API are co-dissolved in a volatile water-miscible solvent (e.g. acetone) (Tyrrell et al. 2010; Gaucher et al. 2005). Subsequently, exposure to water under stirring and evaporation of the organic solvent induces preferred arrangement of core segments into micelles. Precipitation is prevented by the gradual change from organic to aqueous solvent (Tyrrell et al. 2010). In the solvent evaporation process, the choice of solvent and ratio of solvent to water can be optimised to maximise final drug load.

6. Freeze-drying/lyophilizing (Fig. 2e; Le Garrec et al. 2004; Fournier et al. 2004)

In this procedure, API and polymer are co-dissolved in a freeze-dryable solvent (e.g. *tert*-butanol). Subsequently, the solution is mixed with water, freeze-dried and reconstituted in an aqueous (isotonic) solvent (e.g. 0.9% sodium chloride (NaCl)). The proportion of organic solvent affects the aggregation number and the size of the micelles both before and after freeze-drying (Fournier et al. 2004). Micelle formation occurs spontaneously upon reconstitution at time of use, making this an elegant approach. However, application is limited to micelles consisting of polymers that are soluble in freeze-dryable solvents, excluding polymers such as PEO (Aliabadi and Lavasanifar 2006).

7. Heating of aqueous polymer solution from below to above the critical micelle temperature (CMT)

The thermosensitivity of certain polymers allows for a manufacturing procedure with limited use of organic solvents. The hydrophilic to hydrophobic conversion of thermosensitive polymers allows solutions of polymers to be converted into micellar dispersions by simply increasing the temperature above the CMT. The use of organic solvents can therefore be limited to small amounts that may be needed to solubilise the API of interest.

Formation upon heating is a competitive process between intrapolymer coil-to-globule transition (collapse) of the thermosensitive segments and interpolymer association (aggregation) of polymers. The heating rate is a critical parameter: faster heating causes rapid dehydration of the thermosensitive segments, and therefore the subsequent collapse of these segments precedes the aggregation between polymers. This results in micelles with a well-defined core-shell structure. Advantages of this method are the ease of preparation, scalability and the limited use of toxic solvents. A disadvantage is that some APIs may have insufficient stability over the required heating temperature range (although heating may be mild and short). However, this problem could be circumvented by addition of the API to preformed polymeric micelles and solubilisation under stirring rather than simultaneous addition with the polymer. Obviously, applicability of this technique is limited to thermosensitive polymers with a CMT below 37 °C. Micelles prepared this way must be stored at temperatures above the CMT (unless they are stabilised by crosslinking after formation), which may be undesired depending on the API encapsulated.

8. Flash nanoprecipitation

Polymer and drug are co-dissolved in an organic solvent and this solution is pumped into a jet simultaneously with a water stream. The rapid change in solvent conditions at the point where both fluids meet induces precipitation with the polymers collapsing around the drug. The dominant parameters that control the process are mixing time and polymer concentration (Tyrrell et al. 2010). Success of this method has been highly variable for different formulations, with loading contents varying between <0.01 and 50 % (Gindy et al. 2008; Kumar et al. 2009).

Depending on the micelle manufacturing method chosen, certain residual solvents, process impurities and non-encapsulated drug must be removed. Precipitated or crystallized non-encapsulated drug can be removed by low-speed centrifugation (Lee et al. 2005a) or filtration. Techniques to purify the micelles from dissolved non-encapsulated drug as well as residual solvents include high-speed centrifugation (Zweers et al. 2004), dialysis, ultrafiltration (tangential flow filtration, also known as cross flow filtration) or size exclusion techniques. High-molecular weight solutes such as free polymer chains or impurities present in the polymer intermediate can be removed using similar techniques, but require higher resolution separation and are therefore more difficult to remove (and to identify and quantify analytically). Volatile solvents can be removed upon evaporation induced by stirring and spontaneous degassing or nitrogen purging. In addition, freeze-drying can be exploited for removal of volatile residual solvents (Fournier et al. 2004; Saez et al. 2000).

Once purified micelles are obtained, the product can be formulated such that osmolality and pH are suitable for the intended use and route of administration. In addition, excipients may be added to improve the freeze-drying process and product stability. An important advantage of polymeric micelles is that their small size allows sterile end-filtration. This is unlike polymeric microspheres or larger sized nanomedicines exceeding 200 nm which require complex and costly aseptic processing. The filtered product can be stored as a liquid or freeze-dried and stored as dry cakes. In the clinic, freeze-dried cakes are reconstituted and both liquid products and reconstituted freeze-dried products may be further diluted in a suitable infusion medium prior to use. This approach requires maintenance of micellar integrity throughout the freezing as well as reconstitution processes.

Control

In Europe, the EMA has issued a 'reflection paper' specifically for the development of block copolymer micelle medicinal products (EMA Committee for Medical Products for Human Use CHMP 2013a). The information for the pharmaceutical development in this document should be read in connection with generally applicable harmonized (US, EU, Japan) guidelines formulated by the International Conference on 'Harmonisation of Technical Requirements for Registration of Pharmaceuticals for Human Use' (ICH) (http://www.ich.org) as well as relevant regional guidelines.

The control strategy for polymeric micelle drug products includes quality control of raw materials, intermediates and final drug product (see Table 3). All raw materials must be pure and safe. Raw materials can be commercially obtained, and their quality can be based on the certificate of analysis provided by the supplier and/or by in house testing. Common techniques to confirm identity of raw materials are similar to those used for conventional drugs and include NMR, LC-UV and (LC-) MS. In general, polymers used to manufacture polymeric micelle drug products are synthesized in house rather than commercially obtained. Important quality aspects of polymers are their size (M_w and M_n) and polydispersity (Mw/Mn), and the length of the individual blocks which can be measured by NMR and GPC. Changes in these parameters can have an impact on critical quality attributes (CQAs) of the final micellar drug product, such as size, polydispersity, loading capacity and stability. Impurities can include non-reacted monomers, homopolymers and catalysts and should be carefully understood and controlled. In cases where impurities are difficult to remove (e.g. homopolymers of similar size to the polymer of interest), especially in large scale processes, or can only be quantified with insensitive assays, a risk assessment should be performed to assess the chance that the impurities are not removed in subsequent manufacturing steps and the impact of remaining impurities on critical quality attributes of the polymeric micelles as well as on direct potential toxicities. Testing of heavy metals at early stages during the synthesis and manufacturing process is cost-effective as it can reduce the need for such tests on the (expensive) drug product.

Characterisation of the final polymeric micelle drug product comprises of an elaborate set of tests of which a selection is discussed below. Testing can be subdivided into assays for characteristics typical to the nanoparticulate nature, characteristics related to the API formulated, characteristics specific to polymers and standard assays for general and microbial testing. An overview of potential test items is presented in Table 3. Which of these tests are required must be decided on a case by case basis, depending on the specific product characteristics and development phase. Early discussion with regulatory authorities is therefore recommended.

Size (Distribution), Morphology and Microstructure Nanoparticle size and size distribution are routinely studied by dynamic light scattering (DLS, also known as quasi-elastic light scattering or photon correlation spectroscopy), a technique that measures intensity fluctuations of scattered light as a function of particle size. This technique is fast and consumes small amounts of sample. However, it is only suitable for rather homogeneous populations and is largely biased towards larger particles in case of heterogeneous samples. Alternatively, microscopy-based techniques such as transmission electron microscopy (TEM) and atomic force microscopy (AFM) can be used to obtain information on single particle level. Moreover, these techniques provide information on the morphology of the particles. Disadvantages

Table 3 Potential test items for the characterisation of raw materials, intermediates and polymeric micelle drug products

Raw materials and intermediates		
Monomers, initiator, API, API-derivative	*Polymer, copolymer, modified copolymer, polymer-drug conjugate*	
Identity	Identity	
Purity	Purity	
Impurities	Impurities	
Residual solvents	Mw, Mn, Mw/Mn	
Heavy metals	Length of hydrophobic and of hydrophilic block	
Water content	Residual solvents	
	Heavy metals	
	Extent of modification	
	Drug:polymer ratio	
	Tg	
	Cloud point	
Micellar drug product		
Particulate-related characteristics	*API-related characteristics*	*Polymer-related characteristics*
Mean size and size distribution	Identity	Identity
cmc/CAC	Total content	Total polymer content
Aggregation number (association number)	Chemically bound API content	Extra-micellar polymer content
Dynamics of micellisation	Physically encapsulated content	Degradation products
(Apparent) molecular weight	Extra-micellar content	Degradation rate
Core crystallinity/fluidity	Loading capacity	Related substances
Volume of entrapment	Entrapment/encapsulation efficiency	Impurities
Stability	*In vitro* release profile	
Morphology	Physical state of API in core	

Table 3 (continued)

Micellar drug product		
Zetapotential	Distribution of API within particle	
Surface properties	Degradation products	
Protein-carrier interactions	Related substances	
	Impurities	
General testing	*Microbial testing*	
Colour	Endotoxin	
Opalescence	Sterility	
Particulate matter	Non-aerobic count	
pH		
Osmolality		
Water content		
Reconstitution time		
Redispersibility		
Residual solvents		
Extractables		
Antimicrobial and preservative content		
Uniformity of dosage units		

Abbreviations: see list of abbreviations

are that they are laborious, have poor statistical power and are sensitive to artefacts induced during sample preparation. A more recent technique that allows analysis at single particle level is nanoparticle tracking analysis (NTA), which relates the rate of Brownian motion to particle size. When using this technique, sample preparation (dilution) is a critical factor. Given the differences in the principles underlying size determination among the various tools, sizes obtained via different methods cannot be easily compared. At current, no consensus exists regarding the technique or standards to be used.

Detailed information on the microstructure of micelles, e.g. dimension and fluidity of the core, aggregation number (increases with increasing length of hydrophobic block), density of hydrophilic coverage of the shell can be obtained using a combination of DLS and static light scattering (SLS) and NMR.

Content Assays for content are typically based on disruption of the polymeric micelles followed by chromatographic separation of the API from other components and quantification through UV absorption detection. In some cases, exposure of the micelles to the mobile phase may be sufficient to separate the API from the polymers. In other cases, sample preparation is required. For example, for drug products in which the API is cross-linked to the polymer (matrix), chemical pre-treatment is required. Such assays require careful development towards conditions that combine quantitative release with absence of degradation. For all content assays, recovery is the critical part. Determination of recovery is challenging, especially for systems in which the API is cross-linked. Ideally, direct methods able to quantify API in intact particles should be used. However, such assays are scarce and complex in nature, and may not be suitable for characterisation of complex polymeric drug products. For example, radio-actively labelled API could be directly quantified, but such assay would not be useful to characterise test items and drug products in which the API is used without label. However, such orthogonal methods can be useful in the development and validation phase to determine recovery of the LC-based assay.

Chemically Bound Versus Physically Encapsulated Versus Extra-Micellar Although the total content represents the dose, additional information regarding the chemical and physical state of the API is required to evaluate potential toxicity and efficacy. At the extremes, 100 % of the total content would be present in the extra-micellar form and would be immediately bioavailable upon administration, or 100 % of the total content would be present chemically cross-linked to the polymer (matrix) resulting in no exposure. In practice, the systemic exposure will be in between these extremes and will depend on the initial composition of the formulation as well as on its chemical and kinetic stability upon exposure to blood. Depending on the micellar architecture, the API can be present in 3 states: (1) chemically bound to polymers which are either self-assembled into micelles or which are covalently cross-linked into nanoparticles, (2) physically encapsulated in the core of non-cross-linked micelles or cross-linked nanoparticles and (3) present as free API outside the micelles. The latter will be referred to as extra-micellar API to discriminate it from terminology of free and bound API used in PK/PD which reflects interaction with blood components rather than the micelle. The extra-micellar fraction represents the fraction which is immediately bioavailable and is therefore critical for assessing immediate toxicological and therapeutic effects. In order to quantify the individual fractions, they must be separated from one another. The extra-micellar fraction can be separated from the other two using size-based separation tools such as dialysis, ultrafiltration or gel permeation chromatography. The sum of extra-micellar plus physically encapsulated API can be obtained upon extraction using a solvent. The chemically bound fraction can either be deducted from the total content minus the sum of the extra-micellar plus the physically encapsulated fraction or be experimentally determined by subsequent extraction of the physically encapsulated phase, followed by size-based isolation of the micelles and subsequent forced release of the chemically bound fraction. The major challenges in developing assays to isolate the individual fractions include resolution and creation of artefacts because the separation itself affects the composition of the formulation (e.g. through extraction of physically encapsulated API).

Encapsulation/Entrapment Efficiency and Loading Capacity Besides the absolute values representing the levels of the API and polymer, two relative values are often used to describe micellar systems. The first is called encapsulation efficiency, or sometimes entrapment efficiency, and is calculated by dividing the encapsulated (or entrapped) content by the total feed content. This parameter is relevant from a cost perspective, as it determines the consumption of API. Loading capacity is calculated by dividing total drug content by total polymer content or total micelle content. This parameter is informative for the manufacturing process as well as a critical quality attribute. As the micellar drug product is dosed based on the API content, the loading capacity provides information on the dose of polymer that will be co-administered. Both encapsulation efficiency and loading capacity should be maximised to obtain the most efficient manufacturing process and minimal polymer exposure for patients.

In Vitro Release (IVR) Except for polymeric micelles developed solely as a solubilizing agent, the release profile of the API is one of the most important product characteristics. From the release profile the overall exposure time and peak levels can be estimated. In addition, the release profile within the first 24 h may help to predict what the main site of drug release in the body will be. The majority of polymeric micellar nanomedicines is developed to exploit the EPR effect for targeting of cytostatics to tumors. This requires the API to stay encapsulated for the time needed until the particles have accumulated in the target tissue. All API released from the micelles during their circulation in the central compartment should then be considered as lost payload. However, this immediately released fraction can actually contribute to the therapeutic effect via systemic exposure.

In vitro release assays can serve two purposes: (1) batch to batch comparison to show reproducibility of manufacturing, and (2) prediction of *in vivo* release kinetics (see above). Overall, components of an IVR assay are (1) incubation of the micelles in a medium of choice, (2) separation of the released API from the micelles and (3) quantification of the extra-micellar fraction. Since the latter two components are covered in the previous paragraph, this paragraph will be restricted to selection of the incubation medium. For the first purpose, simple but well-defined buffer systems can be used. For example, polymeric micelles can be incubated in pH 7.4 phosphate buffers of various molarities and in the presence or absence of salt. However, for an IVR assay to have predictive value for *in vivo* PK, careful attention must be paid to the composition of the incubation medium. The impact of proteins (e.g. albumin) and/or lipids should be taken into consideration.

A challenging aspect in the development of suitable release assays is finding the conditions to reach 100% release while minimizing the degradation of the API, or compensating for potential degradation upon IVR testing.

Endotoxin Endotoxins (lipopolysaccharides) are part of the outer membrane of the cell wall of Gram-negative bacteria with very strong immunostimulatory activity. Because of the risk for severe immune reactions, the dose levels for endotoxins in products for intravenous administration to humans should be <5 EU/kg/h. The criteria for animal studies are less stringent, but also there endotoxin levels should be minimized and controlled to enable discrimination of immunostimulatory effects of a drug product from effects caused by endotoxin contamination.

Although methods exist for endotoxin removal, the primary objective is to minimize introduction of endotoxins and Gram-negative bacteria (and all microbes) into a drug product through careful manufacturing. This is achieved by selecting raw materials (including water) of suitable quality, by using clean(ed) equipment and disposables and by working under clean conditions in restricted areas. The final endotoxin levels are mostly quantified using the *in vitro* limulus amoebocyte lysate (LAL) assay while the biological consequence may still be required to be assessed in an *in vivo* rabbit pyrogen test (RPT) as part of the preclinical safety package.

LAL tests can be performed in different variants, as end-point or kinetic assays and with readout based on colour formation, turbidity or gel-clotting (Neun and Dobrovolskaia 2011). However, these assays must be verified or validated for application to nanoparticle formulations as these can interfere with the assay (Dobrovolskaia et al. 2010). Rather predictable interferences are those of coloured or turbid nanomedicine formulations. However, less predictable interferences were also discovered in the extensive test program by the Nanotechnology Characterization Laboratory (NCL). For example, nanoparticles filtered through cellulose-based filters were shown to generate false-positive results (Dobrovolskaia et al. 2010). Such interferences should be evaluated for each specific product and a decision tree for choosing a suitable LAL-assay published by the NCL can be consulted (Dobrovolskaia et al. 2009). General recommendations from the NCL are (1) to verify assays by including appropriate inhibition and enhancement controls (IEC) and (2) to run two assays in parallel to check consistency and identify potential bias (Crist et al. 2013) and (3) to verify *in vitro* tests by *in vivo* RPT analysis in case two *in vitro* assays give inconsistent results or if only one LAL-assay is valid (Dobrovolskaia et al. 2010).

A concern which is not addressed by the NCL, but is of importance for solid or cross-linked particles is the risk for false negatives due to entrapment of endotoxins inside particles. Given the size of endotoxins, this is of particular relevance for large sized particles such as microspheres. Endotoxin monomers with sizes around or even below 10 kDa however exist, which could fit inside nanoparticulate cores (Anspach 2001). Therefore, special attention should also be paid to adequate sample preparation to ensure release of any endotoxins from the nanoparticles prior to quantification.

Pharmacology, PK/PD

At the time of writing, eight polymeric micelle products have reached clinical testing or marketing status (Table 1). All eight products are formulations of cytostatic agents and were originally developed for cancer, although one product (Paxceed) was further investigated for rheumatoid arthritis, multiple sclerosis, psoriasis and neurological disorders. Four out of eight products (BIND-014, Genexol-PM, NK105 and Paxceed) are formulations of antimitotic cytostatics (the taxanes paclitaxel and docetaxel) that act via binding and stabilization of microtubules, resulting in cell cycle arrest at the G2/M phase and subsequent apoptosis. Two products (NK911, SP1049C) contain the anti-tumor antibiotic doxorubicin, which binds to nucleic acids and interferes with mitotic processes. One product delivers the topo-isomerase I inhibitor SN-38 (NK012), which exerts its cytotoxicity via interfering with breaking and religation of DNA strands during the normal cell cycle. The last product is a polymeric micelle formulation of cisplatin (NC-6004), a platinum-based anticancer drug that binds to DNA and initiates crosslinking of DNA strands and/or DNA with proteins. A summary of the preclinical and clinical data for each of the eight pro-ducts is given below (in alphabetic order).

Examples from the Clinic

BIND-014 (Bind 014, Prostate-Specific Membrane Antigen (PSMA) Targeted Docetaxel Nanoparticles, DTXL-TNP) BIND-014 is an actively targeted PEG-PLA-based polymeric micelle formulation of docetaxel which is being developed for hormone refractory metastatic prostate cancer (phase II), non-small cell lung cancer (phase II), and (unspecified) solid tumors (phase I) (BIND Biosciences 2013a, b, c).

BIND-014 is prepared by self-assembly of particles consisting of 10 wt% docetaxel (physically) encapsulated in 97.5 % PLA-PEG (16-kD PLA, 5-kD-PEG), and 2.5 % PLA-PEG-ACUPA (also 16-kD PLA, 5-kD-PEG and S,S-2-[3-[5-amino-1-carboxypentyl]-ureido]-pentanedioic acid (ACUPA)). The obtained suspension is diluted with an aqueous polysorbate 80 solution, purified, concentrated and stored as frozen suspension in aqueous 10 % sucrose solution (Lu and Park 2013). This formulation was designed based on *in vitro* and *in vivo* screening of a combinatorial library in which the composition of the copolymers (PLA, PLGA and PEG length, ratio of glycolic to lactic acid) as well as of the micelles (ratio of PLA-PEG-ACUPA, docetaxel, PLA, PLGA, PLA-PEG, PLGA-PEG) was varied (Gu et al. 2008; Shi et al. 2011; Hrkach et al. 2012). The small-molecule ACUPA is exploited for targeting to prostate-specific membrane antigen (PSMA) expressing cells (Hrkach et al. 2012) such as prostate tumor cells and tumor-associated neo-vasculature cells (Chang et al. 1999).

The active targeting concept was confirmed *in vitro* and *in vivo*, based on a 77-fold increase in cell association of a PSMA-targeted RNA aptamer A10 versus a non-targeted formulation (Lupold et al. 2002; Farokhzad et al. 2004) and 3.77-fold increased intra-tumoral levels of the API upon administration of the targeted versus non-targeted formulation in mice bearing human PSMA-positive prostate xenograft tumors (Cheng et al. 2007).

BIND-014 showed an altered PK-profile compared to solvent-based docetaxel (e.g., Taxotere®, a formulation of docetaxel in polysorbate 80) in several *in vivo* preclinical as well as in clinical studies. Interim results of three patients demonstrated that docetaxel plasma levels were two orders of magnitude higher when administered at the 30 mg/m^2 dose level as BIND-014 as compared to solvent-based docetaxel (Hrkach et al. 2012). Docetaxel levels upon administration of BIND-014 exceeded those after solvent-based docetaxel at all except for the 1 h time points and were sustained for at least 48 h (Hrkach et al. 2012). Dose-proportionality was shown in the range between 3.5 and 75 mg/m^2.

Efficacy of BIND-014 is substantiated by *in vitro, in vivo* as well as intermediate clinical data. Increased cytotoxicity *in vitro* and tumor inhibition and prolonged survival in *in vivo* xenograft models were observed for ligand-targeted compared to ligand-lacking docetaxel nanoparticles (Hrkach et al. 2012; Farokhzad et al. 2006). In addition, a lack of difference between ligand-targeted versus ligand-lacking formulations was shown in PSMA-negative xenograft models, further substantiating the active targeting concept (Hrkach et al. 2012). A phase I study in patients with advanced or metastatic cancer was performed with the primary objective to assess the dose limiting toxicities (DLTs) and define the MTD of BIND-014 given intravenously once per 3 weeks (Q3W) or weekly (Q1W) both in a 28-day cycle (BIND Biosciences 2013a). At the Q3W schedule, BIND-014 was shown to be well-tolerated with neutropenia as the dose limiting toxicity and minimal neuropathy, fluid retention, mucositis, rash and nail changes as additional toxicities. In a press release, it was reported that neutropenia was observed to be considerably less for the Q1W schedule compared to the Q3W schedule (BIND Therapeutics 2014). The MTD was defined at 60 mg/m^2 for a Q3W schedule, at which dose an anti-tumor response was shown in 9 out of 28 patients, including 1 complete response, 3 partial responses and 5 patients with stabilized disease (up to at least 4 treatment cycles) (Von Hoff et al. 2013). Disappearance or shrinkage of metastases or lesions was observed in two patients treated at dose levels below 30 mg/m^2 (Hrkach et al. 2012).

Two phase II studies investigating the safety and efficacy of BIND-014 at the Q3W schedule in patients with metastatic castration-resistant prostate cancer or as second-line therapy for patients with lung cancer have recently been initiated (BIND Biosciences 2013b, c).

Genexol-PM (Cynviloq™, Genexol PM, IG 001, IG001, Paclitaxel PM) This product is known under the names Cynviloq, Genexol PM, IG 001, IG001, Paclitaxel PM. Genexol-PM is most commonly used in literature. Genexol-PM is developed for various types of cancer, including metastatic breast cancer, non-small cell lung cancer, pancreatic cancer, ovarian cancer and bladder cancer. It is marketed under the name Genexol-PM in India, the Philippines and Vietnam (metastatic breast cancer and metastatic non-small cell lung cancer) and the Republic of Korea (metastatic breast cancer, metastatic non-small cell lung cancer, ovarian cancer). In the US, it is in development under the name Cynviloq for metastatic breast cancer and non-small cell lung cancer. In 2013, the FDA Division of Oncology Products 1 approved to follow the 505(b)(2) Bioequivalence (BE) regulatory submission pathway for Cynviloq.

Genexol is a formulation of paclitaxel physically encapsulated in mPEG-b-poly(D,L-lactide) (mPEG-PDLLA) copolymeric micelles. Micelles are prepared via the solvent evaporation technique, resulting in particles of 20–50 nm and a drug loading of 16.7 % (Kim et al. 2004). The micelles are sterile filtered and freeze-dried. The cake is reconstituted in saline and further diluted in 5 % dextrose prior to administration (Kim et al. 2001, 2004).

Preclinical *in vitro* and *in vivo* studies proved empty Genexol-PM micelles to be non-toxic and biocompatible, and to lack intrinsic anti-tumor activity (Kim et al. 2001; Burt et al. 1999). *In vitro* activity of paclitaxel-loaded Genexol-PM was shown to be similar to Taxol, as tested in breast, colon, ovarian and NSCLC cells (Kim et al. 2001; Zhang et al. 1997a). Anti-tumor activity was also shown *in vivo* (Kim et al. 2001; Zhang et al. 1997b). Mice with subcutaneous ovarian and breast cancer tumors were treated with Genexol-PM or Taxol at their MTDs. In this comparison, Genexol-PM proved superior. Activity was most impressive in the breast cancer model, in which some complete responses were observed. Plasma AUC upon equimolar dosing was 5-fold lower for Genexol-PM than for Taxol (Zhang et al. 1997b). Organ (including tumor) AUC increased with increasing dose. The low plasma AUC indicates premature release and kinetically unstable micelles. This was further supported by data from a dual-label biodistribution study. The 14C-distribution, indicative of polymer distribution, deviated from that of tritium, indicative for paclitaxel, and both were eliminated via different routes (Burt et al. 1999). This instability is thought to be caused by decomposition of the micelles upon exposure to α- and β-globulins (Chen et al. 2008). When looking at the toxicity profile, Genexol-PM had a favourable profile with an increased MTD. Overall, the safety and efficacy profile served as the basis to proceed to clinical testing.

The first phase I trial was performed in South Korea in 2001 (Kim et al. 2004). Twenty-one patients were treated with Genexol-PM administered as a 3 h infusion once every 3 weeks (Q3W) at doses ranging between 135 and 390 mg/m^2. The MTD was reached at 390 mg/m^2, with neutropenia, sensory neuropathy and myalgia as the dose limiting toxicities (for comparison, the MTDs of reference products Taxol and Abraxane® are 175 and 300 mg/m^2, respectively (Ibrahim et al. 2005)). Hypersensitivity reactions necessitating premedication in the case of Taxol were not observed after administration of Genexol-PM even though premedication was not administered. Three hundred milligrams per meter square as a 3-h infusion every 3 weeks was reported as the recommended dose for phase 2 trials. Evaluation of the PK revealed that plasma AUC for Genexol-PM was lower than that known for Taxol. Regardless of the inferior PK-profile, preliminary signs of clinical efficacy were obtained, with three partial responders and six patients with stable disease.

A second phase I trial was performed to evaluate a weekly dosing schedule (3 out of every 4 weeks) (Lim et al. 2010). Twenty-four patients were administered 80–200 mg/m^2 via a 1 h infusion. In this regimen, the MTD and recommended dose (RD) were set at 200 and 180 mg/m^2, respectively. Incidence of neutropenia was comparable to that in the Q3W schedule, but the Q1W schedule induced less severe non-hematological toxicities. Five partial responders and nine patients with stable disease were reported. Both PK-studies revealed a linear PK profile, distinct from the typical non-linear PK of Taxol which has been attributed to the presence of Cremophor.

Data from phase II trials studying Genexol-PM in metastatic breast cancer (Lee et al. 2008) and in advanced or metastatic pancreatic cancer (Saif et al. 2010), and combination therapy of Genexol-PM with cisplatin in non-small cell lung cancer (NSCLC) (Kim et al. 2007) are available. In the metastatic breast cancer trial, 41 patients were treated with 300 mg/m^2 administered as a 3 h infusion Q3W until disease progression or intolerability. The most important toxicities were haematological (51.2 % grade 3 and 17.1 % grade 4 neutropenia, but no febrile neutropenia; 22 % grade 1–2 thrombocytopenia), sensory peripheral neuropathy (51.2 %), and myalgia (2.4 %). Unlike in the phase I studies, hypersensitivity reactions were reported in 19.5 % of the patients. An overall response rate of 58.5 %, five complete responses and 19 partial responses were reported. The median time to disease progression was 9 months. These response rates are similar to those reported for Abraxane and superior to those reported for Taxol (Gong et al. 2012) at their MTDs, which could be explained by the higher doses and/or increased tumor accumulation.

The same dosing regimen was applied in the pancreatic cancer trial (Saif et al. 2010), but higher doses were administered. After initial treatment of the first 11 patients with 435 mg/m^2, the remaining 45 patients were dosed at 350 or 300 mg/m^2. Neutropenia (40.0 %), fatigue (17.8 %), infection, dehydration, neuropathy (each 13.3 %), and abdominal pain (11.1 %) were reported as the most important grade 3 toxicities. One complete response and two partial responses were observed and the median progression free survival was 2.8 months.

In the NSCLC study (Kim et al. 2007), both Genexol-PM and cisplatin were given Q3W with cisplatin dosed at 60 mg/m^2 and the Genexol-PM dose titrated on an individual basis. The mean Genexol-PM dose used was 252 mg/m^2. The major hematological and non-hematological toxic effects were grade 3/4 neutropenia (29.0 and 17.4 %, respectively), grade 3 peripheral sensory neuropathy (13.0 %) and grade 3/4 arthralgia (7.3 %). Grade 3/4 hypersensitivity reactions were experienced by four patients. The overall response rate was 37.7 %, with a median time to progression of 5.8 months and a median survival period of 21.7 months. Based on comparisons with other trials, superior efficacy was concluded for Genexol-PM combined with cisplatin as compared to the combination of Taxol with cisplatin

(Taxol administered as a 3 h infusion at 175–200 mg/m^2 combined with cisplatin dosed at 75–80 mg/m^2), combination therapy of free paclitaxel with cisplatin and to monotherapy with Abraxane (Kim et al. 2007; Gong et al. 2012).

Based on the results available from the clinical studies performed plus the post-marketing surveillance in outside US territories, the FDA approved pursuance of the 505(b)(2) bioequivalence regulatory pathway using Taxol and Abraxane as the reference drugs. The pivotal bioequivalence study of Cynviloq in metastatic breast cancer patients was initiated in 2014 (IgDraSol 2014).

NC-6004 NC-6004 is a PEG-p(Glu) based polymeric micelle formulation of cisplatin (cis-dichlorodiammineplatinum (II) (CDDP)) under development as a combination therapy with gemcitabine for locally advanced or metastatic pancreatic cancer (III) (Nanocarrier Co. 2014; Orient Europharma Co. 2014). Micelle formation and encapsulation of CDDP are driven by spontaneous assembly and the coordinate complex formation of p(Glu) and CDDP. The resulting product has a cmc below 5.10^{-7} M, a loading capacity of 39 % and an average size of approximately 30 nm with a narrow size distribution (Nishiyama et al. 2003; Uchino et al. 2005). CDDP is an important anti-tumor agent but use of the conventional formulation is hampered by rapid clearance and dose limiting nephrotoxicity (Hartmann and Lipp 2003). NC-6004 was therefore developed to alter the biodistribution and PK profile with the overall aim of increasing intra-tumoral levels of CDDP. NC-6002 was observed to have a decreased cytotoxicity compared to free cisplatin in *in vitro* studies, which presumably is due to the sustained and reduced cisplatin levels (Uchino et al. 2005). *In vivo* however, NC-6004 was observed to be more effective than free cisplatin in a colon adenocarcinoma and a gastric cancer mice model as well as a cisplatin-resistant model (Nishiyama et al. 2003; Uchino et al. 2005; Alami et al. 2006). The PK profile was also studied (Uchino et al. 2005), revealing a 65 and 8-fold increased plasma AUC and C_{max}, respectively, and a 3.6-fold increased tumor AUC for total cisplatin levels after NC-6004 treatment compared to control cisplatin treatment. In addition, NC-6004 was considered less toxic based on absence of body weight losses. Decreased levels in the sciatic nerve as well as reduced neurotoxicity for NC-6002 administered to rats were also reported (Matsumura and Kataoka 2009). NC-6004 had a distinct distribution profile, with extensive distribution into liver and spleen but reduced distribution into the kidneys. This profile is consistent with the decreased nephrotoxicity observed.

A phase I study was performed in the UK in 2006 (Plummer et al. 2011). Seventeen patients were treated with NC-6004 administered once per 3 weeks (Q3W) at a dose ranging from 10 to 120 mg/m^2. In general, NC-6004 was well tolerated with minimal nephrotoxicity and no significant myelosuppression, ototoxicity, emesis, or neurotoxicity. One patient dosed at 90 mg/m^2 experienced a DLT. The unexpected development of severe renal impairment and hypersensitivity reactions (despite prophylactic medication and hydration measures) at all dose levels prevented the MTD as defined in the protocol to be identified. Taking into account the severity of these reactions, the MTD and RD were defined at 120 and 90 mg/m^2. In the PK study, total cisplatin, micellar-encapsulated cisplatin (described as gel-filterable cisplatin) and extra-micellar cisplatin (described as ultrafilterable cisplatin) were discriminated. When comparing total cisplatin levels after NC-6004 to those obtained at equimolar doses of cisplatin, an 11-fold increased AUC and prolonged $t_{1/2}$ and decreased clearance were seen. The PK-profile of the intra-micellar fraction resembled that of the total fraction. For the extra-micellar (active) fraction, a 34-fold decrease in C_{max} combined with an 8.5-fold increased plasma AUC and 230-fold increased $t_{1/2}$ as compared to free cisplatin were observed, supporting the sustained release profile of NC-6004. Stabilisation of disease in seven patients was a first indication of clinical efficacy.

Results of a subsequent phase I/II study of the combination therapy with NC-6004 and gemcitabine in pancreatic cancer (Nanocarrier Co. 2014) reported in a conference abstract (Su et al. 2012) showed an MTD and RD of 120 and 90 mg/m^2 for NC-6004 combined with gemcitabine. One patient had a partial response and ten patients developed stable disease. In this study, prophylactic steroid therapy succeeded in preventing the hypersensitivity reactions that complicated the UK-trial. In January 2014, a phase III study was initiated to study the combination of NC-6004 and gemcitabine versus monotherapy with gemcitabine (Orient Europharma Co. 2014).

NK012 (NK 012) NK012 is currently in phase II development for colorectal cancer (Japan), metastatic breast cancer and small cell lung carcinoma (US) (Matsumura 2011). It is a formulation of SN-38, the active metabolite of irinotecan hydrochloride (CPT-11). CPT-11 is converted into SN-38 by carboxylesterases, but the conversion is incomplete. The metabolite SN-38 itself was shown to be up to 1000-fold more potent against various cancer cells *in vitro*. Therefore, developing a formulation of SN-38 rather than CPT-11 would be highly beneficial. In NK012, SN-38 is covalently entrapped upon conjugation to the P(Glu) segment of PEG-P(Glu) copolymers via a condensation reaction between the carboxylic acid on PGlu and the phenol on SN-38, followed by self-assembly in aqueous media (Koizumi et al. 2006). This procedure yields very small micelles of 20 nm with a loading capacity of 20 wt%. The obtained micelles are freeze-dried and reconstituted in 5% glucose prior to administration. Because the ester bond between SN-38 and P(Glu) is degraded gradually at weak basic condition (PBS, pH 7.4) but not under acidic conditions, the product is considered stable upon storage while the micelles will release SN-38 after administration, preferably after their accumulation in tumors.

Several preclinical studies in various cancer models showed promising anti-tumor activity for NK012, with a significantly enhanced anti-tumor activity compared to CPT-11 Matsumura 2011). In the preclinical studies, it was demonstrated that NK012 exerted significantly more potent antitumor activity with no intestinal toxicity against various orthotopic human tumor xenografts than CPT-11.

Phase I dose escalation studies with similar dosing schemes but different dose ranges were performed in Japan (Hamaguchi et al. 2010) and in the US (Burris et al. 2008). NK012 was administered as a 30 min infusion once per 3 weeks (Q3W). In Japan, the dose was escalated from 2 to 28 mg/m^2 and in the US from 9 to 37 mg/m^2. Protocol definition of MTD was not reached. DLTs were experienced in the first treatment cycle by two out of nine patients dosed at 28 mg/m^2 in the Japanese trial. In the US trial, one out of six patients dosed at 28 mg/m^2 and two out of five patients dosed at 37 mg/m^2 experienced DLTs. The recommended dose for phase II was defined at 28 mg/m^2. Neutropenia or related events were identified as the most important dose limiting toxicity. Non-hematological toxicity was minimal, and cholinergic toxicity typical for CPT-11 was not observed. The PK profiles were similar in both studies and the PK were shown to be linear in the dose range of 2–28 mg/m^2. Interestingly, this PK study included quantification of total (sum of the chemically bound and the extra-micellar fraction), non-covalently bound SN-38, and its glucuronide SN-38-G. Sustained systemic exposure was confirmed based on prolonged plasma levels of these analytes. Observed half-lifes were 36.0–168 h for the chemically bound SN-38 and 70.7–266 h for non-bound SN-38 (Hamaguchi et al. 2010). Consistent with preclinical data, high systemic exposure and slow elimination were observed for NK012 in comparison to data obtained in the CPT-11 phase I trial (Pitot et al. 2000).

Preliminary anti-tumor activity was observed based on 6 and 2 partial responses in the US and Japanese trial, respectively. The recommended dose for phase II was set at 28 mg/m^2. Phase II trials in small cell lung cancer (Nippon Kayaku 2013a) and metastatic breast cancer (Nippon Kayaku 2013b) have been completed, but no study results have been reported. In addition, a phase II trial for colorectal cancer is ongoing in Japan (Matsumura 2011).

NK105 (NK 105, Paclitaxel Micelle) NK105 is a formulation in which paclitaxel is physically encapsulated in PEG-P(Asp) modified with 4-phenyl-1-butanol for increased core hydrophobicity. It is developed as a safer alternative to Taxol, in which paclitaxel is solubilised by Cremophor and ethanol. NK105 is currently in development for metastatic breast cancer (III) (Nippon Kayaku 2014) and stomach cancer (II).

Paclitaxel is a highly effective anticancer agent, but its use comes with serious toxicities such as neutropenia and peripheral neuropathy. In addition, the Cremophor and ethanol that are used to solubilise it in Taxol formulations is associated with severe hypersensitivity reactions and anaphylactic shock. Taxol treatments are therefore accompanied by premedication with anti-allergic agents and administration of granulocyte colony-stimulating factor to prevent or treat neutropenia. However, no treatment exists for the peripheral neuropathy, which is therefore an important dose limiting toxicity of Taxol.

NK105 is manufactured by self-association of the modified PEG-P(Asp) polymers and paclitaxel in solution, resulting in particles with a mean size of 85 nm, a size distribution between 20–430 nm and a drug loading of 23 wt%. The obtained dispersion is freeze-dried and reconstituted in 5 % glucose prior to administration (Hamaguchi et al. 2005).

Preclinical studies have shown an altered PK and biodistribution profile for paclitaxel formulated as NK105 compared to Taxol and Genexol-PM, increased antitumor activity and reduced toxicity. The PK were studied in plasma and tumor after equimolar dosing of NK105 versus paclitaxel at 50 or 100 mg/kg in C26 tumor-bearing mice (Hamaguchi et al. 2005). Plasma levels after 5 min as well as the AUC were significantly increased: 11–20 fold and 50-86-fold, respectively, for NK105 as compared to free paclitaxel. Importantly, a ~3-fold increase in tumor C_{max}, a 25-fold increase in tumor AUC and a prolongation of intra-tumoral paclitaxel levels (up to 72 h as compared to <24 h for free paclitaxel) were reported, indicating increased exposure of the tumor. Efficacy was studied in BALB/c mice bearing subcutaneous HT-29 colon cancer tumors treated with equimolar doses (25–100 mg/kg) of either NK105 or free paclitaxel (Hamaguchi et al. 2005). Dose-dependent and superior anti-tumor activity was observed for NK105 ($p<0.001$). A dose of 25 mg/kg was sufficient to reach a similar tumor growth suppression as for paclitaxel dosed at 100 mg/kg, whereas a single dose of 100 mg/kg resulted in disappearance of tumors and tumor-free survival. At equimolar doses, significantly less weight loss was observed for NK105-treated compared to paclitaxel-treated animals. Neurotoxicity was studied in rats and was significantly reduced for NK105 based on both electrophysiological and histopathological findings (Hamaguchi et al. 2005). In a separate study, a superior radiosensitizing effect of NK105 compared to paclitaxel was shown in Lewis Lung carcinoma bearing mice treated with radiation plus either NK105 or paclitaxel at 45 mg/kg (Negishi et al. 2006).

In a phase I study NK105 was administered once per 3 weeks as a 1 h infusion without anti-allergic pretreatment in a dose escalation scheme from 10 to 180 mg/m^2 (Hamaguchi et al. 2007). Grade 4 neutropenia was observed as the dose limiting toxicity in two patients treated at the 180 mg/m^2 level. Importantly, only grade 1–2 neurotoxicity was observed, indicating a substantial reduction in neurotoxicity for paclitaxel formulated s NK105. Based on this study, the MTD was defined at 180 mg/m^2, the recommended dose for phase II at 150 mg/m^2 and the DLT as neutropenia. In the subsequent phase II study, patients previously treated for stomach cancer were administered 150 mg/m^2 NK105 (Kato et al. 2012). Neutropenia was confirmed as the most important toxicity, with grade 3/4 non-hematological toxicities being infrequent and no grade 3/4 hypersensitivity. Incidence of grade 3 neuropathy was 1.8 %, which is considerably less than the grade 3/4 neuropathies reported for alternative paclitaxel formulations such as Taxol (10 %), Xyotax (15 %) and Abraxane (11 %). The overall response rate was 25 %, with 2 complete responses and 12 partial responses.

NK911 NK 911 is a polymeric micelle formulation of doxorubicin (adriamycin) developed for metastatic pancreatic cancer. NK911 is a special formulation containing both chemically bound and physically bound doxorubicin. The chemically bound doxorubicin, however, is not entrapped with the intention to have it released, but to modify the hydrophobic core to increase the affinity of the core for physically encapsulated doxorubicin. This design was the result of preliminary development experiments with micelles which contained chemically bound as well as physically encapsulated doxorubicin (Yokoyama et al. 1994, 1998, 1999). Studies with this formulation and controls showed that the physically entrapped fraction accounts for the activity, both *in vitro* and *in vivo*. The chemically bound fraction was shown neither to contribute to plasma levels of doxorubicin nor to its activity. The doxorubicin dimer that was observed in some cases appeared to stabilize the micelles, thereby prolonging the release kinetics. However, because the dimer detrimentally affected product stability, the final formulation was chosen without dimer. In the final product (NK911), doxorubicin is conjugated to the P(Asp) block of PEG-P(Asp) at a substitution level of 45% (Nakanishi et al. 2011). The substitution level was observed to affect the loading efficiency of the micelles. Micelles were formed upon direct dissolution of the conjugated polymer in an aqueous medium, after which free doxorubicin was added which was then incorporated in the micellar core (Nakanishi et al. 2011). After freeze-drying and reconstitution in phosphate buffered saline, the micelles had a size of 40 nm (Nakanishi et al. 2011).

Preclinical studies were performed to compare the PK, efficacy and toxicity of NK911 to that of free doxorubicin (Nakanishi et al. 2011). In C26, M5076 and P388 tumor bearing mice a superior anti-tumor activity and reduced toxicity based on body weight loss and survival were obtained for NK911. Non-superior anti-tumor activity was also shown in Lu-24 and MX-1 tumor bearing mice. C_{max} and plasma AUC (0–24 h) were increased 36.4-fold and 28.9-fold, respectively, compared to free doxorubicin. Moreover, tumor drug levels were shown to be increased upon treatment with NK911.

Efficacy was evaluated in CDF1 female mice with subcutaneous mouse colon carcinoma 26 (C26) tumors. The animals were treated once per 4 days for three cycles with either doxorubicin, doxorubicin chemically entrapped or doxorubicin chemically and physically encapsulated (Yokoyama et al. 1998). The anti-tumor activity of polymeric micelle formulations with high contents of chemically conjugated and physically encapsulated doxorubicin was very high, while that of the polymeric micelle containing chemically bound but lacking physically encapsulated doxorubicin showed negligible *in vivo* activity. Anti-tumor activity of micelles containing physically encapsulated doxorubicin was superior to that of free doxorubicin at equivalent doses.

A phase I study was performed in Japan in 2001 (Matsumura et al. 2004). Twenty-three patients were treated with NK911 at doses ranging from 3 to 67 mg/m^2 administered by infusion once per 3 weeks (Q3W). Grade 3 or 4 neutropenia was observed at 50 en 67 mg/m^2. The MTD was defined at 67 mg/m^2, with grade 4 neutropenia lasting more than 5 days as the dose limiting toxicity. Non-hematological (non-dose limiting) toxicities observed were mild alopecia, stomatitis, and anorexia. Importantly, infusion-related reactions common in patients receiving Doxil™ (see Chapter "Liposomes: the science and the regulatory landscape" on liposomes) were not observed. PK evaluation showed a 2.5-fold increased plasma half-life and 2-fold increased AUC for NK911 compared to free doxorubicin. Clearance and volume of distribution were decreased 2.2-fold and 1.6-fold, respectively. When comparing the PK of NK911 to that of Doxil, superior blood stability is seen for Doxil. One patient had a partial response. Based on the good tolerability, the preliminary signs of efficacy and the lack of infusion-related reactions it was decided to proceed with a phase II study. The recommended dose for this study was 50 mg/m^2 given Q3W. No study results have been reported.

Paxceed (Micellar Paclitaxel) Paxceed consisted of mPEG-PDDLA micelles containing physically encapsulated paclitaxel (25 mg paclitaxel/75 mg polymer) prepared by film hydration (Zhang et al. 1996). *In vivo* (Zhang et al. 1997b), a superior safety profile was shown for Paxceed compared to Taxol based on decreased weight losses at equimolar dosing and an increase in MTD from 20 to 25 mg/kg. At equimolar doses of 20 mg/kg, plasma AUC was decreased 5.5-fold for Paxceed as compared to Taxol. Paxceed did not succeed in an improvement in tumor growth inhibition or regression upon intravenous administration, but was shown to be superior to Taxol upon intraperitoneal administration (both dosed at their MTD). After initial anticancer studies, Paxceed was in development for psoriasis (phase II) (Angiotech Pharmaceuticals 2008a), rheumatoid arthritis (phase II) (Angiotech Pharmaceuticals 2008b), neurological disorders (preclinical) (Zhang et al. 2005) and secondary progressive multiple sclerosis (discontinued), but since 2004 no news has been reported.

SP1049C SP1049C is a PEO-PPO-PEO (Pluronic)-based polymeric micelle formulation of doxorubicin. The micelles are spontaneously formed upon reconstitution of doxorubicin in an 0.9% NaCl solution containing 0.25% (w/v) Pluronic L61 which was shown to contribute to the cytotoxic activity and 2% (w/v) F127 for stabilization of the micelles (Danson et al. 2004; Alakhov et al. 1999). In this composition, micelles of 30 nm are obtained. Based on *in vitro* studies, it was discovered that Pluronic L61 has a sensitizing effect on doxorubicin-resistant cells.

In preclinical biodistribution studies (Alakhov et al. 1999), SP1049C was shown to elicit an increased AUC in tumor and brain, but not in liver, kidney, heart, lung and plasma as compared to non-micellar doxorubicin, indicating a tumor targeting effect. In line with the increased exposure, tumor inhibition and survival time were increased for SP1049C. The toxicity profile of SP1049C was similar to that of doxorubicin, with the same MTD, toxicities and vascular irritation observed at the MTD (Alakhov et al. 1999).

Although SP1049C did not alter the PK profile of doxorubicin, its potential to reduce drug resistance and to deliver increased doses in tumors served as the basis to initiate a phase I trial in 1999 (Danson et al. 2004). Twenty-six patients were treated with SP1049C given at doses ranging from 5 to 90 mg/m^2 and administered as an infusion once per 3 weeks for at least six cycles. At a dose of 35 mg/m^2 and above, toxicities as expected for doxorubicin were observed, with neutropenia as the DLT and 90 mg/m^2 and 70 mg/m^2 as the MTD and RD, respectively. The hand-foot syndrome typically observed for Doxil, a liposomal doxorubicin, was not reported. Promising though not long-lasting responses included three patients with complete or partial response and eight patients with stable disease. In a subsequent phase II trial in patients with advanced carcinoma of the esophagus and gastroesophageal junction, patients were treated with SP1049C once every 3 weeks at the MTD (Valle et al. 2011). Neutropenia was confirmed as the principal toxicity of SP1049C. Non-hematological toxicities included mucositis. Cumulative decrements in left ventricular injection fraction (LVEF) were observed, but the severity and clinical relevance of this seem limited and need to be evaluated. Of the 21 patients, none had a complete response, nine had a partial response (confirmed by post-hoc radiological review) and eight had a minor response or stable disease. SP1049C was granted orphan drug designation in the US and the protocol for an international phase III study has been approved by the FDA.

Considerations Polymeric micelle drug products are developed to increase solubility, to reduce toxicity, to enhance efficacy and/or to increase stability of an API inside the body via altered pharmacokinetics and/or biodistribution. In general, the available PK/PD data discussed in this chapter confirm an improved safety profile, as reflected by the increased MTD of the majority of the micellar drug products (Table 4). Interestingly, however, BIND-014 displayed a decreased MTD as compared to Taxotere, a finding for which no explanation is provided.

Whether polymeric micelle formulations result in superior efficacy remains to be established, as the majority of the products have not yet reached phase III trials.

In a theoretic model, (depicted in Fig. 5a) initially 100% of the API would be intra-micellar (chemically bound and/or physically encapsulated) and finally 100% of the API would be extra-micellar (initial free plus released API). In the initial phase, the API is protected from degradation/metabolism, the body is protected from toxic effects, clearance is fully driven by micellar kinetics, and distribution of the API is determined by the micellar distribution. In practice, it is unlikely that the initial fraction of extra-micellar API will be 0% as formulations will typically contain a certain amount present in free form and a burst release may occur upon administration *in vivo*, especially for formulations in which the API is not chemically conjugated. Assuming that the API is released from the micelles in its native form, its PK/PD profile will be identical to that of a reference drug (e.g. formulated in conventional surfactants). Assuming that the API levels are within the range of linear kinetics, clearance of the API released from micelles will also be identical to that of the reference drug. Therefore, if AUC = dose/clearance, the AUC for the API released from micelles cannot exceed that of the reference drug at equimolar dosing. Intra-micellar API levels are however dominated by micellar kinetics and distribution and will differ significantly from those for the reference drug, generally resulting in increased plasma AUC's. Because clearance of intact micelles will result in decreased systemic exposure to released API, the AUC of the released API will usually be decreased compared to that of the reference drug. Therefore, general expectations for AUC's for total API (intra- plus extra-micellar) and free API (released from micelles) measured in plasma or blood are:

$$\mathrm{AUC}_{\text{total API}} > \mathrm{AUC}_{\text{reference drug}}$$

$$\mathrm{AUC}_{\text{free API}} < \mathrm{AUC}_{\text{total API}}$$

$$\mathrm{AUC}_{\text{free API}} < \mathrm{AUC}_{\text{reference drug}}$$

This shows that a micellar delivery system can increase the AUC for the total API present in bioavailable (released) + non-bioavailable fractions (chemically bound

Table 4 Dose limiting toxicities, MTDs and RDs of polymeric micelle drug products and their comparators (in italic) in clinical trials

API	Product	DLT	MTD (mg/m^2)	RD (mg/m^2)	Ref.
Cisplatin					
	Cisplatin	*Nephrotoxicity*	*100 Q3-4 W* *40 mg/m2/d X 5 (200 mg/m2)*	*100 Q3-4 W*	(Bajorin et al. 1987)
	NC-6004	No DLT except one grade 3 fatigue. However, unexpected hypersensitivity and renal impairment were observed.	120 Q3W	90 Q3W	(Plummer et al. 2011)
Docetaxel					
	Taxotere	*Myelosuppression*	*75 Q3W*[a]		
	BIND-014	Myelosuppression, fatigue Neutropenia	60 Q3W		(Von Hoff et al. 2013)
Doxorubicin					
	Doxorubicin	*Cardiomyopathy, myelosuppression*	*40–60 Q3–4W*[a]		
	Doxil	Palmar plantar erythrodysesthesia, mucositis	50 Q4W (70 for single injection)		(Gabison 2001; Uziely et al. 1995)
	NK911	Neutropenia	67 Q3W	50 Q3W	
	SP1049C	Neutropenia	70 Q3W	70 Q3W	(Danson et al. 2004)

Table 4 (continued)

API	Product	DLT	MTD (mg/m²)	RD (mg/m²)	Ref.
Paclitaxel					
	Taxol	*Myelosuppression*		*135–175 Q3W*[a]	(Gianni et al. 1995)
	Abraxane	*Sensory neuropathy, stomatitis, superficial keratopathy*	*300 Q3W*		(Ibrahim et al. 2002)
	Genexol-PM	Neutropenia, sensory neuropathy and myalgia	390 Q3W 200 Q1W	300 Q3W 180 Q1W	(Kim et al. 2004; Lim et al. 2010)
	NK105	Neutropenia	180 Q3W	150 Q3W	(Hamaguchi et al. 2007)
	PAXCEED	Not found	Not found	75 Q4W or 37.5 Q2W[b]	(Angiotech Pharmaceuticals 2008a)
SN-38					
	CPT-11[c]	*Neutropenia and/or diarrhoea*	*240–250 Q3W* *100-150 Q1W*	*100–150 Q1W* *175 Q2W* *350 Q3W*[d]	(Armand 1996)
	NK012	Neutropenia (related events)	Not available	28 Q3W	(Hamaguchi et al. 2010; Burris et al. 2008)

[a] Most commonly used dose
[b] Highest dose used in phase II trials
[c] Precursor of SN-38; no alternative formulations of SN-38 are available
[d] With antidiarrheal-treatment

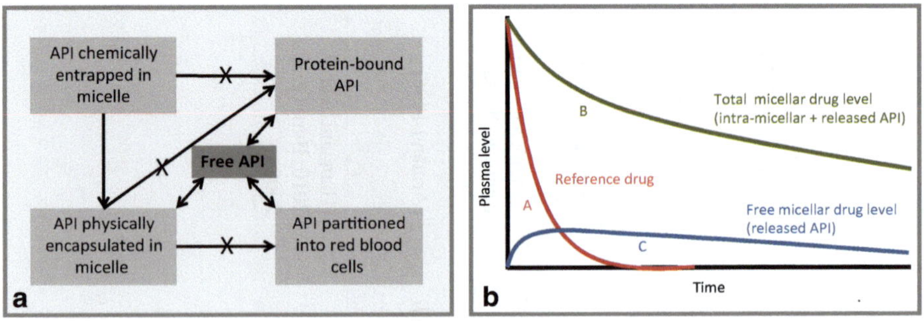

Fig. 5 **a** States in which an API can be present after administration in vivo and their equilibrium and **b** fictional pharmacokinetic profiles for reference drug versus micellar drug

plus physically encapsulated), but not the plasma (or blood) AUC of the bioavailable (free) API itself. If plasma AUC's for free API released from a polymeric micelle are reported to be increased compared to its reference drug at equimolar dosing, the presence of artefacts in the bioanalysis (especially the sample preparation) should be carefully evaluated. The added value of the delivery system lies not in increasing plasma AUC for the free API, but in alteration of the exposure profile by altering the C_{max}, t_{max} and exposure time in plasma/blood (see Fig. 5b). Moreover, the micellar delivery system can change AUCs in target as well as non-target tissues because of the altered distribution profile.

When looking at the available data on pharmacokinetic profiles, the early products such as Paxceed, Genexol-PM and SP1049C failed to show increased plasma AUCs (total API) in comparison to the existing formulations with surfactants, indicating limited stability of these polymeric micelles. The remaining products all showed increased AUCs in plasma and/or organs (including tumors in preclinical studies), providing indirect proof of circulation of intact micelles. However, the data available is almost exclusively obtained based on processing of blood to plasma (or homogenisation of tissues in the presence or absence of albumin) and subsequent extraction using an organic solvent. This sample preparation yields levels of the API that reflect the sum of extra-micellar plus physically encapsulated API and does not allow discrimination between bioavailable, free and non-bioavailable fractions. Also, it does not provide information on red blood cell partitioning of the API in the central compartment. Reporting plasma levels of API therefore provides limited information on the behavior of micellar API and obscures comparisons to non-micellar formulations. It is therefore encouraged to develop assays that do allow separate quantification of bioavailable as well as total drug fractions. The complexity and diversity of these matrices such as plasma, serum, blood, tissues, feces and urine already poses challenges to the bioanalysis of conventional drugs and isolation of micellar versus extra-micellar fractions may therefore not be feasible. In such cases, biodistribution studies with dual labelled formulations should be pursued. Dual labelling of polymer and API enables tracing of released API (no polymer signal detected), of free polymer chains/fragments (no API signal detected) and of intact micelles/drug-polymer conjugates (co-localisation of both signals). Labels and site of labelling should be carefully chosen to prevent artefacts due to altered behavior of the particle, due to loss of signal due to degradation or metabolism or due to generation of non-relevant signals when fragments instead of intact particles are measured.

Regulatory Status

Polymeric micelle drug products are less progressed to the market than other categories in the group of non-complex biological drugs such as liposomes, complex iron nanoparticles and glatiramoids. Currently, no polymeric micelle products have been registered in US or Europe. Outside the US and European territories, Genexol-PM is the only product that has reached marketing status for metastatic breast cancer, non-small cell lung cancer (in India, the Philippines, Republic of Korea and Vietnam) and ovarian cancer (Republic of Korea). At the FDA, in 2012 2 applications and 9 IND's had been submitted for polymeric micelle products. In total, polymeric micelles account for 7 % of all nanotechnology-related platform applications and submissions, a shared 4th position together with superparamagnetic iron oxide (SPIO) after liposomes (39.2 %), nanoparticles (27.2 %) and nanocrystals (13.9 %). Seventy-three percentage of the polymeric micelles are developed for intravenous administration.

Regulatory pathways for the two furthest progressed products Genexol-PM and SP1049C have been identified. Genexol-PM is under development for the treatment of metastatic breast cancer (MBC) and non-small cell lung cancer (NSCLC) in the US under the name Cynviloq. An end of phase II meeting was held by Sorrento Therapeutics, Inc. with the FDA in 2013. Based on data available from the postmarketing surveillance studies conducted outside the US and phase I to III studies in MBC, NSCLC, as well as ovarian, bladder, and pancreatic cancers, the FDA Division of Oncology Products I approved pursuing the 505(b)(2) regulatory submission pathway approach using Abraxane and Taxol as the Reference Listed Drugs. The pivotal study has been initiated in patient with MBC in 2014 (IgDraSol 2014) and the company aims filing an NDA for both MBC and NSCLC in 2015 (Sorrento Therapeutics 2014).

SP1049C is under development for gastrointestinal cancer, colorectal cancer and non-small cell lung cancer. In 2005 and 2008, the FDA granted orphan drug designation for oesophageal cancer and gastrointestinal cancer, respectively. In 2007, the FDA cleared the Investigational New Drug (IND) application for SP1049C for the treatment of metastatic adenocarcinoma of the upper gastrointestinal tract. In the same year, agreement with the FDA was obtained for the design of the study protocol for the randomized Phase III pivotal clinical trial that will compare SP1049C plus Best Supportive Care (BSC) versus BSC alone for the treatment of patients with advanced adenocarcinoma of the esophagus, gastroesophageal junction and stomach who failed adjuvant or 1st or 2nd line chemotherapy. Despite this protocol approval, no development news has been reported since 2008 and the trial is not listed in the ClinicalTrials.gov database.

In Europe, the EMA has issued a 'reflection paper' specifically for the development of block copolymer micelle medicinal products (EMA Committee for Medical Products for Human Use CHMP 2013a). This document provides basic information for the pharmaceutical development, non-clinical and early clinical studies of block copolymer micelle drug products and should be read in connection with generally applicable harmonized (US, EU, Japan) guidelines formulated by the International Conference on 'Harmonisation of Technical Requirements for Registration of Pharmaceuticals for Human Use' (ICH) (http://www.ich.org) as well as relevant regional guidelines. For the US territories, no guideline specific for polymeric micelles is available. Both the EMA and the FDA have issued documents with indirect relevance for the development of polymeric micelle drug products, such as (draft) guidances on liposomes (EMA Committee for Medical Products for Human Use CHMP 2013b, FDA Centre for Drug Evaluation and Rsearch (CDER) 2002), surface coatings (EMA Committee for Medical Products for Human Use CHMP 2013c) and iron-based nano-colloidal products (EMA Committee for Medical Products for Human Use CHMP 2011, 2013b). The EMA has distinguished documents for new nanomedicine and nanosimilars, acknowledging patent expiry of the first generation of nanomedicines such as liposomes, iron-based and nanocrystal-based medicines.

Since guidance on related non-biological complex drugs will be addressed elsewhere in this book, this section will focus on information specifically applicable to polymeric micelles (innovator products).

The EMA 'reflection paper' is focused on products for intravenous administration, which covers the majority of polymeric micelles and all micelles discussed in this book chapter. In this section, selected aspects will be highlighted for discussion whereas for a complete overview the reader is referred to the full text guideline. As a general statement, early discussion with the regulators is strongly encouraged.

Quality Characterization

Thorough evaluation of release kinetics is requested, covering studies on *in vitro* stability, release of the (native) API, degradation of the polymer in plasma or relevant media as well as *in vitro* release studies predictive for release of the API in the circulation and at the target site. In addition, detailed information regarding the physical state and location of the API inside the micelles as well as its surface-association (extra-micellar fraction) is listed. Besides focus on the API, attention is paid to the *in vivo* fate of the polymer, where information on the *in vivo* degradation rate as well as location requested.

This translates into a strong appeal for the development of sophisticated and sensitive (bio)analytical methods. Unfortunately, dedicated research and development on such assay developments is scarce in the public domain. Description of assays used for the characterization and for bioanalysis is often limited to a few sentences in 'materials and methods sections' of papers focused on showing biological proof of concept. An important initiative addressing this gap was the launch of the Nanotechnology Characterization Laboratory (NCL) by the National Cancer Institute (NCI) in 2004, with the aim to accelerate the development of cancer nanomedicines towards the clinic (Nanotechnology Characterization Laboratory 2014; Tinkle et al. 2014). The NCL is a partnership among the NCI, FDA, and the National Institute of Standards and Technology (NIST) and has collaborated with more than 80 companies and academic labs. The NCL has so far developed over 40 protocols that rigorously characterize nanoparticle physicochemical properties, as well as *in vitro* immunological and cytotoxic characteristics and ADME and toxicity profiles in preclinical models. The NCL recommends use of multiple orthogonal methods to measure each physicochemical and performance property of a nanomedicine to compensate for the bias of each individual method. It stresses the importance of thorough characterization to ensure developers to understand their product and to set and justify its specifications.

The protocols developed thus far are largely focused on *in vitro* characterization; physicochemical characterization protocols are limited to size (distribution), chemical composition and zeta-potential assays. With the ongoing efforts it is hoped that the list of protocols will be expanded to cover standardized assays for release studies in various relevant matrices and methodologies for separating encapsulated from non-encapsulated fractions (cf. Chapter "NBCD Pharmacokinetics and Bioanalytical Methods to Measure Drug Release"). Sterility tests are covered under the *in vitro* characterization section and protocols for quantification of endotoxins are available. However, these protocols do not address potential complications due to encapsulation of endotoxins inside cross-linked particles, which would require special sample workup to release the endotoxins prior to their detection.

Manufacturing

The necessity of detailed information on extraction and purification procedures is highlighted. Indeed, these process steps are crucial for the final purity, impurity profile, fraction of extra-micellar API and residual solvents. However, scientific papers describing manufacturing of polymeric micelles are largely focused on the (small scale) micelle formation step, omitting detailed information on downstream processes.

The EMA reflection paper individually addresses components containing block copolymers and/or copolymer-API conjugates, and block copolymer micelle products. Interestingly, a clear distinction is made between micelles formed upon spontaneous self-assembly in aqueous media and those prepared via other routes (as reviewed above). The first is regarded equal to simple dispersion of the block copolymer. The complexity of the different steps in the micelle formation processes calls for careful identification of critical quality attributes (CQAs) and critical process parameters (CPPs) based on a Quality by Design (QbD) approach.

Non-Clinical

In addition to the normal requirements for preclinical PK studies, rate and location of the API release and rate of dissociation of the micelles are listed. For a better understanding of the PK and biodistribution of the API while inside the micelles as well as after its release, validated methods are required for total versus free API in blood/plasma/serum and for total drug levels in organs/tissues.

The PK should describe the C_{max}, AUC and $t_{1/2}$ for free and total drug levels in the central compartment. It is recommended to include the non-micellar drug in PK studies to allow for direct evaluation of the effect of the polymeric micelle on its PK. At a later stage, this could be translated into a recommendation to include the reference product in PK-studies performed with the generic product under development. However, in a discussion on comparability of PK, understanding of bioanalytical assays is essential. NB, the reflection paper defines free API as all API that is outside the micelles, regardless of API-protein-binding or partitioning into red blood cells. Such terminology is confusing, as the terms 'free' and 'total' API are elsewhere used to discriminate the extra-micellar from the intra-micellar API contents in a formulation, and the free drug is commonly used to address the fraction of drug that is not bound to proteins in PK studies. In the case of polymeric micelles which may contain drug chemically bound and/or physically encapsulated, it is suggested to expand the categorisation and specify terminology further into:

- Total: sum of chemically-bound (entrapped), physically encapsulated and extra-micellar API
- Intra-micellar: sum of chemically-bound (entrapped) and physically encapsulated
- Extra-micellar: API that is present outside the micelles due to release/exchange (suggest to restrict the term "free" to reflect non-protein-bound fractions in PK studies)
- Free: extra-micellar, unbound API (not protein-bound)

In addition, one should realize that the free drug will partition between the aqueous plasma and red blood cells whereas the hydrophilic shell of the nanoparticles causes them, together with their payload, to be retained in the plasma fraction. Samples for PK studies are commonly processed by (1) blood sampling followed by (2) processing of blood into plasma, (3) parallel precipitation of plasma proteins and extraction of drug from plasma using an organic solvent (e.g. acetonitrile) and (4) LC-UV based quantification. If such assays are applied to plasma/blood samples from animals (cf. 'reflection paper) after administration of a polymeric micelle drug product, the resulting values would reflect the sum of extra-micellar and physically encapsulated drug. Since the physically encapsulated fraction does not distribute into red blood cells, it will be confined to the plasma fraction. When comparing the data to those obtained for the reference, non-micellar API, plasma levels in micelle-treated animals will greatly exceed those of the reference. Prerequisites for interpretation and comparison of PK data between micellar products and conventional products or other NBCD products are therefore (1) clear recognition of the various physical states of drugs, (2) availability of assays discriminating between these various physical states of drugs, and (3) insight in the red blood cell partitioning.

The EMA reflection document also states that besides evaluation of the fate of the intact API, the metabolic and excretion pathways of the active substance as well as of the micelle constituents should be characterized unless justified otherwise.

Variability in parameters such as size, surface charge and morphology on distribution should be justified in the specifications. However, establishment of such relations would require substantial efforts to (1) create a library of particles in which the parameters are varied in a controllable and reproducible way, (2) screen such libraries *in vivo* and (3) perform analyses to establish the biodistribution based on complex sample preparations or imaging studies. Although the rationale for such studies is acknowledged, realization and financing such studies in early phase development can be challenging, especially for small companies. An example of how such development trajectories could be realized is illustrated by the development of BIND-014 (Gu et al. 2008). Before the final formulation was established, a library of constructs was synthesized and manufactured and, after selection based on *in vitro* characteristics, screened *in vivo*. High throughput screening (allowing creation of large numbers of variables) as well as Design of Experiment (efficient analysis of large numbers of variables) approaches enabled by technological advances offer new opportunities and should be exploited where possible in the development of NBCDs.

For non-clinical pharmacodynamics, safety pharmacology and toxicology, strategies as for conventional drug products and general recommendations in the ICH guidelines should be followed. Special attention should be paid to defining the dose and dosing scheme, which should take into account the PK-profile specific for the micellar drug product. In comparative studies, choices must be made for equimolar versus equitoxic dosing and for relevant intervals in repeated dose studies (to prevent particulate accumulation effects).

Clinical

The EMA reflection paper also discusses considerations for first-in-human studies. Special attention is given to the essence of careful evaluation of non-clinical pharmacokinetic data specific for the micellar product of interest. Sampling time points and duration should be chosen to match the anticipated release profile of the product such that the profiles for free and total API levels can be evaluated. PK parameters of interest are C_{max}, $t_{1/2}$ and AUC of the free as well as the total API fraction in the central compartment (based on plasma, blood or serum). Of special interest are the early time points, which reflect the burst release of API from the micelles. In addition to analysis of the plasma (or related) samples, efforts should be directed at obtaining information on the distribution of the product in target organs and major organs. The preclinical data should be used together with the standard guidelines for selection of the starting dose for first-in-human studies. For studies on dose limiting toxicity, the approach as for conventional drugs can be used, but paying attention to hypersensitivity reactions which may be dose-independent. With regard to product quality, identification and consistency of the critical quality attributes is of importance. Consistency should be shown between non-clinical and clinical products as well as upon changes in the manufacturing process. As for conventional products, test procedures should be validated before initiating clinical testing and the clinical study period must be covered by stability data.

Generics

Given that the first polymeric micelle drug product is yet to be registered, development of *generic* versions will not be announced in the near future and regulatory guidance documentation remains to be written. However, based on resemblances between polymeric micelles and liposomes, future requirements may be derived from the 'Reflection paper on the data requirements for intravenous liposomal products developed with reference to an innovator liposomal product' published by the EMA (EMA Committee for Medical Products for Human Use CHMP 2013b). For development of generic liposomal products an approach is chosen with a strong focus on showing pharmaceutical equivalence prior to any other efforts. The developer is encouraged to apply extensive investigations using state of the art characterization methods to both the innovator and the intended generic product in order to demonstrate with a high level of assurance that the characteristics are comparable. This is of special importance since the developer will likely not have access to procedures used for manufacturing of the innovator product. Thorough characterization and impurity profiling may aid in identifying the manufacturing processes used (e.g. the choice of (co)solvents). Any differences in characteristics identified should be addressed and thoroughly evaluated and justified with respect to safety and efficacy. In addition to the characterization of the normal product, physical and chemical degradation should be compared under stress conditions.

This EMA document on generic versions of liposomes then continues: based on the assessment of pharmaceutical equivalence, the subsequent type and number of non-clinical studies can be defined and justified. Generally, comparative PK (including tissue distribution) would be called for, but it should be decided on a case by case basis which studies could be waived. Interestingly, this EMA reflection paper recommends including assessment of free drug levels in tissues, this in addition to the required total and free levels in blood/serum/plasma and total levels in tissues required for innovator liposomes and indicated in the reflection paper (preclinical phase) for development of block copolymer micelles. Equivalence in non-clinical PD should where possible be evaluated *in vitro* and should include *in vivo* studies at various dose levels. Tissue distribution studies performed as part of the non-clinical PK package are essential for judging equivalence in tissue distribution. Since such evaluations are not possible at the clinical level, the criticality of the preclinical animal studies is such that they should be performed under Good Laboratory Practice (GLP). Tissues of special concern are those related to the safety and efficacy of the drug as well as those involved in significant elimination and processing of the particles. Non-clinical toxicology studies may be waived based on pharmaceutical equivalence and nature of toxicity of the reference product. However, immunoreactogenicity must be addressed. Relevance of such studies for PEGylated polymeric micelles remains to be defined. Clinical PK studies must cover the recommended dose range unless linearity for free, encapsulated and total drug levels is proven. In some cases, quantification of at least 1 metabolite may be used to facilitate assessment of comparability of release rates. In case the elimination rates of unencapsulated and encapsulated drug differ, which is generally the case in all formulations not solely designed as solubilizing agents, clearance, volume of distribution, terminal half-life and partial AUCs should be studied supplementary to the standard PK parameters. The ratio of unencapsulated/encapsulated drug should be followed in time. The need for clinical efficacy studies should be decided on a case by case basis. In case of differences in qualitative composition, additional efforts to prove therapeutic equivalence would be required. It is however clearly stated that this is not the preferred route and exhaustive attempts should be made to reach and demonstrate equivalence of pharmaceutical quality instead. Any discrepancies in pharmaceutical quality between innovator and generic product require justification and are stated to raise serious concerns. This concern was substantiated by the observation that doxorubicin liposomes with a different lipid composition but similar PK profiles differed in their anti-tumor activity in preclinical models (Mamidi et al. 2010).

Prospects, Innovations

For discussion of the prospects for polymeric micelles in drug delivery, three points of focus are selected: (1) increased understanding, (2) increased control, and (3) first product to market.

Increased Understanding

Over the past decades, extensive research and development work was done resulting in a wide range of polymers and polymeric micelles for various indications, with various release mechanisms. The vast body of information is sourced from academic publications. However, with eight products in clinical development and the closing of the gap to marketing and use, products are viewed from a different perspective. Questions, requests and requirements identified during upscaling and GMP manufacturing, design and conduct of preclinical safety packages and comparative studies to test superiority of efficacy, have highlighted the potential of the technology but also shortcomings and blind spots. GMP manufacturing and upscaling drive efforts to detailed characterisation of products and processes, which will in turn lead to an increased understanding of critical process parameters and critical quality attributes. It is hoped that the search for such knowledge by companies developing polymeric micelle drug products also protrudes into the scientific community, leading to systematic studies investigating structure-activity relations. Given the extreme versatility of polymeric micelles, high throughput screening and Design of Experiment approaches are expected to be of great value for such studies, especially when relations between pharmaceutical parameters (e.g. size) and preclinical effects (e.g. distribution, toxicity, efficacy) are included.

Although detailed information is available on procedures for polymer synthesis and micelle formation, an information gap exists between micelle formation and the administration of a drug product. Procedures developed for purification and other downstream processing are hardly available in the public literature. Moreover, description of tools and assays for characterisation and bioanalysis is undervalued and is usually limited to a few sentences as part of the 'materials and methods' section. Given the complexity of the assays required, thorough development and validation as well as a critical assessment of limitations of assays are essential to be able to properly evaluate and interpret data on product characteristics, PK and biodistribution. Protocols are available in the IMPDs for clinical studies, but are not accessible to the public. It could be argued whether publication of pharmaceutical characterisation and bioanalysis procedures should be requested to a same extent as clinical protocols and results. At this point in time, shortcomings of analytical tools and of assays are a bottleneck in the understanding of the *in vivo* behavior of polymeric micelles, as well as other nanomedicine. Initiatives such as the NCL are therefore of utmost importance. When selecting a formulation for commercialization, a confirmatory phase is advised where findings and conclusions drawn in academic setting are verified using higher quality (validated) analytics and test programs (animal models).

Besides being essential for increased understanding, development of relevant and reliable *in vitro* assays can increase efficiency of development programs. Proper study of physicochemical properties and stability under conditions representative for the *in vivo* (and preferably clinical) situation should precede any animal studies for financial as well as ethical reasons.

Increased Control

Innovations in the field of polymeric micelles are focused on increased control over disposition and release of the active compound, as well as expansion of the technology to administration routes less invasive than the intravenous route to which current polymeric micelle products are confined. Increased levels of control are pursued by enhancing the kinetic stability of micelles, targeting the uptake or residence of particles in diseased areas and inducing the release of the bioactive in a given time or location.

The small size and stealth properties of block copolymeric micelles are believed to account for long circulation times and tumor accumulation. In addition to passive accumulation obtained this way, active targeting strategies are pursued. The shell of polymeric micelles can be functionalized and subsequently modified with a targeting ligand selected for specific (diseased) cells or tissues. Polymeric micelles with chemically conjugated monoclonal antibodies (Kabanov et al. 1989), Fab fragments (Torchilin 2004; Torchilin et al. 2003), carbohydrates (Nagasaki et al. 2001; Jule et al. 2003; Cho et al. 2001), peptides (Xiong et al. 2007, 2008; Nasongkla et al. 2004; Yamamoto et al. 1999), folate (Lee et al. 2003a; Yoo and Park 2004; Park et al. 2005) and transferrin (Kursa et al. 2003) have been developed for this purpose. Although this is an elegant approach and safety and altered biodistribution have been shown in some cases, the clinical relevance of targeting ligands for the efficacy of ligand-targeted particulate nanomedicines in general remains to be proven (van der Meel et al. 2013).

Besides aiming to control disposition and/or uptake, strategies to control the time and site of drug release are sought after. Such strategies are based on stimulus-sensitive disassembly and simultaneous drug release or on stimulus-sensitive cleavage of drug molecules from the polymers or intact polymeric micelles (for review see (Cheng et al. 2014; Gao et al. 2013)). Stimuli that are exploited may be passive (biological) or active (externally applied). Examples of passive stimuli that can be exploited include locally decreased pH in tumor and endosomes, elevated glutathione concentrations inside cells or local enzyme overexpression (Ge and Liu 2013; Rijcken et al. 2007b). In case of oral delivery, the changes in pH throughout the gastrointestinal tract offer a further opportunity (Sant et al. 2004; Jones et al. 2003; Sant et al. 2005). pH triggerable polymeric micelles can be categorised in two groups: one uses pH-dependent ionization of weak acid or weak base groups to induce a hydrophobic to hydrophilic conversion of the core-forming block resulting in micellar disassembly; the other one exploits pH-sensitive cleavage of linkers leading to specific release of the API. Successful examples of the first group include micelles with a pNIPAM shell modified with MAA units for intracellular delivery of a photosensitizer (Dufresne et al. 2004), PEG-b-PDMAEMA-b-PDMAEMA triblock polymeric micelles with protonizable tertiary amines (Tang et al. 2003), and PEG-b-PHis + PEG-b-PLLA or biotin-Phis-b-PEG-b-PLLA mixed micelles (Lee et al. 2003a, 2005b) or histidine-grafted PLys containing protonizable histidines (Benns et al. 2000). A typical example of an acid-cleavable linker used for covalent drug entrapment is hydrazine, as used in PEG-b-P(Asp)-hydrazone-doxorubicin (Bae et al. 2005a), doxorubicin-conjugated PLLA-mPEG (Yoo et al. 2002) and core-cross-linked biodegradable polymeric micelles composed of poly(ethylene glycol)-b-poly[N-(2-hydroxypropyl) methacrylamide-lactate) (mPEG-b-p(HPMAm-Lac(n))) (Talelli et al. 2010b). P(Lys)-based polymeric micelles stabilized through disulfide cross-links have been investigated for intracellular disassembly and release of oligonucleotides upon elevated glutathione levels (Kakizawa et al. 1999, 2001; Miyata et al. 2004, 2005). External triggers under investigation are ultrasound (Rapoport 2012) and local heating (Chung et al. 1999, 2000; Kohori et al. 1998; Nakayama et al. 2006; Kim et al. 2000b, c, 2011). Although the previous strategies can be considered as next level control, they are useless if the kinetic stability of micelles is insufficient. The kinetic stability of micelles can be improved by measures, as discussed previously, such as careful matching of the polymeric core and drug chemistry, chemical modification of the core with hydrophobic groups or drug compatible structures and partial crystallisation of the core through stereo-complex formation. In addition to these strategies, micelles can be converted into stable nanoparticles by cross-linking the core and/or shell (for review see Shao et al. 2012). Superior stability for such micelles has been established in several studies *in vitro* and *in vivo*. However, intactness of the polymeric assembly does not guarantee stable residence of the payload inside the assemblies. This was demonstrated in studies with non-cross-linked and cross-linked polymeric micelles of mPEG-b-p(HPMAm-Lacn) (Rijcken et al. 2007a; Soga et al. 2005). The non-cross-linked micelles were shown to be rapidly cleared and to fail to achieve high tumor accumulation, which was ascribed to premature disassembly. Stabilized micelles obtained by end-functionalizing of the polymer lactate groups with methacrylate and subsequent crosslinking were shown to be long-circulating and to achieve enhanced tumor accumulation (Rijcken et al. 2007a). However, when comparing the distribution of labelled paclitaxel to that of labelled polymer, it was revealed that the paclitaxel rapidly leaked out of the intact micelles, resulting in elimination of 95 % of the payload within 30 min after administration (Rijcken 2007). This finding initiated the development of polymeric micelles in which both the core and the drug are cross-linked via covalent but biodegradable links. Tailorable and sustained release kinetics as well as *in vivo* proof of concept of efficacy have been established for various therapeutic agents formulated in these micelles (Coimbra et al. 2012; Crielaard et al. 2012; Quan et al. 2014; Talelli et al. 2010a, b, 2011). The platform is under commercial development under the name CriPec® and CriPec® docetaxel is currently in late stage preclinical development (Cristal Therapeutics 2014).

The development of technologies for ligand-mediated targeting, triggered release, triggered disassembly, thermosensitivity, crosslinking offers a versatile toolbox that enables design of multifunctional polymeric micelles (Xiong et al. 2008; Bae and Kataoka 2009; Bae et al. 2005b; Blanco et al. 2009; Chen et al. 2013; Nasongkla et al. 2006; Torchilin 2009; Yang et al. 2010).

Parallel to the efforts for optimisation of polymeric micelles suitable for intravenous administration, attempts are made to develop polymeric micelles that can increase the bioavailability of poorly water-soluble drugs upon oral delivery through mechanisms such as solubilization, protection against degradation, prolonged residence and bypassing of efflux pumps (Xiong et al. 2012). Preliminary successes have been reported, but the current limited understanding necessitates fundamental studies before specific drug products can be expected.

First Products to Market

In the coming 5 years the first polymeric micelle drug product is anticipated to enter the US market. The first patient in the registration trial of Cynviloq following the 505(b)(2) pathway (testing bioequivalence of Cynviloq against Abraxane as the Reference Listed Drug) has been dosed and the company aims to launch the product in 2016. Follow-up landmarks will be the progression of the first stable polymeric micelle which is not solely functional as a solubilizer, and of the first actively targeted polymeric micelle product. Furthermore, it will be exciting to see which of the next-generation polymeric micelles proves to be worth the added complexity. Lastly, the next era will shed light on the comparative value of polymeric micelles in comparison to established formulations such as liposomes and polymer-drug conjugates.

References

Adams ML, Lavasanifar A, Kwon GS (2003) Amphiphilic block copolymers for drug delivery. J Pharm Sci 92(7):1343–1355

Alakhov V et al (1999) Block copolymer-based formulation of doxorubicin. From cell screen to clinical trials. Colloids Surf B 16(1–4):113–134

Alami N et al (2006) NC-6004, a novel cisplatin-incorporated polymeric micelle, is highly effective against oxaliplatin-resistant tumor models. AACR Meet Abstr 2006(1):133-c-134

Aliabadi HM, Lavasanifar A (2006) Polymeric micelles for drug delivery. Expert Opin Drug Deliv 3(1):139–162

Aliabadi HM, Brocks DR, Lavasanifar A (2005a) Polymeric micelles for the solubilization and delivery of cyclosporine A: pharmacokinetics and biodistribution. Biomaterials 26(35):7251–7259

Aliabadi HM et al (2005b) Micelles of methoxy poly(ethylene oxide)-b-poly(epsilon-caprolactone) as vehicles for the solubilization and controlled delivery of cyclosporine A. J Control Release 104(2):301–311

Aliabadi HM et al (2007) Encapsulation of hydrophobic drugs in polymeric micelles through cosolvent evaporation: the effect of solvent composition on micellar properties and drug loading. Int J Pharm 329(1–2):158–165

Aliferis T, Iatrou H, Hadjichristidis N (2004) Living polypeptides. Biomacromolecules 5(5): 1653–1656

Allen C et al (2000) Polycaprolactone-b-poly(ethylene oxide) copolymer micelles as a delivery vehicle for dihydrotestosterone. J Control Release 63(3):275–286

Angiotech P (2008a) A pilot open-label single-dose study using intravenous micellar paclitaxel for patients with severe psoriasis. In: ClinicalTrials.gov [Internet], Bethesda (MD): National Library of Medicine (US). http://clinicaltrials.gov/ct2/show/NCT00006276?term=paxceed&rank=2&rank=2, NLM Identifier: NCT00006276. Accessed June 2014

Angiotech P (2008b) A phase 2 open-label clinical study using intravenous paxceed™ to treat patients with rheumatoid arthritis. In: ClinicalTrials.gov [Internet], Bethesda (MD): National Library of Medicine (US). http://clinicaltrials.gov/ct2/show/NCT00055133?term=paxceed&rank=1&rank=2, NLM Identifier: NCT00055133. Accessed June 2014

Anspach FB (2001) Endotoxin removal by affinity sorbents. J Biochem Biophys Methods 49(1–3):665–681

Armand JP (1996) CPT-11: clinical experience in phase I studies. Semin Oncol 23(1 Suppl 3): 27–33

Bae Y, Kataoka K (2009) Intelligent polymeric micelles from functional poly(ethylene glycol)-poly(amino acid) block copolymers. Adv Drug Deliv Rev 61(10):768–784

Bae Y et al (2005a) Preparation and biological characterization of polymeric micelle drug carriers with intracellular pH-triggered drug release property: tumor permeability, controlled subcellular drug distribution, and enhanced in vivo antitumor efficacy. Bioconjug Chem 16(1):122–130

Bae Y et al (2005b) Multifunctional polymeric micelles with folate-mediated cancer cell targeting and pH-triggered drug releasing properties for active intracellular drug delivery. Mol Biosyst 1(3):242–250

Bajorin D, Bosl GJ, Fein R (1987) Phase I trial of escalating doses of cisplatin in hypertonic saline. J Clin Oncol 5(10):1589–1593

Benahmed A, Ranger M, Leroux JC (2001) Novel polymeric micelles based on the amphiphilic diblock copolymer poly(N-vinyl-2-pyrrolidone)-block-poly(D, L-lactide). Pharm Res 18(3):323–328

Benns JM et al (2000) pH-sensitive cationic polymer gene delivery vehicle: N-Ac-poly(L-histidine)-graft-poly(L-lysine) comb shaped polymer. Bioconjug Chem 11(5):637–645

BIND Biosciences (2013a) A phase 1 open-label, safety, pharmacokinetic and pharmacodynamic dose-escalation study of BIND-014 (docetaxel nanoparticles for injectable suspension), given by intravenous infusion to patients with advanced or metastatic cancer. In: ClinicalTrials.gov [Internet], Bethesda (MD): National Library of Medicine (US). http://www.clinicaltrials.gov/ct2/show/NCT01300533?term=bind-014&rank=2, NLM Identifier: NCT01300533. Accessed June 2014

BIND Biosciences (2013b) An open label, multicenter, phase 2 study to determine the safety and efficacy of BIND-014 (docetaxel nanoparticles for injectable suspension), administered to patients with metastatic castration-resistant prostate cancer. In: ClinicalTrials.gov [Internet], Bethesda (MD): National Library of Medicine (US). http://www.clinicaltrials.gov/ct2/show/NCT01812746?term=bind-014&rank=1, NLM Identifier: NCT01300533. Accessed June 2014

BIND Biosciences (2013c) An open label, multicenter, phase 2 study to determine the safety and efficacy of BIND-014 (docetaxel nanoparticles for injectable suspension) as second-line therapy to patients with non-small cell lung cancer. In: ClinicalTrials.gov [Internet], Bethesda (MD): National Library of Medicine (US). http://www.clinicaltrials.gov/ct2/show/NCT01792479?term=bind-014&rank=3, NLM Identifier: NCT01792479. Accessed June 2014

BIND Therapeutics (2014) BIND Therapeutics presents clinical data highlighting unique attributes of lead cancer drug candidate, BIND-014, at AACR 2014 Annual Meeting

Blanco E et al (2009) Multifunctional micellar nanomedicine for cancer therapy. Exp Biol Med (Maywood) 234(2):123–131

Burris HA III, Infante JR, Spigel DR et al (2008) A phase I dose-escalation study of NK012. ASCO Annual Meeting Proceedings (Post-Meeting Edition). J Clin Oncol 26(15S):2538

Burt HM et al (1999) Development of copolymers of poly(d,l-lactide) and methoxypolyethylene glycol as micellar carriers of paclitaxel. Colloids Surf B 16(1–4):161–171

Cammas S, Nagasaki Y, Kataoka K (1995) Heterobifunctional poly(ethylene oxide): synthesis of alpha-methoxy-omega-amino and alpha-hydroxy-omega-amino PEOs with the same molecular weights. Bioconjug Chem 6(2):226–230

Chang SS et al (1999) Prostate-specific membrane antigen is produced in tumor-associated neovasculature. Clin Cancer Res 5(10):2674–2681

Chen H et al (2008) Fast release of lipophilic agents from circulating PEG-PDLLA micelles revealed by in vivo forster resonance energy transfer imaging. Langmuir 24(10):5213–5217

Chen YC, Lo CL, Hsiue GH (2013) Multifunctional nanomicellar systems for delivering anticancer drugs. J Biomed Mater Res A 102(6):2024–2038. doi:10.1002/jbm.a.34850

Cheng J et al (2007) Formulation of functionalized PLGA-PEG nanoparticles for in vivo targeted drug delivery. Biomaterials 28(5):869–876

Cheng W et al (2014) Stimuli-responsive polymers for anti-cancer drug delivery. Mater Sci Eng C 45:600–608

Chiappetta DA, Sosnik A (2007) Poly(ethylene oxide)-poly(propylene oxide) block copolymer micelles as drug delivery agents: improved hydrosolubility, stability and bioavailability of drugs. Eur J Pharm Biopharm 66(3):303–317

Cho CS et al (2001) Receptor-mediated cell modulator delivery to hepatocyte using nanoparticles coated with carbohydrate-carrying polymers. Biomaterials 22(1):45–51

Choi JS et al (1999) Poly(ethylene glycol)-block-poly(L-lysine) dendrimer: novel linear polymer/dendrimer block copolymer forming a spherical water-soluble polyionic complex with DNA. Bioconjug Chem 10(1):62–65

Chung JE et al (1999) Thermo-responsive drug delivery from polymeric micelles constructed using block copolymers of poly(N-isopropylacrylamide) and poly(butylmethacrylate). J Control Release 62(1–2):115–127

Chung JE, Yokoyama M, Okano T (2000) Inner core segment design for drug delivery control of thermo-responsive polymeric micelles. J Control Release 65(1–2):93–103

Chung TW et al (2004) Novel micelle-forming block copolymer composed of poly (ε-caprolactone) and poly(vinyl pyrrolidone). Polymer 45(5):1591–1597

Coimbra M et al (2012) Antitumor efficacy of dexamethasone-loaded core-crosslinked polymeric micelles. J Control Release 163(3):361–367

Crielaard BJ et al (2012) Glucocorticoid-loaded core-cross-linked polymeric micelles with tailorable release kinetics for targeted therapy of rheumatoid arthritis. Angew Chem Int Ed Engl 51(29):7254–7258

Crist RM et al (2013) Common pitfalls in nanotechnology: lessons learned from NCI's Nanotechnology Characterization Laboratory. Integr Biol (Camb) 5(1):66–73

Cristal Therapeutics (2014, June) http://www.cristaltherapeutics.com/. Accessed June 2014

Danson S et al (2004) Phase I dose escalation and pharmacokinetic study of pluronic polymer-bound doxorubicin (SP1049C) in patients with advanced cancer. Br J Cancer 90(11):2085–2091

Dimitrov I, Schlaad H (2003) Synthesis of nearly monodisperse polystyrene-polypeptide block copolymers via polymerisation of N-carboxyanhydrides. Chem Commun (Camb) 23:2944–2945

Dobrovolskaia MA, Germolec DR, Weaver JL (2009) Evaluation of nanoparticle immunotoxicity. Nat Nanotechnol 4(7):411–414

Dobrovolskaia MA et al (2010) Ambiguities in applying traditional Limulus amebocyte lysate tests to quantify endotoxin in nanoparticle formulations. Nanomedicine (Lond) 5(4):555–562

Dreher MR et al (2006) Tumor vascular permeability, accumulation, and penetration of macromolecular drug carriers. J Natl Cancer Inst 98(5):335–344

Dufresne MH, Leroux JC (2004) Study of the micellization behavior of different order amino block copolymers with heparin. Pharm Res 21(1):160–169

Dufresne MH et al (2004) Preparation and characterization of water-soluble pH-sensitive nanocarriers for drug delivery. Int J Pharm 277(1–2):81–90

Elhasi S, Astaneh R, Lavasanifar A (2007) Solubilization of an amphiphilic drug by poly(ethylene oxide)-block-poly(ester) micelles. Eur J Pharm Biopharm 65(3):406–413

EMA Committee for Medical Products for Human Use CHMP (2011) Non-clinical studies for generic nanoparticle iron medicinal product applications. http://www.ema.europa.eu/docs/en_GB/document_library/Scientific_guideline/2011/04/WC500105048.pdf. Accessed June 2014

EMA Committee for Medical Products for Human Use CHMP (2013a) Joint MHLW/EMA reflection paper on the development of block copolymer micelle medicinal products. http://www.ema.europa.eu/docs/en_GB/document_library/Scientific_guideline/2013/02/WC500138390.pdf. Accessed June 2014

EMA Committee for Medical Products for Human Use CHMP (2013b) Reflection paper on the data requirements for intravenous liposomal products developed with refernece to an innovator liposomal product. http://www.ema.europa.eu/docs/en_GB/document_library/Scientific_guideline/2013/03/WC500140351.pdf. Accessed June 2014

EMA Committee for Medical Products for Human Use CHMP (2013c) Reflection paper on surface coating: general issues for consideration regarding parenteral administration of coated nanomedicine products. http://www.ema.europa.eu/docs/en_GB/document_library/Scientific_guideline/2013/08/WC500147874.pdf. Accessed June 2014

Engels FK, Mathot RA, Verweij J (2007) Alternative drug formulations of docetaxel: a review. Anticancer Drugs 18(2):95–103

Ernsting MJ et al (2013) Factors controlling the pharmacokinetics, biodistribution and intratumoral penetration of nanoparticles. J Control Release 172(3):782–794

Farokhzad OC et al (2004) Nanoparticle-aptamer bioconjugates: a new approach for targeting prostate cancer cells. Cancer Res 64(21):7668–7672

Farokhzad OC et al (2006) Targeted nanoparticle-aptamer bioconjugates for cancer chemotherapy in vivo. Proc Natl Acad Sci U S A 103(16):6315–6320

FDA Centre for Drug Evaluation and Rsearch (CDER) (2002) Draft guidance on liposome drug products. http://www.fda.gov/downloads/Drugs/GuidanceComplianceRegulatoryInformation/Guidances/ucm070570.pdf. Accessed June 2014
Fournier E et al (2004) A novel one-step drug-loading procedure for water-soluble amphiphilic nanocarriers. Pharm Res 21(6):962–968
Gabizon AA (2001) Pegylated liposomaldoxorubicin metamorphosis of an old drug into a new form of chemotherapy. Cancer Invest 19(4):424–436
Gadelle F, Koros WJ, Schechter RS (1995) Solubilization of aromatic solutes in block copolymers. Macromolecules 28(14):4883–4892
Gao GH, Li Y, Lee DS (2013) Environmental pH-sensitive polymeric micelles for cancer diagnosis and targeted therapy. J Control Release 169(3):180–184
Gaucher G et al (2005) Block copolymer micelles: preparation, characterization and application in drug delivery. J Control Release 109(1–3):169–188
Ge Z, Liu S (2013) Functional block copolymer assemblies responsive to tumor and intracellular microenvironments for site-specific drug delivery and enhanced imaging performance. Chem Soc Rev 42(17):7289–7325
Gelderblom H et al (2001) Cremophor EL: the drawbacks and advantages of vehicle selection for drug formulation. Eur J Cancer 37(13):1590–1598
Gianni L et al (1995) Nonlinear pharmacokinetics and metabolism of paclitaxel and its pharmacokinetic/pharmacodynamic relationships in humans. J Clin Oncol 13(1):180–190
Gindy ME, Panagiotopoulos AZ, Prud'homme RK (2008) Composite block copolymer stabilized nanoparticles: simultaneous encapsulation of organic actives and inorganic nanostructures. Langmuir 24(1):83–90
Gong J et al (2012) Polymeric micelles drug delivery system in oncology. J Control Release 159(3):312–323
Gu F et al (2008) Precise engineering of targeted nanoparticles by using self-assembled biointegrated block copolymers. Proc Natl Acad Sci U S A 105(7):2586–2591
Hamaguchi T et al (2005) NK105, a paclitaxel-incorporating micellar nanoparticle formulation, can extend in vivo antitumour activity and reduce the neurotoxicity of paclitaxel. Br J Cancer 92(7):1240–1246
Hamaguchi T et al (2007) A phase I and pharmacokinetic study of NK105, a paclitaxel-incorporating micellar nanoparticle formulation. Br J Cancer 97(2):170–176
Hamaguchi T et al (2010) Phase I study of NK012, a novel SN-38-incorporating micellar nanoparticle, in adult patients with solid tumors. Clin Cancer Res 16(20):5058–5066
Harada A, Kataoka K (1995) Formation of polyion complex micelles in an aqueous milieu from a pair of oppositely-charged block copolymers with Poly(ethylene glycol) segments. Macromolecules 28(15):5294–5299
Hartmann JT, Lipp HP (2003) Toxicity of platinum compounds. Expert Opin Pharmacother 4(6):889–901
Hong CY, You YZ, Pan CY (2004) Synthesis and characterization of well-defined diblock and triblock copolymers of poly(N-isopropylacrylamide) and poly(ethylene oxide). J Polym Sci, Part A Polym Chem 42:4873–4881
Hrkach J et al (2012) Preclinical development and clinical translation of a PSMA-targeted docetaxel nanoparticle with a differentiated pharmacological profile. Sci Transl Med 4(128):128ra39
Hurter PN, Scheutjens JMHM, Hatton TA (1993a) Molecular modeling of micelle formation and solubilization in block copolymer micelles. 1. A self-consistent mean-field lattice theory. Macromolecules 26(21):5592–5601
Hurter PN, Scheutjens JMHM, Hatton TA (1993b) Molecular modeling of micelle formation and solubilization in block copolymer micelles. 2. Lattice theory for monomers with internal degrees of freedom. Macromolecules 26(19):5030–5040
Ibrahim NK et al (2002) Phase I and pharmacokinetic study of ABI-007, a Cremophor-free, protein-stabilized, nanoparticle formulation of paclitaxel. Clin Cancer Res 8(5):1038–1044
Ibrahim NK et al (2005) Multicenter phase II trial of ABI-007, an albumin-bound paclitaxel, in women with metastatic breast cancer. J Clin Oncol 23(25):6019–6026

ICH http://www.ich.org. Accessed June 2014
IgDraSol (2014) An open-label, randomized, multi-center, single-dose, 2-sequence, 2-period, crossover, comparative bioequivalence study of IG-001 260 mg/m^2 Versus Abraxane® 260 mg/m2 administered intravenously in patients with metastatic or locally recurrent breast cancer. In: ClinicalTrials.gov [Internet], Bethesda (MD): National Library of Medicine (US). http://clinicaltrials.gov/ct2/show/NCT02064829?term=cynviloq&rank=1, NLM Identifier: NCT 02064829. Accessed June 2014
Jette KK et al (2004) Preparation and drug loading of poly(ethylene glycol)-block-poly(epsilon-caprolactone) micelles through the evaporation of a cosolvent azeotrope. Pharm Res 21(7):1184–1191
Jones M, Leroux J (1999) Polymeric micelles—a new generation of colloidal drug carriers. Eur J Pharm Biopharm 48(2):101–111
Jones MC, Ranger M, Leroux JC (2003) pH-sensitive unimolecular polymeric micelles: synthesis of a novel drug carrier. Bioconjug Chem 14(4):774–781
Jule E, Nagasaki Y, Kataoka K (2003) Lactose-installed poly(ethylene glycol)-poly(d,l-lactide) block copolymer micelles exhibit fast-rate binding and high affinity toward a protein bed simulating a cell surface. A surface plasmon resonance study. Bioconjug Chem 14(1):177–186
Kabanov AV, Alakhov VY (2002) Pluronic block copolymers in drug delivery: from micellar nanocontainers to biological response modifiers. Crit Rev Ther Drug Carrier Syst 19(1):1–72
Kabanov AV et al (1989) The neuroleptic activity of haloperidol increases after its solubilization in surfactant micelles. Micelles as microcontainers for drug targeting. FEBS Lett 258(2):343–345
Kabanov AV, Batrakova EV, Alakhov VY (2002) Pluronic block copolymers as novel polymer therapeutics for drug and gene delivery. J Control Release 82(2–3):189–212
Kakizawa Y, Harada A, Kataoka K (1999) Environment-sensitive stabilization of core-shell structured polyion complex micelle by reversible cross-linking of the core through disulfide bond. J Am Chem Soc 121(48):11247–11248
Kakizawa Y, Harada A, Kat K (2001) Glutathione-sensitive stabilization of block copolymer micelles composed of antisense DNA and thiolated poly(ethylene glycol)-block-poly(l-lysine): a potential carrier for systemic delivery of antisense DNA. Biomacromolecules 2(2):491–497
Kang N, Leroux J-C (2004) Triblock and star-block copolymers of N-(2-hydroxypropyl)methacrylamide or N-vinyl-2-pyrrolidone and d, l-lactide: synthesis and self-assembling properties in water. Polymer 45(26):8967–8980
Kang H et al (2002) Self-aggregates of poly(2-hydroxyethyl aspartamide) copolymers loaded with methotrexate by physical and chemical entrapments. J Control Release 81(1–2):135–144
Kang N et al (2005) Stereocomplex block copolymer micelles: core-shell nanostructures with enhanced stability. Nano Lett 5(2):315–319
Kataoka K et al (2000) Doxorubicin-loaded poly(ethylene glycol)-poly(beta-benzyl-L-aspartate) copolymer micelles: their pharmaceutical characteristics and biological significance. J Control Release 64(1–3):143–153
Kato K et al (2012) Phase II study of NK105, a paclitaxel-incorporating micellar nanoparticle, for previously treated advanced or recurrent gastric cancer. Invest New Drugs 30(4):1621–1627
Kim C et al (2000a) Amphiphilic diblock copolymers based on Poly(2-ethyl-2-oxazoline) and Poly(1,3-trimethylene carbonate): synthesis and micellar characteristics. Macromolecules 33(20):7448–7452
Kim IS et al (2000b) Thermo-responsive self-assembled polymeric micelles for drug delivery in vitro. Int J Pharm 205(1–2):165–172
Kim IS et al (2000c) Drug release from thermo-responsive self-assembled polymeric micelles composed of cholic acid and poly(N-isopropylacrylamide). Arch Pharm Res 23(4):367–73
Kim SC et al (2001) In vivo evaluation of polymeric micellar paclitaxel formulation: toxicity and efficacy. J Control Release 72(1–3):191–202
Kim TY et al (2004) Phase I and pharmacokinetic study of Genexol-PM, a cremophor-free, polymeric micelle-formulated paclitaxel, in patients with advanced malignancies. Clin Cancer Res 10(11):3708–3716

Kim DW et al (2007) Multicenter phase II trial of Genexol-PM, a novel Cremophor-free, polymeric micelle formulation of paclitaxel, with cisplatin in patients with advanced non-small-cell lung cancer. Ann Oncol 18(12):2009–2014

Kim SH et al (2011) Thermoresponsive nanostructured polycarbonate block copolymers as biodegradable therapeutic delivery carriers. Biomaterials 32(23):5505–5514

Knop K et al (2010) Poly(ethylene glycol) in drug delivery: pros and cons as well as potential alternatives. Angew Chem Int Ed Engl 49(36):6288–6308

Kohori F et al (1998) Preparation and characterization of thermally responsive block copolymer micelles comprising poly(N-isopropylacrylamide-b-DL-lactide). J Control Release 55(1):87–98

Koizumi F et al (2006) Novel SN-38-incorporating polymeric micelles, NK012, eradicate vascular endothelial growth factor-secreting bulky tumors. Cancer Res 66(20):10048–10056

Kore G et al (2014) Polymeric micelle as multifunctional pharmaceutical carriers. J Nanosci Nanotechnol 14(1):288–307

Kumar N, Ravikumar MN, Domb AJ (2001) Biodegradable block copolymers. Adv Drug Deliv Rev 53(1):23–44

Kumar V et al (2009) Formulation and stability of itraconazole and odanacatib nanoparticles: governing physical parameters. Mol Pharm 6(4):1118–1124

Kursa M et al (2003) Novel shielded transferrin-polyethylene glycol-polyethylenimine/DNA complexes for systemic tumor-targeted gene transfer. Bioconjug Chem 14(1):222–231

Kwon GS (2003) Polymeric micelles for delivery of poorly water-soluble compounds. Crit Rev Ther Drug Carrier Syst 20(5):357–403

Kwon GS, Okano T (1996) Polymeric micelles as new drug carriers. Adv Drug Deliv Rev 21(2):107–116

Kwon GS et al (1993) Biodistribution of micelle-forming polymer-drug conjugates. Pharm Res 10(7):970–974

Kwon GS et al (1995) Physical entrapment of adriamycin in AB block copolymer micelles. Pharm Res 12(2):192–195

Kwon G et al (1997) Block copolymer micelles for drug delivery: loading and release of doxorubicin. J Control Release 48(2–3):195–201

La SB, Okano T, Kataoka K (1996) Preparation and characterization of the micelle-forming polymeric drug indomethacin-incorporated poly(ethylene oxide)-poly(beta-benzyl L-aspartate) block copolymer micelles. J Pharm Sci 85(1):85–90

Lavasanifar A, Samuel J, Kwon GS (2000) Micelles of poly(ethylene oxide)-block-poly(N-alkyl stearate L-aspartamide): synthetic analogues of lipoproteins for drug delivery. J Biomed Mater Res 52(4):831–835

Lavasanifar A, Samuel J, Kwon GS (2001) Micelles self-assembled from poly(ethylene oxide)-block-poly(N-hexyl stearate L-aspartamide) by a solvent evaporation method: effect on the solubilization and haemolytic activity of amphotericin B. J Control Release 77(1–2):155–160

Lavasanifar A, Samuel J, Kwon GS (2002) The effect of fatty acid substitution on the in vitro release of amphotericin B from micelles composed of poly(ethylene oxide)-block-poly(N-hexyl stearate-L-aspartamide). J Control Release 79(1–3):165–172

Le Garrec D et al (2004) Poly(N-vinylpyrrolidone)-block-poly(D, L-lactide) as a new polymeric solubilizer for hydrophobic anticancer drugs: in vitro and in vivo evaluation. J Control Release 99(1):83–101

Lee ES, Na K, Bae YH (2003a) Polymeric micelle for tumor pH and folate-mediated targeting. J Control Release 91(1–2):103–113

Lee ES et al (2003b) Poly(L-histidine)-PEG block copolymer micelles and pH-induced destabilization. J Control Release 90(3):363–374

Lee H et al (2005a) Methoxy poly(ethylene glycol)-block-poly(delta-valerolactone) copolymer micelles for formulation of hydrophobic drugs. Biomacromolecules 6(6):3119–3128

Lee ES, Na K, Bae YH (2005b) Super pH-sensitive multifunctional polymeric micelle. Nano Lett 5(2):325–329

Lee KS et al (2008) Multicenter phase II trial of Genexol-PM, a Cremophor-free, polymeric micelle formulation of paclitaxel, in patients with metastatic breast cancer. Breast Cancer Res Treat 108(2):241–250

Lele BS, Leroux JC (2002) Synthesis and micellar characterization of novel amphiphilic A–B–A triblock copolymers of N-(2-Hydroxypropyl)methacrylamide or N-Vinyl-2-pyrrolidone with Poly(ε-caprolactone). Macromolecules 35(17):6714–6723

Lim WT et al (2010) Phase I pharmacokinetic study of a weekly liposomal paclitaxel formulation (Genexol-PM) in patients with solid tumors. Ann Oncol 21(2):382–388

Lin WJ, Juang LW, Lin CC (2003) Stability and release performance of a series of pegylated copolymeric micelles. Pharm Res 20(4):668–673

Liu L et al (2001) Biodegradable polylactide/poly(ethylene glycol)/polylactide triblock copolymer micelles as anticancer drug carriers. J Appl Polym Sci 80(11):1976–1982

Liu J, Xiao Y, Allen C (2004) Polymer-drug compatibility: a guide to the development of delivery systems for the anticancer agent, ellipticine. J Pharm Sci 93(1):132–143

Liu Z, Wang Y, Zhang N (2012) Micelle-like nanoassemblies based on polymer-drug conjugates as an emerging platform for drug delivery. Expert Opin Drug Deliv 9(7):805–822

Lu Y, Park K (2013) Polymeric micelles and alternative nanonized delivery vehicles for poorly soluble drugs. Int J Pharm 453(1):198–214

Luo L et al (2004) Novel amphiphilic diblock copolymer of low molecular weight Poly(N-vinylpyrrolidone)-block-poly(d,l-lactide): synthesis, characterization, and micellization. Macromolecules 37(11):4008–4013

Lupold SE et al (2002) Identification and characterization of nuclease-stabilized RNA molecules that bind human prostate cancer cells via the prostate-specific membrane antigen. Cancer Res 62(14):4029–4033

Luppi B et al (2002) Poly(vinylalcohol-co-vinyloleate) for the preparation of micelles enhancing retinyl palmitate transcutaneous permeation. Drug Deliv 9(3):147–152

Mamidi RN et al (2010) Pharmacokinetics, efficacy and toxicity of different pegylated liposomal doxorubicin formulations in preclinical models: is a conventional bioequivalence approach sufficient to ensure therapeutic equivalence of pegylated liposomal doxorubicin products? Cancer Chemother Pharmacol 66(6):1173–1184

Matsumura Y (2011) Preclinical and clinical studies of NK012, an SN-38-incorporating polymeric micelles, which is designed based on EPR effect. Adv Drug Deliv Rev 63(3):184–192

Matsumura Y, Kataoka K (2009) Preclinical and clinical studies of anticancer agent-incorporating polymer micelles. Cancer Sci 100(4):572–579

Matsumura Y et al (2004) Phase I clinical trial and pharmacokinetic evaluation of NK911, a micelle-encapsulated doxorubicin. Br J Cancer 91(10):1775–1781

Miyata K et al (2004) Block catiomer polyplexes with regulated densities of charge and disulfide cross-linking directed to enhance gene expression. J Am Chem Soc 126(8):2355–2361

Miyata K et al (2005) Freeze-dried formulations for in vivo gene delivery of PEGylated polyplex micelles with disulfide crosslinked cores to the liver. J Control Release 109(1–3):15–23

Moghimi SM et al (1991) Non-phagocytic uptake of intravenously injected microspheres in rat spleen: influence of particle size and hydrophilic coating. Biochem Biophys Res Commun 177(2):861–866

Moghimi SM, Hunter AC, Murray JC (2001) Long-circulating and target-specific nanoparticles: theory to practice. Pharmacol Rev 53(2):283–318

Molineux G (2002) Pegylation: engineering improved pharmaceuticals for enhanced therapy. Cancer Treat Rev 28(Suppl A):13–16

Nagarajan R, Barry M, Ruckenstein E (1986) Unusual selectivity in solubilization by block copolymer micelles. Langmuir 2(2):210–215

Nagasaki Y et al (1995a) Primary amino-terminal heterobifunctional poly(ethylene oxide). Facile synthesis of poly(ethylene oxide) with a primary amino group at one end and a hydroxyl group at the other end. Bioconjug Chem 6(6):702–704

Nagasaki Y et al (1995b) Formyl-ended heterobifunctional poly(ethylene oxide): synthesis of poly(ethylene oxide) with a formyl group at one end and a hydroxyl group at the other end. Bioconjug Chem 6(2):231–233

Nagasaki Y, Kada T, Scholz C, Iijima M, Kato M, Kataoka K (1998) The reactive polymeric micelle based on an aldehyde-ended poly(ethylene glycol)/Poly(lactide) block copolymer. Macromolecules 31(5):1473–1479

Nagasaki Y et al (2001) Sugar-installed block copolymer micelles: their preparation and specific interaction with lectin molecules. Biomacromolecules 2(4):1067–1070

Nakamura T, Nagasaki Y, Kataoka K (1998) Synthesis of heterobifunctional poly(ethylene glycol) with a reducing monosaccharide residue at one end. Bioconjug Chem 9(2):300–303

Nakanishi T et al (2001) Development of the polymer micelle carrier system for doxorubicin. J Control Release 74(1–3):295–302

Nakayama M et al (2006) Molecular design of biodegradable polymeric micelles for temperature-responsive drug release. J Control Release 115(1):46–56

Nam YS et al (2003) New micelle-like polymer aggregates made from PEI-PLGA diblock copolymers: micellar characteristics and cellular uptake. Biomaterials 24(12):2053–2059

Nanocarrier Co. (2014) Phase I/II study of the combination therapy with NC-6004 and gemcitabine in patients with locally advanced or metastatic pancreatic cancer in Asian countries. In: ClinicalTrials.gov [Internet], Bethesda (MD): National Library of Medicine (US). http://clinicaltrials.gov/ct2/show/NCT00910741?term=nc-6004&rank=2, NLM Identifier: NCT00910741. Accessed June 2014

Nanotechnology Characterization Laboratory (2014, June) http://ncl.cancer.gov/. Accessed June 2014

Nasongkla N et al (2004) cRGD-functionalized polymer micelles for targeted doxorubicin delivery. Angew Chem Int Ed Engl 43(46):6323–6327

Nasongkla N et al (2006) Multifunctional polymeric micelles as cancer-targeted, MRI-ultrasensitive drug delivery systems. Nano Lett 6(11):2427–2430

Negishi T et al (2006) NK105, a paclitaxel-incorporating micellar nanoparticle, is a more potent radiosensitising agent compared to free paclitaxel. Br J Cancer 95(5):601–606

Neradovic D, van Nostrum CF, Hennink WE (2001) Thermoresponsive polymeric micelles with controlled instability based on hydrolytically sensitive N-isopropylacrylamide copolymers. Macromolecules 34(22):7589

Neun BW, Dobrovolskaia MA (2011) Detection and quantitative evaluation of endotoxin contamination in nanoparticle formulations by LAL-based assays. Methods Mol Biol 697:121–130

Nippon Kayaku (2013a) A phase ii study of NK012 in sensitive relapsed and refractory relapsed Small-Cell Lung Cancer (SCLC). In: ClinicalTrials.gov [Internet], Bethesda (MD): National Library of Medicine (US). http://clinicaltrials.gov/ct2/show/NCT00951613?term=nk012&rank=1, NLM Identifier: NCT00951613. Accessed June 2014

Nippon Kayaku (2013b) A phase ii study of NK012 in locally advanced non-resectable and metastatic breast cancer patients with triple negative phenotype. In: ClinicalTrials.gov [Internet], Bethesda (MD): National Library of Medicine (US). http://clinicaltrials.gov/ct2/show/NCT00951054?term=nk012&rank=2, NLM Identifier: NCT00951054. Accessed June 2014

Nippon Kayaku (2014) A multi-national phase III clinical study comparing NK105 versus paclitaxel in patients with metastatic or recurrent breast cancer. In: ClinicalTrials.gov [Internet], Bethesda (MD): National Library of Medicine (US). http://clinicaltrials.gov/ct2/show/NCT01644890?term=nk105&rank=1, NLM Identifier: NCT01644890. Accessed June 2014

Nishiyama N et al (2003) Novel cisplatin-incorporated polymeric micelles can eradicate solid tumors in mice. Cancer Res 63(24):8977–8983

Orient Europharma Co. (2014) A phase iii, open-label, randomized study of the combination therapy with NC-6004 and gemcitabine versus gemcitabine alone in patients with locally advanced or metastatic pancreatic cancer. In: ClinicalTrials.gov [Internet], Bethesda (MD): National Library of Medicine (US). http://clinicaltrials.gov/ct2/show/NCT02043288?term=nc-6004&rank=1, NLM Identifier: NCT02043288. Accessed June 2014

Orienti I et al (2005) Modified doxorubicin for improved encapsulation in PVA polymeric micelles. Drug Deliv 12(1):15–20
Park EK et al (2005) Folate-conjugated methoxy poly(ethylene glycol)/poly(epsilon-caprolactone) amphiphilic block copolymeric micelles for tumor-targeted drug delivery. J Control Release 109(1–3):158–168
Piao L et al (2003) Synthesis and characterization of PCL/PEG/PCL triblock copolymers by using calcium catalyst. Polymer 44(7):2025–2031
Piskin E et al (1995) Novel PDLLA/PEG copolymer micelles as drug carriers. J Biomater Sci Polym Ed 7(4):359–373
Pitot HC et al (2000) Phase I dose-finding and pharmacokinetic trial of irinotecan hydrochloride (CPT-11) using a once-every-three-week dosing schedule for patients with advanced solid tumor malignancy. Clin Cancer Res 6(6):2236–2244
Plummer R et al (2011) A phase I clinical study of cisplatin-incorporated polymeric micelles (NC-6004) in patients with solid tumours. Br J Cancer 104(4):593–598
Popovic Z et al (2010) A nanoparticle size series for in vivo fluorescence imaging. Angew Chem Int Ed Engl 49(46):8649–8652
Quan L et al (2014) Nanomedicines for inflammatory arthritis: head-to-head comparison of glucocorticoid-containing polymers, micelles, and liposomes. ACS Nano 8(1):458–466
Rabinow BE (2004) Nanosuspensions in drug delivery. Nat Rev Drug Discov 3(9):785–796
Rapoport N (2012) Ultrasound-mediated micellar drug delivery. Int J Hyperth 28(4):374–385
Rijcken C (2007) Tuneable and degradable polymeric micelles for drug delivery: from synthesis to feasibility in vivo. From: thesis. Utrecht University, pp 239–260
Rijcken CJ et al (2005) Novel fast degradable thermosensitive polymeric micelles based on PEG-block-poly(N-(2-hydroxyethyl)methacrylamide-oligolactates). Biomacromolecules 6(4):2343–2351
Rijcken CJ et al (2007a) Hydrolysable core-crosslinked thermosensitive polymeric micelles: synthesis, characterisation and in vivo studies. Biomaterials 28(36):5581–5593
Rijcken CJ et al (2007b) Triggered destabilisation of polymeric micelles and vesicles by changing polymers polarity: an attractive tool for drug delivery. J Control Release 120(3):131–148
Rouzes C et al (2000) Surface modification of poly(lactic acid) nanospheres using hydrophobically modified dextrans as stabilizers in an o/w emulsion/evaporation technique. J Biomed Mater Res 50(4):557–565
Saez A et al (2000) Freeze-drying of polycaprolactone and poly(D, L-lactic-glycolic) nanoparticles induce minor particle size changes affecting the oral pharmacokinetics of loaded drugs. Eur J Pharm Biopharm 50(3):379–387
Saif MW et al (2010) Phase II clinical trial of paclitaxel loaded polymeric micelle in patients with advanced pancreatic cancer. Cancer Invest 28(2):186–194
Sant VP, Smith D, Leroux J-C (2004) Novel pH-sensitive supramolecular assemblies for oral delivery of poorly water soluble drugs: preparation and characterization. J Control Release 97(2):301–312
Sant VP, Smith D, Leroux J-C (2005) Enhancement of oral bioavailability of poorly water-soluble drugs by poly(ethylene glycol)-block-poly(alkyl acrylate-co-methacrylic acid) self-assemblies. J Control Release 104(2):289–300
Schmeenk JM, Löwik DWPM, Van Hest JCM (2005) Peptide-containing blockcoplolymers synthesis and potential applications of bio-mimetic materials. Curr Org Chem 9:1115–1125
Shao Y et al (2012) Reversibly crosslinked nanocarriers for on-demand drug delivery in cancer treatment. Ther Deliv 3(12):1409–1427
Shi L, Chapman TM, Beckman EJ (2003) Poly(ethylene glycol)-block-poly(N-vinylformamide) copolymers synthesized by the RAFT methodology. Macromolecules 36(7):2563–2567
Shi J et al (2011) Self-assembled targeted nanoparticles: evolution of technologies and bench to bedside translation. Acc Chem Res 44(10):1123–1134
Shuai X et al (2003) Novel biodegradable ternary copolymers hy-PEI-g-PCL-b-PEG: synthesis, characterization, and potential as efficient nonviral gene delivery vectors. Macromolecules 36(15):5751–5759

Shuai X et al (2004a) Micellar carriers based on block copolymers of poly(epsilon-caprolactone) and poly(ethylene glycol) for doxorubicin delivery. J Control Release 98(3):415–426
Shuai X et al (2004b) Core-cross-linked polymeric micelles as paclitaxel carriers. Bioconjug Chem 15(3):441–448
Slager J, Domb AJ (2003) Biopolymer stereocomplexes. Adv Drug Deliv Rev 55(4):549–583
Soga O et al (2004) Physicochemical characterization of degradable thermosensitive polymeric micelles. Langmuir 20(21):9388–9395
Soga O et al (2005) Thermosensitive and biodegradable polymeric micelles for paclitaxel delivery. J Control Release 103(2):341–353
Sorrento Therapeutics. Cynviloq™ (2014, June) http://sorrentotherapeutics.com/programs/cynviloq/. Accessed June 2014
Sparreboom A et al (1999) Cremophor EL-mediated alteration of paclitaxel distribution in human blood: clinical pharmacokinetic implications. Cancer Res 59(7):1454–1457
Su W-C, Chen L, Chung PL (2012) Phase I/II study of NC-6004, a novel micellar formulation of cisplatin in combination with gemcitabin in patients with pancreatic cancer in Asia ESMO; (Abs# 746 p)
Taillefer J et al (2000) Preparation and characterization of pH-responsive polymeric micelles for the delivery of photosensitizing anticancer drugs. J Pharm Sci 89(1):52–62
Talelli M et al (2010a) Targeted core-crosslinked polymeric micelles with controlled release of covalently entrapped doxorubicin. J Control Release 148(1):e121–e122
Talelli M et al (2010b) Core-crosslinked polymeric micelles with controlled release of covalently entrapped doxorubicin. Biomaterials 31(30):7797–7804
Talelli M et al (2011) Synthesis and characterization of biodegradable and thermosensitive polymeric micelles with covalently bound doxorubicin-glucuronide prodrug via click chemistry. Bioconjug Chem 22(12):2519–2530
Tang Y et al (2003) Solubilization and controlled release of a hydrophobic drug using novel micelle-forming ABC triblock copolymers. Biomacromolecules 4(6):1636–1645
ten Tije AJ et al (2003) Pharmacological effects of formulation vehicles: implications for cancer chemotherapy. Clin Pharmacokinet 42(7):665–685
Teng Y, Morrison ME, Munk P, Webber SE, Prochazka K (1998) Release kinetics studies of aromatic molecules into water from block polymer micelles. Macromolecules 31(11):3578–3587
Tessmar JK, Mikos AG, Göpferich A (2002) Amine-reactive biodegradable diblock copolymers. Biomacromolecules 3(1):194–200
Thompson MS, Vadala TP, Vadala ML, Lin Y, Riffle JS (2008) Synthesis and applications of heterobifunctional poly(ethylene oxide) oligomers. Polymer 49(2):345–373
Tian M et al (1995) Light scattering study of solubilization of organic molecules by block copolymer micelles in aqueous media. J Polymer Sci Part B: Polymer Phys 33(12):1713–1722
Tinkle S et al (2014) Nanomedicines: addressing the scientific and regulatory gap. Ann N Y Acad Sci 1313:35–56
Torchilin VP (2004) Targeted polymeric micelles for delivery of poorly soluble drugs. Cell Mol Life Sci 61(19–20):2549–2559
Torchilin VP (2007) Micellar nanocarriers pharmaceutical perspectives. Pharm Res 24(1):1–16
Torchilin V (2009) Multifunctional and stimuli-sensitive pharmaceutical nanocarriers. Eur J Pharm Biopharm 71(3):431–444
Torchilin VP, Trubetskoy VS (1995) Which polymers can make nanoparticulate drug carriers long-circulating? Adv Drug Deliv Rev 16(2–3):141–155
Torchilin VP et al (2003) Immunomicelles: targeted pharmaceutical carriers for poorly soluble drugs. Proc Natl Acad Sci U S A 100(10):6039–6044
Tyrrell ZL, Shen Y, Radosz M (2010) Fabrication of micellar nanoparticles for drug delivery through the self-assembly of block copolymers. Prog Polymer Sci 35(9):1128–1143
Uchino H et al (2005) Cisplatin-incorporating polymeric micelles (NC-6004) can reduce nephrotoxicity and neurotoxicity of cisplatin in rats. Br J Cancer 93(6):678–687
Uziely B et al (1995) Liposomal doxorubicin: antitumor activity and unique toxicities during two complementary phase I studies. J Clin Oncol 13(7):1777–1785

Valle JW et al (2011) A phase 2 study of SP1049C, doxorubicin in P-glycoprotein-targeting pluronics, in patients with advanced adenocarcinoma of the esophagus and gastroesophageal junction. Invest New Drugs 29(5):1029–1037
van der Meel R et al (2013) Ligand-targeted particulate nanomedicines undergoing clinical evaluation: current status. Adv Drug Deliv Rev 65(10):1284–1298
Van Domeselaar GH et al (2003) Application of solid phase peptide synthesis to engineering PEO–peptide block copolymers for drug delivery. Colloids Surf B 30(4):323–334
van Tellingen O et al (1999) Cremophor EL causes (pseudo-) non-linear pharmacokinetics of paclitaxel in patients. Br J Cancer 81(2):330–335
van Zuylen L, Verweij J, Sparreboom A (2001) Role of formulation vehicles in taxane pharmacology. Invest New Drugs 19(2):125–141
Vangeyte P, Gautier S, Jérôme R (2004) About the methods of preparation of poly(ethylene oxide)-b-poly(ε-caprolactone) nanoparticles in water: analysis by dynamic light scattering. Colloids Surf A: Physicochem Eng Asp 242(1–3):203–211
Vayaboury W et al (2004) Living polymerization of α-amino acid N-carboxyanhydrides (NCA) upon decreasing the reaction temperature. Macromol Rapid Commun 25(13):1221–1224
Von Hoff DDM, Eisenberg P, LoRusso P, Weiss G, Sachdev J, Mita A, Low S, Hrkach J, Summa J, Berk G, Ramanathan R (2013) A phase I study of BIND-014, a PSMA-targeted nanoparticle containing docetaxel, in patients with refractory solid tumors [abstract]. Proceedings of the 104th Annual Meeting of the American Association for Cancer Research, April 2013, Washington, DC: AACR; 2013. Abstract nr LB-203
Woodle MC, Lasic DD (1992) Sterically stabilized liposomes. Biochim Biophys Acta 1113(2):171–199
Xing L, Mattice WL (1997) Strong solubilization of small molecules by triblock-copolymer micelles in selective solvents. Macromolecules 30(6):1711–1717
Xiong XB et al (2007) Conjugation of arginine-glycine-aspartic acid peptides to poly(ethylene oxide)-b-poly(epsilon-caprolactone) micelles for enhanced intracellular drug delivery to metastatic tumor cells. Biomacromolecules 8(3):874–884
Xiong XB et al (2008) Multifunctional polymeric micelles for enhanced intracellular delivery of doxorubicin to metastatic cancer cells. Pharm Res 25(11):2555–2566
Xiong XB et al (2012) Amphiphilic block co-polymers: preparation and application in nanodrug and gene delivery. Acta Biomater 8(6):2017–2033
Yamamoto Y et al (1999) Surface charge modulation of poly(ethylene glycol)–poly(d,l-lactide) block copolymer micelles: conjugation of charged peptides. Colloids Surf B 16(1–4):135–146
Yang X et al (2010) Multifunctional polymeric vesicles for targeted drug delivery and imaging. Biofabrication 2(2):025004
Yokoyama M et al (1990) Characterization and anticancer activity of the micelle-forming polymeric anticancer drug adriamycin-conjugated poly(ethylene glycol)-poly(aspartic acid) block copolymer. Cancer Res 50(6):1693–1700
Yokoyama M et al (1992) Preparation of micelle-forming polymer-drug conjugates. Bioconjug Chem 3(4):295–301
Yokoyama M et al (1994) Improved synthesis of adriamycin-conjugated poly (ethylene oxide)-poly (aspartic acid) block copolymer and formation of unimodal micellar structure with controlled amount of physically entrapped adriamycin. J Control Release 32(3):269–277
Yokoyama M et al (1998) Characterization of physical entrapment and chemical conjugation of adriamycin in polymeric micelles and their design for in vivo delivery to a solid tumor. J Control Release 50(1–3):79–92
Yokoyama M et al (1999) Selective delivery of adriamycin to a solid tumor using a polymeric micelle carrier system. J Drug Target 7(3):171–186
Yoo HS, Park TG (2001) Biodegradable polymeric micelles composed of doxorubicin conjugated PLGA-PEG block copolymer. J Control Release 70(1–2):63–70
Yoo HS, Park TG (2004) Folate receptor targeted biodegradable polymeric doxorubicin micelles. J Control Release 96(2):273–283

Yoo HS, Lee EA, Park TG (2002) Doxorubicin-conjugated biodegradable polymeric micelles having acid-cleavable linkages. J Control Release 82(1):17–27
Yu BG et al (1998) Polymeric micelles for drug delivery: solubilization and haemolytic activity of amphotericin B. J Control Release 53(1–3):131–136
Zeng F, Liu J, Allen C (2004) Synthesis and characterization of biodegradable poly(ethylene glycol)-block-poly(5-benzyloxy-trimethylene carbonate) copolymers for drug delivery. Biomacromolecules 5(5):1810–1817
Zhang X, Jackson JK, Burt HM (1996) Development of amphiphilic diblock copolymers as micellar carriers of taxol. Int J Pharm 132(1–2):195–206
Zhang X et al (1997a) An investigation of the antitumour activity and biodistribution of polymeric micellar paclitaxel. Cancer Chemother Pharmacol 40(1):81–6
Zhang X et al (1997b) Anti-tumor efficacy and biodistribution of intravenous polymeric micellar paclitaxel. Anticancer Drugs 8(7):696–701
Zhang SQJ, Xiong C, Peng Y (2004) Synthesis of end-functionalized AB copolymers. II. Synthesis and characterization of carboxyl-terminated poly(ethylene glycol)-poly(amino acid) block copolymers. J Polym Sci, Part A Polym Chem 42(14):3527–3536
Zhang B et al (2005) Microtubule-binding drugs offset tau sequestration by stabilizing microtubules and reversing fast axonal transport deficits in a tauopathy model. Proc Natl Acad Sci U S A 102(1):227–231
Zuccari G et al (2005) Modified polyvinylalcohol for encapsulation of all-trans-retinoic acid in polymeric micelles. J Control Release 103(2):369–380
Zuccari G et al (2009) Micellar complexes of all-trans retinoic acid with polyvinylalcohol-nicotinoyl esters as new parenteral formulations in neuroblastoma. Drug Deliv 16(4):189–195
Zweers ML et al (2004) In vitro degradation of nanoparticles prepared from polymers based on DL-lactide, glycolide and poly(ethylene oxide). J Control Release 100(3):347–356

Liposomes: The Science and the Regulatory Landscape

Daan J.A. Crommelin, Josbert M. Metselaar and Gert Storm

Contents

Introduction 79
Structure of the Chapter 84
Chemistry and Structure 84
Manufacturing 86
Physicochemical Characterization 87
Pharmacology 91
The Regulatory Landscape 96
Liposomes in New Drug Applications 97
Generic versions or Other Liposomal Drug Products 98
Final Considerations and Reflections 101
Access to All Existing Information 101
Pharmaceutical Aspects of Liposome Design 101
Another Point for Consideration 102
A Bumpy Ride and Surprises 102
In Conclusion 102
References 103

Abstract Liposomes are vesicular (phospho)lipid-based drug carrier systems in the nanometer/micrometer range. The therapeutic performance of these 'composite' drug products heavily depends on their supramolecular structure. They are heterodisperse, difficult to fully characterize by physicochemical means and produced via complex manufacturing processes. This renders them part of the NBCD group.

The first liposome based drug formulation received market authorization over 20 years ago (Ambisome™ 1990). Since then a number of liposome drug products were approved by the FDA and the EM(E)A. The first generic versions of the innovator products are now being registered.

D. J. A. Crommelin (✉) · G. Storm
Department Pharmaceutics, Utrecht Institute for Pharmaceutical Sciences, UIPS, Utrecht University, Utrecht, The Netherlands
e-mail: d.j.a.crommelin@uu.nl

J. M. Metselaar
Enceladus Pharmaceuticals BV, Naarden, The Netherlands

D. J. A. Crommelin, J. S. B. de Vlieger (eds.), *Non-Biological Complex Drugs,* AAPS Advances in the Pharmaceutical Sciences Series 20, DOI 10.1007/978-3-319-16241-6_3

In this chapter the CMC (Chemistry, Manufacturing and Control) aspects of liposomes will be discussed, followed by a short overview of pharmacological aspects (e.g., pharmacokinetics/disposition, dosing, hypersensitivity). The regulatory landscape as it developed over the years is described next, followed by reflections on the future of this family of lipid, vesicular drug carrier systems.

Keywords Liposomes · Doxil™ · Caelyx™ · Ambisome™ · Enhanced Permeability and Retention (EPR) · EMA · FDA · Manufacturing · Formulation · Generic/follow-on liposomes

List of Abbreviations

AA	Anaplastic astrocytoma (grade III)
ABC	Accelerated blood clearance
ADME	Absorption, distribution, metabolism and excretion
ANDA	Abbreviated New Drug Application
API	Active Pharmaceutical Ingredient
AUC	Area under the Curve
CARPA	Complement Activation-Related Pseudo-Allergy
CBG-MEB	College ter Beoordeling van Geneesmiddelen—Medicines Evaluation Board
CHMP	Committee for Medicinal Products for Human Use
Chol	Cholesterol
CMC	Chemistry, Manufacturing and Control
CTD	Common Technical Document
DLS	Dynamic Light Scattering
DMPC	Dimyristoylphosphatidylcholine
DOPC	Dioleylphosphatidylcholine
DOPS	Dioleylphospatidylserine
DPPC	Dipalmitoylphosphatidylcholine
DPPG	Dipalmitoylphosphatidylglycerol
DSC	Differential Scanning Calorimetry
DSPC	Distearoylphosphatidylcholine
DSPE-PEG	Distearoylphosphatidylethanolamine polyethyleneglycol
DSPG	Distearoylphosphatidylglycerol
ELSD	Evaporative Light scattering detection
EPAR	European Public Assessment Report
EPC	Egg Phosphatidylcholine
EPR	Enhanced Permeability and Retention
FACS	Fluorescence Activated Cytometric Cell Sorting
FDA	Food and Drug Administration
FFF	FieldFlow Fractionation
GBM	Glioblastoma Multiforme (grade IV)
HSPC	Hydrogenated Soy Bean Phosphatidylcholine
ICH	International Conference on Harmonization

IVR	In Vitro Release
MPS	Mononuclear Phagocyte System
MS	Mass Spectrometry
NA	Not Assessable
NBCD(s)	Non-Biological Complex Drug(s)
NCE	New Chemical Entity
NCL	Nanotechnology Characterization Lab
NIH	National institutes of Health
NMR	Nuclear Magnetic Resonance
NTA	Nanoparticle Tracking Analysis
PDI	Polydispersity Index
QbD	Quality by Design
ROI	Region of Interest
SAXS	Small Angle X-ray Scattering
SCC	Squamous Cell Cancer
SEC	Size Exclusion Chromatography
SmPC	Summary of Product Characteristics
SPECT	Single Photon Emission Computed Tomography
SPH	Sphingomyelin
TEM	Transmission Electron Microscopy
WHO	World Health Organization

Introduction

Liposomes are a family of (phospho)lipid-based vesicular structures that can be used as drug delivery systems. These vesicles can vary in size, number of bilayers, rigidity and charge of the bilayer. Their morphology and physicochemical characteristics depend on the choice of the lipids and the manufacturing process. These characteristics determine liposome behavior in vivo, such as their circulation time in blood, distribution over different parts of the body and drug release kinetics.

Many bilayer-forming (phospho)lipids have been identified over the years. The phosphatidylcholines, phoshatidylglycerols, phosphatidylethanoamines, phosphatidylserines, and sphingomyelins are well known candidates. They can be used alone or in combination to form liposomes. Often other lipids (e.g. cholesterol, cationic lipids and/or lyso-phosphatidylcholine) are added to render specific properties: stabilization of the bilayer, positive charge or stimulus-sensitivity, respectively. In the scientific literature many examples of surface modification by attaching polyethyleneglycol (PEG) and/or ligands can be found. This wide variety of bilayer building blocks, combined with different manufacturing approaches creates uncountable permutations for liposome final products. Therefore, a selection of (high quality) lipids from reliable sources and the use of a robust manufacturing process are essential to obtain reproducible results in drug delivery.

A drug-containing liposome product consists of a dispersion of vesicles that are difficult to be fully characterized in physicochemical terms. Besides, liposome product properties heavily depend on the manufacturing process. Minor changes in manufacturing protocols can strongly influence final product characteristics. Therefore, they fall under the definition of the Non-Biological Complex Drugs (NBCDs) (cf. Introduction to this book).

The critical importance of exercising full control over the manufacturing process is exemplified by a number of regulatory actions that were reported in recent years. Manufacturing flaws (of undisclosed nature) led to shortage of one of the leading liposome products, brand name Doxil™ in the US/Caelyx™ in the rest of the world) (FDA Ben Venue Laboratories—Voluntary Shutdown 2011; EMA Questions and answers on the supply situation of Caelyx 2012). Other, single case, examples were reported for Depocyte™ (CBG-MEB 2012) and for Ambisome™ (Fierce Pharma Manufacturing 2013). These examples underline the necessity to have both a well-designed and validated protocol, well-trained personnel and a state-of-the-art manufacturing facility in place.

When one realizes the myriad of options to design liposomes for therapeutic use, it is interesting to evaluate what liposome designs actually made it to clinical investigations and market approval. When going through the list of currently (Table 1) registered liposome products in the EU and/or US, only a limited number of (phospho)lipids are being used and the same holds true for manufacturing concepts.

Since the 1960s 24,000 articles and 3500 + patents with 'liposom*' in the title have been published (Scopus). In 1980 250 articles appeared and output increased and reached 1000–1100 publications per year over the last 5 years. When one considers this enormous scientific input and then looks at the 11 approved products and their revenues, the output is rather disappointing. Admittedly, there are a number of recent registrations, but with the exception of Exparel, their revenues are very modest. Other authors drew similar conclusions and propose changes in research strategies (Venditto and Szoka 2013). The present arsenal of liposomal products may be limited, a survey of the clinicaltrial.gov data bank using the search term 'liposomes' scores 670+ hits. Among those clinical trials some are aimed at expanding the indications for existing products. But, there are also a number of new products in development (cf. Kraft et al. 2013; Allen and Cullis 2013). Interestingly, among those are five clinical trials with immunoliposomes, i.e. liposome structures with antibody (fragments) on their surface for targeting purposes. In one study transferrin is used as a targeting ligand (Mamot et al. 2012; van der Meel et al. 2013). Glutathione attached to liposomes may facilitate transport of the liposomes through the blood-brain-barrier. Clinical trials in cancer patients and multiple sclerosis patients are under way (clinicaltrials.gov) (Table 2).

Lipodox is the only generic version of an existing liposome product that was registered so far in the US (not in the EU yet) (Table 1). But outside the EU and US a number of liposome drug products have been registered, e.g. amphotericin liposomes in India under the name: Fungisome™ (Fungisome 2014), Phosome™, Lambin™, Lipholyn™, Amphonex™ (Amphonex 2014). In Taiwan, Argentina,

Table 1 Liposomal formulations approved in EU and US

Liposome	API	Lipids	Formulation	Size range[a]	Introduced	Patent expiration	Revenue in M US$	License holder/ manufacturer
Ambisome	Amphotericin	HSPC, DSPG, chol	Freeze dried	<100 nm	EU 1990[b], US 1997 US	US 2016, EU expired[c]	500[d]	Astellas/Gilead Sciences
DaunoXome	Daunorubicin	DSPC, chol	Aqueous dispersion	40–80 nm	EU 1997, US 1997			Galen/Gilead Sciences
DepoCyt	Cytarabine	DOPC, DPPG	Aqueous dispersion	20 μm	US 1999			Sigma-Tau Pharmaceuticals/ Pacira
DepoDur	Morphine	DOPC, DPPG	Aqueous dispersion	17–23 μm	US, 2004, UK 2006			In UK: Flynn Pharma Ltd/Almac Pharma Services Ltd, Ireland
Doxil/Caelyx	Doxorubicin	HSPC, chol, DSPE-PEG	Aqueous dispersion	100 nm	US 1995, EU 1996	Expired[d]	600[d]	Jansen-Cilag
Exparel	Bupivacaine	DOPC, DPPG, tricaprylin	Aqueous dispersion	24–31 μm	US 2011		100[d]	Pacira
Lipodox[c]	Doxorubicin	HSPC, chol, DSPE-PE	Aqueous dispersion	100 nm	US 2013			Sun Pharma
Marqibo	Vincristine	SPH, chol	Freeze dried	100 nm	US 2012			Talon Therapeutics
Mepact	Mifamurtide	DOPC, DOPS	Freeze dried	?	EU 2009			Takeda
Myocet	Doxorubicin	EPC, chol	Freeze dried	80–90 nm	EU 2000			TEVA

Table 1 (continued)

Liposome	API	Lipids	Formulation	Size range[a]	Introduced	Patent expiration	Revenue in M US$	License holder/ manufacturer
Visudyne	Verteporfin	EPC, DMPC	Freeze dried	18–104 nm[f]	US 2000, EU 2000	Expired US 2016,1016[g]	35[h]	Valeant

Chol cholesterol, *DMPC* dimyristoylphosphatidylcholine, *DOPC* dioleylphosphatidylcholine, *DOPS* dioleylphospatidylserine, *DPPC* dipalmitoylphosphatidylcholine, *DPPG* dipalmitoylphosphatidylglycerol, *DSPC* distearoylphosphatidylcholine, *DSPE-PEG* distearoylphosphatidylcholine polyethyleneglycol, *DSPG* distearoylphosphatidylglycerol, *EPC* egg phosphatidylcholine, *HSPC* hydrogenated soy bean phosphatidylcholine, *SPH* sphingomyelin

[d] TLC report 2013, data for 2012

[h] 2011 Valeant

[e] Gaspani and Milani 2013

[g] Visodyne patent report

[c] Generic version of Doxil/Caelyx;from 2013 the major product in the US

[b] National regulatory path (Gaspar 2010)

[f] Chassu et al. 2013

[a] cf. Kraft et al. 2013

Table 2 Overview of ligand-targeted lipid-based nanomedicines undergoing clinical evaluation. adapted from van der Meel et al. 2013 (for abbreviations: see original publication)

Product name	Company	Approx. size (nm)	Payload	Ligand	Target	Clinical indication	Clinical phase
Lipid-based nanomedicines							
MBP-426	Mebiopharm	50–200	Oxaliplatin	Protein	Transferrin receptor	Metastatic gastric, gastro esophageal junction, esophageal adenocarcinoma	Phase II
SGT-53	SynerGene Therapeutics	90	p53 plasmid DNA	Antibody fragment (scFv)	Transferrin receptor	Solid tumors	Phase Ib
SGT-94	SynerGene Therapeutics	90	RB94 plasmid DNA	Antibody fragment (scFv)	Transferrin receptor	Solid tumors	Phase I
MM-302	Merrimack Pharmaceuticals	75–110	Doxorubicin	Antibody fragment (scFv)	ErbB2 (HER2)	Breast cancer	Phase I
Lipovaxin-MM	Lipotek		Melanoma antigens and IFNγ	Single domain antibody (dAb) fragment (VH)	DC-SIGN	Melanoma vaccine	Phase I
Anti-EGFR ILs-DOX	University Hospital Basel	85	Doxorubicin	Antibody fragment (Fab’)	EGFR	Solid tumors	Phase I
2B3-101	to-BBB Technologies		Doxorubicin	Protein	Gluthathione transporters	Solid tumors	Phase I/IIa
MCC-465	Mitsubishi Pharma Corporation	140	Doxorubicin	Antibody fragment (F(ab)’2)	Not characterized	Advanced gastric cancer	Phase I (discontinued)

China and South Korea generic versions of Ambisome or Doxil are also registered or under development (TLC 2013).

In this chapter attention will be paid to the regulatory processes that were developed, or are under development to attain approval of liposome products. Both the regulatory landscapes for novel liposome-based products and for generic versions are discussed. When looking at the requirements for these two groups of liposome drug products, some general observations can be made. The CMC (Chemistry, Manufacturing and Control) part of the dossiers contains the same list of items to be dealt with. For the innovator's product the major objectives are to find a comprehensive set of specifications that assure the quality of the product and indicate the allowed batch-to-batch variations, leading to optimal therapeutic reproducibility. For generic versions, the prime objective is to demonstrate close similarity (based on 'sameness' principle) of the product characteristics with those of the innovator product. In that sense the CMC part of the dossier for generics is at least as detailed as that of the innovator's product. The same holds for the 'pharmacokinetics' part, where bioequivalence has to be shown. However, the sections on nonclinical animal studies and on the clinical trials (to show efficacy and safety) to be performed before market approval are much more elaborate for innovative drug products than for generic versions. Recent EMA documents for Caelyx (Caelyx EPAR product information EMA 2014) and for Lipodox (not-approved; CHMP assessment report Doxorubicin SUN EMA 2011) exemplify these conclusions).

The focus in this chapter will be on the EU and US practices as—unfortunately—only little information regarding practices in other parts of the world could be found.

Structure of the Chapter

Before we deal with the regulatory landscape, attention will be paid to topics that are part of a regulatory dossier. The Chemistry, Manufacturing and Control (the CMC part of the dossiers) of liposomes will be discussed in the sections: Chemistry and Structure, followed by: Manufacturing and Physicochemical Characterization. These sections are then followed by a section on Pharmacology, including considerations regarding Toxicology, Disposition and Pharmacokinetics of liposomes.

Chemistry and Structure

Bilayer Composition In the liposome drug formulations that are presently on the market and listed in Table 1. The major lipid component of the bilayer is typically a member of the phosphatidylcholine (PC) family. The acyl chain length and degree of unsaturation may vary, but most products are based on saturated phospholipids with long acyl chain lengths. Besides the relatively high transition temperatures of the bilayer that leads to increased physical stability, also the excellent chemical stability explains the preference for these phospholipids. Marqibo (see Table 1) is the only approved product based on sphingomyelin instead of phosphatidylcholine. To further stabilize the bilayer cholesterol is usually added in significant quantities. A phospholipid conjugated to poly(ethylene glycol) (PEG) is added to the bilayer of some liposome products, e.g. Doxil. The name of this approach, 'Stealth technology', refers to the alleged capacity of this polymer to shield the liposome from opsonization and premature clearance. The choice of these lipids and the composition of the lipid bilayer affects bilayer charge, rigidity, stability, and—immediately upon i.v. injection—the interaction with blood elements such as opsonins and circulating cells. The particle size of the different products varies widely, between 20 nm and 30 µm (cf. Table 1).

An important consideration to choose a lipid is the guarantee of high quality supply, existing drug master files (DMF), presence in approved products and acceptable costs. Over the many years that liposomes were studied in preclinical and clinical tests a number of phospholipid suppliers 'surfaced' who met the strict market requirements.

Stability About half of the liposome products listed in Table 1 are stored as an aqueous dispersion in the refrigerator, the other half in freeze dried form. Preferably, pharmaceutical products have a shelf life at ambient temperature of at least 2 years. Loosely speaking there are two stability concerns: (1) chemical degradation of the bilayer components and/or the associated drug, and (2) physical instability, i.e. aggregation/fusion, and drug leakage.

Re 1: Hydrolysis and oxidation: as phospholipids such as PC have four ester bonds, hydrolysis may occur. In particular the fatty acid-glycerol esters are sensitive to hydrolysis. The hydrolysis rate depends on the temperature, presence of an aqueous medium, its pH and ionic strength and the charge of the bilayer. Oxidation is a concern when cholesterol and/or unsaturated fatty acids are present in the bilayer forming phospholipids. Adding antioxidants such as vitamin E, filling and finishing under argon and/or freeze drying helps to minimize the problem. Grit et al. (1993) and later Zuidam published extensively on this topic of hydrolysis/oxidation and the consequences for maintaining liposome integrity (see review Barenholz and Crommelin 1994; Zuidam et al. 2003). Doxil is a liquid formulation with a shelf life of 18 months at temperatures between 2 and 8 °C (EMEA scientific discussion Caelyx 2005) indicating that under the chosen conditions the lyso-phospholipid content was within specifications over 18 months.

Re 2: Aggregation, fusion and leakage: these are in particular challenging issues when the formulation has to be freeze-dried to meet shelf life requirements. After reconstitution, aggregation and fusion may occur and part of the liposome-associated drug may be released. Successful freeze drying of liposome dispersions (maintain integrity throughout the drying-reconstitution process) depends on the careful selection of process parameters such as freezing rate, sublimation temperature/pressure (primary drying), secondary drying temperature and time, residual water content and lyoprotectants, usually sugars such as sucrose, maltose or trehalose (van

Winden 2003; Arshinova et al. 2012). A robust freeze drying protocol for manufacturing is of critical importance to minimize batch-to-batch variation.

Manufacturing

For small-scale liposome production a large number of preparation methods have been described over the years. They are all based on a few principles, basically three steps: hydrate lipids, apply a controlled sizing procedure and remove non-liposome associated drug (Barenholz and Crommelin 1994; Lasch et al. 2003). For a number of weak base or acid compounds active loading principles can be used. The literature on small-scale manufacturing is rich (reviewed e.g. by Fenske et al. 2003) and studies on the rational design of liposome drug product formulations have been published. Quality by design (QbD) approaches to screen critical variables and define the design space have been reported by Xu et al. 2011, 2012a, b.

In contrast, very little can be found in the peer-reviewed literature on upscaling and large-scale production. However, a scheme on the production steps for Doxil/Caelyx is available (Fig. 1).

The 100+ patents that can be found through the Scopus data bank combining the search terms 'liposome*' and 'manufacturing' do not give new basic insights. Large scale, GMP production relies on strict and robust protocols, excellent technical facilities and well-trained personnel. This conclusion is not to be taken lightly as

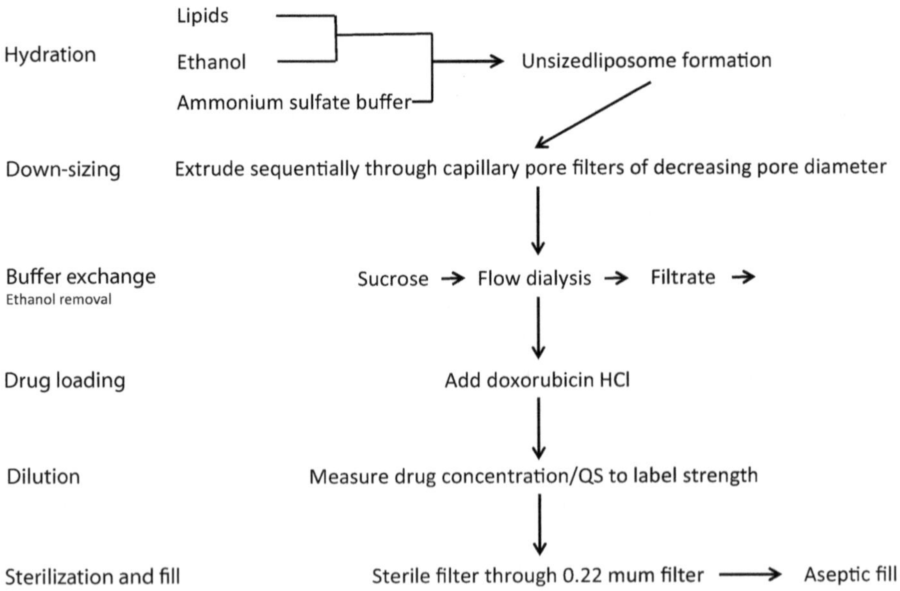

Fig. 1 Manufacturing process of Doxil/Caelyx (with permission from Frank Martin)

is proven by the earlier mentioned supply problems with Doxil/Caelyx, DepoCyte and Ambisome.

The FDA draft guidance document that was issued in 2002 (FDA 2002) (http://www.fda.gov/downloads/Drugs/GuidanceComplianceRegulatoryInformation/Guidances/ucm070570.pdf) provides information on a number of issues that should be properly dealt with when designing the manufacturing process, including identifying critical parameters. It also discusses how to deal with changes in the manufacturing process of existing products (comparability studies). Depending on the nature of these changes it states: 'In vivo studies may be warranted to demonstrate that the changed product is equivalent to the original product with respect to safety and efficacy'.

Physicochemical Characterization

To ensure reproducible therapeutic outcomes, the physicochemical characteristics of liposomal drug products have to be constant or—more precisely—stay within an earlier established design space. That holds for batch-to-batch variation of 'fresh' batches and for batches during storage. This notion was recognized in the period 1985–1995 when the first liposomal products (Ambisome and Doxil) received their IND and NDA status (FDA system). Barenholz and Crommelin (1994) listed 'quality control assays' for characterization of liposomal drug products. This list has expanded over time as new analytical techniques and insights became available (Zuidam et al. 2003; Jiang et al. 2011; van den Hoven 2012, Table 3, 4). For instance, the introduction of the cryo-TEM technique facilitated the establishment of liposome morphology (lamellarity and shape), and visualization of contents and size. Dynamic light scattering (DLS) techniques evolved over the years and became more powerful to the point of monitoring single particles (Nanosight™). FFF (field flow fractionation) offers another way of elucidating liposome size distributions, as does FACS (fluorescence activated cell sorting technology) (Jiang et al. 2011) and SEC (size exclusion chromatography).

Besides particle size, also particle size distribution is a critical determinant for in vivo performance. While a product consisting of two separate particle size populations may look similar to a monodisperse product in terms of average diameter, its in vivo performance will likely be inferior with the larger particles (that contain most of the drug) cleared prematurely when administered intravenously. With DLS as the standard way of gauging particle size, the polydispersity index (PDI) that is usually reported along with the particle diameter (mostly reported as Z-average) has become the most convenient and straightforward way of assessing particle size distribution. However, it is has been recognized that small quantities of aggregates and/or very large particles are sometimes not sufficiently reported by the PDI value and can go undetected. DLS at a 20 ° angle of detection has been proposed as a novel way of detecting large aggregates and seems to be successful to some extent. The combination with new particle diameter assessment methods, including Nanosight NTA and qNano, is nowadays generally being recommended to collect more de-

Table 3 Test items for the characterization of liposomal drug formulations. From van den Hoven 2012

Liposomal components	Purity API	Liposome characteristics	Size
	Crystallinity API		Size distribution
	Encapsulated API (including salt forms)		Presence of aggregates
	Unencapsulated API		Internal volume
	Lipid contents/ Lipid composition (ratio)		Biological activity
	Degradation products of the lipid components Drug to lipid ratio		Immunochemical properties and interactions
	Buffer composition		In vitro (plasma) stability and release Long term toxicity profile
Bilayer characteristics	Morphology Lamellarity	Liposomal dispersion	Appearance Residual solvents
	Surface area		Uniformity of dosage units
	Folding regimen of surface structures		pH
	Charge		Sterility/Microbial limits
	Thickness membrane layer		Bacterial endotoxins
	Phase transition temperature		Particulate matter
			Water content
			Reconstitution time
			Antimicrobial and antioxidant preservative content
			Osmolality
			Extractables
			Stability upon storage

tailed information about the liposome size distribution. However, to our knowledge, these new methods have not led to new specifications to particle size distributions that are considered acceptable by authorities and/or the peers in the field.

Particle size distribution and the presence of large aggregates/particles has been associated with the occurrence of hypersensitivity reactions. To predict hypersensitivity reactions, often seen when injecting liposomes intravenously, in vitro complement binding assays were proposed, as discussed by Szebeni et al. 2011) (see also under 'Pharmacology').

Table 4 Analytical techniques used for the characterization of liposomal drug formulations. Adjusted from van den Hoven (2012). For references: see thesis

Analytical technique	Liposomal property	Liposomal property has an influence on:	Characteristics of
HPLC (UV, MS detection), or other suitable assay method	Content and purity API	Efficacy and toxicity	Liposomal components
HPLC (UV, MS, ELSD detection)	Content and purity lipid components	Efficacy and toxicity, liposomal properties, release profile of the encapsulated drug	Liposomal components
	Type of surface structures and their folding regimen (indicative)	Circulation time, clearance, distribution, cellular uptake	Liposomal bilayer
	Thickness of the liposomal bilayer (indicative)	Aggregation in the formulation and interactions with proteins and cells in vivo	Liposomal bilayer
Small angle scattering	Thickness of the liposomal bilayer	Aggregation in the formulation and interactions with proteins and cells in vivo	Liposomal bilayer
Zeta potential measurement	Surface charge	Interaction and uptake by target cells and MPS, toxicity by rupture of cell membranes	Liposomal bilayer
TEM	Lamellarity, morphology and thickness of the liposomal bilayer	Release profile of the encapsulated drug	Liposomal bilayer
	Drug crystallinity	Correct salt form, shape of the precipitate	Liposomal components
DSC	Phase transition temperature	Permeability of the liposomal membrane, release profile of the encapsulated drug	Liposomal bilayer
DLS, Nanosight, SEC, FFF or TEM	Size	Biological interactions, biodistribution	Liposomal system
	Size distribution	Indicates absence of aggregates or agglomerates	Liposomal system
	Surface area (indicative, related to size)	Interactions with cells, tissues, organ systems, proteins and other macromolecules	Liposomal bilayer
	Internal volume	Drug content (indicative) and release profile (indicative)	Liposomal system

Table 4 (continued)

Analytical technique	Liposomal property	Liposomal property has an influence on:	Characteristics of
FACS	Detection of aggregates, even when very low numbers of aggregates are present	Aggregates could activate complement, resulting in hypersensitivity reactions, influencing the safety of the formulation	Liposomal system
Entrapment fluorescent probe	Internal volume	Drug content (indicative) and release profile (indicative)	Liposomal system
Complement assay	Immunochemical properties	Complement induced hypersensitivity reactions	Liposomal system
Release testing method	In vitro release and stability	Release profile of the encapsulated drug, membrane stability in vivo (indicative)	Liposomal system

MS mass spectrometry, *ELSD* evaporative light scattering detection, *TEM* transmission electron microscopy, *DSC* differential scanning calorimetry, *DLS* dynamic light scattering, *SEC* size exclusion chromatography, *FFF* field-flow fractionation, *FACS* fluorescence activated cell sorting

Some of the bilayer characteristics listed are not so easy to measure. For instance, Barenholz (2012) points out the important difference between surface potential and zeta-potential measurements. This difference is in particular relevant when dealing with PEGylated liposomes. The surface potential is difficult to measure directly and one may question what its relevance is for understanding liposome behavior. For the zeta-potential measurements, DLS equipment is available with special features to monitor particle mobility in an electric field. The PEG content can be quantified colorimetrically (Nag et al. 1997) or by NMR (Vernooij et al. 1999). Information on the structure of the PEG-layer can be obtained by SAXS (Varga et al. 2012). The complex nature of the SAXS measurements and data analysis makes this approach not suitable for routine measurements. Yoshino et al. describe an anion exchange chromatographic technique to provide information on the liposome surface condition of low-percentage PEG-containing liposomes. Basically, this technique depends on the screening of the negative zeta-potential by PEG. As a caveat, this negative potential is partially induced by hydrolyzed neutral phospholipids and is in that respect dependent on the purity and chemical stability of the bilayer-forming phospholipids (Yoshino et al. 2012). An important characterization assay, which appears to be especially important to the FDA, is the In Vitro Release (IVR) assay. Burgess et al. (Bhardwaj and Burgess 2010; Xu et al. 2012c) received 'FDA critical path funding' to develop experimental protocols to establish release characteristics under different experimental conditions. In this assay liposomes are exposed to well-controlled stress conditions, such as a high temperature, the presence of serum and the addition of surfactants or water-miscible solvents. The underlying concept is that a robust performance assessed with the IVR assay is a predictor for minimal variations in terms of bioavailability, efficacy and safety when given to humans. The design of a suitable IVR assay can pose a challenge to liposomal product developers. Indeed the FDA continues to be concerned regarding the development of accurate and appropriate IVR assays, as is demonstrated by a recent call from the FDA for research proposals aiming at the design of more accurate IVR assays for liposomal products ((FDA Grants 2014) http://grants.nih.gov/grants/guide/rfa-files/RFA-FD-14-016.html).

The FDA and EMA both request the assessment of free and liposome-associated drug when monitoring the pharmacokinetic profile of drug-liposome combinations. For Doxil, the free drug concentration/liposome associated concentration ratio upon i.v. injection can be in the order of 1:100. Validated assays to reliably determine free doxorubicin concentrations among relatively large quantities of liposome-associated doxorubicin are necessary and a number of such assays have been described. They use ion-exchange (Druckmann et al. 1989), solid state extraction (Thies et al. 1990; Griese et al. 2002), and ultrafiltration (Mayer and St-Onge 1995). A capillary electrophoresis assay with separation of free doxorubicin and liposome associated doxorubicin in the capillary was recently described by Kim and Wainer 2010) (Also see Chap. 8 by Stern in this book).

Pharmacology

General Aspects

The liposomes that are mentioned in Table 1 are all parenterally administered, mainly via the intravenous route but also via the intrathecal (DepoCur) or epidural route (DepoDur). Since the early 1980s a large number of studies have been undertaken to find out what happens to liposomes upon administration and a wealth of information has been collected. The end conclusion is that 'the' liposome does not exist. The biological performance (pharmacokinetics, tissue distribution, efficacy, and side effects) of liposomes in vivo is controlled by a complex combination of interrelated physicochemical and biological factors. The former factors include liposome-related properties such as size distribution and surface characteristics determined by bilayer composition and/or surface modifications with polymers (e.g. PEG) and targeting ligands. The latter factors include anatomical, physiological and immunological barriers, but also opportunities offered by pathological conditions which can be exploited by liposomes (e.g. the Enhanced Permeability and Retention, (EPR) effect, overexpression of certain receptors). Obviously, it is important to understand the factors affecting the PK/PD profile of liposomal drug formulations and the mechanisms responsible for the occurrence of considerable intra- and inter-individual variability.

The literature on the pharmacological effects of liposome-associated drugs versus the free drug has grown over the years forming a large knowledge base that can be used when designing new liposome formulations (Gregoriadis 2006). In general, one can say that the liposome-associated drug does not exert a pharmacological effect as long as it is liposome associated. Basically, the released drug shows the same mode of action as the free drug, but the change in disposition (reaching other parts of the body) may lead to different pharmacological outcomes compared to the free drug. Differences in pharmacological and toxicological effects are directly related to the change in disposition of the free drug. A selection of observations relevant for an understanding of the pros and cons of liposomal delivery will be discussed in the following paragraphs.

Enhanced Permeability and Retention (EPR)

The present generation of intravenously administered liposomal drug products (Table 1) consists of liposomes that show minimal drug release while the liposomes are circulating in the blood. That implies that the main drug dose follows the disposition of these carrier systems in the body. Both in animals and humans the main sites of uptake are the MPS cells in liver (Kupffer cells) and spleen. PEGylation slows down uptake by these phagocytic cells and then—dependent on the AUC—the skin may become a site of uptake together with the MPS. There is an ongoing debate about the extent of uptake of intact liposomes by target sites such as tumors and inflammatory sites: the enhanced permeation and retention (EPR) effect (Bae and Park 2011, Park 2013, Crommelin and Florence 2013). In fast growing tumors in animals, this EPR effect definitely helps to accumulate a fraction of the liposomal drug dose in the tumors. In Kaposi's Sarcoma patients, often AIDS patients, also preferential accumulation of doxorubicin liposomes in tumors occurs. However, Harrington et al. (2001) showed that for a number of frequently diagnosed (other) solid tumors in humans, accumulation in primary tumors and/or metastases (EPR-effect) only occurs in a subset of patients. This group used radiolabelled liposomes to assess tumor uptake (Table 5). It would be logical to first screen patients for tumor uptake before using EPR-dependent delivery systems (Lammers et al. 2012; Crommelin and Florence 2013).

Opsonization

One should realize that physicochemical characteristics are often changed upon contact with the bloodstream (Chonn et al. 1992). The composition of the resulting protein coat on the liposomes is expected to be dynamic in time regarding amount and protein types. Blood protein adsorption onto the liposome surface may affect the biological performance substantially, by inducing drug release and aggregation, with large aggregates being trapped in the first capillary bed encountered after intravenous administration, in the lungs. In addition, deposition of proteins can opsonize the liposome surface, facilitate MPS recognition and consequently accelerate clearance from the bloodstream (Moghimi and Szebeni 2003). Therefore, the protein

Table 5 Patient details, stage, and results of gamma camera imaging and estimated tumor uptake from ROI analysis. Harrington et al. 2001

Patient	Tumor	Stage (cf. the web)	Whole body scan	SPECT	Total % injected dose[a]	%ID/kg[b]
1	SCC bronchus	T4N0M0	Positive	Positive	1.7	12.5
2	SCC bronchus	T4N0M0	Positive	Positive	1.6	25.4
3	Breast (ductal)	T4N2M1	Negative	Negative		
4	SCCHN	T3N2M0	Positive	Positive	3.5	46.8
5	Breast (ductal)	T4N3M0	Positive	Positive	0.3	2.7
6	Breast (ductal)	T4N2M1	Positive	Positive	1.5	3.9
7	Breast (ductal)	T3M2N0	Positive	Positive	1.7	9.5
8	SCCHN	T4N0M0	Positive	Positive	0.7	24.2
9	SCCHN	T3N1M0	Positive	Positive	1.0	32.0
10	SCC cervix	FIGO IIIB	Negative	Positive	NA	NA
11	Breast (ductal)	T4N2M0	Positive	Positive	1.4	5.2
12	SCC bronchus	T2NoM1	Negative	Negative		
13	SCCHN	T3N2M0	Positive	Positive	0.6	9.0
14	SCCHN	T3N0M0	Positive	Positive	1.6	53.0
15	SCC bronchus	T3N0M1	Positive	Positive	2.6	16.7
16	Glioma (AA)	Inoperable	Negative	Positive	NA	NA
17	Glioma (GBM)	Inoperable	Negative	Positive	NA	NA

[a] Tumor uptake determined from region of interest (ROI) on 72 h whole body scan
[b] Percentage injected dose/kg calculated from estimated tumor volumes
SCC (HN) squamous cell cancer (head and neck), *AA* anaplastic astrocytoma (grade III), *GBM* glioblastoma multiforme (grade IV), *NA* not assessable

coat is a key determinant of pharmacokinetics even though not all the deposited proteins have an impact.

Dose and Dosing Interval Effects

From a pharmacokinetic viewpoint, it is useful to discriminate between MPS-'directed' (short-circulating) and MPS-'avoiding' (long-circulating) liposomes. The use of the latter is certainly preferred if the drug target is not located within the MPS. The former have relatively short circulation times affected by their size and lipid composition which determines bilayer fluidity and charge. In general, increasing liposome size and charge results in increased clearance from the blood whereas increasing bilayer rigidity has the opposite effect (Drummond et al 2008). MPS-directed liposomes are cleared from the circulation in a lipid dose-dependent saturable manner (Allen and Hansen 1991). Circulation times of these liposomes increase proportionally with increasing lipid dose. The decreased clearance by the MPS at higher lipid doses is considered to be the consequence of a combination of

MPS saturation and depletion of serum opsonins. In particular small (< 100 nm), neutral and rigid (i.e., composed of saturated phospholipids below their phase transition and a high cholesterol content) can exhibit prolonged circulation at high lipid doses (Storm et al. 1995). In the latter example, opsonic factors likely experience difficulty to attach to such bilayer structures. This knowledge has been exploited in generating commercial MPS-avoiding liposomes containing daunorubicin (DaunoXome) and amphotericin B (AmBisome). However, the major current approach to design MPS-avoiding liposomes is to modify the liposome surface with hydrophilic polymers, with clearly polyethyleneglycol (PEG) being the most popular and effective in prolonging the circulation time of liposomes (Woodle and Lasic 1992). PEG surface modification (with PEG of 1000–5000 mol.wt. at a 5 mol% density) has been shown to have important advantages over other methods to obtain prolonged circulation. One of the principal advantages is that PEG-liposomes possess dose-independent, nonsaturable, log-linear pharmacokinetics over a large lipid dose range (4–400 µmol/kg). This permits dose escalation without complications arising from changes in pharmacokinetic profile. Another advantage is the versatility in choice of lipid composition, as this provides an opportunity to optimize the liposomal formulation characteristics for drug entrapment and stability without risking a change in prolonged circulation and tissue distribution characteristics (Woodle et al. 1992).

Although the pharmacokinetics of PEG-liposomes have shown to be much less dependent on lipid dose than their corresponding non-PEGylated counterparts, some unexpected pharmacokinetic irregularities have been observed at lower lipid doses. The first striking observation was made when liposomes were studied for scintigraphic imaging applications. In such studies, very low lipid doses of PEG-liposomes (< 1 µmol/kg), not relevant for drug targeting applications, are applied. A complete loss of the long circulation property of PEG-liposomes was observed, in various animal species and in humans (Carstens et al. 2007). The rapid clearance pattern of PEG-liposomes at such very low lipid doses suggests that a limited amount of some type of opsonic protein(s) is present in the circulation causing this. Above a critical threshold dose, this opsonin pool is depleted, leading to the appearance of the long circulation characteristic at higher doses. The second striking observation of altered clearance of PEG-liposomes was made upon repeated administration (Oussoren and Storm 1999; Dams et al. 2000). A second dose of PEG-liposomes, given 5 days up to 4 weeks after the first injection, yielded a dramatically decreased circulation time and elevated hepatosplenic uptake in rats. The effect was most pronounced at a dosing interval of one week. At subsequent weekly injections, the intensity of this so-called 'accelerated blood clearance' (ABC) effect attenuated. The ABC phenomenon was seen in a number of animal species (Laverman et al. 2001), but is clinically not confirmed (yet). Notwithstanding the lack of clinical evidence, the ABC effect deserves proper attention in case of clinical applications requiring repeated dosing schemes of PEG-liposomes. Enhanced blood clearance could compromise the therapeutic efficacy and the increased uptake by liver and spleen could cause toxic effects towards these organs. The exact mechanism underlying the ABC effect is still under debate.

Release Rate

The rate of in vivo drug release is the key parameter affecting the bioavailability and therefore efficacy and toxicity of any liposomal drug formulation. Therefore, it is important to assess the PK of liposome-encapsulated and released drug. Mechanisms of in vivo drug release are not exactly known but can involve diffusion from intact particles, extra- and intracellular biodegradation of the liposomal structures, and/or uptake by MPS macrophages and subsequent secretion by these cells. After the drug is released from the liposomes, the PK and disposition of the drug will not follow that of the liposome particles anymore. Simple in vitro incubations in the presence of plasma or serum may yield a first rough impression but are certainly inadequate predictors of the bioavailability. While accurate methodologies are available to monitor the PK of the liposomal particles (e.g. gamma scintigraphy), translational scientists involved in liposomal drug development are often still facing a large analytical gap to accurately assess the PK of the released drug. In particular, assays to separate and distinguish between co-existing fractions of liposome-bound and released drug need to be carefully validated (see 2.3. and Chapter in this book by Stern)

Bioequivalence

In the Introduction the point was made that there is a difference between the extent of clinical efficacy and safety testing for innovator's liposome drug products and generic versions. Innovator products undergo larger test programs (see e.g., EMA (2012) summary of product characteristics Myocet; EMA (2014) summary of product characteristics Caelyx). But, in terms of pharmacokinetics, the generic version has to meet the challenge of showing bioequivalence to the innovator. The dossier of the first generic version of Doxil (following the sameness approach) failed (i.a.) on the ground of not showing bioequivalence. The plasma levels of free doxorubicin from Caelyx and Lipodox in a bioequivalence trial can be found in this document: CHMP assessment report Doxorubicin SUN EMA (2011). In tissues, including tumor tissue, such discrimination between liposome-bound and free drug is very difficult, if not impossible, to achieve. One approach followed in the blood compartment relates to the situation that when the clearance rate of the released drug is sufficiently faster than the clearance rate of the liposomal particle, the rate of drug release from circulating liposomes can be estimated by simply monitoring the plasma drug-to-liposomal lipid ratio in time.

Complement Activation-Mediated Hypersensitivity Reaction (CARPA)

As pointed out above, the interaction of liposomes with the cellular arm of the nonspecific (innate) immune system is dominating the in vivo fate of liposomal drug particles. The humoral innate response to liposomes, manifested by activation of the

Table 6 Clinical Picture of CARPA (adapted from Szebeni et al 2011). Frequency: 1–40%, depending on drug and recipient factors, Death: ~1–10/10,000 (0.01–0.15%)

Organ system	Human	Animals
Hemodynamic	Hypo/hypertension	Pulmonary hypertension/systemic hypo/hypertension
Cardiopulmonary	Dyspnoea, chill, tachy/bradycardia, chest pain, back pain	Tachy-bradycardia/ arrhythmia, ischemic ECG
Skin	Flushing, rash	Flushing, rash
Blood	Thrombopenia, leukopenia,/ leukocytosis	Thrombopenia, leukopenia,/ leukocytosis

complement system, has also been early recognized, but this effect got much less attention than the interaction of liposomes with phagocytic MPS cells. This situation changed with the manifestation of complement activation-mediated hypersensitivity reactions upon infusion of PEG-liposomes in up to 30% of patients in clinical studies (Szebeni et al. 2011). These reactions usually occur at the start of the infusion and include symptoms like cardiopulmonary distress, hypo- or hypertension, dyspnea, tachypnea, facial edema and pain in the chest and back (Table 6)

Most of these reactions are mild and transient, but life-threatening reactions also have been documented. Since these reactions occur at the first exposure to the liposomal drug particles (without prior sensitization), they are often referred to as 'pseudoallergy'. The phenomenon has therefore been termed 'complement activation-related pseudoallergy (CARPA)' (Szebeni 2005). Hypersensitivity reactions have been observed for practically all marketed liposomal drugs but are not limited to the use of liposome formulations as they also occur with monoclonal antibodies, micellar drugs, PEGylated proteins, radio and ultrasound contrast agents, therapeutic enzymes and other, small molecule drugs (e.g. Taxol). Because of its potentially fatal outcome, the phenomenon is considered a safety issue in nanopharmacotherapy (Szebeni et al. 2011). Assessment of CARPA was recently recommended by the European Medicines Agency as a preclinical immune toxicity tests in the development of (generic) liposomal drugs (see earlier mentioned EMA Reflection Paper 2013). However, at present, there is neither a standard test, nor a validated battery of testing procedures for evaluating the CARPA-genic activity of i.v. administered nanomedicines (Szebeni 2011).

The Regulatory Landscape

This section is divided in two parts. The first deals with the regulatory experience regarding the market approval of *innovator's* liposome drug products, the second regarding market approval of *generic* versions.

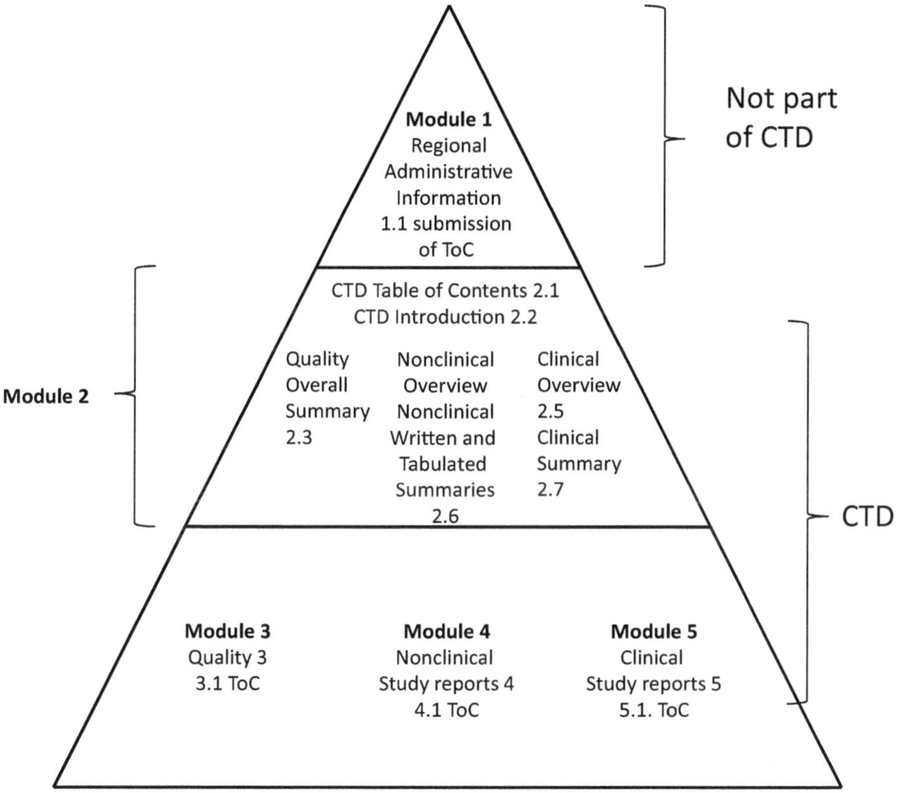

Fig. 2 From: EMA ICH Topic M 4: Common Technical Document for the Registration of Pharmaceuticals for Human Use –Organisation CTD

Liposomes in New Drug Applications

For Europe The Common Technical Document (CTD) with a lay out as described in the ICH topic 4 document is used for submission of a dossier requesting marketing authorization of new liposomal drugs via the 'central procedure' (EMA (2006) Volume 2B Notice to applicants medicinal products for human use; EMA ICH Topic M 4 (2004)). This means that the full dossier as described in the ICH topic 4 document (Fig. 2) has to be submitted, with 5 modules: regional administrative information (module 1), summaries (module 2), quality aspects (module 3), non-clinical study reports (module 4) and clinical reports (module 5).

Upon issuing a marketing authorization, EMA provides a SmPC (Summary of Product Characteristics) to inform healthcare professionals about 'the why and how to use it'. This is a 'living document' that is updated, e.g. by adding new indications. SmPCs of e.g. Caelyx, Myocet, Daunoxome, Mepact and Visodyne can be found on the EMA website.

For the US For new drug-liposome combinations the section 505(b)(1) protocol—a full NDA—has to be followed, even if the drug (without the liposome) has been in use for a long time. This means that to support approval, a. o. two controlled clinical studies have to be performed to show the expected positive therapeutic outcome. One could consider to follow the 505(b)(2) route here instead of the full NDA (505(b)(1)) in the US and hybrid/mixed market application procedures in Europe. Such regulatory approval procedures form intermediates between generic and new drug applications. They allow referral to existing dossiers of approved active pharmaceutical ingredients (API) given in other dosage forms, significantly reducing the number of clinical trials needed and making liposome product development way more cost effective than developing a new chemical entity (NCE). E.g. Doxil was approved through the 505(b)(2) route referring to the doxorubicin dossier, and Depodur to the morphine sulfate dossier.

In 2002 a draft guidance document on liposomal drug products was issued by the FDA (FDA 2002 Guidance for Industry: Liposome Drug Products). Today it is still in this draft form. In this document recommendations are made regarding the necessary CMC actions and work to be submitted on 'human pharmacokinetics and bioavailability', comparing liposome drug and non-liposome drug pharmacokinetics, including mass balance outcomes, absorption, distribution, metabolism and excretion (ADME) and in vivo integrity.

Since the introduction of this guidance document in 2002 a number of liposome products received market approval in the USA. It would be interesting to learn to what extent the recommendations in the draft guidance were followed and whether more or less studies had to be performed and where new insights impacted the protocols. For instance, regarding the use of QbD approaches to fine tune the final lipid composition and the manufacturing process and to set specifications (see above). Which new analytical techniques were introduced/requested and which insights were gained around evaluating complement binding effects (Szebeni et al. 2011)?

Generic versions or Other Liposomal Drug Products

Doxil or Ambisome: For Europe

Up to now, no generic versions of liposomes have been approved by the EMA. An application for a generic version (Lipodox) of Doxil/Caelyx was not accepted by the EMA (see above and CHMP assessment report Doxorubicin SUN EMA (2011)). According to the assessment report 'there are outstanding major non-clinical and clinical objections which preclude a recommendation for marketing authorization at the present time'. TLC (Taiwan Liposome Company) is now conducting bioequivalence studies in Europe with her Doxil generic (Doxisome™)

with the aim to gain EU market approval at the end of 2015. It will continue pursuing bioequivalence and market approval of Doxisome in the US immediately after. Ambil™ is the Ambisome generic developed by the Taiwan Liposome Company (TLC) and locally approved for the treatment of systemic infections. TLC expects to file the EMA Ambil generic application before the end of 2014 with marketing clearance for EU countries potentially taking another 6–9 months of review time (TLC 2013).

In 2013 the EMA published a 'Reflection paper on the data requirements for intravenous liposomal products developed with reference to an innovator liposomal product'. In this document, pharmaceutical quality aspects, non-clinical and clinical requirements for approval of a generic liposome drug product are discussed. But, it also mentions that these 'reflections' may be useful for those who wish to submit a dossier for a novel liposome drug product. The agency recommends the generic version to be similar to the innovator's product in qualitative and quantitative terms and follows the philosophy that 'In the comprehensive evaluation of the new liposomal product the body of evidence obtained in quality, non-clinical and clinical studies must be considered as a whole'. Final decisions regarding the full dossier requirements will be taken on a case-by-case basis. This also concerns the need for clinical efficacy studies. No full scale clinical safety trials are anticipated and reference is made to EU pharmacovigilance guidelines. The regulatory philosophy behind this 'reflection paper' is similar to the one expressed in the documents published by the FDA on the subject of liposomal drug products.

Doxil or Ambisome: For the US

In 2010 the FDA issued a 'Draft guidance on doxorubicin hydrochloride liposomes' containing non-binding recommendations for sponsors of generic versions of reference list doxorubicin containing liposomes (Doxil)(FDA Draft Guidance on Doxorubicin Hydrochloride (2010)). The recommendations are meant for generic products where the test and reference PEGylated liposome products: '(1) have the same drug product composition and (2) are manufactured by an active liposome loading process with an ammonium sulfate gradient and (3) have equivalent liposome characteristics including liposome composition, state of encapsulated drug, internal environment of the liposome, liposome size distribution, number of lamellae, grafted PEG at the liposome surface, electrical surface potential or charge, and in vitro leakage rates.' In vitro release studies have to be performed and a set of recommended experimental conditions (pH, temperature) is provided. The FDA advises to use QbD (see above, Xu et al. 2012a, b) approaches to identify critical process parameters and material attributes to guide optimization of the manufacturing process. This document does not mention a requirement to run clinical efficacy comparisons.

The FDA approved the market authorization of Lipodox™, a generic version of Doxil™ in 2013 through the Abbreviated New Drug Application (ANDA) pathway (the EMA still has not followed suit). First FDA issued a temporary permission (until 4 February 2013) for the use of Lipodox™ because of the shortage of supply of Doxil™. Lipodox™ is also marketed in India.

Doxil/Ambisome: For the rest of the world

Outside the EU and US amphotericin liposomes other than Ambisome have received marketing authorization. The same is true, mutatis mutandis, for doxorubicin liposomes and Doxil. It is not always clear whether these companies have aimed for a generic version (sameness principle) of Ambisome or Doxil or whether a different, new formulation has been introduced.

As mentioned above, several (generic) versions of amphotericin-liposomes (reference product: Ambisome™?) are on the market under names such as, Fosome™—Cipla, India, Lambin Liposome™—Sun Pharma India (personal communications)(cf. http://www.medindia.net/drug-price/amphotericin-b/fungisome.htm). We were not able to find more information about these products than what was mentioned by Balasegaram et al. 2012. In Taiwan Ambil™ is marketed by TLC (Taiwan Trade Center, Toronto (2013)).

In India Fungisome™—Lifecare India (Fungisome 2014) (amphotericin containing soy phosphatidylcholine/cholesterol liposomes) is marketed. This amphotericin-lipid formulation needs to be sonicated before administration, which is a somewhat striking feature in the field of parenteral liposome products. Several clinical studies have been performed and published (Kshirsagar et al. 2005). Anfogen™ is an amphotericin liposome with a similar lipid composition as Ambisome but it is produced with a different manufacturing process. Anfogen has been approved by the Argentinean authorities and is sold by Genpharma S.A., Argentina. Olson et al. (2008) compared the physicochemical properties, antifungal and toxicity properties in vitro and in animals. Their study shows that Anfogen and Ambisome differ in their physicochemical properties. Therefore, Anfogen should not be considered to be a generic version of Ambisome. Gaspani and Milani (2013) mention that the FDA is looking at generic version of Ambisome, but no further information became available since. To stimulate and increase the access to affordable amphotericin liposome products throughout the world Gaspani and Milani propose the WHO to play an active role in the development of protocols to ensure high quality, generic versions of amphotericin liposomes. The existing innovator's formulation is highly active against visceral leishmaniasis, a serious, even deadly, disease in parts of Asia and Africa. Maybe the format of an assessment of generics alternatives using the model of the World Health Organization (WHO) prequalification system could be considered (http://apps.who.int/prequal/)(Gaspani and Milani 2013). The mission of this WHO program is: 'In close cooperation with national regulatory agencies and partner organizations, the Prequalification Programme aims to make quality priority medicines available for the benefit of those in need. This is achieved through its evaluation and inspection activities, and by building national capacity for sustainable manufacturing and monitoring of quality medicines.'

Final Considerations and Reflections

Access to All Existing Information

Above, information on liposomal drug formulations has been provided with a focus on regulatory issues that are specific for NDA and ANDA applications (and their European variants). But there is more information that could help to facilitate dossier formation of novel and generic liposome submissions for marketing approval. Unfortunately, the present rules do not allow all documentation used for registration dossiers to be published. Thus, what is in the public domain and accessible to us, authors, is only part of a much larger knowledge data bank. We also realize that we used search engines with English as leading language and by that we could, and probably, have missed material in other languages from countries outside the US and EU territories.

If clinical studies are requested for generic product registrations, then how extensive should these be? It is obvious that generic liposomal drug products need to fulfill pharmaceutical equivalence and bioequivalence criteria. However, it is currently under debate whether clinical studies can be limited to pharmacokinetics studies only, as usually is the case with conventional oral dosage forms such as tablets, or whether additionally (pre)clinical efficacy and/or safety studies should be performed to show similarity. In this context, the discussion on the consequences of following the' sameness' or 'not-sameness' approach has not been finished. An example where the 'sameness/not-sameness' topic is brought up (although not explicitly mentioned by the authors) is in the following study: A group from ALZA Johnson and Johnson (Mamidi et al. 2010) emphasized the importance of the selection of the phospholipid bilayer composition. They studied doxorubicin liposomes with different, but closely related bilayer structures and found in some cases in animals (mice and monkeys were used) the same pharmacokinetic profiles but different safety or efficacy outcomes (likely related to different in vivo drug release profiles). The authors conclude 'that plasma pharmacokinetics and systemic exposure of doxorubicin did not correlate well with the antitumor activity and toxicity profiles for PEGylated liposomal doxorubicin products'. One might argue that this conclusion cannot be drawn for cases where the 'sameness' principle (the bilayer composition, loading principle and particle size of the reference product and generic version are the same) is followed for the design and production of the generic liposome formulations.

Pharmaceutical Aspects of Liposome Design

A number of practical questions have to be addressed again and again when trying to design a new drug-liposome product. E.g. there is a great need for validated assays for free/liposome encapsulated drug determination in vivo/the clinic. What are the best separation protocols? What are validated in vitro release assays predictive for in vivo behavior? What are the best (and validated) protocols to establish sterility and absence of pyrogens? And: what are relevant impurities and what are the specifications for their presence in the final drug product?

In this context, the recommended choice for determination of particle size/aggregate size and sensitivity of the variation therein for clinical performance (domain space) should be given attention. Regularly, new techniques for particle sizing are introduced, but it takes years before the full potential benefit for liposome characterization is assessed. Here an institution such as the Nanotechnology Characterization Lab (NCL, NIH) is extremely helpful. An European initiative, as proposed in the white paper of the European Technology Platform (NANOMEDICINE 2020: Contribution of Nanomedicine to Horizon 2020 (2013)) could help to carry the burden.

Another Point for Consideration

Quite a few products use semi-natural phospholipids (notably hydrogenated soy bean phosphatidylcholine (HSPC), e.g. in Doxil). Is there any indication whether the nature of the fatty acids is affected by the genetic changes the soy bean plants will go through in the future? Would that be a reason to prefer synthetic lipids? However, HSPC has the advantage of being much cheaper than dipalmitoylphosphatidylcholine (DPPC) or distearoylphosphatidylcholine (DSPC).

A Bumpy Ride and Surprises

This chapter would be incomplete without mentioning the article on the ‘bumpy ride’ of the development of Doxil by one of its visionary drivers, Chezy Barenholz (Barenholz 2012). In 2012 Barenholz asked the question: ‘how come there is still no generic PEGylated liposomal doxorubicin-like product approved by the FDA or EMA?’ And he presented an analysis of the hurdles to be taken by generic products when maneuvering through the ANDA process. In that analysis he frequently quoted the excellent paper by Jiang et al. 2011 (FDA officers). Surprisingly, one year later Lipodox, the generic version of Doxil, was approved by the FDA.

In Conclusion

The development of novel and generic liposomal drug products poses challenges as discussed in this chapter. But, we as ‘liposomologists’ active in the academic, industrial or regulatory field, have learned a lot over the last 30+ years about these complex systems. And this know-how, particularly if efficiently brought together, can help novel, therapeutically beneficial liposome-based products and generic versions of existing ones to reach the market and the patient faster.

References

Allen TM, Cullis PR (2013) Liposomal drug delivery systems: from concept to clinical applications. Adv Drug Deliv Rev 65:36–48

Allen TN, Hansen CB (1991) Pharmacokinetics of stealth versus conventional liposomes: effect of dose. Biochim Biophys Acta 1068:133–141

Amphonex (2014) http://www.bharatserums.com/product/neurology/AMPHONEX%2050mg%20%28Lyo%29%20Pack%20Insert%20for%20Domestic_2013_04_06_07_45_20_893.pdf

Arshinova OY, Sanarova EV, Lantsova AV, Oborotova NA (2012) Drug synthesis methods and manufacturing technology: lyophilization of liposomal drug forms. Pharm Chem J 46:228–233

Bae YH, Park K (2011) Drug delivery to tumors: myths, reality and possibility. J Controlled Release 153:198–205

Balasegaram M, Ritmeijer K, Lima MA, Burza S, Genovese GO, Milani B, Gaspan S, Potet J, Chappuis F (2012) Liposomal amphotericin B as a treatment for human leishmaniasis. Expert Opin Emerg Drugs 17:493–510

Barenholz Y (2012) Doxil®—The first FDA-approved nano-drug: lessons learned. J Control Release 160:117–134

Barenholz Y, Crommelin DJA (1994) Liposomes as pharmaceutical dosage forms. In: Swarbrick J, Boylan JC (eds) Encyclopedia of pharmaceutical technology, vol 9. Marcel Dekker, Inc, NY, pp 1–39

Bhardwaj U, Burgess DJ (2010) A novel USP apparatus 4 based release testing method for dispersed systems. Intl J Pharm 388:287–294

Carstens MG, Romberg B, Oussoren C, Storm G (2007) Observations on the disappearance of the stealth property of PEGylated liposomes: effects of lipid dose and dosing frequency. In: Gregoriadis G (ed) Liposome technology, vol III, 3rd edn. CRC Press, Boca Raton pp 79–94

CBG-MEB (2012) http://www.cbg-meb.nl/NR/rdonlyres/E2508764-9FC9-4C59-BE29-B02DB4361E91/0/1208DHPCDepocyteEN.pdf. Accessed 26 April 2015

CHMP assessment report Doxorubicin SUN EMA (2011) http://www.ema.europa.eu/docs/en_GB/document_library/Application_withdrawal_assessment_report/human/002049/WC500112957.pdf. Accessed 26 April 2015

Chassu D, Holt K, Chader GJ (2013) Nanoparticle-based therapeutics, an overview. In: Chassu D, Chader GJ. (eds) Ocular drug delivery systems: barriers and application of nanoparticulate systems. CRC Press, Boca Raton, pp 3–16

Chonn A, Semple SC, Cullis PR (1992) Association of blood proteins with large unilamellar liposomes in vivo. Relation to circulation lifetimes. J Biol Chem 267:18759–18765

Crommelin DJA, Florence AT (2013) Towards more effective advanced drug delivery systems. Int J Pharm 454:496–511

Dams ET, Laverman P, Oyen WJ, Storm G, Scherphof GL, Van der Meer JW, Corstens FHM, Boerman OC (2000) Accelerated blood clearance and altered biodistribution of repeated injections of sterically stabilized liposomes. J Pharmacol Exp Ther 292:1071–1079

Druckmann S, Gabizon A, Barenholz Y (1989) Separation of liposome-associated doxorubicin from non-liposome-associated doxorubicin in human plasma: implications for pharmacokinetic studies. Biochim Biophys Acta 980:381–384

Drummond DC, Noble CO, Hayes ME, Park JW, Kirpotin DB (2008) Pharmacokinetics and in vivo drug release rates in liposomal nanocarrier development. J Pharm Sci 97:4696–4740

EMA ICH, Topic M 4 (2004) Common technical document for the registration of pharmaceuticals for human use –organisation CTD. http://www.ema.europa.eu/docs/en_GB/document_library/Scientific_guideline/2009/09/WC500002721.pdf. Accessed 26 April 2015

EMA Questions and answers on the supply situation of Caelyx (2012) http://www.ema.europa.eu/docs/en_GB/document_library/Medicine_QA/2013/04/WC500142510.pdf. Accessed 26 April 2015

EMA Reflection paper on the data requirements for intravenous liposomal products developed with reference to an innovator liposomal product (2013) http://www.ema.europa.eu/docs/en_GB/document_library/Scientific_guideline/2013/03/WC500140351.pdf. Accessed 26 April 2015

EMA Summary of product characteristics Myocet (EPAR) (2010) http://www.ema.europa.eu/docs/en_GB/document_library/EPAR__Product_Information/human/000297/WC500031811.pdf. Accessed 26 April 2015

EMA Summary of product characteristics Caelyx (EPAR) (2014) http://www.ema.europa.eu/docs/en_GB/document_library/EPAR__Product_Information/human/000089/WC500020180.pdf. Accessed 26 April 2015

EMA Volume 2B Notice to applicants medicinal products for human use (2006) http://ec.europa.eu/health/files/eudralex/vol-2/b/update_200805/ctd_05-2008_en.pdf. Accessed 26 April 2015

EMEA Scientific discussion Caelyx (2005) http://www.ema.europa.eu/docs/en_GB/document_library/EPAR_-_Scientific_Discussion/human/000089/WC500020175.pdf. Accessed 26 April 2015

FDA (2002) Guidance for Industry: Liposome Drug Products http://www.fda.gov/downloads/Drugs/GuidanceComplianceRegulatoryInformation/Guidances/ucm070570.pdf. Accessed 26 April 2015

FDA Ben Venue Laboratories—Voluntary Shutdown (2011) http://www.fda.gov/Drugs/DrugSafety/ucm281782.htm. Accessed 26 April 2015

FDA Draft Guidance on Doxorubicin Hydrochloride (2010) http://www.fda.gov/downloads/Drugs/…/Guidances/UCM199635.pdf. Accessed 26 April 2015

FDA Grants (2014) http://grants.nih.gov/grants/guide/rfa-files/RFA-FD-14-016.html. Accessed 26 April 2015

Fenske DB, Maurer N, Cullis PR (2003) Encapsulation of weakly-basic drugs, antisense oligonucleotides and plasmid DNA within large unilamellar vesicles for drug delivery applications. In: Torchilin VP, Weissig V (eds) Liposomes, a practical approach, second edition. Oxford University, Oxford, pp 173–180

Fierce Pharma Manufacturing (2013) http://www.fiercepharmamanufacturing.com/story/gilead-recalls-20-batches-Ambisome-due-possible-contamination/2013-06–19. Accessed 26 April 2015

Fungisome (2014) http://www.lifecareinnovations.com/. Accessed 26 April 2015

Gaspani S, Milani B (2013) Access to liposomal generic formulations: beyond AmBisome and Doxil/Caelyx. GaBI J 2:60–62

Gaspar R (2010) Therapeutic products: regulating drugs and medical devices. In: Hodge GA et al. (eds) International handbook on regulating nanotechnologies. Edward Elgar, Cheltenham, pp 291–320

Gregoriadis G (2006) Liposome technology, vol I, II and III, 3rd Revised edn. Taylor & Francis Inc, Hoboken

Griese N, Blaschke G, Boos J, Hempel G (2002) Determination of free and liposome-associated daunorubicin and daunorubicinol in plasma by capillary electrophoresis. J Chromatogr A 979:379–388

Grit M, Zuidam N, Crommelin DJAC (1993) Analysis and hydrolyis kinetics of phospholipids in liposome dispersions. In: Gregoriadis G (ed) Liposome technology, vol III, 2nd edn. CRC Press, Boca Raton, pp 455–485

Harrington KJ, Mohammadtaghi S, Uster PS, Glass D, Peters AM, Vile RM, Stewart JSW (2001) Effective targeting of solid tumors in patients with locally advanced cancers by radiolabeled pegylated liposomes. Clin Cancer Res 7:243–254

Jiang W, Lionberger R, Yu LX (2011) In vitro and in vivo characterizations of PEGylated liposomal doxorubicin. Bioanalysis 3:333–344

Kim HS, Wainer IW (2010) Simultaneous analysis of liposomal doxorubicin and doxorubicin using capillary electrophoresis and laser induced fluorescence. J Pharm Biomed Anal 52:372–376

Kraft JC, Freeling JP, Wang Z, Ho RJY (2013) Emerging research and clinical development trends of liposome and lipid nanoparticle drug delivery systems. J Pharm Sci 103:1520–6017
Kshirsagar NA, Pandya SK, Kirodian BG, Sanath S (2005) Liposomal drug delivery system from laboratory to clinic. J Postgrad Med 51:5–15
Lammers T, Rizzo LY, Storm G, Kiessling F (2012) Personalized nanomedicine. Clin Cancer Res 18:4889–4894
Lasch J, Weissing V, Brandi M (2003) Preparation of liposomes. In: Torchilin VP, Weissing V (eds) Liposomes, a practical approach, 2nd edn. Oxford University Press, Oxford, pp 267–286
Laverman P, Carstens MG, Boerman OC, Dams ET, Oyen WJ, van Rooijen N, Corstens FH, Storm G (2001) Factors affecting the accelerated blood clearance of polyethylene glycol-liposomes upon repeated injection. J Pharmacol Exp Ther 298:707–612
Mamidi RNVS, Weng S, Stellar S, Wang C, Yu N, Huang T, Tonelli AP, Kelley MF, Angiuoli A, Fung MC (2010) Pharmacokinetics, efficacy and toxicity of different pegylated liposomal doxorubicin formulations in preclinical models: is a conventional bioequivalence approach sufficient to ensure therapeutic equivalence of pegylated liposomal doxorubicin products? Cancer Chemother Pharmacol 66:1173–1184
Mamot C, Ritschard R, Wicki A, Stehle G, Dieterle Th, Bubendorf L, Hilker C, Deuster S, Herrmann R Rochlitz C (2012) Tolerability, safety, pharmacokinetics, and efficacy of doxorubicin-loaded anti-EGFR immunoliposomes in advanced solid tumours: a phase 1 dose-escalation study. Lancet Oncol 13:1234–1241
Mayer LD, St-Onge G (1995) Determination of free and liposome-associated doxorubicin and vincristine levels in plasma under equilibrium conditions employing ultrafiltration techniques. Anal Biochem 232:149–157
Moghimi SM, Szebeni J (2003) Stealth liposomes and long circulating nanoparticles: critical issues in pharmacokinetics, opsonization and protein-binding properties. Prog Lipid Res 42:463–478
Nag A, Mitra G, Ghosh PC (1997) A colorimetric estimation of polyethyleneglycol-conjugated phospholipid in stealth liposomes. Anal Biochem 250:35–43
NANOMEDICINE 2020: Contribution of Nanomedicine to Horizon 2020 (2013) http://www.etp-nanomedicine.eu/public/press-documents/publications/etpn-publications/etpn-white-paper-H2020
Olson JA, Adler-Moore JP, Jensen GM, Schwartz J, Dignani MC, Proffitt RT (2008) Comparison of the physicochemical, antifungal, and toxic properties of two liposomal amphotericin b products. Antimicrob Agents Chemother 52:259–268
Oussoren C, Storm G (1999) Effect of repeated i.v. administration on the circulation kinetics of poly(ethyleneglycol)-liposomes in rats. J Liposome Res 9:349–355
Park K (2013) Facing the truth about nanotechnology in drug delivery. ACS Nano 7:7442–7447
Storm G, Belliot SO, Daemen T, Lasic DD (1995) Surface modification of nanoparticles to oppose uptake by the mononuclear phagocyte system. Adv Drug Delivery Rev 17:31–48
Szebeni J (2005) Complement activation-related pseudoallergy: a new class of drug-induced immune toxicity. Toxicology 216:106–121
Szebeni J, Muggi F, Gabizon A, Barenholz Y (2011) Activation of complement by therapeutic liposomes and other lipid excipient-based therapeutic products: prediction and prevention. Adv Drug Delivery Rev 63:1020–1030
Taiwan Trade Center, Toronto (2013) http://toronto.taiwantrade.com.tw/news/detail.jsp?id=12817&lang=En-US_US. Accessed 26 April 2015
Thies RL, Cowens DW, Cullis PR, Bally MB, Mayer LD (1990) Method for rapid separation of liposome-associated doxorubicin from free doxorubicin in plasma. Anal Biochem 188:65–71
TLC (2013) http://www.tlcbio.com/upload/media/investor/%E6%B3%95%E8%AA%AA%E6%9C%83%E7%B0%A1%E5%A0%B1/Company%20Yearly%20Update%20for%20201403_v5.pdf. Accessed 26 April 2015
van den Hoven JM (2012) Liposomal glucocorticoids: pharmaceutical, preclinical and clinical aspects, Thesis Utrecht University. http://dspace.library.uu.nl/handle/1874/256300

van der Meel R, Vehmeijer LJC, Kok RJ, Storm G, van Gaal EVB (2013) Ligand-targeted particulate nanomedicines undergoing clinical evaluation: current status. Adv Drug Deliv Rev 65:1284–1298

van Winden EC (2003) Freeze-drying of liposomes: theory and practice. Methods Enzymol 367:99–110

Varga Z, Wacha A, Vainio U, Gummel J, Bóta A (2012) Characterization of the PEG layer of sterically stabilized liposomes: a SAXS study. Chem Phys Lipids 165:387–392

Venditto VJ, Szoka FC (2013) Cancer nanomedicines: so many papers and so few drugs! Adv Drug Delivery Rev 65:80–88

Vernooij EA, Gentry CA, Herron JN, Crommelin DJ, Kettenes-van den Bosch JJ (1999) 1H NMR quantification of poly(ethylene glycol)-phospatidylethanolamine in phospholid mixtures. Pharm Res 16:1658–1661

Woodle MC, Lasic DD (1992) Sterically stabilized liposomes. Biochim Biophys Acta 1113:171–199

Woodle MC, Matthay KK, Newman MS, Hidayat JE, Collins LR, Redemann C, Martin FJ, Papahadjopoulos D (1992) Versatility in lipid composition showing prolonged circulation with sterically stabilized liposomes. Biochim Biophys Acta 1105:193–200

Xu X, Khan MA, Burgess DJ (2011) Formulation, processing design and risk assessment. Int J Pharm 419:52–59

Xu X, Khan MA, Burgess DJ (2012a) A quality by design (QbD) case study on liposomes containing hydrophilic API: II. Screening of critical variables, and establishment of design space at laboratory scale. Int J Pharm 423:543–553

Xu X, Costa AP, Khan MA, Burgess DJ (2012b) Application of quality by design to formulation and processing of protein liposomes. Int J Pharm 434:349–359

Xu X, Khan MA, Burgess DJ (2012c) A two-stage reverse dialysis in vitro dissolution testing method for passive targeted liposomes. Int J Pharm 426:211–218

Yoshino K, Taguchi K, Mochizuki M, Nozawa S, Kasukawa H, Kono K (2012) Novel analytical method to evaluate the surface condition of polyethylene glycol-modified liposomes. Colloids Surf A Physicochem Eng Aspects 397:73–79

Zuidam NJ, Van Winden E, De Vrueh R, Crommelin DJA (2003) Stability, storage, and sterilization of liposomes. In: Torchilin VP, Weissing V (eds) Liposomes, a practical approach, 2nd edn. Oxford University Press, Oxford, pp 149–165

Glatiramoids

Vera Weinstein, Rivka Schwartz, Iris Grossman, Benjamin Zeskind and J. Michael Nicholas

Contents

Introduction 110
Glatiramer Acetate 111
Chemistry, Manufacturing, and Control (CMc) 112
Chemistry: Drug Substance and Drug Product 112
Manufacturing: Standard Techniques 113
Control: Glatiramoid Identity and Quality 114
Pharmacology 125
Mechanisms of Action 126
Biological Activity: Glatiramoid-Mediated Effects on Gene Expression 134
Regulatory Status 138
Prospects 140
References 142

Abstract Glatiramoids are non-biologic complex drugs (NBCDs) comprising four naturally occurring amino acids in a complex copolymeric mixture. The first and most thoroughly studied glatiramoid, glatiramer acetate (Copaxone®, Teva Pharmaceutical Industries, Ltd.) is approved for treatment of relapsing-remitting forms of multiple sclerosis, an autoimmune disorder characterized by neuroinflammation and progressive neurodegeneration. Glatiramoid mixtures comprise a potentially incalculable number of structurally closely related active peptide moieties that cannot be isolated, quantified, or identified using even the most sophisticated available multidimensional separation techniques. Numerous studies have demonstrated that the glatiramer acetate in Copaxone® modulates innate and adaptive immune cell responses to promote antiinflammatory and neuroprotective activities; however, the

V. Weinstein (✉) · R. Schwartz · I. Grossman
Teva Pharmaceutical Industries, Ltd., Petach Tikva, Israel
e-mail: vera.weinstein@teva.co.il

B. Zeskind
Immuneering Corporation, Cambridge, USA

J. Michael Nicholas
Teva Pharmaceuticals, Kansas City, MO, USA

D. J. A. Crommelin, J. S. B. de Vlieger (eds.), *Non-Biological Complex Drugs,* AAPS Advances in the Pharmaceutical Sciences Series 20, DOI 10.1007/978-3-319-16241-6_4

active epitopes in Copaxone® are unknown and the precise mechanisms of immunomodulatory activity responsible for its therapeutic efficacy are not entirely elucidated. The identity, quality, and consistency of a glatiramoid are inexorably linked to its own manufacturing process. Several manufacturers now market glatiramoids in various countries that are purported to be generic or follow-on versions of Copaxone®; at this writing, no full set of peer-reviewed long term safety and efficacy data for these products is available in the medical literature. Sophisticated analysis techniques, though unable to completely characterize glatiramoid mixtures can differentiate among them based on physicochemical features and biological activities. Comparative gene expression studies have demonstrated important differences between the reference drug (Copaxone®) and purported generic glatiramer acetate products that may have significant implications for the safety and efficacy of the purported generic products. Currently, there is no globally agreed defined pathway for regulatory approval of follow-on and generic glatiramoid products. In the interest of patient safety and well-being, there is an urgent need for regulatory agencies to come to consensus regarding criteria needed to establish therapeutic equivalence among members of the glatiramoid class. Scientific approaches discussed in this chapter may be helpful when evaluating glatiramoid formulations in the framework of equivalence testing.

Keywords Copaxone® · Glatiramer acetate · Glatiramoids · Multiple sclerosis (RRMS) · Non-biological complex drug (NBCD) · Copolymeric mixture · Colloidal suspension · Nanoparticles · Purported generic GA · Size exclusion chromatography · Peptide mapping · Capillary electrophoresis (CE) · Dynamic light scattering (DLS) · Atomic force microscopy (AFM) · Ion mobility mass spectrometry (IMMS) · Polimunol GTR® · Glatimer® · Escadra® · Probioglat® · TV-5010 · Microarray · Gene expression · Genomics · MRNA · Pathways · Safety · Efficacy · Adverse events · Follow-on glatiramoids · Human monocytes · THP-1 · Mouse splenocytes · Mechanism of action · Biological activity · Biological analysis

Abbreviations

AA	Amino acids
ACTRIMS	Americas committee for treatment and research in multiple sclerosis
AFM	Atomic force microscopy
ALA	L-alanine
APC	Antigen-presenting cells
APL	Altered peptide ligand
APP	Amyloid precursor protein
ARR	Annual relapse rate
BBB	Blood Brain Barrier
BDNF	Brain-derived neurotrophic factor
CCR7	Chemokine receptor type 7

CDMS	Clinically definitive multiple sclerosis
CNS	Central nervous system
Co-stim	Costimulatory
DLS	Dynamic light scattering
EAE	Experimental autoimmune encephalomyelitis
ECTRIMS	European committee for treatment and research in multiple sclerosis
GA	Glatiramer acetate
GdE	Gadolinium-enhancing
GLU	L-glutamic acid
HLA	Human leukocyte antigen
IEF	Isoelectric focusing
IL	Interleukin
IMMS	Ion mobility mass spectrometry
IFN-γ	Interferon gamma
LYS	L-lysine
MBP	Myelin basic protein
MHC	Major histocompatibility complex
MOG	Myelin oligodendrocyte glycoprotein
MRI	Magnetic resonance imaging
MRS	Magnetic resonance spectroscopy
MS	Multiple sclerosis
MW	Molecular weight
NBCD	Non-biological complex drug
NK	Natural killer
PBMC	Peripheral blood mononuclear cells
PD	Pharmacodynamic
PD1	Programmed death receptor 1
PFS	Pre-filled syringe
PK	Pharmacokinetic
PLP	Proteolipid protein
RP-HPLC	High-performance liquid chromatography
RRMS	Relapsing-remitting multiple sclerosis
SC	Subcutaneous
SEC	Size exclusion chromatography
TCR	T cell receptor
TGF-β	Transforming growth factor beta
TIV	Total intensity value
TNBS	2,4,6-trinitrobenzenesulfonic acid
TNF-α	Tumor necrosis factor alpha
TYR	L-tyrosine
UV	Ultraviolet
WFI	Water for injection
YFP	Yellow fluorescent protein

Introduction

Glatiramoids are non-biological complex drug (NBCD) products for which the active ingredient is a copolymeric mixture of four synthetic, naturally occurring amino acids (AA), L-alanine, L-glutamic acid, L-lysine, and L-tyrosine, in a constant molar ratio (Varkony et al. 2009). Copaxone®, the prototype and most intensively studied glatiramoid, is an injectable colloidal suspension containing the active ingredient, glatiramer acetate (GA, formerly known as Copolymer-1). Copaxone® is now approved and used in 57 countries worldwide as an immunomodulator to reduce the frequency of relapses in ambulatory patients with relapsing-remitting multiple sclerosis (RRMS) and in patients who have experienced a clinically isolated syndrome and are determined to be at high risk of developing clinically definitive multiple sclerosis (CDMS) (COPAXONE® 1996).

GA and its basic biological properties were discovered in the 1960s by scientists at the Weizmann Institute in Israel and later developed by Teva Pharmaceutical Industries into the medicinal product, Copaxone®, to treat MS, a chronic, degenerative autoimmune disease. Discovery of Copaxone® was serendipitous in that the scientists were trying to develop an encephalitogenic mixture of copolymers in order to induce experimental autoimmune encephalomyelitis (EAE, an animal model of MS) in guinea pigs; however, GA proved to be protective against EAE and to ameliorate severity of established disease (Teitelbaum et al. 1971). Since then, decades of research and clinical use have shown Copaxone® to be safe and effective; however, the individual active components in GA have not been identified (and cannot be, give the limits of current technology) and the exact mechanism of action of GA is not completely elucidated. In recent years, there has been growing interest in the pharmaceutical industry to develop new glatiramoids for treatment of MS. Teva Pharmaceutical Industries pursued the development of TV-5010 and currently, a phase II clinical study of a "second generation" co-polymer, plovamer acetate (also called PI-2301; Merck Serono) in MS patients is underway; this agent has three of the same amino acid constituents as GA but replaces glutamic acid with phenylalanine. In addition to development of new glatiramoids, attempts have been made to manufacture (purported) generic versions of GA. At this writing, products purported to be generic versions of GA are marketed in India and the Ukraine (Glatimer®, Natco Pharma, Ltd., Hyderabad, India), in Mexico (Probioglat®, Probiomed S.A. de C.V.; Mexico City, Mexico), and two in Argentina (Escadra®, Raffo Laboratories, Buenos Aires, Argentina), and (Polimunol GTR®, Synthon). Moreover, at this writing, a generic to Copaxone®, Glatopa by Sandoz was approved in US, however not on the market yet.

Currently, there are no peer-reviewed data of the clinical efficacy of all but one of these purported generic products in the medical literature. In September 2014, data was presented at the Joint ACTRIMS-ECTRIMS Meeting (see list of abbreviations) from a clinical trial ("GATE") sponsored by Synthon B.V., the Netherlands, to evaluate the safety and efficacy of a GA-like product ("GTR") compared with Copaxone® (http://www.synthon.com/Corporate/News/PressReleases/Synthon-announces-successful-outcome-of-the-PhaseIII-GATE-study-with-its-generic-glatiramer-acetate?sc_lang=en). The GATE study investigators reported the key result that "The estimated geometric mean numbers of GdE (gadolinium-enhanced) lesions were 0.42 (GTR) and 0.39 (Copaxone®, COP), resulting in an estimated GTR/COP GdE lesion ratio of 1.097 with a 95 % confidence interval of 0.884–1.362, which is within the pre-defined equivalence margin" (Cohen et al. 2014). During the question and answer period following the presentation at the meeting, participants raised substantive questions regarding the study design and reported outcomes. Most importantly, it was pointed out that the ARR observed in GATE for Copaxone® (0.41) was essentially the same as for the placebo (0.39). This finding is inconsistent with two decades of clinical findings for Copaxone. For instance, in the Comi et al. (2001) study (which was also 9 months in duration, and was cited by the investigators as the model adopted for GATE) Copaxone® decreased ARR substantially (0.81 for Copaxone® versus 1.21 for placebo). At the Joint ACTRIMS-ECTRIMS meeting, the presenter, Dr. Jeffery Cohen, responded to the question by stating that he does not know why the effect observed on MRI in GATE was not reflected in ARR.

In addition to the questions raised about the GATE study, Teva has initiated physicochemical, biological and genomic characterization studies of the Synthon product marketed in Argentina since May 2014. Results of these analyses indicate significant differences between the two glatiramoids and will be published in the peer-reviewed literature upon completion.

Purported generic drug products should be the "same" or at minimum, "highly similar" to Copaxone®. However, important differences from Copaxone® have been detected using sophisticated state-of-the-art technologies (Bakshi et al. 2013; Towfic et al. 2014). As described here, information gleaned by comparing Copaxone® with some of the purported generic GA products is contributing to a better understanding of the complexity of glatiramoids, and of the crucial relationship between drug manufacture and drug identity and quality for members of the glatiramoid class furthermore, the biological and clinical implications of lack of sameness have been examplified in Teva's multiple publications and citizen petitions. However, no testing in humans has been required from the first generic approved by FDA less than two weeks from publishing these lines.

Glatiramer Acetate

GA is a first-generation nanomedicine comprising a multitude of immunogenic synthetic polypeptides that are not fully characterized in a colloidal suspension (EMEA/CHMP/79769/2006 2006). Unlike most small-molecule drugs, GA is a heterogeneous mixture of potentially millions of distinct polypeptides, each containing up to 200 AA, with higher order (secondary) structural elements and polypeptide sizes comparable to those of proteins. The complexity of GA is amplified by the fact that its exact mechanism of action is unknown. Like other NBCDs, GA does not have a homo-molecular structure, and the GA in Copaxone® is defined in large part by its well-controlled manufacturing process, which was developed and has been optimized by the manufacturer (Teva Pharmaceutical Industries, Ltd.) over the last few decades (Duncan and Gaspar 2011; Nicholas 2012; Schellekens et al. 2011; Varkony et al. 2009). The importance of the stringent manufacturing method used to create GA is exemplified by development of a second generation GA glatiramoid known as TV-5010 that the same manufacturer prepared by making slight changes to the GA manufacturing process conditions at downstream stages (Ramot et al. 2012). The product had the same molar ratio of AA as GA but a higher average molecular weight (MW). Despite similarities in some physicochemical parameters, the *in vivo* safety profile of TV-5010 differed dramatically from that of GA, even though both products were produced from the same upstream intermediate mixture. As described below, TV-5010 showed severe toxicity in long-term preclinical studies that was never seen with GA (Ramot et al. 2012).

Routine and advanced analytical techniques and assessment of biological activities indicate critical differences between Copaxone® and purported generic products that highlight the unique challenges of creating new glatiramoids or replicating Copaxone®. Although common nonspecific analytical methods can indicate similarities between Copaxone® and purported generic GA products, they also identify important differences in physicochemical properties, including mixture composition, polypeptide sizes, and charge distribution. Similarly, gene expression studies show profoundly different gene transcription profiles from GA-activated mouse splenocytes when re-activated *ex vivo* with GA, from transcription profiles when splenocytes are re-activated with purported generic GA product (Bakshi et al. 2013; Towfic et al. 2014; Citizens petition 2013, 2014). Moreover, gene expression analysis has shown that at least one purported generic GA product (Glatimer®) with multiple different marketed batches ($N > 10$) demonstrates very poor batch-to-batch consistency of biological effects (Bakshi et al. 2013; Nicholas 2012; Towfic et al. 2014; Citizen Petition 2009).

The active AA sequences (biological epitopes) in glatiramoids are believed to act, in part, as altered peptide ligands (APLs) of encephalitogenic epitopes within myelin basic protein (MBP), an autoantigen implicated in MS (Aharoni et al. 1999). As antigen-based therapeutics, glatiramoids are highly immunogenic. Decades of clinical use demonstrate that Copaxone® does not contain encephalitogenic epitopes and does not induce auto-reactive antibodies (Johnson 2010). Copaxone® induces anti-GA antibodies that are not neutralizing and do not affect clinical efficacy or safety (Teitelbaum et al. 2003). However, the same cannot be assumed for purported generic products. Experts in the field of MS agree that the immunogenicity profile of follow-on glatiramoid products cannot be assumed and should be established for each new glatiramoid (Cohen et al. 2008). Recommendations for establishing the safety, efficacy, and immunogenicity of new and generic glatiramoid products before market access approval are described below.

Chemistry, Manufacturing, and Control (CMc)

Chemistry: Drug Substance and Drug Product

Copaxone® (GA) is a copolymeric mixture comprising L-alanine (ALA), L-glutamic acid (GLU), L-lysine (LYS) and L-tyrosine (TYR), in a defined molar ratio of 0.43: 0.14: 0.34: 0.09, calculated relative to the total number of moles of AA. Copaxone® is an injectable product composed of nano-sized peptide particles of

GA suspended in an aqueous solution with mannitol (COPAXONE® 1996). The GA dose is 20 mg GA and 40 mg mannitol suspended in 1 mL water for injection [WFI] in a pre-filled syringe [PFS] administered by daily SC injection, or 40 mg GA suspended in the same mannitol solution administered by PFS three times weekly. The size of the polypeptide structures, as individual molecules or in the form of aggregates, in Copaxone® makes it a colloidal suspension; i.e., small peptide entities, 1 to 500 nm in size, are distributed in a continuous medium and do not settle out under the influence of gravity. The average MW of GA is 5000 to 9000 Daltons, whereas most of the polypeptides in the GA mixture fall within a MW distribution range of approximately 2500 to 20,000 Daltons. Glatiramoids are characterized by the molecular formula below, in which X represents an anion (e.g., acetate or any other pharmaceutically acceptable salt), the superscripts represent the relative molar ratios of AA, the subscript "n" relates to the polymeric chain length, and "m" is the molar quantity of counter ions.

$$(\text{L-Glu}^{13-15}, \text{L-Ala}^{39-46}, \text{L-Tyr}^{8.6-10}, \text{L-Lys}^{30-37})_n \cdot m\text{X}$$

The variety of AA sequences and the size distribution of the polypeptide nanoparticles in GA are what makes the mixture so complex. Although the final GA composition is not entirely random, no specified AA sequences are generated during synthesis. No assays are available to determine the sequences of, or quantify, all the individual polypeptides produced, because there are too many to isolate individually. A polypeptide with a representative composition of 60 AA (about 7000 Daltons) will contain on average 8 GLU, 26 ALA, 6 TYR and 20 LYS residues, and the estimated number of possible AA sequence combinations could reach as high as 10^{29} (Carter and Keating 2010). As the molecular composition of a particular glatiramoid is linked inexorably to its own manufacturing process, the GA in Copaxone® is unique to the specific, proprietary manufacturing process used by its manufacturer (Krull and Cohen 2009).

Although purported generic GA products are manufactured using the same component AA as Copaxone® as starting materials, minor alterations in the reaction conditions used in their manufacture have resulted in polypeptides mixtures with distinctly different compositional arrangements (described below) and with potentially different immunogenicity, efficacy and safety profiles from those of Copaxone®. Attempts to correlate the clinical effects of purported generic products with those of Copaxone®, based on similarity of basic chemical parameters only, are confounded by the fact that the actual active sequences or structures within the complex GA polypeptides mixture that are responsible for drug efficacy and safety are unknown.

Manufacturing: Standard Techniques

GA is prepared from N-carboxy-α-amino acid anhydrides (monomers) using diethylamine as the polymerization initiator. The bifunctional AA are protected (the δ-NH2 of LYS is protected by a trifluoroacetyl group and the γ-COOH of GLU is protected by a benzyl group); therefore, polymerization occurs through the growth of linear chains of monomers without crosslinking between the polymer chains. The MW of the polymer is then reduced by acidolytic cleavage and deprotection. The AA sequences of the resulting polypeptides are not completely random and the entire complex polypeptide mixture composition is highly reproducible, but only under strictly controlled, specific reaction conditions. For the polymerization step (establishment of the primary structure), the following factors are critical: the relative concentration of activated AA in the reaction mixture, the reactivity of the AA, the amount of initiator used, the reaction conditions, and the quality of the AA and the solvent used. For the acidolytic cleavage step (which establishes size distribution), the concentration and purity of the reagents, as well as the reaction time and temperature, are the most critical factors.

Control: Glatiramoid Identity and Quality

Because GA comprises a potentially incalculable number of structurally closely related active peptide moieties that cannot be isolated, quantified, or identified using even the most sophisticated available multidimensional separation techniques, it is impossible to identify and characterize its active AA sequences (Varkony et al. 2009). Nevertheless, while there is/are no reliable characterization technique(s) to show that two glatiramoid mixtures are identical, there are cutting-edge analytical methods by which two glatiramoid mixtures can be evaluated, compared, and differentiated.

The identity and consistency of complex glatiramoid mixtures depend on a number of important physicochemical characterization parameters: composition, sequence of structurally similar constituents in the mixture, higher order structures/conformation, aggregation, impurities, and degradation products. Additional considerations are related to essential requirements for nano/colloidal drug suspensions: the size and shape of the molecular entities, relative size distribution and agglomeration, and charge distribution.

Similarities between Copaxone® and purported generic GA products may be observed when using common, nonspecific analytical methods (e.g., MW distribution by size exclusion chromatography [SEC] or AA ratio determination). The ability of such techniques to ensure physicochemical similarity or to establish pharmaceutical equivalence between glatiramoid mixtures is limited, however, because of their inability to differentiate the vast number of structurally related, but essentially different, constituents.

The only reliable way to compare key attributes of glatiramoids is to evaluate them in intact form. Analyses of proteolyzed materials widely used in proteomics to identify AA sequences on protein chains are inadequate to characterize a glatiramoid mixture. Chemical or enzymatic cleavage of polypeptides in a glatiramoid produces a mixture that is less structurally complex and that is informative about only the building blocks comprising the polypeptide sequences, not the original complex structures of the parent sequences. It is not possible to reverse engineer the components in the parent mixture from the identified fragments because the nature of the original complex material is irreversibly lost during fragmentation. Figures 1–3 show the loss of discriminatory capacity of fragmentation (i.e., peptide mapping) when comparing two glatiramoid mixtures—first in the intact form, and

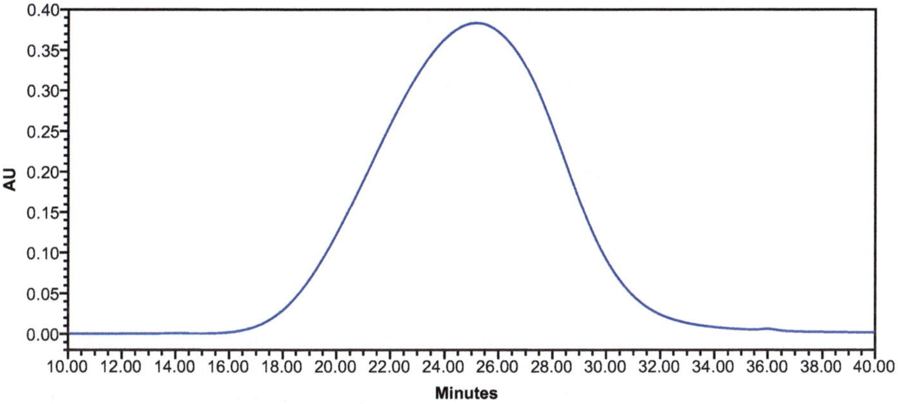

Fig. 1 Typical UV size exclusion chromatogram (SEC) of a glatiramoid (Sample A) showing characteristic MW distribution. (*AU* absorption units)

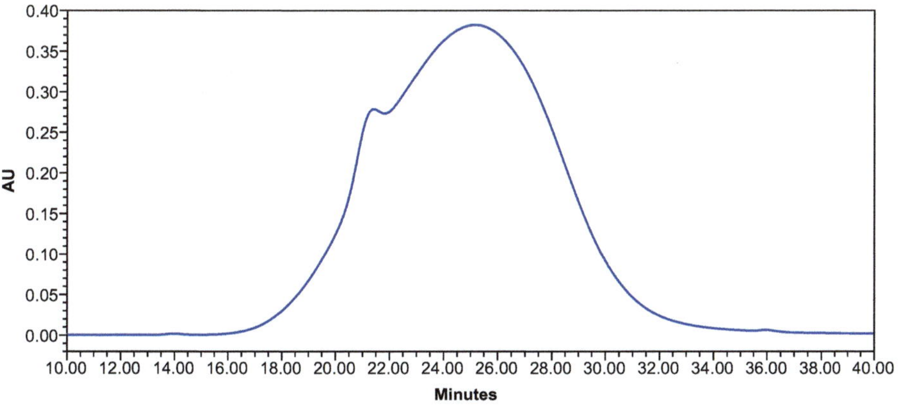

Fig. 2 The same glatiramoid as in Fig. 1 but now spiked with 2 % foreign peptide comprising the same AA as the original glatiramoid in the same ratio (Sample B). The spike is clearly detected on the chromatogram

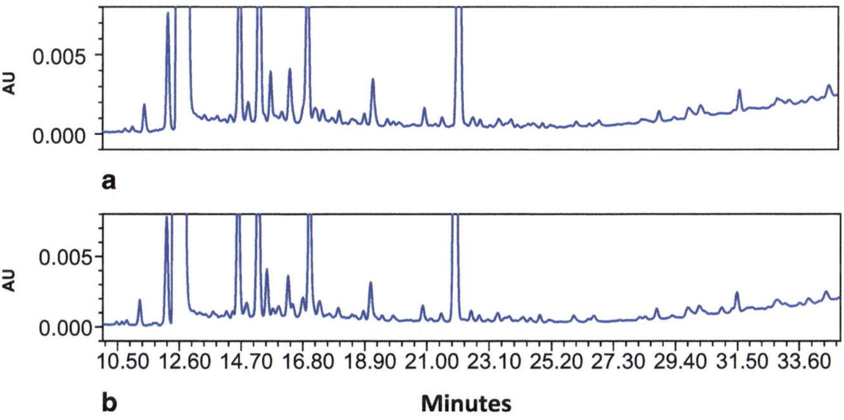

Fig. 3 Peptide mapping of Samples A and B after submitting to enzymatic hydrolysis with pronase and separated by reverse-phase high-performance liquid chromatography (RP-HPLC)

then, after cleavage to shorter entities. Figure 1 shows the characteristic MW distribution of a glatiramoid mixture (Sample A). In Fig. 2, the same glatiramoid (Sample A) has been spiked with a foreign peptide containing the same AA in the same molar ratio as the original glatiramoid sample (now Sample B). As Fig. 3 illustrates, the peptide maps of Sample A and Sample B are identical. The profound difference observed between the intact glatiramoids in Figs. 1 and 2 was masked when the mixtures were fragmented and analyzed using a conventional nonspecific method.

Further analysis showed that the more exhaustive the extent of cleavage, the weaker the correlation between the digested fragments and the parent molecules. Hence, analysis of peptide fragments is not sensitive enough to indicate small differences in the composition and quality of unknown glatiramoid mixtures manufactured under different conditions. This same analysis was performed comparing Copaxone® with the purported generic GA products, Glatimer®, Escadra®, and Probioglat® (Fig. 4).

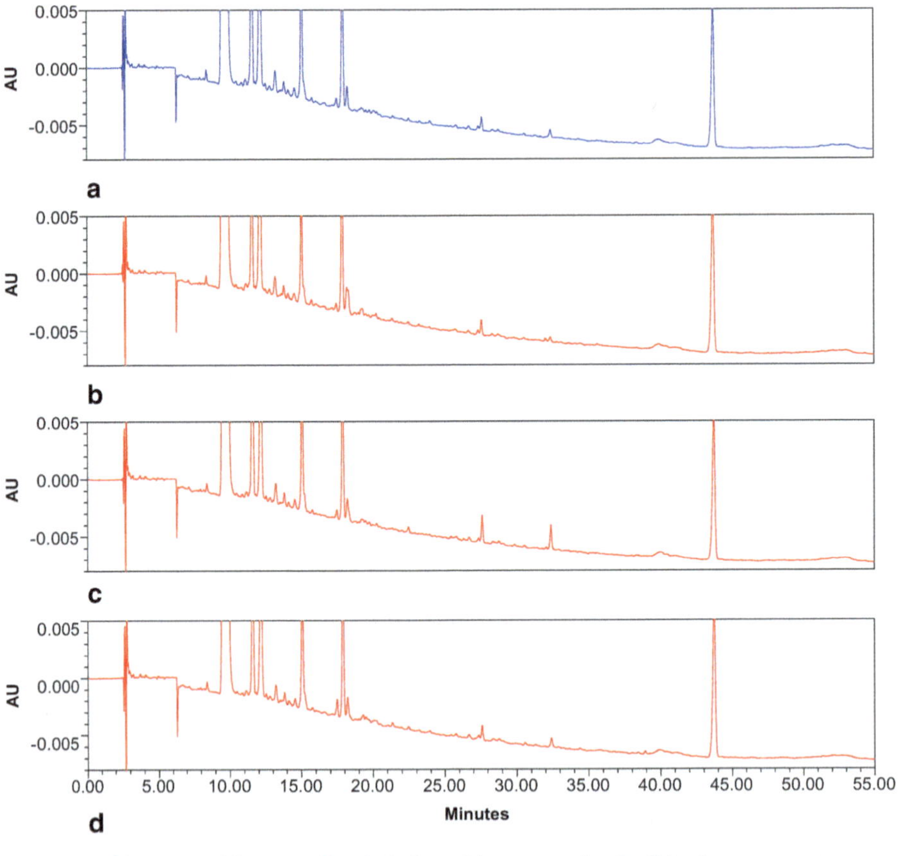

Fig. 4 Peptide maps of Copaxone® and glatiramoids purported to be GA after exposure to enzymatic digestion. The top chromatogram (**a**, *blue*) is Copaxone® and three other chromatograms (*red*) are (**b**) Glatimer® (**c**) Escadra®, and (**d**) Probioglat®. The peptide maps are practically identical, although the purported generic products are shown to be compositionally different from Copaxone® by other methods

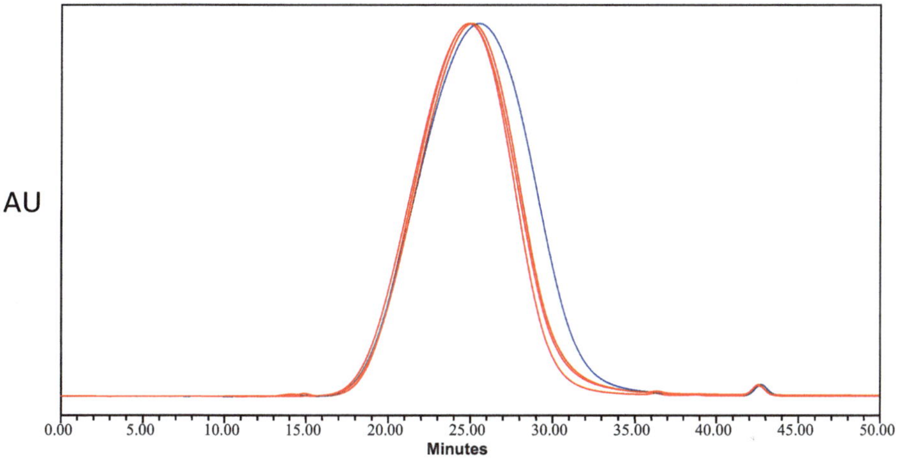

Fig. 5 Size exclusion chromatographic (SEC UV) MW distribution profile of Copaxone® (*blue*) and three purported generic GA products (Glatimer®, Probioglat®, and Escadra®; *red*). (*AU* absorption units)

Similarly, a nonspecific general method used to analyse intact glatiramoid mixtures, MW distribution determination by SEC, is also only partially able to reliably reflect characteristic physicochemical parameters of complex glatiramoid mixtures. While MW distribution might suggest an almost perfect correlation between the GA in Copaxone® and purported generic GA products in terms of their hydrodynamic size distribution, it does not indicate their structural equivalence. As shown in Fig. 5, the MW distribution profiles of Copaxone® and three purported generic GA products are practically overlaid; nevertheless, they have been demonstrated to be compositionally different in assessments that employ more sophisticated analytical techniques.

More sensitive analytical methods can discriminate among glatiramoids of different origin (i.e., made by different manufacturers). The charge distribution within the glatiramoid mixture of polypeptide chains that are rich in charged residues (LYS and GLU) is an important and specific drug characteristic as it reflects the primary structure through arrangement of the charges. Capillary isoelectric focusing (IEF) electrophoresis is a sensitive discriminatory technique to detect variations among glatiramoid batches produced by different manufacturing processes. The charge distribution of the purported generic GA products differs from that of Copaxone® when measured by capillary IEF. This technique showed excellent batch-to-batch consistency of charge distribution among several Copaxone® batches (Fig. 6), indicating a well-controlled and robust manufacturing process. The IEF profiles for the generic products reveal dissimilarities from Copaxone® in polypeptide primary structures, charge distribution, and sequence composition (Fig. 7).

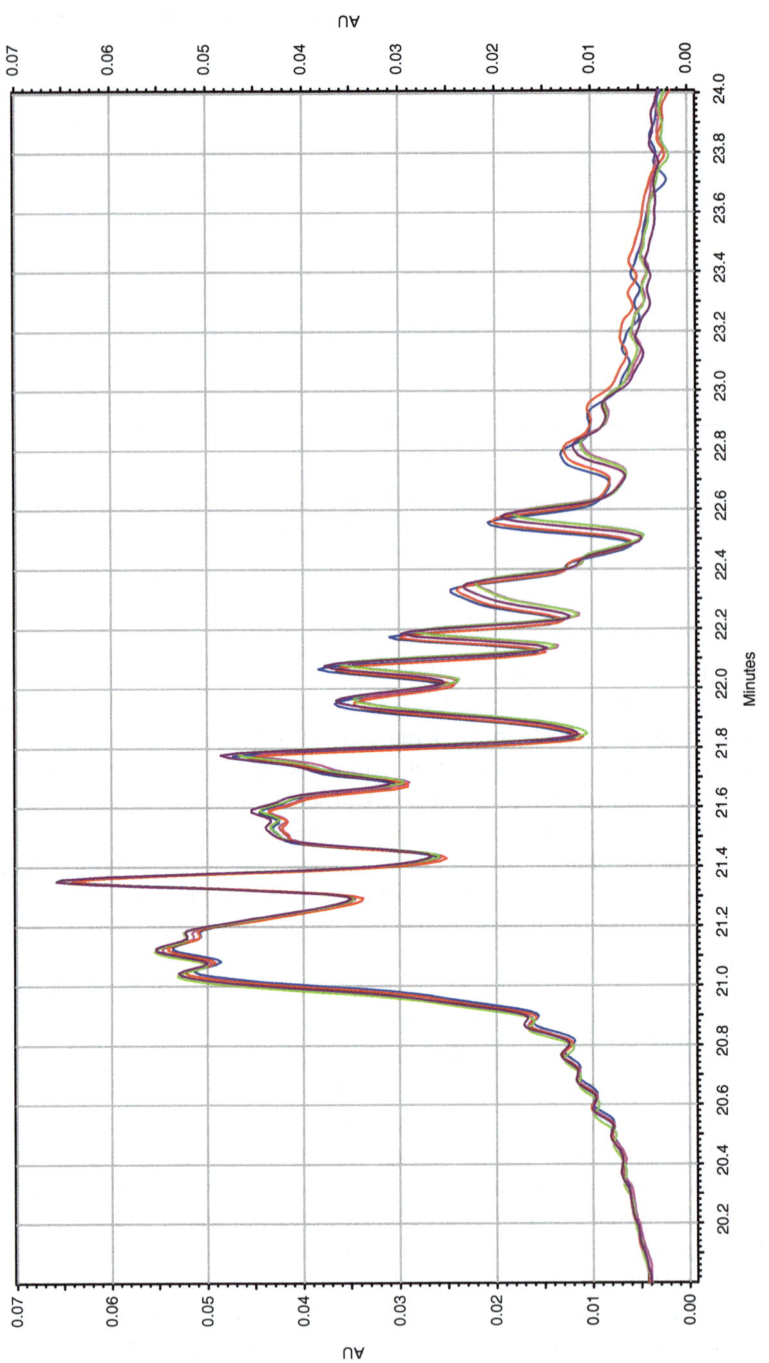

Fig. 6 Charge distribution pattern of 5 randomly chosen Copaxone® batches shows tight consistency based on capillary electrophoresis (CE) testing. (x-axis—retention time; y-axis—absorbance units (*AU*)

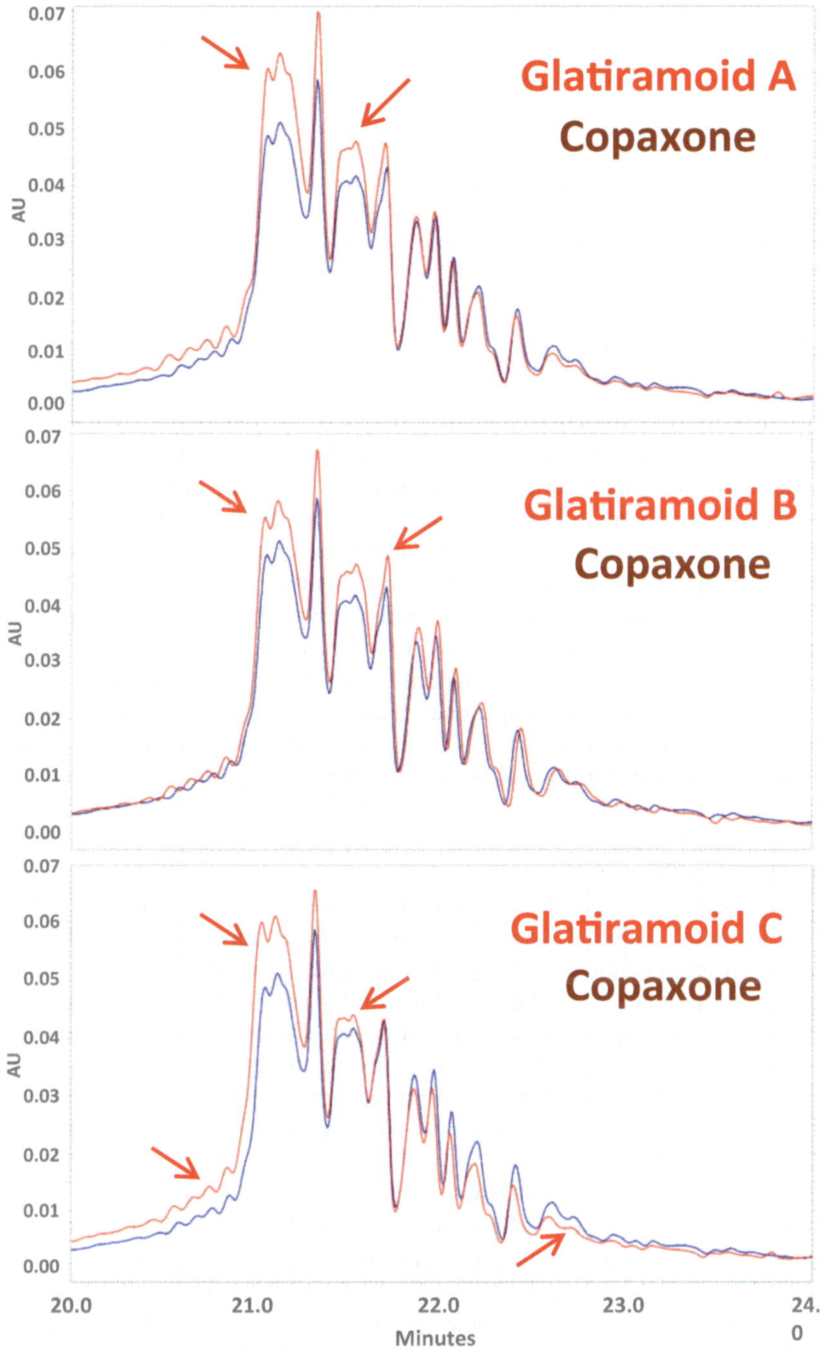

Fig. 7 The charge distribution profile indicated by isoelectric focusing (IEF) electrophoresis is different for each glatiramoid. The Copaxone® profile (*blue*) is compared with profiles for the purported generic products, Glatimer®, Escadra®, and Probioglat®. *Red arrows* point to important differences in the charge distribution (i.e., composition) of peptide subpopulations. (*AU* absorption units)

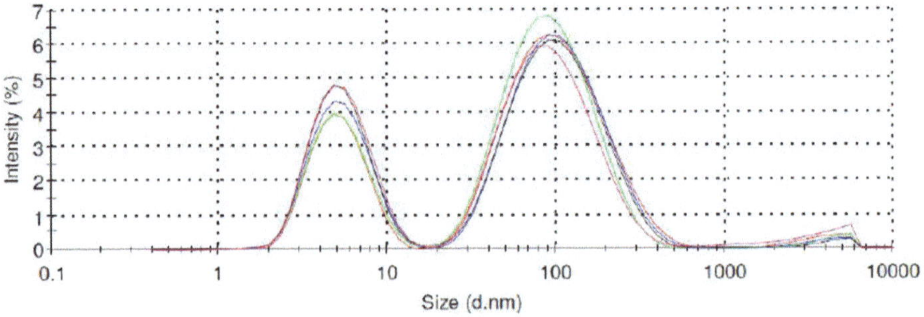

Fig. 8 Particle size distributions of 10 batches of Copaxone® by dynamic light scattering (DLS)

Another useful analytical technique to characterize a glatiramoid is dynamic light scattering (DLS), which assesses particle size distribution. DLS analysis showed that the Copaxone® nanoparticle mixture consists of two main polypeptide populations. The first population is characterized by a distribution of particle sizes in the range of 1 to 15 nm, whereas the second population contains particles in the range of 20–500 nm. As shown in Fig. 8, the first population likely represents "monoparticles," or separated molecules with a size of about 6 nm. The second population comprises larger entities of around 100 nm in size (e.g., labile intermolecular associates) that may be formed by interactions among the polypeptide chains.

Again, analysis of many Copaxone® batches indicated highly consistent DLS size distribution profiles (Fig. 8). In contrast, substantial differences in aggregate sizes in the colloidal solutions of purported generic GA products were found when compared with Copaxone®—as well as inconsistencies among batches from the same supplier. The purported generic GA products also differed considerably from each other (Fig. 9).

Differences in the distribution of particle sizes may also affect the nature of aggregates in the colloidal suspensions. Atomic force microscopy (AFM) is a sensitive and useful method to study morphological equivalence and presentation of nanosized particles in colloidal glatiramoid suspensions. AFM analysis of the morphology of aggregates in the Copaxone® solution and in the purported generic GA products reveal inconsistencies in aggregate appearance. Copaxone® samples exhibit consistent structures with linear shapes (strings) (Fig. 10), whereas large globular particles and nonhomogenous structures are found in the generic glatiramoid products (Fig. 11).

A cutting-edge technology that has been used to assess the consistency and reproducibility of the composition of complex, structurally similar glatiramoid mixtures is ion mobility mass spectrometry (IMMS). IMMS is capable of differentiating between closely related moieties, such as isomeric peptides, and provides separation at a sensitivity level that other chromatographic techniques cannot achieve. The use of HDMS Compare software (Waters Corporation, Milford, MA, USA) amplifies visual differentiation and clearly shows differences between otherwise apparently

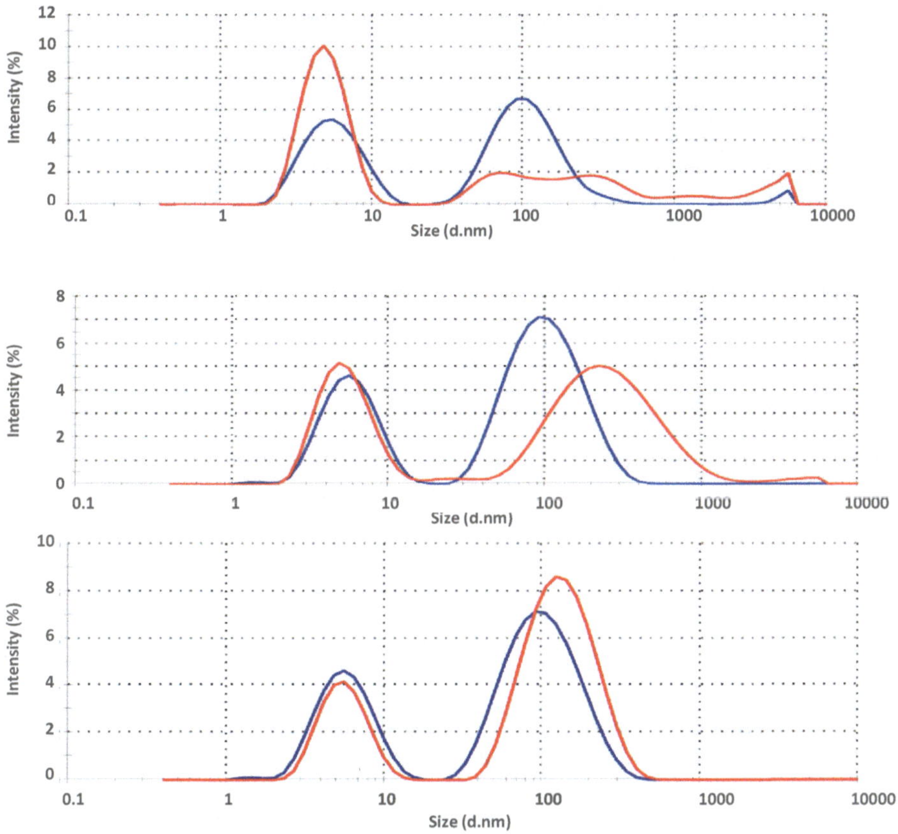

Fig. 9 Comparative dynamic light scattering (DLS) scans showing Copaxone® (*blue*) and purported generic GA products (*red*). Glatimer® in the top panel, Escadra® in the middle panel, and Probioglat® in the lower panel

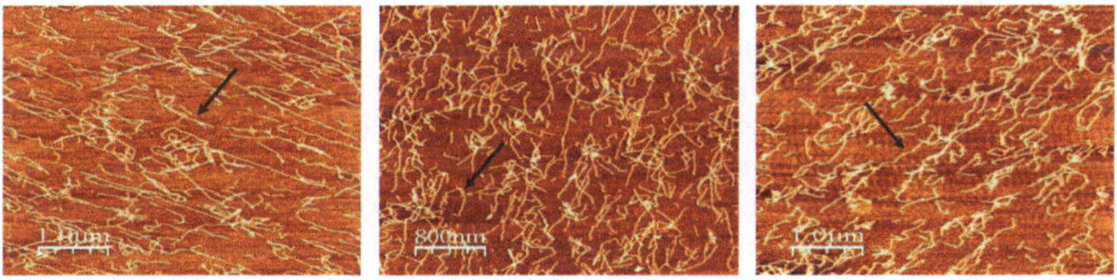

Fig. 10 Morphology of aggregates: dried samples from different batches of Copaxone® analyzed by atomic force microscopy (*AFM*) show consistent folded linear structures (strings, examples indicated by *arrows*)

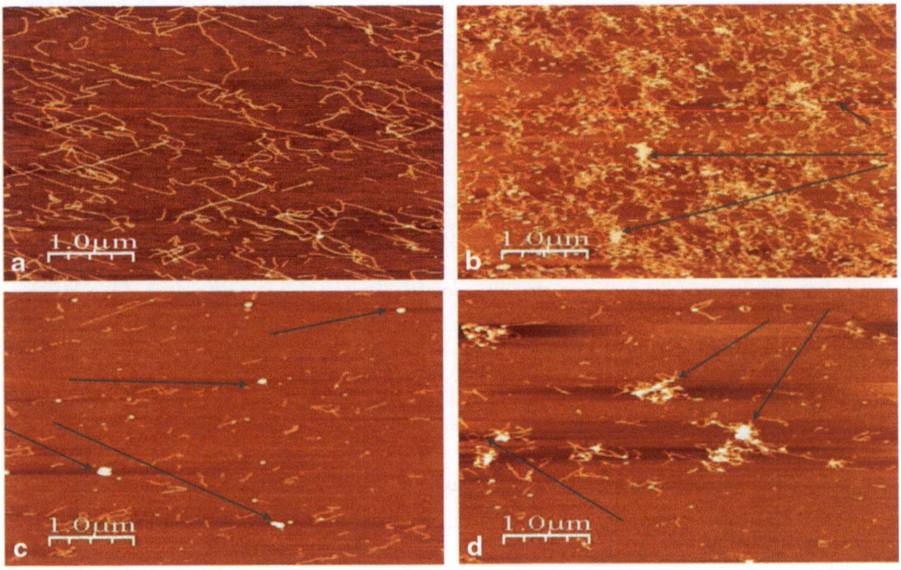

Fig. 11 Morphology of aggregates in purported generic GA products was different from that of GA (Copaxone®). Arrows point to nonhomogenous aggregates: **a** Copaxone®, **b** Glatimer®, **c** Probioglat®, **d** Escadra®

identical samples (Berkowitz et al. 2012). IMMS allows for 2-dimensional separation of ionized molecules based on molecular size, shape, and mass/charge. Ion mobility measures the time it takes an ion to traverse a nitrogen-filled tube ("drift time"), which is dependent on the size and shape of the ionized molecules. After exiting the tube, ions are then collected and pulsed into the mass spectrometer for a second phase of the separation based on mass/charge. This technique produces a 3-dimensional heat map to highlight intensity (a marker of abundance) differences among peptides at various mass/charge and drift times. The heat maps are compared using software that evaluates each pixel on the heat map to generate a comparison map that reflects the difference in an illustrative way (Fig. 12). A quantitative assessment of the intensity values within the highlighted areas (reflecting the differences in composition) produces a total intensity value (TIV). This technique is also extremely useful in analyses of complex mixtures such as crude oil and polymer blends to determine differences in composition.

IMMS was used to analyze 15 randomly chosen batches of Copaxone® and a few batches of each of the purported generic GA products mentioned above. All samples were compared to a randomly chosen representative batch of Copaxone®. Theoretically, if the composition of a tested generic GA product was exactly the same as the composition of the reference Copaxone®, the comparative heat map would have no highlighted areas and the resulting TIV would be zero. Conversely, comparative heat maps with more colored areas signify greater differences from the reference and would have a higher TIV. The comparative heat maps generated

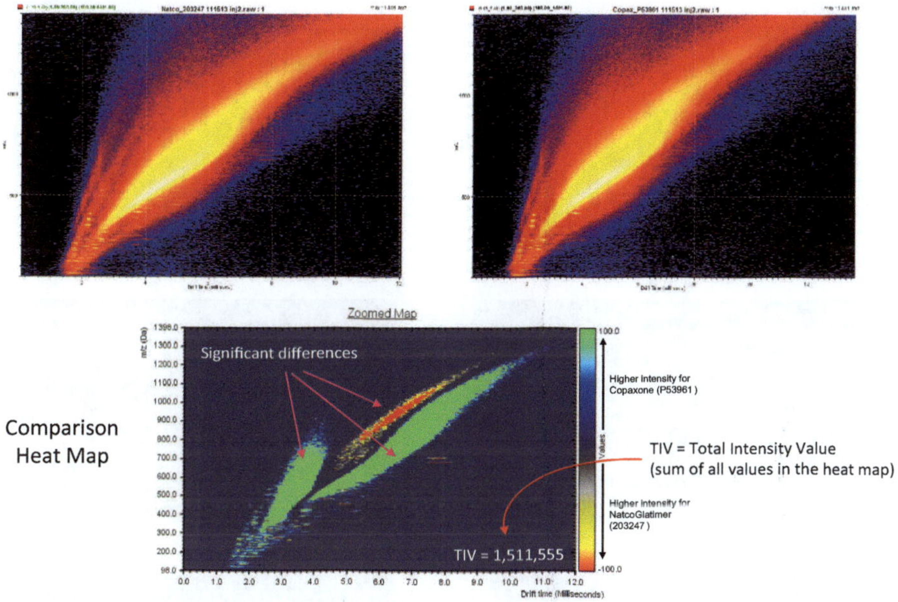

Fig. 12 Comparing heat maps to calculate the total intensity value (*TIV*), a comparative measure of peptide abundance based on drift time and mass/charge between samples

for the purported generic GA products showed highly intensely colored areas when compared with the Copaxone® lots tested (Fig. 13).

As can be seen, the variability in the comparative heat maps for Copaxone® batches is minimal, while more profound and significant variability is observed when comparing Copaxone® with the purported generic GA products. This is also reflected by the calculated TIVs for the comparative maps. The average TIV for comparisons among the Copaxone® batches was 117,245 (range 22,010–255,265). When the generic products were compared with Copaxone® the TIVs were 8- to 13-fold higher (range 943,110–1,511,555). Thus, IMMS analysis showed minimal batch-to-batch variability in Copaxone®- Copaxone® comparisons as reflected by small TIVs, in contrast to the high TIVs recorded when comparing Copaxone® with the purported generic GA products (Fig. 14).

As is true for other NBCDs, strict manufacturing protocols ensure that the quality of the Copaxone® complex mixture is within the established range of physicochemical parameters and that it has appropriate biological activity and safety. Even slightly different synthesis conditions, which are almost certain to occur when glatiramoids are made by different manufacturers, can create generic GA products with peptide composition that vary significantly from that of Copaxone®, as demonstrated in the previous series of analyses. Variability in glatiramoids purporting to be the same substance (the samples in the analyses described here all purport to be GA) appear to arise from alterations in AA sequences; polypeptide size, shape, or charge; and aggregate material in the various products.

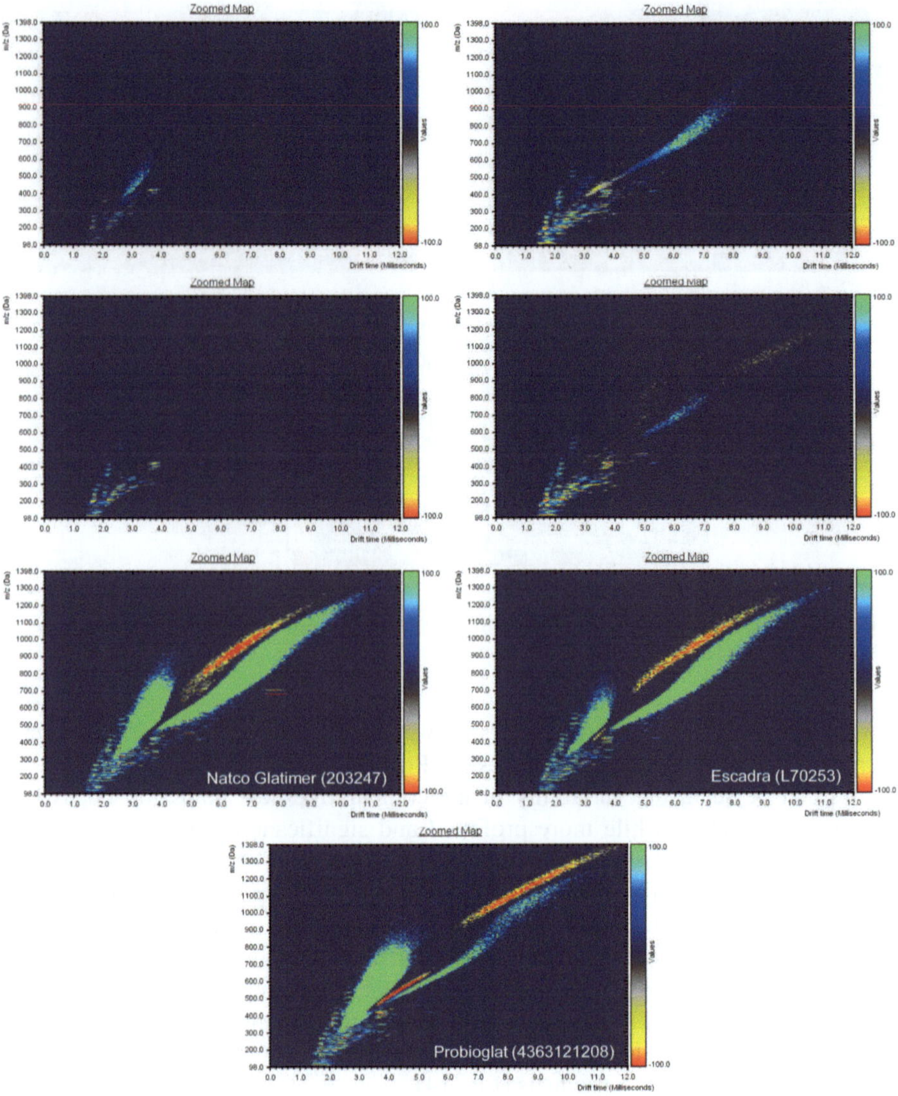

Fig. 13 Comparing heat maps between Copaxone® batches (upper 4 heat maps) and between Copaxone® and purported generic products (lower three heat maps)

Key Points:

- Glatiramoids are non-biological complex mixtures of synthetic polypeptides that are not fully characterizable using available analytical tools.
- Glatiramoids comprise nano-sized complex peptide entities suspended in a colloidal solution.
- No polypeptide sequences related to therapeutic activity of Copaxone® have been identified, isolated or quantified.

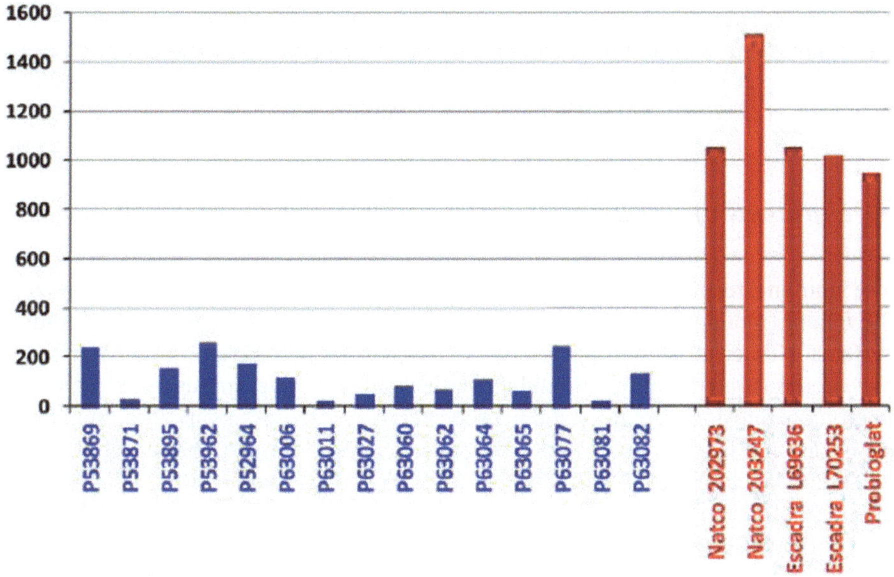

Fig. 14 Calculated total intensity value (*TIV*) among Copaxone® batches (*blue bars*) and TIVs when Copaxone® was compared with purported generic GA products (*red bars*)

- No reliable analytical technology exists to demonstrate sameness, similarity, or equivalence among glatiramoid mixtures. Cutting-edge analytical techniques may be used to evaluate, compare, and differentiate between glatiramoids. Meaningful comparisons are best made between glatiramoids in the intact state.
- Nonspecific and highly specific analytical techniques indicate that the well-controlled proprietary manufacturing method used to create Copaxone® provides a highly consistent product.
- Slight deviations in manufacturing procedures lead to different compositions of glatiramoid mixtures, thereby creating unique therapeutic entities.
- Sensitive techniques show dissimilarities between Copaxone® and purported generic GA products in the composition of constituent polypeptides and in physicochemical attributes.

Pharmacology

Copaxone® acts, in part, as an antigen-based therapeutic vaccine, presenting "safe" antigenic epitopes that do not induce autoimmunity to lymphocytes. The synthetic peptides in Copaxone® mimic the structure of epitopes for myelin-reactive T cells (i.e., peptide fragments of myelin antigens, including MBP, generated by antigen processing). Repetitive antigen encounter following Copaxone® administration simulates the effect of chronic infection, and promotes GA-induced regulatory cells that suppress disease and secrete anti-inflammatory cytokines (Aharoni et al. 1997, 1999, 2003, 2011; Allie et al. 2005; Arnon and Aharoni 2004; Vieira et al. 2003; Ziemssen et al. 2002, 2005; Ziemssen and Schrempf 2007).

Mechanisms of Action

The immune response to acute and chronic CNS damage includes cells of both the innate (monocytes and dendritic cells) and adaptive (T and B cells) immune systems. Numerous studies published in the scientific literature over recent decades have demonstrated that GA modulates innate and adaptive immune cell responses to promote anti-inflammatory and neuroprotective activities in various animal models of chronic inflammatory and neurodegenerative diseases (Aharoni et al. 2000, 2003, 2005a, b, c, 2008, 2010; Azoulay et al. 2005; Dhib-Jalbut 2003; Gilgun-Sherki et al. 2003; Hafler 2002; Hong et al. 2005; Johnson 2010; Jung et al. 2004; Kala et al. 2010, 2011; Kipnis et al. 2000, 2003; Kipnis and Schwartz 2002; Putheti et al. 2003; Sand et al. 2009; Schori et al. 2001; Schwartz 2003; Stern et al. 2008; Teitelbaum et al. 1971, 1988). The exact mechanisms of GA action in humans remains uncertain. Some of the proposed immunomodulatory activities of GA thought to contribute to its therapeutic effects are shown in Fig. 15. They include (Arnon and Aharoni 2004; Carter and Keating 2010; Dhib-Jalbut 2003; Kala et al. 2011; Lalive et al. 2011; Schrempf and Ziemssen 2007; Sela and Teitelbaum 2001; Weber et al. 2007a; Wolinsky 2004; Ziemssen and Schrempf 2007):

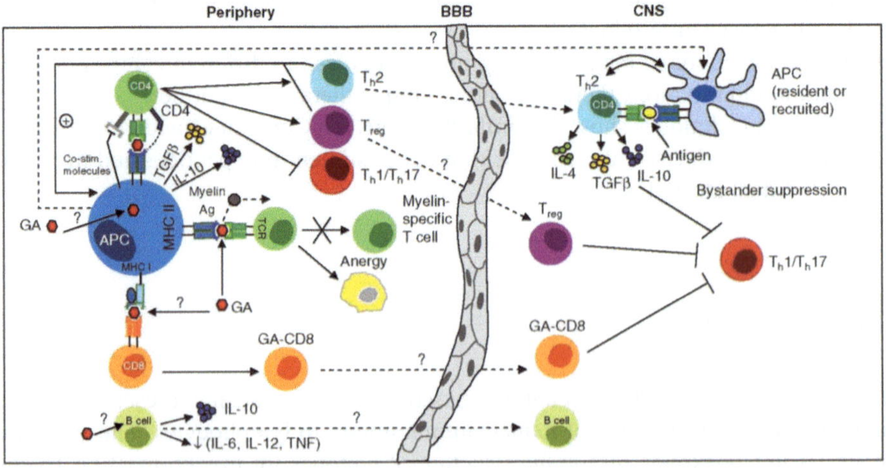

Fig. 15 Proposed immunomodulatory mechanisms of GA activity (Lalive et al. 2011). (*Co-stim* costimulatory, *TCR* T cell receptor, *TNF* tumour necrosis factor,? indicates an assumed mechanism, ↓ indicates decrease

- High-affinity binding to antigen-presenting cells (APCs)
- Development of type II APCs
- Interplay between GA-reactive APCs and T cells
- Induction of GA-reactive T cells
- Migration of GA-reactive T cells to the site of damage in the central nervous system (CNS)
- Modulation of B-cell function
- Modulation of natural killer (NK) cell function
- Neuroprotective effects (e.g., promotion of neurotrophic factors)
- Immunoglobulin response.

High-Affinity Binding to Antigen-Presenting Cells

GA is administered subcutaneously. A substantial fraction of GA peptides is hydrolyzed locally, while a fraction of the injected material (either intact or partially hydrolyzed) is presumed to enter the lymphatic circulation, enabling it to reach regional lymph nodes, and another fraction may enter the systemic circulation in intact form (Carter and Keating 2010; COPAXONE® 1996). Although the active sequences in the GA mixture have not been identified, it has been recognized that the "multi-epitopal" nature of GA is an important feature of its activity, because to respond to the immense inherent heterogeneity of the human immune system, a medicine that possesses an equally heterogeneous character may be required. Binding of GA antigens to major histocompatibility complex (MHC) proteins on APCs is a prerequisite step that initiates the immunomodulatory effects of GA. The antigenic sequences in GA compete with myelin antigens and preferentially bind to MHC II molecules with high avidity and without antigen processing or any obvious allelic preference (Fridkis-Hareli and Strominger 1998; Kipnis and Schwartz 2002; Lalive et al. 2011; Citizen Petition 2009). Initial binding of GA peptides to MHC class II molecules on APCs in the periphery (e.g., monocytes, dendritic cells, and macrophages) most likely occurs immediately upon GA administration at the site of injection, and in the lymphatic circulation (COPAXONE® 1996).

The multiple sequences of GA peptides allow binding to many different alleles of MHC class II molecules of different animal species and strains, and HLA class II (DR) molecules from MS patients with different genetic backgrounds (Fridkis-Hareli and Strominger 1998). A study revealed that GA also induces HLA class I-restricted T-cell responses, suggesting that GA may also bind to HLA class Ia or class Ib molecules. These additional mechanisms may require cross-presentation or nonclassical HLA molecules (e.g., HLA-E) (Tennakoon et al. 2006).The "promiscuous binding" capacity of GA peptides has been shown to successfully prevent or suppress EAE induced by different encephalitogenic antigens in a wide variety of animal species with different genetic backgrounds (Aharoni et al. 2011; Fridkis-Hareli and Strominger 1998; Kala et al. 2011; Sela and Teitelbaum 2001).

Development of Type II APCs

Recent studies indicate that GA may suppress CNS autoimmune disease in an antigen-nonspecific manner by modulating APC function (Jung et al. 2004; Kim et al. 2004; Sanna et al. 2006; Vieira et al. 2003; Weber et al. 2007b). In fact, APCs may be the primary target of immunomodulation by GA (Weber et al. 2007b). Studies in EAE have demonstrated that GA treatment promotes development of anti-inflammatory type II monocytes that secrete anti-inflammatory cytokines Burger et al. 2009; Sanna et al. 2006; Weber et al. 2007a, b). Type II monocytes direct differentiation of T cells toward an anti-inflammatory Th2 phenotype and promote the generation of $CD_4^+CD25^+FoxP3^+$ regulatory T cells, independent of antigen specificity (Weber et al. 2007b). Interestingly, GA did not influence T cell polarization when added to naïve Th cells activated in an APC-free system, suggesting that APCs are essential to facilitating the well-studied GA-mediated shift from Th1 to Th2/3 cells (described further below) (Vieira et al. 2003).

Dendritic cells are the most effective APCs and play an important role in MS. In EAE and MS pathogenesis, dendritic cells present myelin antigens to autoaggressive myelin-specific T cells and naïve T cells, leading to the induction of proinflammatory Th17 responses (Sanna et al. 2006). *In vitro*, GA was shown to inhibit the production of the proinflammatory cytokines, IL-12 and TNF-α, by human dendritic cells (Sanna et al. 2006), and in animal models, GA increased production of anti-inflammatory IL-10 by bone marrow-derived dendritic cells (Jung et al. 2004).

Interplay Between APCs and T Cells

The interplay between APCs and T cells is fundamental to GA-mediated immune modulation. APCs present GA peptides on MHC molecules to T cell receptors, and subsequent T cell activation induces expansion of GA-reactive Th2 cells and promotes generation of $CD_4^+CD25^+FoxP3^+$ regulatory T cells (Aharoni et al. 2003, 2010, 2011; Kim et al. 2004; Vieira et al. 2003). In a positive feedback loop, Th2 cells reciprocally modify APC function, pushing APCs toward the anti-inflammatory type II phenotype (Kim et al. 2004; Vieira et al. 2003; Weber et al. 2007a). Accordingly, upon exposure to GA, human dendritic cells induce IL-4-secreting Th2 cells and increased levels of anti-inflammatory IL-10 (Vieira et al. 2003); and T-cell-mediated deviation of APC toward a type II cytokine pattern has been demonstrated in human microglial cells exposed to GA-reactive T cells from MS patients (Kim et al. 2004).

Induction of GA-Reactive T Cells

The immunomodulatory activity of GA is mainly attributed to its ability to induce GA-reactive T cells that have both anti-inflammatory and neuroprotective activity (Kipnis and Schwartz 2002). Research in animal models has shown that adoptive transfer of GA-specific T cells can block induction of EAE by the encephalitogen, MBP, indicating that the immunomodulatory effect of GA is mediated, at least in part, by these GA-specific immune cells (Aharoni et al. 1993). GA exhibits low affinity for auto-aggressive MBP-specific T cells (Aharoni et al. 1999; Hafler 2002; Ziemssen and Schrempf 2007). Repeated injections of GA induce a moderate loss of T-cell responsiveness to antigenic peptides due to T cell receptor degeneracy, accompanied by a shift to a Th2 type of $CD4^+$ T cell (Aharoni et al. 1999; Arnon and Aharoni 2004; Duda et al. 2000). Thus, upon encountering GA-reactive T cells, self-antigens such as MBP may be recognized as weak agonists, much like APLs (Aharoni et al. 1997, 1999; Arnon and Aharoni 2004; Hafler 2002; Ziemssen and Schrempf 2007). GA-reactive T cells then secrete suppressive Th2/Th3 cytokines, thereby restricting local inflammation (Aharoni et al. 1997, 2000, 2003, 2010; Hafler 2002; Kipnis and Schwartz 2002; Putheti et al. 2003; Stern et al. 2008; Ziemssen et al. 2005). This is known as "bystander suppression" and it might explain the ability of GA to simulate the T-cell response to self-antigens without invoking the inflammatory response associated with autoimmune diseases (Ziemssen and Schrempf 2007).

$CD4^+$ T-Cell Responses

Evidence suggests that $CD4^+$ T-cell lines generated at the initiation of treatment with GA secrete both pro-inflammatory Th1 (IL-2 and interferon gamma [IFN-γ]), and anti-inflammatory Th2 cytokines (IL-4, IL-5) (Kantengwa et al. 2007; Miller et al. 1998; Neuhaus et al. 2000; Weder et al. 2005). However, studies in MS patients and EAE models have demonstrated that continued exposure to GA induces a shift from a primarily Th1-type cytokine profile to a Th2-type profile, characterized by increased secretion of IL-5 and IL-13 (Aharoni et al. 1997; Chen et al. 2001; Duda et al. 2000; Franciotta et al. 2003; Miller et al. 1998; Neuhaus et al. 2000; Sanna et al. 2006; Weder et al. 2005). GA-specific Th2 cells are also the source of anti-inflammatory cytokines such as IL-4, IL-6, IL-10, (Aharoni et al. 1997) and brain-derived neurotrophic factor (BDNF) (Aharoni et al. 1997, 2000; Azoulay et al. 2005; Miller et al. 1998; Putheti et al. 2003; Weber et al. 2007a; Ziemssen et al. 2005). A similar shift to anti-inflammatory cytokines is also observed in the colons of mice with 2,4,6-trinitrobenzenesulfonic acid (TNBS)-induced colitis (Aharoni et al. 2005c). Following daily treatment with GA, reduced secretion of tumor necrosis factor alpha (TNF-α, a proinflammatory cytokine) and increased secretion of transforming growth factor beta (TGF-β, generally an anti-inflammatory cytokine and mediator of bystander suppressive effects) was observed in cultures of isolated mesenchymal lymph cells activated with colon extracts (Aharoni et al. 2005c).

GA therapy also alters chemokine-receptor expression on GA-reactive T cells, which may have a significant effect on trafficking of these T-cell populations. Chemokine receptors expressed by GA-reactive T cells are predominantly Th2-biased. Prominent expression of the CC chemokine receptor type 7 (CCR7) by GA-reactive

T cells may promote their recruitment to the lymphoid tissue, with subsequent regulatory effects in the immune compartment (Allie et al. 2005).

$CD8^+$ T-Cell Responses

$CD8^+$ T cells have regulatory activity in that they can suppress proliferation of autoreactive $CD4^+$ T cells. Upon initial exposure to GA, $CD8^+$ T-cell proliferative responses are significantly lower in peripheral blood mononuclear cells (PBMC) of untreated MS patients compared with healthy controls. However, continued GA treatment up-regulates $CD8^+$ T cell proliferation, with restoration to levels observed in healthy subjects (Karandikar et al. 2002). GA therapy also enhances the suppressive activity of $CD8^+$ T cells (Tennakoon et al. 2006). Over time, GA treatment is associated with a decrease in $CD4^+$ GA-reactive T cells and an increase in $CD8^+$ GA-reactive T cells (Allie et al. 2005; Kipnis et al. 2002).

Regulatory T-Cell Responses

The ability to maintain healthy immune surveillance without developing autoimmune disease depends on the activity of naturally occurring $CD4^+CD25^+FoxP3^+$ regulatory T cells (Haas et al. 2007; Hong et al. 2005; Schwartz and Kipnis 2002). Significant deficiencies in the number and/or function of these regulatory cells have been found in several autoimmune diseases, including MS (Haas et al. 2005, 2007; Venken et al. 2008; Viglietta et al. 2004). Treatment with GA induces the formation of $CD4^+CD25^+FoxP3^+$ regulatory T cells and increases their suppressive function (Haas et al. 2009). In GA-treated SJL mice, GA-specific regulatory T cells secreted high levels of the anti-inflammatory cytokines IL-10 and IL-13 and small amounts of IL-4, and virtually no IL-17, IL-6, IFN-γ, or TNF-α, which are associated with inflammation (Stern et al. 2008). Alterations in subclasses of regulatory T cells have also been implicated in MS. For example, in patients with MS, the proportion of $CD4^+CD25^+FoxP3^+CD31^+$ regulatory T cells is reduced, whereas the proportion of $CD31^-$ memory regulatory T cells are reciprocally expanded (Haas et al. 2009). GA therapy was shown to expand $CD4^+CD25^+FoxP^3+CD31^+$ regulatory T cells and decrease the proportion of memory regulatory T cells (Haas et al. 2009). GA has also been shown to alter the proportion of regulatory T cells that express the programmed death receptor 1 (PD1) on their cell surface (Saresella et al. 2008). PD1 is expressed intracellularly in the majority of regulatory T cells ($PD1^-$) and requires antigen stimulation for surface expression ($PD1^+$). A recent study demonstrated that $PD1^-$ regulatory T cell concentrations are significantly increased in peripheral blood of MS patients with stable disease compared with blood of MS patients with acute disease activity. In this study, $PD1^-$ regulatory T cells were significantly augmented in MS patients who responded to GA therapy (Saresella et al. 2008). Thus, GA has beneficial effects on multiple regulatory T cells subpopulations by helping to restore the homeostatic composition of regulatory T cell subsets to normal levels (Haas et al. 2009).

Migration of GA-Reactive T Cells into the Site of Damage

Cells that participate in the immune response to acute and chronic damage within the CNS include activated microglia, infiltrating macrophages, and T lymphocytes. Studies in EAE animal models have shown that adoptively transferred GA-reactive Th2 cells from GA-treated mice can be identified in the CNS of untreated mice, indicating that GA-reactive T cells can cross the blood-brain barrier and accumulate in the CNS. Once in the CNS, they are stimulated *in situ* by myelin auto-antigens such as MBP (Aharoni et al. 2000, 2003, 2005a; Arnon and Aharoni 2007). GA-reactive Th2 cells that secrete anti-inflammatory cytokines within the CNS reduce inflammation via bystander suppression and promote secretion of neurotrophic factors known to be pivotal for neuronal survival and tissue repair (Aharoni et al. 2003, 2005a; Azoulay et al. 2005; Hammarberg et al. 2000; Moalem et al. 2000; Muhallab et al. 2002; Ziemssen et al. 2002, 2005).

Modulation of B-Cell Function

Recent findings implicate B cells in both the pathogenesis and inhibition of MS and EAE (Bettelli et al. 2006; Chang et al. 1999, 2003; Duddy et al. 2007; Fillatreau et al. 2002; Hjelmstrom et al. 1998; Matsushita et al. 2008; Wolf et al. 1996; Yanaba et al. 2008). Contradictory activities by different B-cell subsets include demyelination (Bettelli et al. 2006), antibody production against myelin proteins (Breij et al. 2008; Franciotta et al. 2008; Hedegaard et al. 2009; Lassmann 2010), and protective mechanisms involving IL-10 secretion (Aharoni et al. 2011; Duddy et al. 2007; Fillatreau et al. 2002; Mann et al. 2007; Matsushita et al. 2008). As an antigen-based therapy, GA interacts with B cells and modulates B-cell function. Studies in EAE and MS patients have shown that GA can modulate cytokine secretion in specific subsets of B cells, increasing expression of anti-inflammatory cytokines, such as IL-10, IL-4, and TGF-β, which may in turn diminish production of proinflammatory cytokines, reduce autoreactive T-cell proliferation, and promote the generation of regulatory T cells (Kala et al. 2010, 2011; Matsushita et al. 2008; Yanaba et al. 2008). Studies in a murine model of EAE have also shown that B cells from GA-treated mice reduced expression of CD80 and CD86, costimulatory molecules on B cells that promote T cell activation (Kala et al. 2010, 2011). In addition, purified B cells adoptively transferred from GA-treated mice suppressed EAE in recipient mice and inhibited the proliferation of autoreactive T cells and the development of Th1 and Th17 cells (Kala et al. 2010). Whether B cells enter the CNS remains to be determined (Lalive et al. 2011).

Modulation of NK-Cell Function

Cytokines released by NK cells influence the initiation and development of adaptive immune responses by T cells and B cells. Several studies have documented decreased numbers of NK cells and impaired NK-cell function in patients with MS (Baxter and Smyth 2002; Benczur et al. 1980; Flodstrom et al. 2002; French and Yokoyama 2004; Grunebaum et al. 1989). Similarly, depletion of NK cells prior to EAE induction in mouse models results in clinically more severe relapsing EAE (Zhang et al. 1997). A recent study showed that GA treatment enhanced cytolysis by human NK cells against autologous and allogeneic immature and mature monocyte-derived dendritic cells (Sand et al. 2009). Moreover, increased secretion of IFN-γ by NK cells promotes autoreactive Th1 responses, whereas impaired capacity of NK cells to release IFN-γ is found to be a major mechanism underlying resistance to EAE. GA treatment inhibited the release of IFN-γ but increased the release of TNF-α from activated NK cells (Sand et al. 2009).

Neuroprotective Effects of GA

Studies in animals and humans have demonstrated potential neuroprotective activity by GA (Aharoni et al. 2003, 2005a, b; Weder et al. 2005). Neuroprotective effects of GA were examined in a rat model of EAE. GA treatment elicited protective effects on retinal ganglion cells as assessed by electroretinogram measurement of neurodegeneration and neuronal function (Maier et al. 2006). Similar effects of GA have been demonstrated in an optic-nerve injury model in which GA-reactive T cells accumulated at the site of injury and prevented the secondary degeneration of axons (Kipnis et al. 2000). BDNF, a critical factor in the differentiation and survival of neurons and various glial cell functions, is constitutively expressed in inflammatory cells of the innate and adaptive immune systems (monocytes, T cells, and B cells) (Kerschensteiner et al. 1999; Moalem et al. 2000; Muhallab et al. 2002; Ziemssen et al. 2002). In mice with EAE, adoptive transfer of GA-reactive T cells (Aharoni et al. 2003) or daily GA injections (Aharoni et al. 2005a) resulted in an increase in BDNF in the CNS *in situ*. In another study, GA-treated mice showed less axonal damage as revealed by a reduction in non-phosphorylated neurofilaments (SMI-32) and amyloid precursor protein (APP) staining compared with untreated mice (Gilgun-Sherki et al. 2003). The beneficial effects of GA have also been observed *in vivo* in mice with EAE using a yellow fluorescent protein (YFP) transgenic mouse model, which selectively express YFP on their neuronal population (Aharoni et al. 2005a). Compared with untreated mice, GA treatment led to a reduction in neuronal/axonal damage as assessed by quantification of expression of various neuronal antigens. Cell proliferation, migration, and differentiation were augmented in the GA-treated mice, endorsing a direct association between GA treatment and immunomodulation, neurogenesis, and *in situ* therapeutic benefits (Aharoni et al. 2005a).

The putative neuroprotective effects of GA have also been demonstrated in patients with MS. Decreased BDNF levels in MS patients have been restored to normal levels during GA treatment (Azoulay et al. 2005). Moreover, neuroprotection by GA is suggested by findings of axonal metabolic recovery and protection from axonal injury measured by magnetic resonance imaging (MRI) and magnetic resonance spectroscopy (MRS) (Arnold et al. 2013; Filippi et al. 2001; Khan et al. 2005). GA-treated patients had fewer new brain lesions evolve into MRI-identified "black holes," in which severe axonal disruption has occurred, than placebo-treated patients (Filippi et al. 2001).

Consistent with its protective effects on neurons and axons, GA treatment has been shown to decrease demyelination and increase myelin repair in an EAE model via enhanced proliferation, differentiation, and survival of oligodendrocyte progenitor cells, and to promote their recruitment into injury sites (Aharoni et al. 2008). Furthermore, GA treatment in two different mouse models of EAE (proteolipid protein [PLP]-induced relapsing-remitting and myelin oligodendrocyte glycoprotein [MOG]-induced chronic EAE) revealed reduced white matter lesion size, increased axonal density, a higher prevalence of normal-appearing axons, decreased demyelination and degradation, and reduced inflammation within the CNS compared with untreated mice (Aharoni et al. 2011). Quantitative analysis of remyelination relative to demyelination also revealed significant augmentation of remyelination after GA treatment (Aharoni et al. 2011).

Although findings in patients with MS and animals models have provided a convincing argument for the beneficial effects of GA on neuroprotection and neurogenesis, it is not yet clear whether these effects associated are a direct consequence of neurotrophic factor release or, alternatively, a physiological result of abrogation of inflammatory processes. These processes occur simultaneously, making it difficult to delineate the precise therapeutic pathways involved.

GA-Mediated Effect on Immunoglobulin Response

In addition to the cellular immune response, GA treatment has also been shown to induce a humoral response in MS patients (Brenner et al. 2001). GA treatment in MS patients induces a significant and sustained increase of the type II antibody, IgG4. This is in contrast to untreated MS patients who may demonstrate an unprimed humoral response against GA characterized by the production of IgM, IgG1, and IgG2 isotypes (Farina et al. 2002). The preferential secretion of IgG4 antibodies in GA-treated patients may occur secondarily to the induction of GA-reactive Th2 cells, as isotype switching to IgG4 is regulated by the Th2 cytokine, IL-4 (Basile et al. 2006). This finding is in keeping with the Th1 to Th2 shift that occurs during GA treatment. Although most IgG4 antibodies display strong neutralizing activity, serum from GA-treated patients containing anti-GA antibodies did not inhibit the ability of GA to stimulate GA-reactive T cells, indicating that anti-GA Ig had no neutralizing effect *in vitro* (Teitelbaum et al. 1971). Interestingly, MS patients treated with GA who remained relapse-free displayed higher titers of anti-GA antibodies than patients with active disease during GA treatment (Brenner et al. 2001), suggesting that GA-specific antibodies may have therapeutic effects in the

CNS. This hypothesis was supported in a mouse model of demyelinating disease, in which GA-specific antibodies were shown to promote remyelination of spinal cord axons, an effect that may contribute to the neuroprotective properties of GA in MS (Ure and Rodriguez 2002).

Key Points:

- Upon injection, GA antigens compete with myelin antigens for preferential binding to MHC molecules on APCs in the periphery. GA stimulates development of anti-inflammatory type II APCs, which promote naïve T cell differentiation toward development of Th2 cells and $FoxP3^+$ regulatory T cells. Type II APCs and Th2/T regulatory cells may facilitate each other's development via a positive feedback mechanism by which T-cell-derived anti-inflammatory cytokines further induce type II differentiation of APCs.
- GA blocks activation of myelin-reactive T cells or renders these cells anergic, and induces GA-reactive $CD8^+$ T cells with suppressive function.
- In animal models, GA-reactive Th2 cells cross the blood-brain barrier to become reactivated within the CNS and secrete anti-inflammatory cytokines and neurotrophic factors, dampening the surrounding proinflammatory milieu via bystander suppression. Within the CNS, GA-reactive T-cell-derived cytokines may promote type II differentiation of resident or recruited APCs.
- GA demonstrates neuroprotective effects in EAE animal models and reduces axonal injury in MS patients.
- GA has an immunomodulatory effect on B cells that results in decreased activation of autoreactive T cells.
- As an antigen-based therapy, GA induces anti-GA non-neutralizing antibodies in treated subjects.

Biological Activity: Glatiramoid-Mediated Effects on Gene Expression

Traditional biological characterization methods (e.g. ELISA, secreted cytokine screens) are insufficient to fully characterize Copaxone®'s mode of action. Recent experiments with gene expression arrays provide a broader and more comprehensive analysis of immune pathways modulated by Copaxone®. Experiments on mouse splenocytes and human THP-1 monocytes have been done using state-of-the-art gene expression arrays and data analysis (Bakshi et al. 2013; Towfic et al. 2014; Citizens petition 2013, 2014). In these studies, activated splenocytes from Copaxone®-treated mice were re-activated *ex vivo* with Copaxone® or with purported generic GA products. Copaxone® was shown to up-regulate or down-regulate more than 1400 genes, and as such, significantly affects over 100 functional pathways (Bakshi et al. 2013). Genes differentially expressed upon Copaxone® re-activation of the splenocytes were significantly enriched for membership in many relevant and overlapping pathway lists with some shared functionality that further strengthens the type and role of the impacted biology. The pathways were determined by employing the National Institute of Health's DAVID enrichment tool (Huang et al. 2009a, b) and included: regulation of cytokine production, regulation of adaptive immune responses, positive regulation of T cell activation, positive regulation of lymphocyte activation, regulation of B cell-mediated immunity, and positive regulation of B cell proliferation (Bakshi et al. 2013; Towfic et al. 2014; Zeskind et al. 2014).

These studies also demonstrated key differences between Copaxone® and purported generic GA products that are likely to influence drug safety and efficacy (Bakshi et al. 2013; Towfic et al. 2014; Citizens petition 2014). For example, there were marked differences, often in opposite directions (e.g., reversals of up-regulation and down-regulation) on gene expression by activated splenocytes re-activated by the purported generic GA product, Glatimer®, compared with reactivation by Copaxone® (Fig. 16) (Towfic et al. 2014).

As noted, $FoxP3^+$ regulatory T cells suppress autoreactive T cells and are implicated in the reduction of MS disease activity (Hong et al. 2005; Sela and Teitelbaum 2001). FoxP3 induction is important for regulatory T cell activation and mouse studies have shown that Copaxone® treatment can increase $FoxP3^+$ regulatory T cells threefold in the total pool of $CD3^+$ T cells (Aharoni et al. 2010). This is

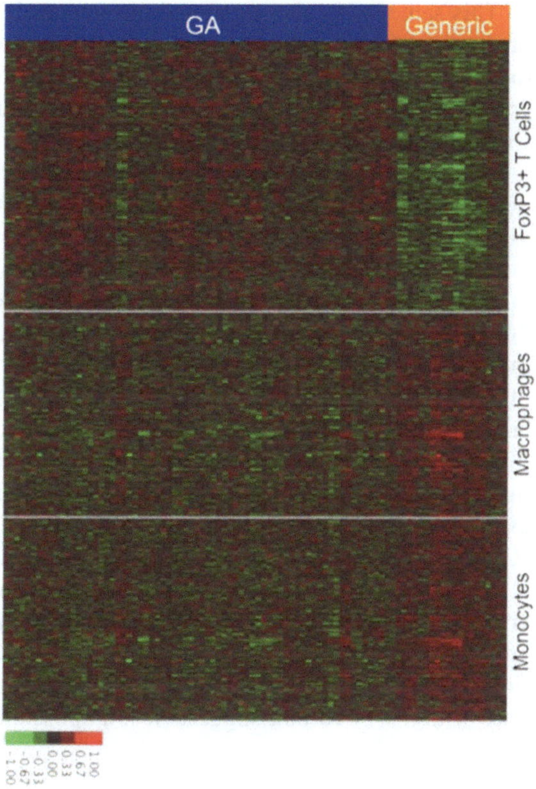

Fig. 16 Cell-type specific differences in the biological impact of Copaxone®and the purported generic GA product, Glatimer®. The heat map depicts relative expression of specific genes in Copaxone®-activated samples and Glatimer®-activated samples. Each of the rows within the Treg, macrophage, and monocyte sections represent a gene with high cell-type specificity scores for the respective cell types. Overall, Copaxone® induces higher expression of Treg-associated genes than Glatimer®, while Glatimer® induces higher expression of macrophage- and monocyte-associated genes than Copaxone®(Towfic et al. 2014). (*GA* Copaxone®, *"generic"* Glatimer)

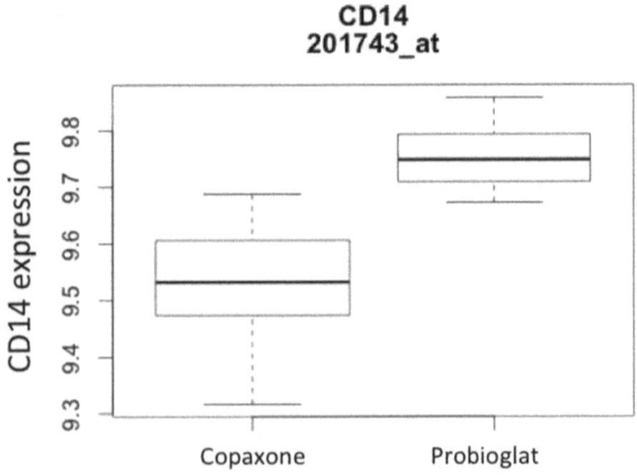

Fig. 17 CD14 expression is significantly higher with stimulation by Probioglat® compared with Copaxone® (a single CD14 probe-set on the chip is shown, 201743_at) in human monocytes (Citizens petition 2014)

consistent with gene expression data showing Copaxone® induced FoxP3 expression to a significantly greater extent than Glatimer® (Towfic et al. 2014). Given these findings, it is reasonable to hypothesize that Copaxone® may have greater therapeutic efficacy than Glatimer® by suppressing harmful cytotoxic cells more effectively, although this hypothesis warrants further investigation.

This same study also examined differences in global variability across relevant probes to determine whether the biological impact of Glatimer® was as consistent (across 5 batches) as of Copaxone® (across 30 batches). There was four times greater variability of biologic impact among the Glatimer® batches compared with the Copaxone® batches (Towfic et al. 2014).

Gene expression studies also show differences between Copaxone® and purported generic GA products that are relevant to the potential safety/toxicology of glatiramoids. In the experiment above, lipopolysaccharide (LPS)-response pathway genes were significantly enriched among genes upregulated by Glatimer® (LPS is an endotoxin that promotes proinflammatory cytokine secretion) (Towfic et al. 2014). Similarly, CD14 is a key monocyte inflammation marker, and monocytes may serve as prominent contributors to neuroinflammation in MS (Bar-Or et al. 2003). Human monocytes stimulated by Probioglat®expressed significantly higher CD14 levels than samples treated with Copaxone®(Fig. 17) (Citizens petition 2014). Potential safety issues related to increased monocyte activation by Glatimer® or Probioglat® may be reflected as increased adverse inflammatory events, such as increased injection-site reactions, and inducing, rather than preventing, relapse (Citizens petition 2014; Gutierrez 2013; Muller 2013).

Results of these gene expression studies demonstrate that glatiramoid products created from the same 4 AA combined in the same molar ratios can be quite different, and key differences between glatiramoids can go undetected using characterization methods that measure only a few drug features. This is supported by experience with TV-5010 that suggests it is impossible to predict the safety or efficacy of a glatiramoid mixture in the absence of clinical testing in MS patients. TV-5010 was synthesized by the originator of Copaxone® (Teva Pharmaceutical Industries, Inc.) by introducing small changes in the Copaxone® manufacturing process (Ramot et al. 2012; Varkony et al. 2009). The product was similar to Copaxone® in AA ratio and in physical properties. Moreover, TV-5010 was identified by Copaxone®-specific antibodies, signalling that they shared similar B cell epitopes and likely, a similar mode of action. In general, TV-5010 showed an improved efficacy profile in preclinical studies, as it was more effective than Copaxone® in blocking EAE. No toxicities were observed in short-term studies (13 weeks) of TV-5010 in rodents and, based on this finding, it was assumed that the toxicity profile of TV-5010 would be similar to that of Copaxone®. Good safety and tolerability was shown in RRMS patients in two small 9-months clinical trials (De Stefano et al. 2009). However, long-term toxicity studies in monkey (52 weeks) and rats (26 weeks) unexpectedly revealed an unfavorable long-term toxicity profile, including, but not limited to, fibrosis in rats and eosinophilia in monkeys (Ramot et al. 2012; Varkony et al. 2009). Development of TV-5010 was halted. Because Copaxone® and TV-5010 had many similarities, the same splenocyte method described in the studies above was used to determine whether genes differentially expressed in response to reactivation with Copaxone® or TV-5010 could have predicted toxicity. Of genes with significantly different expression levels (fold changes >1.5) between the two drugs, the gene with the lowest (most significant) p value was MMP14 (Citizens petition 2013). MMP14 expression, which was significantly higher in response to TV-5010 than in response to Copaxone®, has been associated with fibrosis and eosinophil-related disorders in animals and in humans (Beppu et al. 2013; Zhou et al. 2004; Citizens petition 2013).

Key Points:

- While similar in some traditional features of glatiramoid characterization, e.g. amino acid distribution, purported generic and follow-on GA products can induce significantly different gene expression profiles associated with key immunological pathways in MS than those induced by Copaxone®
- TV-5010 experience indicates that even minor changes in glatiramoid manufacture can cause differences in biological activities that can influence toxicology, and likely leading to clinical safety and efficacy concerns.
- Differences from Copaxone® in the gene expression profiles of TV-5010 and the purported generic GA products indicate that if the manufacture of a glatiramoid differs from that of Copaxone®, it cannot be assumed to have the same efficacy and safety properties as Copaxone®.
- The clinical significance of differences in biological activities between Copaxone® and purported generic GA products are unclear. No full set of peer-reviewed efficacy and safety data has been published for purported generic GA products at this time. Copaxone® is the only glatiramoid with a number of long term clinical studies to support its long-term efficacy and safety.

Regulatory Status

The European Medicines Agency (EMA) and the US Food and Drug Administration (FDA) regulatory statutes make a distinction between small molecules and biological medicinal products, but neither agency has devised a dedicated pathway for follow-on compounds and generic versions of NBCDs, nor are there specific guidelines for the glatiramoid class. Conventional pharmacokinetic (PK) studies cannot demonstrate glatiramoid bioequivalence due to rapid hydrolysis of GA at the site of injection and uptake by local APCs, and because the active epitopes in GA are not identified. Moreover, there is a lack of pharmacodynamic (PD) markers suitable to predict clinical outcomes of glatiramoid therapy, and their essential immunomodulatory mode of action is not fully elucidated (Varkony et al. 2009). Furthermore, because they are injectable colloidal solutions, glatiramoids are subject to problems when trying to establish bioequivalence, "…because differences in particle size, polymorphic structure of the suspended active ingredient, or the suspension formulation can significantly affect the rate of release and absorption…" (ORANGE BOOK 2013). It is crucial to ensure the quality of the glatiramoid injection in terms of morphology and particle size; i.e., to test higher order structures and charge distribution in suspension, as described above.

Nevertheless, as noted, purported generic GA products are marketed in India (Glatimer® Natco), Mexico (Probioglat® Probiomed), and Argentina (Escadra®, Raffo and Polimunol, Synthon) despite the fact that the efficacy and long-term safety of these purported generic products are unknown. Thus, there is an urgent need for general consensus regarding the criteria needed to establish therapeutic equivalence among members of the glatiramoid class, as minor changes from the well-studied GA in Copaxone® could have an important impact on efficacy and patient safety. The currently marketed generic products mentioned above display only a degree of similarity to Copaxone® that is restricted, at best, to broad characteristics such as MW distribution of the polypeptide components in the mixture and AA content. Given that the efficacy and safety of glatiramoids are dependent on cytokine induction and immunogenicity to exert their therapeutic biological activity, and that sensitive analytical techniques can detect important differences between glatiramoids made by different manufacturers, it is necessary to find consensus among innovators, manufacturers, and regulatory agencies worldwide to define a comprehensive set of regulations before generic and follow-on glatiramoid products can make claims of therapeutic equivalence.

Of particular importance to the future regulatory pathway of new and generic glatiramoids is the fact that these products are antigenic immunomodulatory agents with inherent immunogenic activity, which can have an enormous impact on both safety and efficacy. Characterizing the immunogenicity of a glatiramoid should be a key consideration in the regulatory approval process. Purported generic and follow-on GA products cannot be judged to be equivalent to the originator or to each other based on bulk physicochemical characteristics, even if subjected to a battery of "overlapping" criteria, because even small changes in the primary structure might lead to different antigenic epitopes. In turn, different epitopes can markedly influence the product's immunogenicity profile, resulting in an anti-glatiramoid antibody repertoire with different specificities and affinities from those of anti-Copaxone® antibodies. Different epitopes, no matter how slight the difference, can induce antibodies that neutralize drug efficacy, interfere with recognition of foreign antigens in a way that undermines healthy tumor surveillance and defense against infections, or cause undesirable host responses. Moreover, because of the chronic nature of MS and of MS treatment regimens, continued alteration of humoral and cellular immunity with a putative generic GA product with uncharacterized immunogenicity could allow progression of neurologic disability and related safety problems that would not be detected in the short term, such as:

- Decreased efficacy associated with subclinical persistent inflammation
- Risks associated with immunogenicity, including: immune-complex formation leading to deposits in the kidney, hypersensitivity reactions, additional autoimmune disorders causing further demyelination and exacerbation of disease progression, general immune suppression leading to increased risk of opportunistic infections, and drug-related eosinophilia, which can have major clinical consequences, including death.

Based on current knowledge and available data, a recent paper summarized discussions among 25 scientific experts from industry, academia, and regulatory bodies regarding the regulation of NBCDs and their follow-on versions (Schellekens et al. 2014; Crommelin et al. 2014; Holloway et al. 2012; Mühlebach et al. 2013). For glatiramoids, there was a general understanding that because of the physicochemical differences seen between the prototype, Copaxone®, and purported generic versions of GA, and because the clinical implications of these differences are as yet unclear, the only meaningful way to evaluate product efficacy and safety is to perform comparative clinical trials in patients with MS (Schellekens et al. 2014). Such trials should include both placebo and Copaxone® as an active comparator, enroll a sufficient number of patients, provide for long-term monitoring (at least 2 years), evaluate appropriate clinical endpoints (i.e., relapse rate), and thoroughly characterize and document the immunogenicity of the product.

This position appears to be endorsed by regulatory authorities. The New Drug Division at FDA, for instance, refused to approve a supplemental new drug application (NDA) for a new formulation of Copaxone® (prepared by the same manufacturer as the original formulation), in which the same amount of GA would be given in a reduced injection volume (0.5 mL rather than 1.0 mL). The FDA's stated opinion was that a clinical trial would be necessary to demonstrate that the effectiveness of the product was not compromised by reducing the injection volume.

The EMA guidelines for biosimilar drugs released in 2005 stipulates that for marketing approval it must be shown that follow-on products closely related to reference medicines do not have meaningful differences from the reference in quality, safety, and efficacy (EMEA/CHMP/437/04 2005). For a glatiramoid, it is virtually impossible to comply with these criteria. State-of the art technologies can be used to differentiate glatiramoids made by different suppliers, or to indicate the degree of consistency (or lack thereof) of product characteristics among glatiramoid batches made by a specific manufacturer, but cannot completely characterize the structure of a glatiramoid. Thus, the only valid way to establish whether any glatiramoid product has similar quality, safety, and efficacyis to establish therapeutic equivalence in a clinical trial as a route to approval. Consistent with this approach, the manufacturer of a follow-on GA product is currently conducting a 24-month comparative clinical trial with placebo and Copaxone® arms, based on scientific advice from the EMA (http://www.synthon.com/Corporate/News/PressReleases/Synthon-announces-successful-outcome-of-the-PhaseIII-GATE-study-with-its-generic-glatiramer-acetate?sc_lang=en).

Prospects

At the 2013 American Association of Pharmaceutical Sciences (AAPS) annual meeting, it was argued that the foundation of drug similarity assessment is built on analytical, physicochemical, biochemical and/or functional studies. Consequently, there can be no reduced requirement for clinical studies if molecular and functional similarities are not demonstrated. Alternately, a speaker at the meeting stated, "You can't use clinical studies to test a protein into biosimilarity" (sic). In terms of glatiramoids, this would indicate that both analytical and clinical characterization should be prerequisites to glatiramoid marketing approval. Because currently available purported generic GA products are not equivalent to Copaxone® with respect to physicochemical features and/or biological activity, these glatiramoid products should be considered unique chemical entities that require a full battery of evaluations, including full-scale nonclinical testing and controlled clinical trials with validated endpoints. It is clear that significant differences in the active ingredients of Copaxone® and purported generic GA products call for a case-by-case approach to similarity demonstration.

Over time, it may become possible to determine a set of mandatory tests to screen purported generic glatiramoid products to elucidate any deviances from the originator. Substantial advances have been made in modern technologies capable of evaluating the consistency of glatiramoid manufacture and to identify differences between glatiramoid products. The newer approaches discussed in this chapter may guide authorities to establish scientifically based rules and requirements for evaluating glatiramoid formulations in the framework of equivalence testing between the reference drug and generic and follow-on products. Evaluation of these highly complex mixtures containing multiple, closely related structures should be made using specific cutting-edge techniques to assess the composition of intact drug. Gene expression analysis has proven to be a sensitive tool to evaluate drug effects on down- and up-regulation of genes in order to indicate functional pathways of drug activation, to determine differences in biological activities among glatiramoids with potential consequences for safety and efficacy, and to assess batch-to-batch equivalence of as yet unapproved glatiramoids. In activated mouse splenocytes, Copaxone® up-regulated genes associated with beneficial regulatory T cells linked to the suppression of myelin-reactive T cells, while a purported generic GA product up-regulated genes associated with the myeloid cell linage linked to enhanced inflammatory responses (Bakshi et al. 2013; Towfic et al. 2014). While these effects were obtained in mouse splenocyte studies, additional studies of gene expression from human monocytes are underway to further elucidate potential differences in gene expression profiles among glatiramoids.

With respect to clinical outcomes associated with substitution and interchangeability of glatiramoid products, only clinical data from a randomized controlled trial with a cross-over design will provide evidence that switching glatiramoid formulations will not prove detrimental to MS patients. Comparative gene expression studies suggest that the immune system is extremely sensitive to changes in the antigenic nature of GA (Bakshi et al. 2013; Towfic et al. 2014; Citizens petition 2013, 2014). Immune function can vary widely from one MS patient to the next due to differences in genetic backgrounds and environmental exposures during the course of their lives. Because the immune response appears to be exquisitely sensitive to specific antigenic structures, the therapeutic components of Copaxone® may vary from patient to patient. This has implications for changes in efficacy and managing safety risks if members of the glatiramoid class are switched or substituted for one another for individual MS patients. Switching Copaxone® for one of the purported generic products may increase risk of harm to patients. This potential is supported by experience reported by MS patients in Mexico, when the National Social Service began to replace Copaxone® with Probioglat® or to alternate the two glatiramoids in early 2013 (Gutierrez 2013; Muller 2013). At 2 to 4 months after switching to Probioglat®, relapses increased more than 50% at some health clinics and relapse-related hospitalizations increased by 200% in 2013 compared with the previous year, when these patients had well-controlled disease using only Copaxone® (Citizens petition 2014).

As the primary concern of regulatory authorities is to make products available that are both efficacious and safe, specific guidelines and approval requirements for the glatiramoid class should be based on the most recent scientific findings. For biosimilar drugs, marketing authorization by the EMA does not imply that a product is interchangeable with the reference drug, and the FDA requires that clinical effects of interchangeability must be shown in specific studies (Schellekens et al. 2014). A similar requirement is needed for the glatiramoid class of NBCDs. Each new purported generic or follow-on glatiramoid product should be considered a new entity and evaluated with the same scrutiny and rigor.

Acknowledgement The authors sincerely wish to thank the following persons for their advice and contribution to this book chapter and especially the help of those at CMC and R&D departments of Teva Pharmaceuticals: Arthur Komlosh, Tal Hasson, Tatiana Molotsky, Anna Kogan, Revital Krispin, Dalia Pinkert, Galia Papir, Kevin Wells-Knecht, Mehran Yazdanian, Jill Conner, Shlomo Bakshi, Olga Beriozkin and Wim Weyenberg.

References

Aharoni R, Teitelbaum D, Arnon R (1993) T suppressor hybridomas and interleukin-2-dependent lines induced by copolymer 1 or by spinal cord homogenate down-regulate experimental allergic encephalomyelitis. Eur J Immunol 23:17–25

Aharoni R, Teitelbaum D, Sela M et al (1997) Copolymer 1 induces T cells of the T helper type 2 that crossreact with myelin basic protein and suppress experimental autoimmune encephalomyelitis. Proc Natl Acad Sci U S A 94:10821–10826

Aharoni R, Teitelbaum D, Arnon R et al (1999) Copolymer 1 acts against the immunodominant epitope 82-100 of myelin basic protein by T cell receptor antagonism in addition to major histocompatibility complex blocking. Proc Natl Acad Sci U S A 96:634–639

Aharoni R, Teitelbaum D, Leitner O et al (2000) Specific Th2 cells accumulate in the central nervous system of mice protected against experimental autoimmune encephalomyelitis by copolymer 1. Proc Natl Acad Sci U S A 97:11472–11477

Aharoni R, Kayhan B, Eilam R et al (2003) Glatiramer acetate-specific T cells in the brain express T helper 2/3 cytokines and brain-derived neurotrophic factor in situ. Proc Natl Acad Sci U S A 100:14157–14162

Aharoni R, Arnon R, Eilam R (2005a) Neurogenesis and neuroprotection induced by peripheral immunomodulatory treatment of experimental autoimmune encephalomyelitis. J Neurosci 25:8217–8228

Aharoni R, Eilam R, Domev H, et al (2005b) The immunomodulator glatiramer acetate augments the expression of neurotrophic factors in brains of experimental autoimmune encephalomyelitis mice. Proc Natl Acad Sci U S A 102:19045–19050

Aharoni R, Kayhan B, Arnon R (2005c) Therapeutic effect of the immunomodulator glatiramer acetate on trinitrobenzene sulfonic acid-induced experimental colitis. Inflamm Bowel Dis 11:106–115

Aharoni R, Herschkovitz A, Eilam R et al (2008) Demyelination arrest and remyelination induced by glatiramer acetate treatment of experimental autoimmune encephalomyelitis. Proc Natl Acad Sci U S A 105:11358–11363

Aharoni R, Eilam R, Stock A et al (2010) Glatiramer acetate reduces Th-17 inflammation and induces regulatory T-cells in the CNS of mice with relapsing-remitting or chronic EAE. J Neuroimmunol 225:100–111

Aharoni R, Vainshtein A, Stock A et al (2011) Distinct pathological patterns in relapsing-remitting and chronic models of experimental autoimmune encephalomyelitis and the neuroprotective effect of glatiramer acetate. J Autoimmun 37:228–241

Allie R, Hu L, Mullen KM et al (2005) Bystander modulation of chemokine receptor expression on peripheral blood T lymphocytes mediated by glatiramer therapy. Arch Neurol 62:889–894

Arnold D, Narayanan S et al (2013) Neuroprotection with glatiramer acetate: evidence from the PreCISe trial. J Neurol 260(7):1901–1906

Arnon R, Aharoni R (2004) Mechanism of action of glatiramer acetate in multiple sclerosis and its potential for the development of new applications. Proc Natl Acad Sci U S A 101(Suppl 2):14593–14598

Arnon R, Aharoni R (2007) Neurogenesis and neuroprotection in the CNS-fundamental elements in the effect of glatiramer acetate on treatment of autoimmune neurological disorders. Mol Neurobiol 36:245–253

Azoulay D, Vachapova V, Shihman B et al (2005) Lower brain-derived neurotrophic factor in serum of relapsing remitting MS: reversal by glatiramer acetate. J Neuroimmunol 167:215–218

Bakshi S, Chalifa-Caspi V, Olaschkes I et al (2013) Gene expression analysis reveals function pathways of glatiramer acetate activation. Expert Opin Ther Targets 17(4):351–362

Bar-Or A, Nuttall R, Duddy M et al (2003) Analyses of all matrix metalloproteinase members in leukocytes emphasize monocytes as major inflammatory mediators in multiple sclerosis. Brain 126(Pt 12):2738–2749

Basile E, Gibbs E, Aziz T et al (2006) During 3 years treatment of primary progressive multiple sclerosis with glatiramer acetate, specific antibodies switch from IgG1 to IgG4. J Neuroimmunol 177(1–2)161–166

Baxter AG, Smyth MJ (2002) The role of NK cells in autoimmune disease. Autoimmunity 35:1–14

Benczur M, Petranyl GG, Palffy G et al (1980) Dysfunction of natural killer cells in multiple sclerosis: a possible pathogenetic factor. Clin Exp Immunol 39:657–662

Beppu L, Anilkumar AA, Dohil R et al (2013) MMP-14 is elevated in pediatric subjects with eosinophilic esophagitis. J Allergy Clin Immunol 131(2):Abstract 132

Berkowitz S, Engen J, Mazzeo J et al (2012) Analytical tools for characterizing biopharmaceuticals and the implications for biosimilars. Nat Rev Drug Discov 11(7):527–540. doi:10.1038/nrd37

Bettelli E, Baeten D, Jager A et al (2006) Myelin oligodendrocyte glycoprotein-specific T and B cells cooperate to induce a Devic-like disease in mice. J Clin Invest 116:2393–2402

Breij EC, Brink BP, Veerhuis R et al (2008) Homogeneity of active demyelinating lesions in established multiple sclerosis. Ann Neurol 63:16–25

Brenner T, Arnon R, Sela M et al (2001) Humoral and cellular immune responses to Copolymer 1 in multiple sclerosis patients treated with Copaxone. J Neuroimmunol 115:152–160

Burger D, Molnarfi N, Weber MS et al (2009) Glatiramer acetate increases IL-1 receptor antagonist but decreases T cell-induced IL-1beta in human monocytes and multiple sclerosis. Proc Natl Acad Sci U S A 106:4355–4359

Carter NJ, Keating GM (2010) Glatiramer acetate: a review of its use in relapsing-remitting multiple sclerosis and in delaying the onset of clinically definite multiple sclerosis. Drugs 70(12):1545–1577

Chang TT, Jabs C, Sobel RA et al (1999) Studies in B7-deficient mice reveal a critical role for B7 costimulation in both induction and effector phases of experimental autoimmune encephalomyelitis. J Exp Med 190:733–740

Chang TT, Sobel RA, Wei T et al (2003) Recovery from EAE is associated with decreased survival of encephalitogenic T cells in the CNS of B7-1/B7-2-deficient mice. Eur J Immunol 33:2022–2032

Chen M, Gran B, Costello K et al (2001) Glatiramer acetate induces a Th2-biased response and crossreactivity with myelin basic protein in patients with MS. Mult Scler 7:209–219

Cohen B, Oger J, Gagnon A et al (2008) The implications of immunogenicity for protein-based multiple sclerosis therapies. J Neurol Sci 275:7–17

Cohen JA et al (2014) Generic glatiramer acetate is equivalent to copaxone on effcacy and safety: results of the randomized double-blind GATE trial in multiple sclerosis, Abstract FC1.2, 2014 Joint ACTRIMS-ECTRIMS Meeting (MSBoston 2014): Oral Presentations. Mult Scler 20:14–66

COPAXONE® (glatiramer acetate) solution for subcutaneous injection: Full Prescribing Information, FDA-approved labeling. Teva Neuroscience Inc. Initial U.S. Approval: 1996. Revised [2/2009]

Crommelin DJ, de Vlieger JS, Weinstein V et al (2014) Different pharmaceutical products need similar terminology. AAPS J 16(1):11–14. doi:10.1208/s12248-013-9532-0

De Stefano N, Filippi M, Confavreux C et al (2009) The results of two multicentre, open-label studies assessing efficacy, tolerability and safety of protiramer, a high molecular weight synthetic copolymeric mixture, in patients with relapsing-remitting multiple sclerosis. Mult Scler 15(2):238–243

Dhib-Jalbut S (2003) Glatiramer acetate (Copaxone) therapy for multiple sclerosis. Pharmacol Ther 98(2):245–255

Duda PW, Schmied MC, Cook SL et al (2000) Glatiramer acetate (Copaxone) induces degenerate, Th2-polarized immune responses in patients with multiple sclerosis. J Clin Invest 105:967–976

Duddy M, Niino M, Adatia F et al (2007) Distinct effector cytokine profiles of memory and naive human B cell subsets and implication in multiple sclerosis. J Immunol 178:6092–6099

Duncan R, Gaspar R (2011) Nanomedicines under the microscope. Mol Pharm 8(6):2101–2141

Farina C, Vargas V, Heydari N et al (2002) Treatment with glatiramer acetate induces specific IgG4 antibodies in multiple sclerosis patients. J Neuroimmunol 123:188–192
Filippi M, Rovaris M, Rocca MA et al (2001) Glatiramer acetate reduces the proportion of new MS lesions evolving into "black holes". Neurology 57:731–733
Fillatreau S, Sweenie CH, McGeachy MJ et al (2002) B cells regulate autoimmunity by provision of IL-10. Nat Immunol 3:944–950
Flodstrom M, Shi FD, Sarvetnick N, Ljunggren HG (2002) The natural killer cell—friend or foe in autoimmune disease? Scand J Immunol 55:432–441
Franciotta D, Zardini E, Bergamaschi R et al (2003) Interferon gamma and interleukin 4 producing T cells in peripheral blood of multiple sclerosis patients undergoing immunomodulatory treatment. J Neurol Neurosurg Psychiatry 74:123–126
Franciotta D, Salvetti M, Lolli F et al (2008) B cells and multiple sclerosis. Lancet Neurol 7:852–858
French AR, Yokoyama WM (2004) Natural killer cells and autoimmunity. Arthritis Res Ther 6:8–14
Fridkis-Hareli M, Strominger JL (1998) Promiscuous binding of synthetic copolymer 1 to purified HLA-DR molecules. J Immunol 160:4386–4397
Gilgun-Sherki Y, Panet H, Holdengreber V et al (2003) Axonal damage is reduced following glatiramer acetate treatment in C57/bl mice with chronic-induced experimental autoimmune encephalomyelitis. Neurosci Res 47:201–207
Grunebaum E, Malatzky-Goshen E, Shoenfeld Y (1989) Natural killer cells and autoimmunity. Immunol Res 8:292–304
Haas J, Hug A, Viehover A et al (2005) Reduced suppressive effect of CD4+CD25high regulatory T cells on the T cell immune response against myelin oligodendrocyte glycoprotein in patients with multiple sclerosis. Eur J Immunol 35:3343–3352
Haas J, Fritzsching B, Trubswetter P et al (2007) Prevalence of newly generated naïve regulatory T cells [Treg] is critical for Treg suppressive function and determines Treg dysfunction in multiple sclerosis. J Immunol 179:1322–1330
Haas J, Korporal M, Balint B et al (2009) Glatiramer acetate improves regulatory T cell function by expansion of naive CD4(+)CD25(+)FOXP3(+)CD31(+) T cells in patients with multiple sclerosis. J Neuroimmunol 216:113–117
Hafler DA (2002) Degeneracy, as opposed to specificity, in immunotherapy. J Clin Invest 109:581–584
Hammarberg H, Lidman O, Lundberg C et al (2000) Neuroprotection by encephalomyelitis: rescue of mechanically injured neurons and neurotrophin production by CNS-infiltrating T and natural killer cells. J Neurosci 20:5283–5291
Hedegaard CJ, Chen N, Sellebjerg F et al (2009) Autoantibodies to myelin basic protein (MBP) in healthy individuals and in patients with multiple sclerosis: a role in regulating cytokine responses to MBP. Immunology 128:e451–e461
Hjelmstrom P, Juedes AE, Fjell J et al (1998) B-cell-deficient mice develop experimental allergic encephalomyelitis with demyelination after myelin oligodendrocyte glycoprotein sensitization. J Immunol 161:4480–4483
Holloway C, Mueller-Berghaus J, Lima BS et al (2012) Scientific considerations for complex drugs in light of established and emerging regulatory guidance. Ann N Y Acad Sci 1276:26–36. doi:1111/j.1749-6632.2012.06811
Hong J, Li N, Zhang X et al (2005) Induction of CD4+CD25+regulatory T cells by copolymer-I through activation of transcription factor Foxp3. Proc Natl Acad Sci U S A 102:6449–6454
Huang DW, Sherman BT, Lempicki RA (2009a) Systematic and integrative analysis of large gene lists using DAVID Bioinformatics Resources. Nature Protoc 4(1):44–57
Huang DW, Sherman BT, Lempicki RA (2009b) Bioinformatics enrichment tools: paths toward the comprehensive functional analysis of large gene lists. Nucleic Acids Res 37(1):1–13
Johnson KP (2010) Glatiramer acetate and the glatiramoid class of immunomodulator drugs in multiple sclerosis: an update. Expert Opin Drug Metab Toxicol 6(5):643–660. doi:10.1517/17425251003752715

Jung S, Siglienti I, Grauer O et al (2004) Induction of IL-10 in rat peritoneal macrophages and dendritic cells by glatiramer acetate. J Neuroimmunol 148(1–2):63–73
Kala M, Rhodes SN, Piao WH et al (2010) B cells from glatiramer acetate-treated mice suppress experimental autoimmune encephalomyelitis. Exp Neurol 221:136–145
Kala M, Miravalle A, Vollmer T (2011) Recent insights into the mechanism of action of glatiramer acetate. J Neuroimmunol 235:9–17
Kantengwa S, Weber MS, Juillard C et al (2007) Inhibition of naive Th1 CD4+ T cells by glatiramer acetate in multiple sclerosis. J Neuroimmunol 185:123–129
Karandikar NJ, Crawford MP, Yan X et al (2002) Glatiramer acetate (Copaxone) therapy induces CD8(+) T cell responses in patients with multiple sclerosis. J Clin Invest 109:641–649
Kerschensteiner M, Gallmeier E, Behrens L et al (1999) Activated human T cells, B cells, and monocytes produce brain-derived neurotrophic factor in vitro and in inflammatory brain lesions: a neuroprotective role of inflammation? J Exp Med 189:865–870
Khan O, Shen Y, Caon C et al (2005) Axonal metabolic recovery and potential neuroprotective effect of glatiramer acetate in relapsing-remitting multiple sclerosis. Mult Scler 11:646–651
Kim HJ, Ifergan I, Antel JP et al. (2004) Type 2 monocyte and microglia differentiation mediated by glatiramer acetate therapy in patients with multiple sclerosis. J Immunol 172:7144–7153
Kipnis J, Schwartz M (2002) Dual action of glatiramer acetate (Cop-1) in the treatment of CNS autoimmune and neurodegenerative disorders. Trends Mol Med 8:319–323
Kipnis J, Yoles E, Porat Z et al (2000) T cell immunity to copolymer 1 confers neuroprotection on the damaged optic nerve: possible therapy for optic neuropathies. Proc Natl Acad Sci U S A 97:7446–7451
Kipnis J, Mizrahi T, Hauben E et al (2002) Neuroprotective autoimmunity: naturally occurring CD4+CD25+ regulatory T cells suppress the ability to withstand injury to the central nervous system. Proc Natl Acad Sci U S A 99:15620–15625
Kipnis J, Nevo U, Panikashvili D et al (2003) Therapeutic vaccination for closed head injury. J Neurotrauma 20:559–569
Lalive PH, Neuhaus O, Benkhoucha M et al (2011) Glatiramer acetate in the treatment of multiple sclerosis: emerging concepts regarding its mechanism of action. CNS Drugs 25:401–414
Lassmann H (2010) Axonal and neuronal pathology in multiple sclerosis: what have we learnt from animal models. Exp Neurol 225:2–8
Maier K, Kuhnert AV, Taheri N et al (2006) Effects of glatiramer acetate and interferon-beta on neurodegeneration in a model of multiple sclerosis: a comparative study. Am J Pathol 169:1353–1364
Mann MK, Maresz K, Shriver LP et al (2007) B cell regulation of CD4+CD25+ T regulatory cells and IL-10 via B7 is essential for recovery from experimental autoimmune encephalomyelitis. J Immunol 178:3447–3456
Matsushita T, Yanaba K, Bouaziz JD et al (2008) Regulatory B cells inhibit EAE initiation in mice while other B cells promote disease progression. J Clin Invest 118:3420–3430
Miller A, Shapiro S, Gershtein R et al (1998) Treatment of multiple sclerosis with copolymer-1 (Copaxone): implicating mechanisms of Th1 to Th2/Th3 immune-deviation. J Neuroimmunol 92:113–121
Moalem G, Gdalyahu A, Shani Y et al (2000) Production of neurotrophins by activated T cells: implications for neuroprotective autoimmunity. J Autoimmun 15:331–345
Muhallab S, Lundberg C, Gielen AW et al (2002) Differential expression of neurotrophic factors and inflammatory cytokines by myelin basic protein specific and other recruited T cells infiltrating the central nervous system during experimental autoimmune encephalomyelitis. Scand J Immunol 55:264–273
Mühlebach S, Vulto A, de Vlieger J et al (2013) The authorization of non-biological complex drugs (NBCDs) follow-on versions: specific regulatory and interchangeability rules ahead? GaBi J 2(4):204–207
Neuhaus O, Farina C, Yassouridis A et al (2000) Multiple sclerosis: comparison of copolymer-1-reactive T cell lines from treated and untreated subjects reveals cytokine shift from T helper 1 to T helper 2 cells. Proc Natl Acad Sci U S A 97:7452–7457

Nicholas J (2012) Complex drugs and biologics: scientific and regulatory challenged for follow-on products. Drug Inf J 46(2):197–206

Putheti P, Soderstrom M, Link H et al (2003) Effect of glatiramer acetate (Copaxone) on CD4+CD25high T regulatory cells and their IL-10 production in multiple sclerosis. J Neuroimmunol 144:125–131

Ramot Y, Rosenstock M, Klinger E et al (2012) Comparative long-term preclinical safety evaluation of two glatiramoid compounds (glatiramer acetate, Copaxone®, and TV-5010, protiramer) in rats and monkeys. Toxicol Pathol 40:40–54. doi:10.1177/019263311424169

Sand KL, Knudsen E, Rolin J et al (2009) Modulation of natural killer cell cytotoxicity and cytokine release by the drug glatiramer acetate. Cell Mol Life Sci 66:1446–1456

Sanna A, Fois ML, Arru G et al (2006) Glatiramer acetate reduces lymphocyte proliferation and enhances IL-5 and IL-13 production through modulation of monocyte-derived dendritic cells in multiple sclerosis. Clin Exp Immuno 143:357–362

Saresella M, Marventano I, Longhi R et al (2008) CD4+CD25+FoxP3+PD1 regulatory T cells in acute and stable relapsing-remitting multiple sclerosis and their modulation by therapy. FASEB J 22:3500–3508

Schellekens H, Klinger E, Mühlebach S et al (2011) The therapeutic equivalence of complex drugs. Regul Toxicol Pharmacol 59:176–183. doi:10.1016/yrtph.210.09.021

Schellekens H, Stegemann S, Weinstein V et al (2014) How to regulate nonbiological complex drugs (NBCD) and their follow-on versions: Points to consider. AAPS J 16(1):15–21. doi:10.1208/s12248-013-9533-z

Schori H, Kipnis J, Yoles E et al (2001) Vaccination for protection of retinal ganglion cells against death from glutamate cytotoxicity and ocular hypertension: implications for glaucoma. Proc Natl Acad Sci U S A 98:3398–3403

Schrempf W, Ziemssen T (2007) Glatiramer acetate: mechanisms of action in multiple sclerosis. Autoimmun Rev 6:469–475. doi:10.1016/j.autrev.2007.02.03

Schwartz M (2003) Neuroprotection as a treatment for glaucoma: pharmacological and immunological approaches. Eur J Ophthalmol 13(Suppl 3):S27–S31

Schwartz M, Kipnis J (2002) Autoimmunity on alert: naturally occurring regulatory CD4(+) CD25(+) T cells as part of the evolutionary compromise between a 'need' and a 'risk'. Trends Immunol 23:530–534

Sela M, Teitelbaum D (2001) Glatiramer acetate in the treatment of multiple sclerosis. Expert Opin Pharmacother 2:1149–1165

Stern JN, Keskin DB, Zhang H et al (2008) Amino acid copolymer-specific IL-10 secreting regulatory T cells that ameliorate autoimmune diseases in mice. Proc Natl Acad Sci U S A 105: 5172–5176

Teitelbaum D, Meshorer A, Hirshfeld T et al (1971) Suppression of experimental allergic encephalomyelitis by a synthetic polypeptide. Eur J Immunol 1:242–48

Teitelbaum D, Aharoni R, Arnon R et al (1988) Specific inhibition of the T-cell response to myelin basic protein by the synthetic copolymer Cop 1. Proc Natl Acad Sci U S A 85:9724–9728

Teitelbaum D, Brenner T, Abramsky O et al (2003) Antibodies to glatiramer acetate do not interfere with its biological functions and therapeutic efficacy. Mult Scler 9:592–599 B99

Tennakoon DK, Mehta RS, Ortega SB et al (2006) Therapeutic induction of regulatory, cytotoxic CD8+ T cells in multiple sclerosis. J Immunol 176:7119–7129

Towfic F, Funt JM, Fowler KD et al (2014) Comparing the biological impact of glatiramer acetate with the biological impact of a generic. PLoS ONE 9(1):e83757. doi:1130.1371/journal.pone.0083757

Ure DR, Rodriguez M (2002) Polyreactive antibodies to glatiramer acetate promote myelin repair in murine model of demyelinating disease. FASEB J 16:1260–1262 doi:10.1096/fj.01-1023fje:1260-1262

Varkony H, Weinstein V, Klinger E et al (2009) The glatiramoid class of immunomodulator drugs. Expert Opin Pharmacother 10(4):657–68

Venken K, Hellings N, Broekmans T et al (2008) Natural naïve CD4+CD25+CD127 low regulatory T cell (Treg) development and function are disturbed in multiple sclerosis patients: recovery of memory Treg homeostasis during disease progression. J Immunol 180:6411–6420
Vieira PL, Heystek HC, Wormmeester J et al (2003) Glatiramer acetate (copolymer-1, Copaxone) promotes Th2 cell development and increased IL-10 production through modulation of dendritic cells. J Immunol 170:4483–4488
Viglietta V, Baecher-Allan C, Weiner HL et al (2004) Loss of functional suppression by CD4+CD25+ regulatory T cells in patients with multiple sclerosis. J Exp Med 199:971–979
Weber MS, Hohlfeld R, Zamvil SS (2007a) Mechanism of action of glatiramer acetate in treatment of multiple sclerosis. Neurotherapeutics 4:647–653
Weber MS, Prod'homme T, Youssef S et al (2007b) Type II monocytes modulate T cell-mediated central nervous system autoimmune disease. Nat Med 13:935–943
Weder C, Baltariu GM, Wyler KA et al (2005) Clinical and immune responses correlate in glatiramer acetate therapy of multiple sclerosis. Eur J Neurol 12:869–878
Wolf SD, Dittel BN, Hardardottir F et al (1996) Experimental autoimmune encephalomyelitis induction in genetically B cell-deficient mice. J Exp Med 184:2271–2278
Wolinsky JS (2004) Glatiramer acetate for the treatment of multiple sclerosis. Expert Opin Pharmacother 5:875–891
Yanaba K, Bouaziz JD, Haas KM et al (2008) A regulatory B cell subset with a unique CD1dhiCD5+ phenotype controls T cell-dependent inflammatory responses. Immunity 28: 639–650
Zeskind B et al (2014, April 28) Gene expression studies comparing glatiramer acetate and proposed generics (P1.212). AAN Annual Meeting. http://www.neurology.org/content/82/10_Supplement/P1.212. Accessed 28 April 2015
Zhang B, Yamamura T, kKondo T et al (1997) Regulation of experimental autoimmune encephalomyelitis by natural killer (NK) cells. J Exp Med 186:1677–1687
Zhou X, Hovell CJ, Pawley S et al (2004) Expression of matrix metalloproteinase-2 and -14 persists during early resolution of experimental liver fibrosis and might contribute to fibrolysis. Liver Int 24:492–501
Ziemssen T, Schrempf W (2007) Glatiramer acetate: mechanisms of action in multiple sclerosis. Int Rev Neurobiol 79:537–570
Ziemssen T, Kumpfel T, Klinkert WE et al (2002) Glatiramer acetate-specific T-helper 1- and 2-type cell lines produce BDNF: implications for multiple sclerosis therapy. Brain-derived neurotrophic factor. Brain 125:2381–2391
Ziemssen T, Kumpfel T, Schneider H et al (2005) Secretion of brain-derived neurotrophic factor by glatiramer acetate-reactive T-helper cell lines: Implications for multiple sclerosis therapy. J Neurol Sci 233:109–112

Regulatory References

Citizen Petition (2009) Document ID: FDA-2009-P0555-0001. http://www.regulations.gov/.gov/#!documentDetail;D=FDA-2009-P-0555-0001. Accessed 14 Jan 2014
Citizens petition (2013) Docket No. FDA-2013-P-1641. http://www.regulations.gov/#!documentDetail;D=FDA-2013-P-1641-0001. Accessed 10 Aug 2014
Citizens petition (2014) Docket No. FDA 2014-P-0933. http://www.regulations.gov/#!documentDetail;D=FDA-2014-P-0933-0001. Accessed 10 Sept 2014
EMEA/CHMP/437/04 (2005) Guideline on similar biological medicinal products. http://www.ema.europa.eu/docs/en_GB/document_library/Scientific_guideline/2009/09/WC500003517.pdf. Accessed 28 April 2015

EMEA/CHMP/79769/2006 (2006) Reflection paper on nanotechnology-based medicinal products for human use. http://www.ema.europa.eu/docs/en_GB/document_library/Regulatory_and_procedural_guideline/2010/01/WC500069728.pdf. Accessed 10 Aug 2014

Krull I, Cohen S (2009) The complexity of glatiramer acetate and the limits of current multidimensional analytical methodologies in the attempt to characterize the product. Letter in reference to Citizen Petition FDA-2008-P-0529 to the Dockets Management Branch, Food and Drug Administration. 16 January 2009

ORANGE BOOK (2013) Drug products with therapeutic equivalence evaluations 34th Edition. Active ingredients and dosage forms with potential bioequivalence problems. http://www.fda.gov/ucm/groups/fdagov-public/@fdagov-drugs-gen/documents/document/ucm071436.pdf. Accessed 15 Sept 2014

Media References

Comi G, Filippi M, Wolinsky JS (2001) European/Canadian multicenter, double-blind, randomized, placebo-controlled study of the effects of glatiramer acetate on magnetic resonance imaging–measured disease activity and burden in patients with relapsing multiple sclerosis. European/Canadian Glatiramer Acetate Study Group Ann Neurol 49, 290–297

Gutierrez R (2013, Oct 12) IMSS replaces a drug with a generic drug and patients' health is declining. http://www.radiogrupo.com.mx/index.php/local-movil/34-principales-locales/12519-imss-les-cambio-medicamento-por-uno-generico-y-pacientes-empeoraron-su-salud. Accessed 29 April 2014

Muller AA (2013, December 11) IMSS gives a generic drug that affects multiple sclerosis patients. http://www.televisaregional.com/aguascalientes/noticias/IMSS-da-medicamento-generico-que-afecta-a-pacientes-con-esclerosis-235464281.html. Accessed 29 April 2014

Synthon announces successful outcome of the Phase III GATE study with its generic glatiramer acetate. http://www.synthon.com/Corporate/News/PressReleases/Synthon-announces-successful-outcome-of-the-PhaseIII-GATE-study-with-its-generic-glatiramer-acetate?sc_lang=en. Accessed 28 April 2015

Iron Carbohydrate Complexes: Characteristics and Regulatory Challenges

Stefan Mühlebach and Beat Flühmann

Contents

Iron Deficiency/Iron Deficiency Anemia and its Treatment 151
IV Iron Preparations and Manufacturing 151
Pharmaceutical Quality/Stability (In Vitro) 156
Pharmacology Including (Non-Clinical) Biodisposition and Toxicity 160
Clinical Pharmacokinetics, Efficacy and Safety of MPs 162
The Regulatory Landscape: New Drug Applications (NDA) and Generic Versions 164
Conclusion: Final Considerations and Reflections 166
References 167

Abstract Iron carbohydrate complexes for IV therapy consist of nanosize range particles with a polynuclear Fe(III)-oxyhydroxide core and a carbohydrate shell. They are pro-drugs. The iron complexes are stable on the shelf and are modified upon intravenous administration. The iron carbohydrate nanoparticles interact with cells of the innate immune system for uptake and release of iron into the physiological iron metabolic pathways: i.e. phagocytosis by cells of the RES, cleavage of the carbohydrate shell from the iron core which has to deliver the iron to physiological pools after release into and transport through the blood. They are non-biological complex drugs (NBCDs) i.e. showing polydispersity (non-homomolecular structures), cannot be fully characterized, and are highly dependent on a well-controlled manufacturing process.

Iron carbohydrate complexes are nanomedicines or nanocolloidals. Absorption, distribution, metabolism, and excretion (ADME) profiles have to be investigated and defined by appropriate in vivo test systems. The "nanoparticular" properties

S. Mühlebach (✉) · B. Flühmann
Department of Global Regulatory Affairs, Vifor Pharma Ltd., Glattbrugg, Switzerland
e-mail: stefan.muehlebach@viforpharma.com

S. Mühlebach
Department of Pharmaceutical Sciences, Division of Clinical Pharmacy & Epidemiology, University of Basel, Basel Switzerland

D. J. A. Crommelin, J. S. B. de Vlieger (eds.), *Non-Biological Complex Drugs*, AAPS Advances in the Pharmaceutical Sciences Series 20, DOI 10.1007/978-3-319-16241-6_5

are key to their specific disposition which also influences pharmacodynamics and therefore efficacy and safety.

Sensitive biomarkers to correlate fate, efficacy and safety of the products have to be defined and used to ensure therapeutic equivalence when comparing products, especially also follow-on versions developed with reference to an innovator product. These follow-on versions, called nanosimilars, have to undergo a stepwise similarity approach to establish their therapeutic equivalence.

Keywords IV iron carbohydrates · Nanocolloidals · NBCDs, polydispersity · Innate immune system · Biodisposition and in vivo performance · Stepwise similarity approach (for follow-on versions)

Abbreviations

ADME	Absorption, distribution, metabolism and excretion
AFM	Atomic force microscopy
BW	Body weight
EDQM	European directorate for the quality of medicines
HMWD	High molecular weight dextran
ICH	International Conference on Harmonisation of Technical Requirements for Registration of Pharmaceuticals for Human Use
IS	Iron sucrose
ISS	Iron sucrose similar
LMWD	Low molecular weight dextran
MP	Medicinal product
NBCD(s)	Non biological complex drug(s)
NDA	New drug application
NME	New molecular entity
NTBI	Non transferrin bound iron
PD	Pharmacodynamics
PET	Positon emission tomography
PK	Pharmacokinetics
pn	Polynuclear
RES	Reticulo endothelial system
RLD	Reference listed drug
RMP	Risk minimization plan
TE	Therapeutic equivalence
TEM	Transmission electron microscopy
Tf	Transferrin
XRD	X ray diffraction

Iron Deficiency/Iron Deficiency Anemia and its Treatment

According to the World Health Organization (WHO), more than 1 billion people show iron deficiency, the most common nutritional deficiency. About 700 million are anemic (WHO 2008). Iron-deficiency anemia is associated with reduced quality of life, decreased physical and cognitive performance, but also with adverse clinical outcome (Muñoz et al. 2014; Baker and Greer 2010). Iron deficiency anemia is a major finding for chronically ill patients and represents a major complicating factor for women of childbearing age and during pregnancy. Generally, a balanced composition of food including essential nutrients with the necessary bioavailable iron ensures a sufficient iron supply of about 1–2 mg iron to compensate for physiological losses, despite the relatively low absorption rate of dietary iron. The normal total body iron content in adults is between 3 and 4 g, of which 15–30% is bound in storage form (mostly in liver). A very efficient homeostasis of iron exists with only minimal loss from the body (Crichton et al. 2008). In the event of extensive iron loss as a result of bleeding or impaired iron utilization in connection with chronic disease and the accompanying inflammation, iron deficiency or even iron deficiency anemia can quickly develop and supplemental iron has to be provided (Mircescu et al. 2013). To treat iron deficiency three therapeutic options exist: oral iron, IV iron and, in selected cases, blood transfusion. In chronic disease, in particular, guidelines increasingly recommend parenteral administration of iron as the method of choice for anemia therapy (e.g. KDIGO 2012).

Iron deficiency, depending on the level of severity, can often lead to an iron deficit of more than 1000 mg iron, a quarter or more of the normal body stores. Oral and intravenous (IV) iron replacement products are available for the treatment of iron deficiency and iron deficiency anemia. However, oral administration is restricted due to the limited absorption of iron via the small intestine (a few mg iron per day) and gastrointestinal intolerance (Huch and Schäfer 2006; Schrier et al. 2013).

Correction of anemia with oral iron can therefore take weeks or even months before iron stores are replenished and requires strict therapy compliance of the patient. In contrast, large quantities of iron can be administered fast and safely, "bypassing" the intestine, using IV iron products (Auerbach and Ballard 2010; Krikorian et al. 2013; Muñoz et al. 2014).

IV Iron Preparations and Manufacturing

Fe(III) salts are not stable in an aqueous solution at $pH > 3$ and will precipitate by forming iron oxyhydroxides. Fe(III) complexes represent the most prominent physiological iron state for transport (transferrin-bound iron) and storage (ferritin-bound iron). Iron oxyhydroxides as formed from iron salts may differ considerably in their structure and (polymeric) size depending on the procedures used in their preparation. It is possible to prepare polynuclear iron(III) oxyhydroxides with any number of iron atoms from 6 to some 10,000. Around the physiological pH of

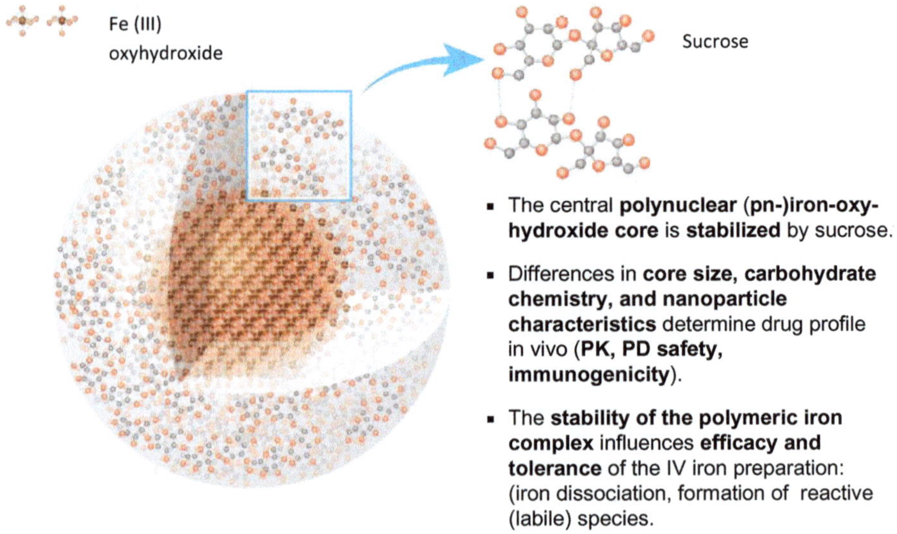

Fig. 1 Colloidal IV iron sucrose structure, indicative drawing

approximately 7, iron hydroxides are insoluble and generally precipitate from the solution. However, by means of surface interactions with organic molecules such as proteins in the body or different carbohydrates, polynuclear ($_P$) units with the composition $[Fe(OH)_3]_p$ or $[FeO(OH)]_p$ can be kept in solution as colloidal particles (Schneider 1988). Therefore, IV iron preparations are designed as colloidal dispersions consisting of non-homomolecular iron carbohydrate complexes made of polynuclear Fe(III)-oxyhydroxide cores surrounded by a carbohydrate ligand (Fig. 1), that stabilizes the complex and prevents the nanoparticles from further polynuclearization and precipitation (Auerbach and Ballard 2010). A wide variety of carbohydrates have been employed for this shell, among others, monosaccharide derivatives, disaccharides, oligosaccharides, and polysaccharides. The type of carbohydrate affects pharmacokinetics and immunogenicity (Wysowski et al. 2010). Drug products currently on the market in different countries include iron sucrose, ferric carboxymaltose, sodium ferric gluconate, low and high molecular weight iron dextran, ferumoxytol and iron isomaltoside 1000.

The choice of the carbohydrate influences the conformation of the polynuclear Fe(III)-oxyhydroxide complex core and the carbohydrate shell. The carbohydrate is important not only to stabilize the polymeric iron complex on the shelf but also for the fate of the whole product in the body (PK, immunogenicity, stability/reactivity in vivo and finally also the tolerance especially when escalating the dose). In Table 1 some characteristics and differences of the most widely used IV iron carbohydrate complexes are listed. In the past, so called iron sucrose similar (ISS) preparations, i.e. copies of the originator iron sucrose (Venofer®), have been introduced in different markets outside the US via a generic regulatory pathway underestimating the high complexity of the product and its manufacturing leading to nano-sized

Table 1 IV iron carbohydrate colloidal brands. (Adapted from Muñoz and Martin-Montanez 2012)

	Iron gluconate	Iron sucrose	HMWID	LMWID	Ferric carboxy-maltose	Iron isomaltoside 1000	Ferumoxytol
Brand name	Ferrlecit®	Venofer®	Dexferrum®	Cosmofer®	Ferinject® Injectafer® (US)	Monofer®	FeraHeme®
Manufacturer	Sanofi-Aventis	Vifor	Watson	Pharma-Cosmos	Vifor	Pharma-Cosmos	AMAG Pharmaceuticals
Carbohydrate shell	Gluconate (mono-saccha-ride)	Sucrose (di-saccharide)	Dextran (branched poly-saccharide)	Dextran (branched poly-saccharide)	Carboxy-maltose (branched poly-saccharide)	Isomaltoside (linear oligo-saccharide)	Polyglucose sor-bitol (branched poly-saccharide)
Molecular weight [kDa]	37.5	34–60	265	165	150	69	185
"Plasma half-life" [h]	1	6	60	30	7–12	20	15
Iron content [mg/mL]	12.5	20	50	50	50	100	30
Maximal single dose [mg]	125	300	20 mg/kg BW	20 mg/kg BW	20 mg/kg BW (max.1000)	20 mg/kg BW	510

HMWID high molecular weight iron dextran, *LMWID* low molecular weight iron dextran, *BW* body weight

MP. However, recent clinical studies indicated a lack of therapeutic and safety equivalence between the innovator's product and ISS (Lee et al. 2013; Martin-Malo et al. 2012; Rottembourg et al. 2011; Stein et al. 2012; Aguera et al. 2014). Also non-clinical investigations showed non-comparability between the reference and the follow-on products and in some cases that was also true with physicochemical (e.g. polarographic) comparisons (Fig. 2) used as a kind of fingerprint of the iron core complex (Toblli et al. 2012).

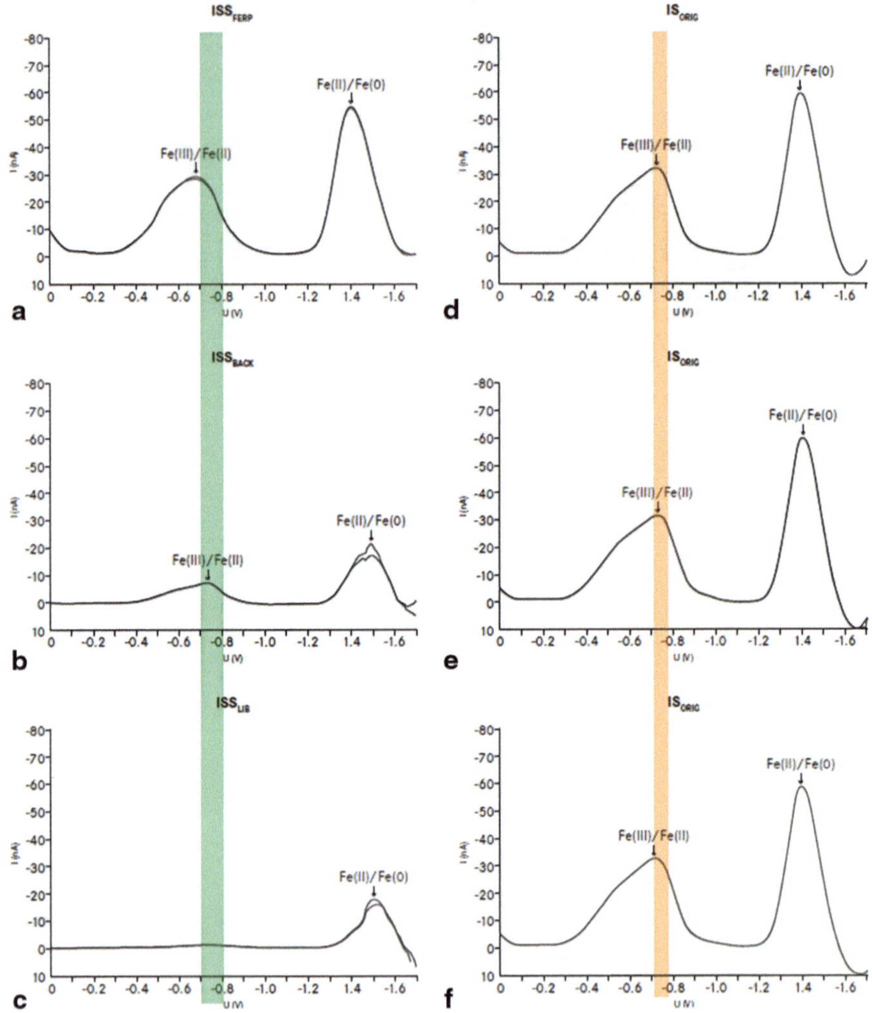

Fig. 2 Polarograms of different ISS compared to different lots of IS_{orig}. (Adapted from Toblli et al. 2012). Polarogram of a potential [V] vs. current [nA] plot. **a**–**c** are from different marketed ISS, **d**–**f** from three randomly chosen lots of IS_{orig}. The *green* and *red* bar mark the peak voltage ranges for the reduction potential Fe(III) → Fe(II) by USP (iron sucrose injection solution -0.750 ± 0.050 V) and the resulting one from the analyzed originator batches (-0.750 ± 0.025 V), respectively

Table 2 Sensitive parameters for IV iron carbohydrate manufacturing

Parameter	Details
Starting materials	Composition including impurities heavy metals, electrolytes
Specific conditions of the multistep synthesis	Concentration of reagents, pH, Temperature, Reaction time, Other reaction conditions
Purification processes (frequency, extent, materials)	Washing, Filtration
Terminal product manufacturing	Quality of container materials, Mixing/Dilution conditions, Final antimicrobial treatment

The manufacturing of such polymeric products goes through numerous steps. The critical details of these processes are often proprietary and not disclosed in the public domain. The manufacturing typically consists of a number of key steps starting with the controlled polymerization and precipitation of the polynuclear Fe(III)-oxyhydroxide (pn-FeOOH) cores upon reaction of a Fe(III)-salt with a base. Next a sophisticated purification step takes place and the precipitate is washed with water. By reaction of this polynuclear Fe(III)-oxyhydroxide with carbohydrate ligands under specific reaction conditions (e.g. temperature and pH) and reaction time a stabilized colloidal solution is obtained. Only rigorous control of starting materials such as the source of iron (e.g. potential heavy metal contamination) and the base, as well as detailed reaction conditions during all manufacturing steps (Table 2) by multiple in- and end-process controls can guarantee the production of a nano-medicinal product with an appropriate pharmaceutical quality, minimal batch to batch variation and with the desired in vivo performance. In a last step the colloidal solution is adjusted for concentration and pH and filled into the primary packaging materials such as ampoules or vials. This final product should be considered the medicinal product, since the specific conditions of the adjustments and filling process, including the characteristics of the chosen primary packaging materials, have a significant impact on the physicochemical properties of the final product. This is also one of the reasons why in pharmacopoeias product monographs and not an API monograph are established or under discussion for in vitro quality issues (British Pharmacopoeia 2015; EDQM 2013; USP 2014). As a consequence, not only IV iron formulations with different carbohydrates show important differences in their characteristics and in vivo performance, but subtle differences in the manufacturing process of potential follow-on versions can lead to clinically meaningful differences as observed in the case of ISS (Lee et al. 2013; Martin-Malo et al. 2012; Rottembourg et al. 2011; Stein et al. 2012; Aguera et al. 2014). Therefore, the exact composition of parenteral iron products is largely defined by the manufacturing process as it determines the in vivo performance (Fig. 3). This is extremely important as the full array of physicochemical techniques to characterize these iron dispersions is still not able to fully predict clinical performance, as discussed in the next section. As a consequence, the therapeutic equivalence of two iron-carbohydrate products cannot be established using the classical generic approach (Schellekens et al. 2011).

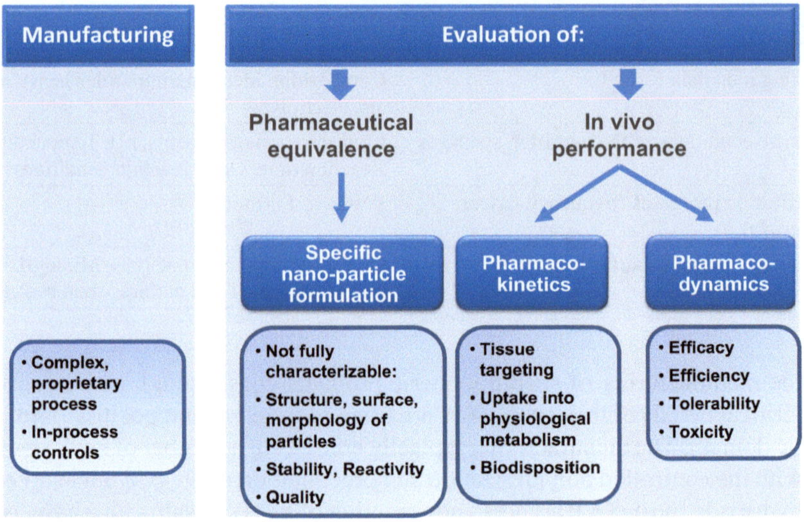

Fig. 3 NBCDs: Therapeutic equivalence from manufacturing to efficacy and safety. (Adapted from Mühlebach et al. 2013)

Pharmaceutical Quality/Stability (In Vitro)

Several parameters are listed in the specifications of the various parenteral iron products as well as the respective pharmacopeia monographs currently under revision. In 2012 the EDQM has set up a working party on non-biological complexes to further define the quality parameters for the characterization of this type of products and to evaluate additional methods suitable for the specification of the iron carbohydrate complex and the nanoparticle properties (Table 3). The challenge is to integrate reliable and robust methods to assess the (nano-)size distribution and also surface properties of such complex MP affecting relevant differences in the quality and impacting the in vivo performance. Today the question is to identify clinically meaningful quality attributes with a known and understood influence on disposition, safety or efficacy of the product. For instance, the fraction of labile iron in the IV iron-based colloidal pro-drug formulation may have an impact on the safety, and efficacy. But, there are no data that definitively associate this quality attribute to a clinical outcome. An example of the lack of correlation is the amount of labile iron measured in these IV iron products which strongly depends on the used analytical methods to quantify it (Balakrishnan et al. 2009; Jahn et al. 2007, 2011; Stefansson et al. 2011; Van Wyck 2004) which directly limits the comparability of published data.

Figure 2 gives an example where polarograms of an IS and ISS are shown. Differences in the starting materials or reaction conditions (e.g. pH, temperature, reaction times, clean-up procedures as mentioned before) have a significant impact on the physicochemical composition and properties of the final product (Toblli et al.

Table 3 Physicochemical tests on IV iron carbohydrates: the iron sucrose example. Adapted from an EDQM working document for the monograph elaboration

Current assays for specification		Assays under discussion	
Parameter	Method	Parameter	Method
Identification of Iron		Particle size and distribution e.g. D10/D50/D90 and span or Z average and PI	DLS
Identification of Sucrose			
Molecular Weight Determination	Gel Permeation Chromatography		
Reduction potentials	Polarography	labile iron' e.g. method as describrd in Jahn-paper	In vitro method
Fe(III)/Fe(II)			
Fe(II)/Fe(0)		Microscopic method	TEM, (AFM)
Turbidity point	Turbidimetric titration	Magnetic properties of the core	Mössbauer, NMR
Impurities		Chemical structure of iron core	XRD, XANES
Chloride	Titration	Degradation path	Degradation by macrophages
Osmolarity	Freezing point depression		
Particulate matter	Light obscuration measurement, Ph. Eur. 2.9.19 "Particulate contamination"		
Bacterial endotoxins	LAL Test, Ph. Eur. 2.6.14		
Sterility	Sterility test		
Ferrous iron	Polarography		
Absence of low molecular weight complexes	Polarography		
Assay			
Iron	Complexometric titration or Atomic Absoption Spectroscopy		
Sucrose	HPLC		

Molecular weight determination includes a polydispersity calculation

2012). The authors try to relate these differences to in vivo performance parameters of the products in animals and found considerable differences in oxidative stress and inflammatory reactions but again there was no correlation. Even a product almost in spec in physicochemical testing showed increased oxidative stress compared to the originator.

As for all nanomedicines, it can be assumed that the properties of the nanoparticle surface strongly impact biodisposition and hence the product's safety and efficacy profile (EMA 2011). When injected into the bloodstream, the iron carbohydrate complexes will be modified, dependent on the stability of the complex (Geisser and Burckhardt 2011). Upon injection, the different environment in the body versus storage conditions will change the size and surface properties of the complexes. This is not only important for the uptake by the monocytes/macrophages of the RES and the distribution processes. It also induces product dependent, direct release of iron from the injected complex (pro-drugs), impacting both the safety and efficacy of the MPs. Therefore, size and morphology attributes influence the biodistribution of iron nanoparticles and have to be assessed. Two techniques [i.e. atomic-force microscopy (AFM) and transmission electron microscopy (TEM)] can conceivably be applied. However, the results of these analyses strongly depend on the sample preparation. AFM is qualitative in nature and is difficult to standardize and validate (Kudasheva et al. 2004). TEM may be better suited to determine particle morphology, but it does not offer any advantage in terms of standardization and validation of the method. The authorities acknowledge in their guidance paper that the suitability (accuracy and precision) of the methods used for morphology and particle diameter assessment have to be demonstrated (FDA 2013a).

In addition to the size, the polymorphic form of the polynuclear iron core is an important parameter for the characterization of nanoparticles. However, this morphological measurement can also be challenging as it depends on the size and degree of crystallinity of the compounds. Data in the literature show that certain tests such as X-ray Diffraction (XRD) yield conflicting and, thus, inaccurate results. Indeed, the core structure of iron sucrose was assigned as pure ferrihydrite mixed with possibly other structures such as lepidocrocite, or akaganeite or as pure akaganeite (Funk et al. 2001; Fütterer et al. 2013; Jahn et al. 2011; Kudasheva et al. 2004).

The difficulties in determining some of the more specific nano-related parameters also led to disagreements in the scientific literature e.g. for the determination of the particle size of iron sucrose. While Kudasheva et al. (2004) report a particle size of 22 nm for the iron sucrose originator product, others have suggested that the particle size is 8 nm (Jahn et al. 2011). As the determination of the particle size depends on the method and conventions applied, a standardized method for sample preparation and determination of the particle size has to be established and validated.

The majority of the parameters mentioned as important for the characterization of test products are not well defined and many of the analytical methods available to determine the physicochemical parameters requested by authorities have not been validated (Schellekens et al. 2014). Very little regulatory guidance is provided as to how the MPs should be properly physico-chemically characterized.

Therefore, it may be difficult to draw conclusions about the clinical impact of identified physicochemical differences (Stefansson et al. 2011; Bailie et al. 2013a, b). Assays like "kinetic studies of Fe(III) reduction by acid degradation" (Erni et al. 1984) are performed in vitro under conditions significantly different from the physiological situation within the human body and are therefore difficult or impossible to interpret for relevancy. Thus again, it will not be possible to predict the impact of any physicochemical parameter difference obtained with these tests on the clinical safety and efficacy of the iron-carbohydrate product.

In contrast, the absence of a certain physicochemical difference does not exclude non-clinical or clinical meaningful differences (Toblli et al. 2012; Lee et al. 2013). This leads to the conclusion that for the evaluation of product quality and stability of these colloidal dispersions of iron carbohydrates, a state of the art extensive, sensitive and –most important- validated physicochemical quality assessment program has to be developed to correlate at best structure to performance (EMA CHMP/SWP 2013; Ehmann et al. 2013; FDA 2013a). Such tests would also be helpful to control the manufacturing process –batch consistency -, and to exclude nanosimilar products with analysis outcomes which fall outside the predetermined specifications from further (expensive) (non)-clinical evaluation.

The difficulties and limitations linked to these physicochemical tests to ensure the pharmaceutical quality, highlight the need to convincingly demonstrate that the applied methods and obtained data are robust, sensitive and indicative for relevant characteristics of the MPs.

To conclude on the issue on comparability of nanosimilar products: it is necessary to continue after the successful physicochemical assessment in a stepwise approach with the next levels of regulatory evaluation, namely the non-clinical and clinical assessment (Fig. 4), to finally end up with a totality of evidence for similarity (Ehmann 2013).

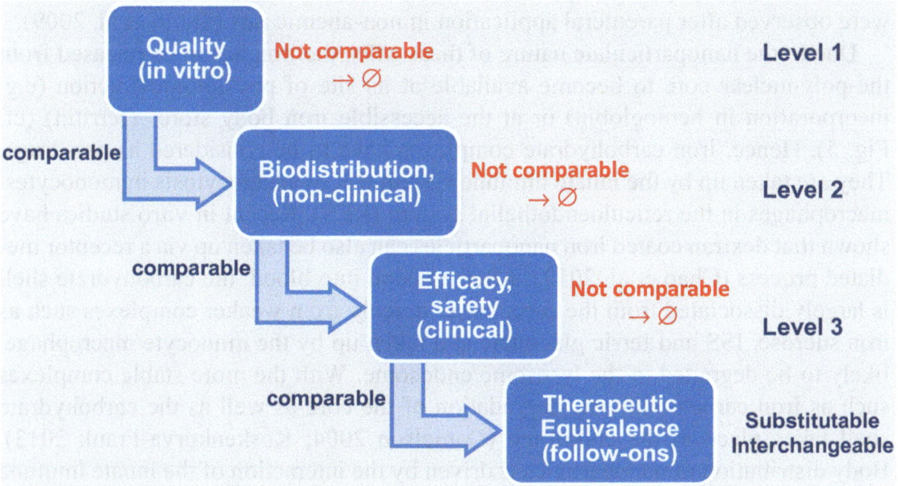

Fig. 4 The three level approach for NBCD follow-ons (algorithm, decision tree)

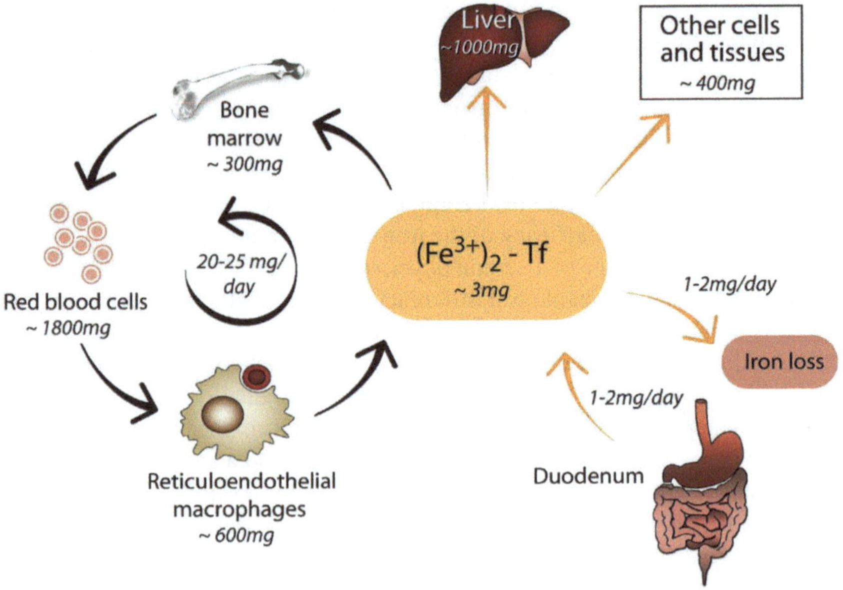

Fig. 5 Iron homeostasis. (From Hentze et al. 2004)

Pharmacology Including (Non-Clinical) Biodisposition and Toxicity

As concluded above, physicochemical tests alone are not sufficient to ensure full comparability between a test and a reference product. An in vivo performance assessment between a test and reference medicinal product has to be added. In several non-clinical studies product specific differences in the biodisposition of the IV iron were observed after parenteral application in non-anemic rats (Toblli et al. 2009).

Due to the nanoparticulate nature of these MPs, the iron has to be released from the polynuclear core to become available at its site of physiological action (e.g. incorporation in hemoglobin) or at the accessible iron body stores (ferritin) (cf. Fig. 5). Hence, iron carbohydrate complexes have to be considered as pro-drugs. They are taken up by the innate immune system e.g. by phagocytosis in monocytes/macrophages in the reticuloendothelial system (RES). Recent in vitro studies have shown that dextran coated iron nanoparticles can also be taken up via a receptor mediated process (Chao et al. 2012). After infusion into blood, the carbohydrate shell is largely dissociated from the core, rather quickly from weaker complexes such as iron sucrose, ISS and ferric gluconate and taken up by the monocyte/macrophage, likely to be degraded in the lysosome/endosome. With the more stable complexes such as iron carboxymaltose, degradation of the core as well as the carbohydrate shell takes place in the endosome (Danielson 2004; Koskenkorva-Frank 2013). Body distribution of nanoparticles is driven by the interaction of the innate immune system with the particle surface. Alterations at the particle surface may thus lead to inefficient organ disposition resulting in a lack of efficacy and increased toxicity (Nyström and Fadeel 2012). Therefore, animal data from a well-defined model are expected to give the necessary information on the biodistribution of colloidal IV iron carbohydrate complexes, which is important for a comparability evaluation (EMA CHMP/SWP 2013) (Fig. 4). In addition, the fate of the MP also needs to be assessed on a cellular level like the uptake by the liver Kupffer cells and by the hepatocytes. These cells may influence the availability of iron from these iron stores and thereby toxicity. The challenge is to develop an appropriate and robust animal model with in-depth characterization of the species and defined sensitivity and reproducibility. From the considerable amount of pre-clinical studies carried out with the iron sucrose originator and its similars, it is clear that these models are not only difficult to standardize but subsequently also difficult to compare regarding their results (Meier et al. 2011; Toblli et al. 2009, 2011). Various non-clinical studies by Toblli et al. demonstrated that, compared to the iron sucrose originator, increased oxidative stress levels were induced by iron sucrose similar (ISS) products. This was even true for an ISS that met the United States Pharmacopeia's physicochemical reference values for iron sucrose injection (Toblli et al. 2012) (Table 3 and Fig. 2a polarography).

Physicochemical properties of the iron carbohydrate complexes, such as molecular weight distribution, particle size, thermodynamic and kinetic stability of the iron complex in blood also define the amount of labile iron initially released from the core. This released iron needs to be immediately bound to transferrin to protect sensitive structures like membranes, proteins or even DNA from interaction with labile iron. Non-transferrin bound iron (NTBI) includes per definition also carbohydrate bound iron. In case of a stable and robust complex the release of iron is slow and as a consequence the acute toxicity is low. This slow release leads to a longer terminal elimination phase/lower clearance rate of the complex and the corresponding half-life will be long (Table 1). The iron released from the lysosomal pool (degraded iron (carbohydrate) complexes after RES uptake) may then also be bound to the still existing, not fully saturated transferrin and yield a high transferrin saturation. In the presence of a weak iron carbohydrate complex the released iron reacts with the transferrin and could lead to full transferrin saturation dependent on the IV iron carbohydrate dose. The released, free iron could also react with other plasma components inhibiting their transportation to the right compartments in the tissues. Such reactive NTBI species (Brissot et al. 2012) would induce oxidative stress as reported in non-clinical and clinical studies (Toblli et al. 2009, 2011; Martin-Malo et al. 2012). Importantly, non-clinical studies have shown that the labile iron is not directly related to, or not the only cause of oxidative stress caused by IV iron preparations. The nature of the complex and/or the core plays an important role as well (Bailie et al. 2013a, b; Koskenkorva-Frank et al. 2013) and other unknown mechanisms may be relevant. In addition the (chronic) disease causing the iron deficit might also modify the inflammatory response to oxidative stress, e.g. in end stage chronic kidney disease (Beguin and Jaspers 2014; Mircescu et al. 2013). To ensure appropriate uptake of the iron complex by macrophages of the RES, the maximal therapeutic dose and the time needed for administration is mainly limited by the complex reactivity and the immediate iron release. The physiological mechanisms and the physicochemical parameters relevant for the uptake of the iron nanoparticles into the RES are still not fully identified and also present a scientific challenge to define appropriate regulatory evaluation (Schellekens et al. 2011, 2014).

These non-clinical studies demonstrate that similar physicochemical properties do not necessarily ensure similar biodistribution and ultimately, efficacy and toxicological effects. But to extrapolate these findings in animals to the treatment of patients, efficacy and safety and eventually therapeutic equivalence between products has to be assessed in clinical studies: the third level (Fig. 4).

Clinical Pharmacokinetics, Efficacy and Safety of MPs

A pharmacokinetics study in healthy volunteers similar to that conducted by Danielson et al. (1996) for Venofer® using a Michaelis-Menten compartmental model may be a reasonable study design as a starting point for assessing the pharmacokinetic characteristics of colloidal iron products (e.g. an ISS compared to the iron sucrose originator). It can however, not replace an appropriate therapeutic equivalence study in patients because the underlying chronic disease may influence iron metabolism. Therefore, a clinical assessment in a sensitive patient population has to be demanded for obtaining marketing authorization of an iron-carbohydrate new molecular entity (NME) or follow-on nanosimilars (Tinkle et al. 2014).

To establish therapeutic equivalence, the last and most important level of evaluation of comparability has to be reached (Fig. 4). This implies for these iron-based colloidal products, that both their pharmacokinetic profile (PK) as well as the performance in safety/efficacy tests (PD) has to be evaluated. This requires that clinical trials with patients must be performed to demonstrate similarity and exclude any clinically meaningful differences that may not show up in physicochemical and non-clinical analyses due to the nanoparticular NBCD characteristics (Borchard et al. 2012; FDA 2013a, 2014; Vifor Pharma Ltd. FDA comment 2014).

A classical bioequivalence assessment for such products is not possible although initially the whole dose of the IV administered iron carbohydrate pro-drug is injected into the blood. This is due to the nano-colloidal properties, the lack of knowledge of the biodistribution and of the relevant central compartment controlling the overall pharmacokinetics (Desai 2012) but also because of the reactivity of the iron prodrug complex affecting the clearance (Geisser and Burckhardt 2011). The amount of iron in the serum is only a small portion of the body iron. It represents the iron that is transferred to the site of action or storage. This total serum iron concentration is not directly related to a given iron dose nor does it allow to calculate an AUC for bioavailability as it only indicates the result of the rates of iron transfer from the uptake pool into (invasion) and from the serum (evasion) into the storage and/or erythropoiesis pool (Fig. 5). Therefore, the direct link between PK and efficacy and safety (PD) of these products doesn't exist and a combined assessment of PK and

PD (biomarkers) is needed, but not yet defined (FDA 2013). As an example, transferrin saturation, a highly regulated biomarker to indicate metabolically available iron (indicates the saturation of the available iron transport capacity in the blood), is often used in clinical practice. However, it is known to be influenced by diurnal variation as well as the dynamic processes of iron metabolism and utilization. The latter is also modified by the patient's underlying disease state, iron status, use of erythropoiesis-stimulating agents, and hemoglobin level. In addition, the measurement of transferrin saturation is influenced by the iron dosing regimen and the time of measurement in relation to the iron dose given (Wish 2013). This also means that other biomarkers have to be defined and related to the disposition and a comparative "bioequivalence" at the site of actions of the iron carbohydrate MP. These multiple factors modulating uptake, distribution, storage, incorporation into hemoglobin and other body processes including re-utilization from physiological breakdown products (Andrews 2008; Hentze et al. 2004; Mircescu et al. 2013) (Fig. 5) request comparative trials in defined patients to finally conclude on therapeutic equivalence and substitutability or interchangeability of a follow-on nanosimilar to a RLD. With the originator's iron sucrose and ferric carboxymaltose products uptake of iron into target organs such as liver and bone marrow and finally red blood cells, and iron utilization over time was demonstrated with ferrokinetics studies using positron emission tomography (PET) after injection of radioactively labeled (^{59}Fe) products (Beshara et al. 1999, 2003).

Evidence of clinically relevant efficacy and safety differences between similar colloidal IV iron sucrose products authorized has been shown by several investigators in various, defined *patient* populations when switching from iron sucrose originator to an ISS (Lee et al. 2013; Rottembourg et al. 2011; Stein et al. 2012; Aguera et al. 2014). In addition, another clinical study (Martin-Malo et al. 2012) also clearly demonstrated significant differences in oxidative stress, cell activation and apoptosis between an iron sucrose originator product and an ISS. Despite the small patient group size the oxidative stress parameters in the monocytes cultivated from the patients were significantly different for the originator IS vs. the ISS, but not between different originator iron carbohydrate MPs investigated. These findings indicate how important the proprietary well-controlled manufacturing processes are for the in vivo performance of the NBCDs (Table 2).

From the clinical and non-clinical evidence mentioned it became obvious that for comparability testing of the safety and efficacy, i.e. assessing therapeutic equivalence, of colloidal iron follow-on products one needs to include patients on chronic treatment who have reached an optimal steady state in their disease condition, as seen e.g. in hemodialysis patient who have established a tight control over the hemoglobinemia. Clinical assessment of comparability of colloidal iron products is wrought with variability and hence often difficult to interpret. For example, the non-linear kinetics of the ADME processes of nanomedicines are not well understood. Therefore, a potential dose dependence of the pharmacokinetic profile may be observed and should therefore be taken into consideration when comparing such products (FDA 2013a). Only the totality of evidence in a stepwise approach allows to assess comparability and the extent of therapeutic equivalence (Fig. 4).

The Regulatory Landscape: New Drug Applications (NDA) and Generic Versions

NDA/NME: a number of different iron-carbohydrate formulations have been approved in the USA and in different countries in Europe. Considering the complex and also nanoparticular nature with the pharmaceutical, pharmacological and toxicological intricacies inherent to this family of NBCD products, these NME should follow the centralized procedure (cf. chapter Pita.) and undergo a drug development program according to the ICH (international Conference on Harmonization of Technical Requirements for Registration of Pharmaceuticals for Human Use) Common Technical Document.

Generic versions: Iron carbohydrates for parenteral use have been regulated within the conventional regulatory framework (Schellekens et al. 2011; EMA 2011). Therefore, a number of iron sucrose follow-on products have been approved through the generic pathway and treated as low molecular weight 'simple' drug molecules. The complexity of the products and the difficulty to establish similarity were not recognized (Schellekens et al. 2014). Clinical findings demonstrating differences in safety and efficacy of these follow-on products led to scientific discussions in expert groups as is reflected in numerous, peer-reviewed publications in well-respected journals discussing proposals on how to resolve the problem (Borchard et al. 2012; Ehmann et al. 2013; Mühlebach et al. 2013; Nicholas 2014; Schellekens et al. 2011, 2014; Shah et al. 2013; Walson et al. 2014). FDA and EMA have been contributing to this scientific exchange discussing the arisen regulatory challenges by issuing draft guidance and reflection papers (EMA CHMP/SWP 2013; FDA 2013a).

Awareness has been created by the evidence of non-equivalence of such follow-on versions (Stein et al. 2012; Toblli et al. 2009, 2012; Martin-Malo 2012). Given the complexity of this class of nanomedicinal products, specific requirements should be addressed. The reflection paper by EMA (EMA CHMP/SWP 2013) indicates a three step approach, moving from step (I) evaluating in vitro pharmaceutical quality/stability (2.1.1) to step II) assessing non-clinical biodisposition (2.1.2) to the final step III: determining clinical pharmacokinetics, efficacy and safety of the MPs (2.1.3) (Fig. 4).

Non-clinical data should demonstrate similar tissue targeting as well as biodistribution in the expected cellular compartments for the products in a standardized animal model with potentially competing compartments as proposed by the EMA in its reflection paper (EMA 2011) (Table 4). In contrast to invasive techniques the use of (non-invasive) imaging techniques like MRI can be an alternative method for appropriate paramagnetic IV iron carbohydrates to trace the fate of such compounds in animals at an organ level, whereas cellular distribution of the iron needs to be assessed by histological studies. The methods have to be validated in the relevant non-clinical settings. In addition, the specific "paramagnetic" properties of the different iron products are key and must be specified in order to obtain the necessary sensitivity that is required to define a time-dependent distribution of the iron, including the

Table 4 Relevant compartments for the distribution of parenteral iron nanoparticles

Plasma
RES: macrophages e.g. in spleen, lymph nodes, liver (Kupffer cells)
Target tissues and not-target tissues
Pharmacological target tissues e.g. bone marrow
Toxicological target tissues e.g. kidney, liver (hepatocytes), lung, heart

degradation profile. Therefore, there is a need for basic validation studies to define sensitivity, variability, and detection limits of such methods.

This insight into the comparability of such colloidal iron products at the non-clinical biodistribution level (Fig. 4, second level) obtained with a well-defined biological model suitable for this assessment, adds to the necessary totality of evidence for the comparability of a product developed with reference to authorized iron colloidals. Then this non-clinical assessment also allows to compare the potential of these products to induce oxidative stress and inflammation as shown by Toblli, Martin-Malo, and coworkers (Toblli et al. 2012; Martin-Malo et al. 2012), and to compare properties indicative of different tissue/cell targeting and biodistribution with an impact on safety/toxicity. Demonstrating the physicochemical and non-clinical similarity between the products will then allow to check safety and efficacy in a clinical comparability study in humans. It is therefore important to realize, that the above mentioned non-clinical studies can potentially identify meaningful differences between such products, but are not sufficient on their own to prove similarity of the colloidal iron products. They will exclude those MP which do not reach this level of comparability as can be concluded from Toblli et al. (2009, 2012) (Section "Pharmacology Including (Non-Clinical) Biodisposition and Toxicity").

It has to be stressed that non-clinical assessments, although contributing to the comparability of MPs, do not allow a direct extrapolation into humans, as e.g. metabolism of iron in humans differs substantially from iron metabolism in animals. Moreover, the performed tests in animal models are generally short- to mid-term. Patients with a chronic disease condition may be exposed for a much longer period of time to such MPs (Borchard et al. 2012).

Although the FDA IV iron guidance (FDA 2013a, b) does not address efficacy and safety, but mainly the pharmacokinetics (bioequivalence), PK and PD can hardly be dissected and a combined assessment of drug concentration over time together with markers for PD has to be used for such complex MP and their complex kinetics. As mentioned, plasma concentrations of iron are not indicative for the biodistribution profile of IV iron products as they interact with the highly controlled physiological or pathophysiological iron homeostasis (EMA 2011; Desai 2013). Clinical evaluation requires a study showing the effective delivery of iron from the MP (pro-drug) to its site of action, e.g. the hemoglobin (Desai 2013). Only a sufficiently powered head-to-head clinical investigation in an appropriate patient population will provide the necessary data to fully evaluate the properties and characteristics of the products and to assess therapeutic equivalence.

This final level (step III, Fig. 4) of enough similarity is needed to conclude on substitutability or interchangeability of the compared products (Schellekens et al. 2014) by demonstrating therapeutic equivalence. It is interesting to note that the FDA has recently scheduled a re-evaluation of an already authorized iron gluconate follow-on product to assess the validity of their authorization data requirements and tools (FDA 2013b).

A clinical assessment of therapeutic equivalence and substitutability of the follow-on colloidal iron product in a comparative clinical trial is recommended by Schellekens (Schellekens et al. 2011). A good guidance for the design of the study and potential waivers can be found in the recent Drafts on Guidance for Industry "Scientific Considerations in Demonstrating Biosimilarity to a Reference Product" issued by the FDA (2012, 2014). A clinical trial for comparison is recommended, considering to use different dosing regimens (amount and timing) to show therapeutic and safety equivalence in relevant patient groups.

Conclusion: Final Considerations and Reflections

The advent of nanoparticular MP may revitalize pharmaceutical R&D and provide solutions to so far unmet clinical needs by targeting drugs towards identified tissues and cells important to cure or prevent a disease. This benefit could also be brought about by inducing less side effects and toxicities often limiting the use of medications. Nanomedicines are complex drugs where the manufacturing process defines the product characteristics relevant for the in vivo performance. The control over product characteristics is difficult and the resulting products contain typically non-homomolecular structures. These products, either of biological or non-biological origin, cannot be fully characterized by physicochemical tests. Small but unknown differences between similar products may impact the clinical performance. Therefore a 'similarity' approach (cf. EU and FDA legislation) for comparability with a follow-on version is recommended (cf. Fig. 1 in the Introduction chapter).

IV iron carbohydrate formulations represent a group of nanoparticular NBCDs that have been successfully used since decades (Beguin and Jaspers 2014). Some follow-on versions were authorized using the national route in the EU without awareness of the complexity of these type of products. New evidence and learnings regarding the evaluation of physicochemical, non-clinical and clinical data in follow-on MP protocols may assist to further define and harmonize science based regulatory evaluation approaches mainly for follow-on versions. This will allow a case-by-case approach in a defined class of products within the nanomedicines group and NBCDs. The idea to authorize nanosimilars by a stepwise approach through quality, non-clinical and clinical assessment may direct the creation of these guidelines. The biosimilar approach, although not set up for NBCDs, might also assist to avoid re-inventing the wheel.

To get the necessary information on the in vivo performance of complex products in 'real patients', an appropriate post-marketing surveillance (labeling, tracking, identification) and a corresponding risk management plan are key to identify potential differences in safety and efficacy for the MPs in 'real life situations' and to ultimately provide high quality and lower priced follow-on MP to patients. Because one cannot establish full pharmaceutical equivalence nor bioequivalence, the pharmacovigilance program might be of even greater importance for such similar follow-on versions compared to earlier programs for generic versions of small molecule drug products. Here again, international harmonization of the protocols for post-marketing surveillance is highly desirable (Bailie 2012).

References

Aguera ML, Martin-Malo A, Álvares de Lara MA et al (2014) The treatment with generic iv iron needs a higher dose of iron and ESA to keep hemoglobin stable. Nefrologia 34(S1):abstr 319

Andrews NC (2008) Forging a field: the golden age of iron biology. Blood 112:219–230

Auerbach M, Ballard H (2010) Clinical use of intravenous iron: administration, efficacy, and safety. Hematology Am Soc Hematol Educ Program 2010:338–347

Bailie GR (2012) Comparison of rates of reported adverse events associated with IV iron products in the United States. Am J Health Syst Pharm 69(4):310–320

Bailie GR, Schuler C, Leggett RE, Levin R (2013a) Oxidative effect ferumoxytol and iron dextran on urinary bladder contraction and impact of antioxidants. Free Radic Antioxid 3(1):7–10

Bailie GR, Schuler C, Leggett RE, Li H, Li HD, Patadia H, Levin R (2013b) Oxidative effect of several intravenous iron complexes in the rat. Biometals 26(3):473–478

Balakrishnan VS, Rao M, Kausz AT, Brenner L, Pereira BJ, Frigo TB, Lewis JM (2009) Physicochemical properties of ferumoxytol, a new intravenous iron preparation. Eur J Clin Invest 39(6):489–496

Baker RD, Greer FR (2010) Diagnosis and prevention of iron deficiency and iron-deficiency anemia in infants and young children (0–3 years of age). Pediatrics 126(5):1040–1050

Beguin Y, Jaspers A (2014) Iron sucrose—characteristics, efficacy and regulatory aspects of an established treatment of iron deficiency anemia in a broad range of therapeutic areas. Expert Opin Pharmacother 15(14):2087–2103

Beshara S, Lundqvist H, Sundin J, Lubberink M, Tolmachev V, Valind S, Antoni G, Langström B, Danilson BG (1999) Pharmacokinetics and red cell utilization of iron(III)-hydroxide-sucrose complex in anaemic patients—a study using positron emission tomography. Br J Haematol 104:296–302

Beshara S, Sorensen J, Lubberink M et al (2003) Pharmakokinetics and red cell utilization of 52Fe/59Fe-labelled iron polymaltose in anemic patients using positron emission tomography. Br J Haematol 120:853–859

Borchard G, Flühmann B, Mühlebach S (2012) Nanoparticle iron medicinal products–requirements for approval of intended copies of non-biological complex drugs (NBCD) and the importance of clinical comparative studies. Regul Toxicol Pharmacol 64:324–328

Brissot P, Roperta M, Le Lana C, Loréala O (2012) Non-transferrin bound iron: a key role in iron overload and iron toxicity. Biochim Biophys Acta (BBA)—Gen Subj 1820(3):403–410

British Pharmacopoeia (2015) Monograph Iron Sucrose Injection. http://www.pharmacopoeia.co.uk/2015/access.htm Accessed Nov 2014

Chao Y, Karmali P, Simberg D (2012) Role of carbohydrate receptors in the macrophage uptake of dextran-coated iron oxide nanoparticles. Adv Exp Med Biol 733:115–123

Crichton RR, Danieslon BG, Geisser P (2008) Iron therapy with special emphasis on intravenous administration, 4th edn. Uni-Med, International Medical Publishers, London

Danielson B (2004) Structure, chemistry, and pharmacokinetics of intravenous iron agents. Am J Soc Nephrol 15:S93–S98

Danielson BG, Salmonson T, Derendorf H, Geisser P (1996) Pharmacokinetics of iron(III)-hydroxide sucrose complex after a single intravenous dose in healthy volunteers. Arzneimittelforschung 46(6):615–621

Desai N (2012) Challenges in development of nanoparticle-based therapeutics. AAPS J 14(2):282–295

Ehmann F, Sakai-Kato K, Duncan R et al (2013) Next generation nanomedicines and nanosimilars: EU regulators' initiatives relating to the development and evaluation of nanomedicines. Nanomedicine 8(5):849–856

EMA (2011) Reflection paper on non-clinical studies for generic nanoparticle iron medicinal product applications. http://www.ema.europa.eu/ema/pages/includes/document/open_document.jsp?webContentId=WC500105048. Accessed July 2014

EMA CHMP/SWP (2013) Draft reflection paper on the data requirements for intravenous iron-based nano-colloidal products developed with reference to an innovator medicinal product. EMA/CHMP/SWP/620008/2012, July 25, 2013. London, UK, European Medicines Agency. http://www.ema.europa.eu/ema/doc_index.jsp?curl=pages/includes/document/document_detail.jsp?webContentId=WC500149496&murl=menus/document_library/document_library.jsp&mid=0b01ac058009a3dc. Accessed July 2014

Erni I, Oswald N, Rich HW, Schneider W (1984) Chemical characterization of iron (III)-hydroxide-dextrin complexes. A comparative study of commercial preparations with alleged reproductions used in the examination of bioavailability. Arzneimittelforschung 34(11):1555–1559

European Directorate for the Quality of Medicines & HealthCare. The European Pharmacopoeia in brief. EDQM brochure (2013) p 10. http://www.edqm.eu/en/Search-EDQM-website-519.html?bStat=1&sChaineRecherchee=non-biological+complexes. Accessed Nov 2014

FDA (2012) Quality Considerations in Demonstrating Biosimilarity to a Reference Protein Product. http://www.fda.gov/downloads/drugs/guidancecomplianceregulatoryinformation/guidances/ucm397017.pdf. Accessed Sept 2014

FDA (2013a) Draft guidance on iron sucrose. http://www.fda.gov/downloads/Drugs/GuidanceComplianceRegulatoryInformation/Guidances/UCM297630.pdf. Accessed May 2014

FDA (2013b) Solicitation on the therapeutic equivalence of generic iron complex product. https://www.fbo.gov/index?s=opportunity&mode=form&id=592788989854da145c8e7b6d103c898d&tab=core&tabmode=list&. Accessed July 2014

FDA (2014) Guidance for Industry. Clinical Pharmacology Data to Support a Demonstration of Biosimilarity to a Reference Product

Funk F, Long GJ, Hautot D, Büchi R, Christl I, Weidler PG (2001) Physical and chemical characterization of therapeutic iron containing materials: a study of several superparamagnetic drug formulations with the β-FeOOH or ferrihydrite structure. Hyperfine Interact 136(1–2):73–95

Fütterer S, Andrusenko I, Kolb U, Hofmeister W, Langguth P (2013) Structural characterization of iron oxide/hydroxide nanoparticles in nine different parenteral drugs for the treatment of iron deficiency anaemia by electron diffraction (ED) and X-ray powder diffraction (XRPD). J Pharm Biomed Anal 86:151–160

Geisser P, Burckhardt S (2011) The pharmacokinetics and pharmacodynamics of iron preparations. Pharmaceutics 3(1):12–33

Hentze MW, Muckenthaler MU, Andrews NC (2004) Balancing acts: molecular control of mammalian iron metabolism. Cell 117(3):285–297

Huch R, Schäfer R (2006) Iron deficiency and iron deficiency anemia. Georg Thieme Medical, Stuttgart

Jahn MR, Mrestani Y, Langguth P, Neubert RH (2007) CE characterization of potential toxic labile iron in colloidal parenteral iron formulations using off-capillary and on-capillary complexation with EDTA. Electrophoresis 28(14):2424–2429

Jahn MR, Andreasen HB, Fütterer S, Nawroth T, Schünemann V, Kolb U, Hofmeister W, Muñoz M, Bock K, Meldal M, Langguth P (2011) A comparative study of the physicochemical properties of iron isomaltoside 1000 (Monofer), a new intravenous iron preparation and its clinical implications. Eur J Pharm Biopharm 78(3):480–491

KDIGO (2012) KDIGO clinical practice guideline for anemia in chronic kidney disease. Chapter 2: Use of iron to treat anemia in CKD. Kidney Int Suppl 2(4):292–298

Koskenkorva-Frank TS, Weiss G, Koppenol WH, Burckhardt S (2013) The complex interplay of iron metabolism, reactive oxygen species, and reactive nitrogen species: insights into the potential of various iron therapies to induce oxidative and nitrosative stress. Free Radic Biol Med 65:1174–1194

Krikorian S, Shafai G, Shamim K (2013) Managing iron deficiency anemia of CKD with IV iron. http://www.uspharmacist.com/content/d/featured%20articles/c/42386/. Accessed July 2014

Kudasheva DS, Lai J, Ulman A, Cowman MK (2004) Structure of carbohydrate-bound polynuclear iron oxyhydroxide nanoparticles in parenteral formulations. J Inorg Biochem. 98(11):1757–1769

Lee ES, Park BR, Kim JS, Choi GY, Lee JJ, Lee IS (2013) Comparison of adverse event profile of intravenous iron sucrose and iron sucrose similar in postpartum and gynecologic operative patients. Curr Med Res Opin 29(2):141–147

Martin-Malo A, Merino A, Carracedo J et al (2012) Effects of intravenous iron on mononuclear cells during the haemodialysis session. Nephrol Dial Transplant 27(6):2465–2471

Meier T, Schropp P, Pater C, Leoni AL, Khov-Tran VV, Elford P (2011) Physicochemical and toxicological characterization of a new generic iron sucrose preparation Arzneimittelforschung 61(2):112–119

Mircescu G, Critchon RR, Geisser P (2013) Iron therapy in renal anaemia, 1st edn. Uni-Med, International Medical Publishers, London

Mühlebach S, Vulto A, de Vlieger JSB, Weinstein V, Flühmann B, Shah VP (2013) The authorization of non-biological complex drugs (NBCDs) follow-on versions: specific regulatory and interchangeability rules ahead? Generics Biosimilars Initiat (GaBi) J 2(4):204–207

Muñoz M, Martin-Montanez E (2012) Ferric carboxymaltose for the treatment of iron-deficiency anemia. Expert Opin Pharmacother 13:907–921

Muñoz M, Gomez-Ramirez S, Liumbruno GM, Grazzini G (2014) Intravenous iron and safety: is the end of the debate on the horizon? Blood Transfus 12:287–289

Nicholas M (2014) Clinical development, immunogenicity, and interchangeability of follow-on complex drugs. Generics Biosimilars Initiat (GaBi) J 3(2):71–78

Nyström AM, Fadeel B (2012) Safety assessment of nanomaterials: implications for nanomedicine. J Controll Release 161:403–408

Rottembourg J, Kadri A, Leonard E, Dansaert A, Lafuma A (2011) Do two intravenous iron sucrose preparations have the same efficacy? Nephrol Dial Transplant 26(10):3262–3267

Schellekens H, Klinger E, Mühlebach S, Brin J, Storm G, Crommelin DJ (2011) The therapeutic equivalence of complex drugs. Regul Toxicol Pharmacol 59(1):176–183

Schellekens H, Stegemann S, Weinstein V et al (2014) How to regulate nonbiological complex drugs (NBCD) and their follow-on versions: points to consider. AAPS J 16(1):15–21

Schneider W (1988) Iron hydrolysis and the biochemistry of Iron—the interplay of hydoxide and biogenic ligands. Chimia 42:9–20

Schrier SL, Auerbach M, Mentzer WC, Landaw SA (2013) Treatment of anemia due to iron deficiency. http://www.uptodate.com/contents/treatment-of-the-adult-with-iron-deficiency-anemia#top. Accessed July 2014

Shah VP, Mühlebach S, Vulto A (2013) Workshop on the challenges in substitution of nonbiological complex drugs. Am J Health-Syst Pharm 70:1879–1880

Stefansson BV, Haraldsson B, Nilsson U (2011) Acute oxidative stress following intravenous iron injection in patients on chronic hemodialysis: a comparison of iron-sucrose and iron-dextran. Nephron Clin Pract 118(3):c249–256

Stein J, Dignass A, Chow KU (2012) Clinical case reports raise doubts about the therapeutic equivalence of an iron sucrose similar preparation compared with iron sucrose originator. Curr Med Res Opin 28(2):241–243

Tinkle S, McNeil SE, Mühlebach S, Bawa R, Borchard G, Barenholz YC et al (2014) Nanomedicines: addressing the scientific and regulatory gap. Ann N Y Acad Sci 1313(1):35–56

Toblli JE, Cao G, Oliveri L, Angerosa M (2009) Differences between original intravenous iron sucrose and iron sucrose similar preparations. Arzneimittelforschung 59(4):176–190

Toblli JE,Cao G, Oliveri L, Angerosa M (2011) Assessment of the extent of oxidative stress induced by intravenous ferumoxytol, ferric carboxymaltose, iron sucrose and iron dextran in a nonclinical model. Arzneimittelforschung 61(7):399–410

Toblli JE, Cao G, Oliveri L, Angerosa M (2012) Comparison of oxidative stress and inflammation induced by different intravenous iron sucrose similar preparations in a rat model. Inflamm Allergy Drug Targets 11(1):66–78

U.S. Pharmacopoeial Convention. Monographs in need of modernizaition, Iron sucrose injection (no. 207). Update Sept. 2014. http://www.usp.org/sites/default/files/usp_pdf/EN/USPNF/2014-09-25_monographs_needing_modernization.xlsx. Accessed Nov 2014

Van Wyck D, Anderson J, Johnson K (2004) Labile iron in parenteral iron formulations: a quantitative and comparative study. Nephrol Dial Transplant 19(3):561–565

Vifor Pharma Ltd. Comment (2014) to the FDA 2013 guidance from Vifor Pharma Ltd, Sidley Austin LLP. http://www.regulations.gov/#!documentDetail;D=FDA-2007-D-0369-0289. Accessed July 2014

Walson P, Mühlebach S, Flühmann B (2014) First Asia-Pacific educational workshop on non-biological complex drugs (NBCDs), Kuala Lumpur, Malaysia, 8 October 2013. Generics Biosimilars Initiat J 3(1):30–33

WHO (2008) Worldwide prevalence of anaemia 1993–2005. http://www.who.int/vmnis/publications/anaemia_prevalence/en/. Accessed July 2014

Wish JB (2006) Assessing iron status: beyond serum ferritin and transferrin saturation. Clin J Am Soc Nephrol Suppl 1:S4–8

Wysowski DK, Swartz L, Borders-Hemphill BV, Goulding MR, Dormitzer C (2010) Use of parenteral iron products and serious anaphylactic-type reactions. Am J Hematol 85(9):650–654

Drug Nanocrystals

Gerrit Borchard

Contents

Drug Nanocrystals 172
Chemistry, Structure, Manufacture 173
Characterization of Nanocrystals and their Formulations 179
Biopharmaceutical Aspects of Nanocrystals 182
Regulatory Status 184
Perspectives for Nanocrystal Technology 184
References 185

Abstract Drug nanocrystals are nanosized particles of pharmacologically active substances, obtained through bottom-up (e.g., precipitation) or top-down (e.g., milling) methods, or by a combination of such processes. Suspensions of nanocrystals contain excipients adsorbing to the nanocrystal surface stabilizing the suspension against aggregation by steric hindrance and electrostatic repulsion. The complexity of this type of non-biological complex drugs (NBCDs) is given, among others, by the ratio of amorphous to crystalline drug form contained in the nanoparticles, the particle size distribution, and the extent of stabilizer adsorption to the nanocrystal surface. These parameters have to be controlled during the manufacturing process to assure product quality, clinical performance and safety. A number of nanocrystal products already have obtained marketing authorization by EMA and FDA through a conventional regulatory framework, which may change with increasing sensitivity towards issues related to nanomedicines. No "generic" or "similar" products of nanocrystal formulations have been developed so far.

Keywords Nanocrystals · Nanosuspension · Nano-comminution · Poorly soluble drugs · Developability Classification System (DCS) · NanoEdge® · Dissocubes® · Microfluidizer® · Nanopure®

G. Borchard (✉)
School of Pharmaceutical Sciences Geneva-Lausanne, University of Geneva, University of Lausanne, 1211 Geneva, Switzerland
e-mail: gerrit.borchard@unige.ch

D. J. A. Crommelin, J. S. B. de Vlieger (eds.), *Non-Biological Complex Drugs,* AAPS Advances in the Pharmaceutical Sciences Series 20, DOI 10.1007/978-3-319-16241-6_6

Abbreviations

AFM	Atomic force microscopy
API	Active pharmaceutical ingredient
ASD	Amorphous solid dispersion
BU	Bottom up
CP	Co-precipitation
CT	Combination technology
DCS	"Developability" classification system
DLS	Dynamic laser light scattering
DSC	Differential scanning calorimetry
FIB	Focused ion beam
HF FIFFF	Hollow fiber flow field-flow fractionation
HPH	High-pressure homogenization
HPMC	Hydroxypropyl methylcellulose
LD	Laser diffractometry
MM	Media milling
PCS	Photon correlation spectroscopy
PDI	Polydispersity index
PEG	Poly(ethylene glycol)
PTIR	Photothermal-induced resonance
PVP	Polyvinylpyrrolidone
RES	Reticuloendothelial system
SCF	Supercritical fluid
SDS	Sodium dodecyl sulfate
SEM	Scanning electron microscopy
TD	Top-down
TEM	Transmission electron microscopy
WBM	Wet ball milling
XRPD	X-ray powder diffraction

Drug Nanocrystals

The discovery of new pharmacologically active principles is often faced with the challenge of their unfavorable properties of low solubility and resulting difficult "developability" (Fig. 1). About 40 % of marketed products, and up to 70 % of candidates currently under development are estimated to show poor solubility (Di et al. 2009). Several strategies to address this challenge, e.g., by optimization of the formulation, the use of co-solvents and cyclodextrins, and size reduction (Fig. 1) have been proposed (Williams et al. 2013). Reduction in size or "nanonizing" (Junghanns and Müller 2008) of poorly soluble class IIa drugs results in nanosized crystals ("nanocrystals") of much enlarged surface area and increased surface energy. As a

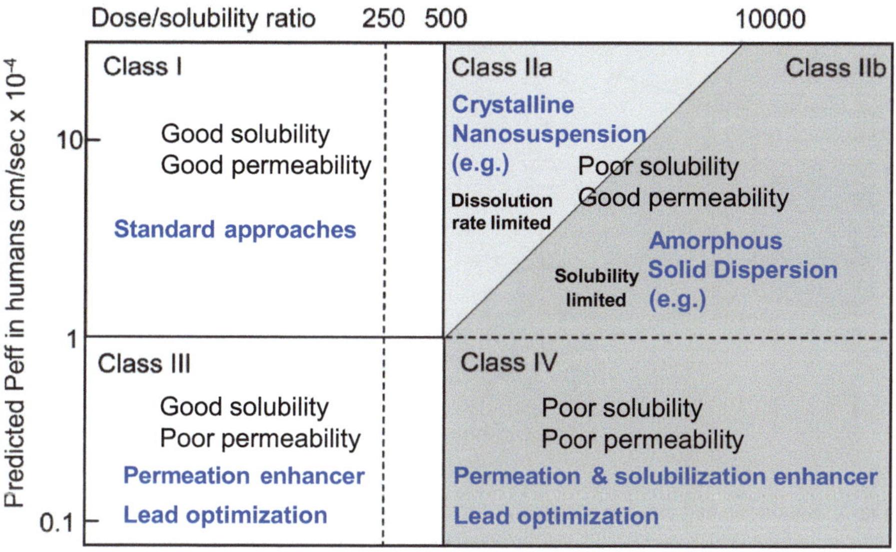

Fig. 1 "Developability" Classification System (DCS) for drug candidates. (Modified from Butler and Dressman 2010, with permission)

result, nanocrystals show increased dissolution velocity and therefore enhanced oral drug bioavailability (Gao et al. 2013a).

Chemistry, Structure, Manufacture

Nanocrystals consist of nanoparticles of pure active principle (Möschwitzer 2013) and are considered intermediates between the amorphous and crystalline states of the active pharmaceutical ingredient (API) (Rabinov 2004). Nanocrystals may be manufactured either by nanoprecipitation techniques in a "bottom-up" (BU) process, or by a "top-down" (TD) approach by nano-comminution, i.e. size reduction of larger particles (crystalline powders) by attrition forces (Fig. 2).

Bottom-up Procedures: Solvent Evaporation and Controlled Evaporation of Droplets

Nanoprecipitation as a bottom-up process was first developed to produce nanocrystals. The first such process was developed by Sucker and co-workers (Gassmann and Sucker 1992) in which the drug is dissolved in an organic solvent preferably miscible with water. This primary drug solution is successively mixed with a non-solvent (e.g., water), resulting in a fine precipitate of nanocrystals (Gassmann et al. 1994). During the crystallization process, the precipitate needs to be stabilized

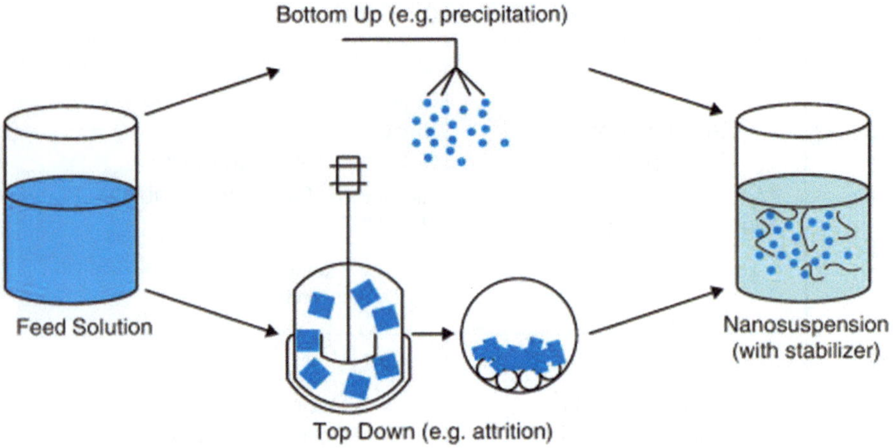

Fig. 2 Bottom-up and top-down strategies for drug nanocrystal manufacturing, followed by formulation as nanosuspension using stabilizers. (Adapted from Wang et al. 2012, with permission)

against agglomeration, as this would produce particles beyond the nanoscale. Rapid mixing in combination with stabilizing agents such as citric acid, ethyl cellulose or gelatin (Auweter et al. 1998) allows for the production of a large number of nanocrystals at a narrow size distribution, and minimization of Ostwald ripening (List and Sucker 1988). Aqueous suspensions of nanocrystals need to be stabilized against aggregation by the addition of electrostatic or steric stabilizers such as hydroxypropyl methylcellulose (HPMC), polyvinylpyrrolidone (PVP), polyethylene glycol (e.g., PEG 400), or copolymers (e.g., Pluronic® F127 and F68). Ionic surfactants such as sodium dodecyl sulfate (SDS) and non-ionic surfactants such as Tween® 20 and Tween® 80 are also in use (Jiang et al. 2012). These stabilizers are suggested to attach to the surface of nanocrystals, imparting electrostatic repulsion and/or steric hindrance to agglomeration. Further stabilization may be achieved by extraction of residual organic solvent through rotary evaporation, by lyophilization or by spray-drying (Chan and Kwok 2011). This extraction technique, however, may not be available to a large percentage of drug candidates that are poorly soluble in organic solvents miscible with water.

Generation of nanocrystals following a bottom-up approach is also achieved by crystallization of the drug crystals in a confined working volume. The use of microfluidic devices (Ali et al. 2009) allows for the nanoprecipitation process (drug supersaturation, nucleation and particle growth) to take place in the central diffusion layer between the solvent and the anti-solvent stream flowing in a laminar fashion in a microreactor. Through adjustment of the rate of antisolvent flow, the resulting particle size and size distribution may be selected. The generated nanocrystal suspension may successively be introduced into a stabilizer solution under sonication, as shown, e.g., for hydrocortisone nanocrystals (Ali et al. 2011).

In addition to methods using solvent precipitation, the controlled evaporation of droplets has been adapted for nanocrystal formation. As an example, spray-drying of drug solutions uses the principle of crystal formation at the droplet surface through the rapid evaporation of the solvent, resulting in a powder whose properties are dependent on process parameters such as drug solution feed and drying kinetics. By successive freeze-drying of the sprayed drug solution in a process called spray freeze-drying, spherical nanocrystals of monodisperse size distribution are obtained (Vehring 2008). In contrast to the other techniques mentioned above, this bottom-up process of spray freeze-drying is considered to be suitable for large scale production of nanocrystals (Waard et al. 2009).

Other procedures, using either the principle of solvent precipitation or controlled evaporation of droplets, are being developed. Among solvent precipitation methods, procedures based on high-gravity controlled precipitation, flash nanoprecipitation, sonoprecipitation and processes adapting supercritical fluid (SCF) technologies are found. Aerosol flow reactor method and electrospraying of drug solutions are among the controlled evaporation methods. An excellent overview of these partially innovative methods has been published by Chan and Kwok (Chan and Kwok 2011).

Top-down Procedures: Nano-Comminution by Milling and Homogenization

Top-down processes, i.e. reducing the size of drug powders to the nanoscale, are considered to be more suitable for industrial-scale production of nanocrystals, as they avoid the challenges (absence of suitable solvent, removal of residual organic solvent, difficult process control, overall costs) to be faced in the application of bottom-up procedures. Generally, two procedures can be distinguished, i.e. milling and high pressure homogenization (Eerdenbrugh et al. 2008). As nano-sized powders may not be obtained by dry milling (e.g., jet milling), in wet ball milling (WBM) slurries of coarse drug powders are used for reduction of particle size down to the nano range (Niwa et al. 2011; Merisko-Liversidge and Liversidge 2011). WBM, in which shear forces are generated by the movement of milling media, was first developed by Liversidge et al. (Liversidge and Cundy 1995). The NanoCrystal® technology successfully used today in the pharmaceutical industry is based on this development. Commonly used milling media in WBM are pearls of various sizes made from ceramics, stainless steel, glass or polystyrene resin. A potential problem, however, is represented by the erosion of the milling beads, resulting in milling product contamination. Wet ball milling times can last from 30 min to several days, basically depending on factors such as drug hardness, slurry viscosity, temperature, size of the milling beads, energy input and choice of stabilizers used. Nanocrystal sizes obtainable by WBM are reported to be in the range of 128 nm (itraconazole for i.v. injection) (BeirowskiJ et al. 2011) to 800 nm (1,3-dicyclohexylurea) (Chan and Kwok 2011).

The second well-established top-down process uses high-pressure homogenization (HPH) of coarse drug powders dispersed in a suitable medium for the generation of nanocrystals (Keck and Müller 2006). Basically, the drug suspension is repeatedly passed through a high pressure homogenizer of different designs, while the pressure is successively increased to production pressure (typically 1000 to 2000 bar). Comminution is caused by cavitation, shear forces and collision of drug particles in three distinguishable set-ups, i.e. Microfluidizer® (Bruno and McIlwrick 1999), Dissocubes® (Müller et al. 2003) and Nanopure® (Bushrab and Müller 2003) technologies. Microfluidizer® technology uses repetitive (50 to 100 cycles) frontal collision of two fluid streams in a Y- or Z-shaped collision chamber, e.g., by jet stream homogenization. Obtained particle dispersions in the submicron range need to be stabilized by the presence of surfactants, which will adsorb to the surface of the nanocrystals.

By contrast, Dissocubes® technology uses piston-gap homogenizers that reduce particle size of drug powders dispersed in an aqueous surfactant solution by passing it through a homogenization gap (5 to 20 μm) at high pressures (1500 to 2000 bar) (Müller et al. 2000). Particle size is reduced by cavitation, i.e. the collapsing of air bubbles created by high dynamic pressure and reduction in static pressure below the vapor pressure of the aqueous phase. Resulting nanocrystals were shown to have cuboid or irregular shapes due to their crystalline nature.

The Nanopure® technology is a variation of the piston-gap approach, performed at low temperatures, using non-aqueous liquids (oils, polyethylene glycol) or water-reduced media (such as glycerol/water or ethanol/water mixtures). Under these conditions, due to the low vapor pressures of the media used, cavitation is almost absent. Size reduction, however, caused by shear forces and drug particle collisions, was shown to be sufficient to achieve nanosized drug particles. Due to the low temperatures applied, Nanopure® technology is especially useful in the preparation of nanocrystals of temperature sensitive drugs. Table 1 shows pharmaceutical nanocrystal products available on the market, together with their year of FDA approval, indication and technology applied.

Table 1 Pharmaceutical nanocrystal products on the market. (Adapted from Weissig et al. 2014; Junyaprasert and Morakul 2014; Hafner et al. 2014)

Name	Drug	Approval/indication	Technology
Cesamet®	Nabilone	FDA 2006, anti-emetic (oral)	BU, CP
Emend®	Aprepitant	FDA 2003, anti-emetic (oral)	TD, MM
Gris-PEG®	Griseofulvin	FDA 1975, antifungal (oral)	BU, CP
Megace ES®	Megestrol acetate	FDA 2005, hypercholesterolemia, hypertriglyceridemia (oral)	TD, MM
Rapamune®	Rapamycin, formulated in tablets	FDA 2002, immunosuppression, (oral)	TD, MM
Tricor®	Fenofibrate as nanocrystals	FDA 2004, hypercholesterolemia, hypertriglyceridemia (oral)	TD, MM
Triglide®	Fenofibrate as non-soluble drug delivery microparticles	FDA 2004, hypercholesterolemia, hypertriglyceridemia (oral)	TD, HPH
Xeplion®	Paliperidone	EMA 2011, schizophrenia (i.m.)	TD, MM
Zypadhera®	Olanzapine	EMA 2008, schizophrenia (i.m.)	TD, MM

BU bottom-up, *TD* top-down, *CP* co-precipitation, *HPH* high-pressure homogenization, *MM* media milling, *i.m.* intramuscular

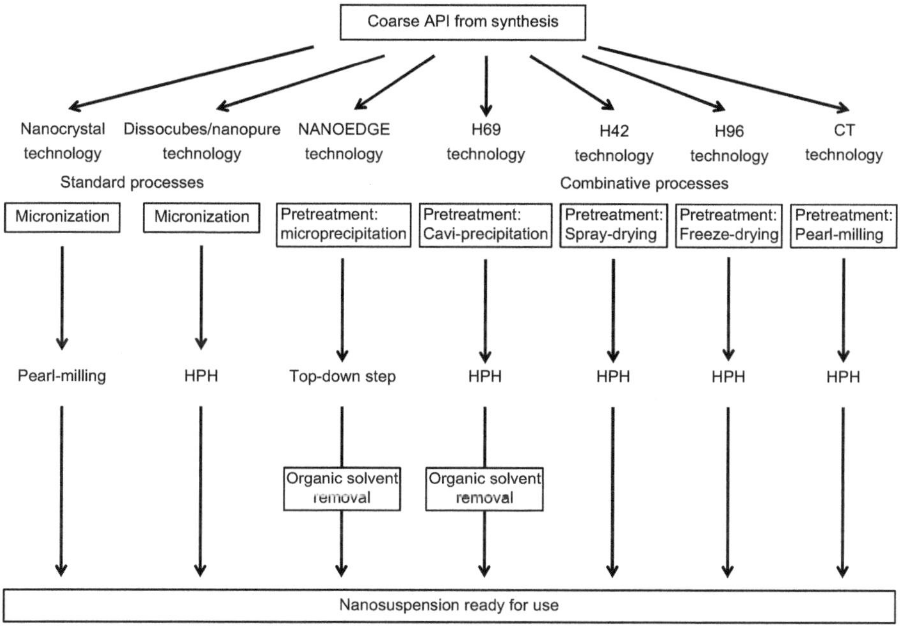

Fig. 3 Standard and combinative technologies for size reduction of nanocrystals. (Adapted from Salazar et al. 2014)

Combinative Technologies for Nanocrystal Production

A combination of two top-down processes is represented by the Combination Technology (CT) approach. CT combines primary low-energy pearl milling of the coarse drug slurry or macrosuspension to a particle size of about 600 to 1500 nm. This step is followed by high pressure homogenization, which improves the homogeneity of the nanosuspension with regard to particle size and reduction of larger crystals (Al Shaal et al. 2010). CT offers the possibility to work at reduced HPH pressures (100–500 bar) to achieve smaller-sized nanocrystals than at higher pressures (usually about 1500 bar), and to reduce the processing time in terms of homogenization cycles (e.g., from usually 20 to 5). As an example, nanocrystals of the plant flavon apigenin, used as an antioxidant, were prepared by CT (Table 1). The volume weighted diameter (d(v)50 %) of 164 ± 15 nm (measured by laser diffractometry, LD) was achieved by HPH at 300 bar, after primary a bead-milling step (7 passages) had reduced the d(v)50 % to 231 ± 18 nm (Al Shaal et al. 2011).

Combinations of bottom-up followed by top-down procedures are also in use, e.g., Baxter's NanoEdge® technology using drug microprecipitation followed by annealing by HPH, and the H69, H42 and H96 technologies using combinations of cavi-precipitation, spray drying of organic solutions or freeze-drying, respectively, with top-down methods such as HPH (Salazar et al. 2014). These "combinative" technologies (Fig. 3) offer the advantages of shorter processing times, improved physical stability of the resulting nanosuspensions, and overall smaller particle sizes favorable for the improvement of solubility and thus bioavailability of poorly soluble drugs. A summary of nanocrystals produced by such combinative technologies is shown in Tables 2, 3 and 4).

Table 2 Particle sizes of nanocrystals for cosmetic products by the Combined Technology approach (pearl milling/HPH). (Salazar et al. 2014)

Drug	Smallest mean particle size [nm]	Administration route
Rutin	604	Topical/oral
Hesperidin	599	Topical/oral
Apigenin	275	Topical/oral

Table 3 Particle sizes of nanocrystals for pharmaceutical products by the NanoEdge® approach (microprecipitation/HPH). (Salazar et al. 2014)

Drug	Smallest mean particle size [nm]	Administration route
Paclitaxel	200	i.v.
Nabumetone	930	i.v.
Prednisolone	640	i.v.
Carbamazepin	400	i.v.
Itroconazol	581	i.v.
Itroconazol[a]	177	i.v.

[a] Produced by microprecipitation/sonication

Table 4 Particle sizes of nanocrystals for oral pharmaceutical products by H69 (cavi-precipitation/HPH), H42 (spray-drying/HPH) and H96 (freeze-drying/HPH) approach. (Salazar et al. 2014)

Process	Drug	Smallest mean particle size [nm]
H69	Ibuprofen	170
H69	Hydrocortisone acetate	787
H69	Resveratrol	150
H69	Omeprazol	921
H69	Prednisolone	22
H42	Amphotericin B	172
H42	Glibenclamide	236
H42	Hydrocortisone acetate	281
H42	Ibuprofen	636
H42	Resveratrol	200
H96	Amphotericin B	62
H96	Glibenclamide	164
H96	Cyclosporin A	440
H96	Hydrocortisone acetate	414

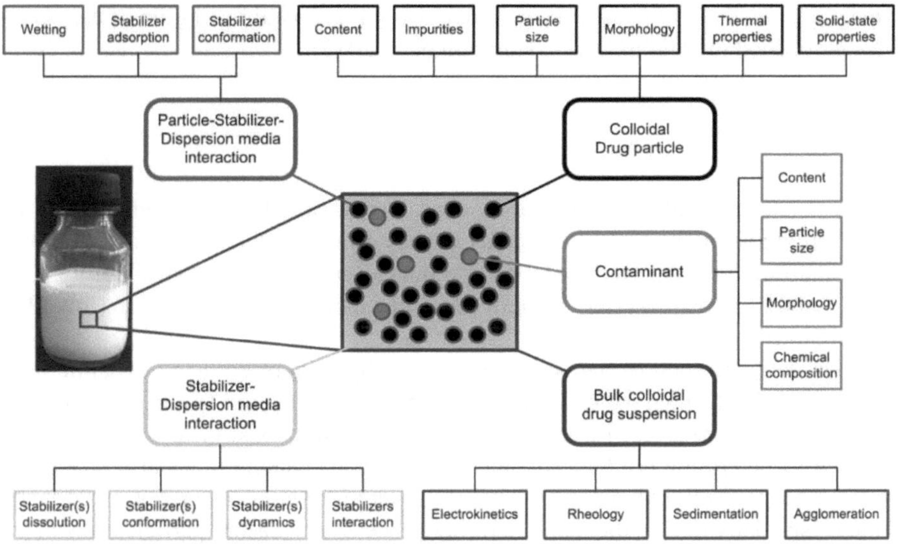

Fig. 4 Different parameters affecting the quality of nanocrystal formulations. (Adapted from Juhnke and John 2014)

Characterization of Nanocrystals and their Formulations

Nanocrystals and their formulations are truly complex drugs, whose biological performance depends on a variety of physicochemical properties. These, in turn, are influenced by the manufacturing and formulation process applied. The different parameters to be considered are depicted in Fig. 4, and the techniques applied to determine the various parameters are briefly discussed below.

Particle Structure and Morphology

Well-established techniques for the imaging of nanocrystals are light microscopy, scanning and transmission electron microscopy (SEM and TEM) (Raghava Srivalli and Mishra 2014), In addition, cryogenic TEM (cryo-TEM) has for example been described for the imaging of magnetic iron oxide nanocrystals (Yu et al. 2006) and for the calcium antagonist, felodipine (Lindfors et al. 2007).

Other suggested techniques, although not routinely used in drug nanocrystal imaging, include the use of focused ion beam (FIB) in conjunction with SEM and TEM. This technique allows for the resolution of structures down to about 1 nm by using a stream of heavy ions (e.g., gallium) instead of an electron beam (Mulders and Day 2005; Wirth 2008). Another analytical technique is atomic force microscopy (AFM) (Gao et al. 2013b), which may eventually be used in connection with photothermal-induced resonance (PTIR) technology. PTIR does not only allow for imaging and sizing of nanostructures, but also for chemical resolution of nanocrystals embedded in solid matrices (Harrison et al. 2013).

Particle Size

The particle size and size distribution, expressed as polydispersity index (PDI) are the defining parameters of nanocrystals, as they directly influence physical stability, drug saturation solubility, dissolution and thus ultimately drug bioavailability (Gao et al. 2008). For nanocrystals, a small size and a narrow size distribution (low PDI) are obviously desired properties.

Size determination of nanocrystals is typically performed using dynamic laser light scattering (DLS), which is also called photon correlation spectroscopy (PCS). In DLS, the diffraction of a laser beam measured at a fixed or at variable angles is used to determine the hydrodynamic diameter of nanoparticles at a size range from 3 nm to 3 µm (Mhatre et al. 2012). In addition, laser diffraction (LD) is applied, especially during the manufacturing process, to measure particle sizes in the range of 50 nm up to 2000 µm (Sahoo et al. 2011). Other techniques, also applied to measure aggregates in formulations of therapeutic proteins, may also be applied to determine the size and size distribution of nanocrystals, e.g., Coulter counter analysis, nanoparticle tracking analysis (NTA), as well as hollow-fiber flow field-flow fractionation (HF FIFFF). These techniques have recently been described and reviewed in (Lapresta-Fernández and Salinas-Castillo 2014).

Particle Surface Charge (zetapotential)

The surface charge and thus the resulting repulsion forces between individual particles is of great importance in the stabilization of suspensions of nanocrystals (Wu et al. 2011). In addition, stabilization can be increased by addition of stabilizers providing steric hindrance against aggregation (Lee et al. 2005). As the surface charge itself cannot directly be measured (Nel et al. 2009), the electrical potential (zetapotential) at the surface of the electric double layer, formed by the Stern layer and the diffusion layer of oppositely charged ions, is measured. A potential of + or −30 mV is commonly regarded to be needed to stabilize nanodispersions against aggregation, if a steric stabilizer is utilized a value of + or −20 mV has been mentioned to be sufficient (Jacobs and Müller 2002). The extent of adsorption of stabilizer molecules to the particle surface as well as strength of binding are, of course, also determined by the zetapotential.

The zetapotential can be measured by measuring the electrophoretic mobility of the particles in an electric field by laser light scattering (Deshiikan and Papadopoulos 1998), or by electroacoustic means (O'Brien et al. 1995). The advantage of the latter method is that it is not restricted to measurements in diluted dispersions, as is the method using laser light diffraction (Hunter 2001).

Particle Crystallinity

The solid state of nanocrystals, i.e. the presence of the drug in its crystalline and/or amorphous state, is decisive for the physical stability of nanocrystals. The drug in its amorphous state is thermodynamically unstable and would therefore show a faster dissolution rate and thus higher bioavailability (Fakes et al. 2009), whereas the crystalline form confers higher storage stability. The amorphous form of the drug is metastable and tends to convert to the crystalline form during storage, which can be slowed by the addition of suitable stabilizers or co-precipitates in a bottom-up process (Yoshioka et al. 1995). In general, top-down manufacturing processes will produce primarily nanocrystals in the crystalline state, as shear forces break away parts of the bigger particles at crystal lattice imperfections (Gao et al. 2011). Bottom-up techniques, such as the controlled evaporation of droplets of drug solution as described above, favors the formation of amorphous nanocrystals (Hancock and Parks 2000). A combinative approach of a bottom-up and top-down technique generally also results in drug particles in their crystalline form.

The state of crystallinity of nanocrystals may be assessed by X-ray powder diffraction (XRPD) (Pinna 2005) and differential scanning calorimetry (DSC) (Gill et al. 2010). While by the former technique characteristic diffraction patterns are obained, the latter measures the heat flow related to the transition of the drug particles from a crystalline to an amorphous state. Both techniques, XRPD and DSC may be combined (Van Eerdenbrugh et al. 2007).

Recently, terahertz spectroscopy, showing specific absorption spectra for crystalline polymorphs, has been introduced to measure crystallinity as a rapid and non-destructive method suitable to be applied in up-scaled manufacturing processes (Otsuka et al. 2010).

Dissolution Testing of Nanocrystals

The dissolution process of drug particles can be described by the Noyes-Whitney equation (Noyes and Whitney 1897) (Eq. 1).

$$\frac{dX}{dt} = \frac{DA}{h_D} \times (c_s - c_t) \quad (1)$$

where $dXI\ dt$ is defined as the dissolution rate, D is the drug diffusion coefficient, A is the dissolution surface area, h_D the diffusion layer thickness, c_s the saturation solubility in a given medium, and $0c_t$ the drug concentration around the nanocrystals at a given time point.

Nanosizing of drug particles therefore increases the surface area A, decreases the diffusion layer thickness h_D (Mosharraf and Nystrom 1995), and increases the saturation solubility c_s (Kesisoglou et al. 2007), resulting in an overall faster dissolution rate. The increase in saturation solubility is described by the Freundlich-Ostwald equation (Müller and Peters 1998) (Eq. 2).

$$S = S_{\infty} exp\left(\frac{2\gamma M}{r\rho RT}\right) \quad (2)$$

where S is the saturation solubility of the drug, S_{∞} the saturation solubility of an infinitely large drug crystal, γ describes interfacial tension between the crystal surface and the dispersion medium, M is the molecular weight of the drug, r the drug particle radius, ρ the density, R a gas constant and T the temperature. From the Freundlich-Ostwald equation the positive impact of surfactants used in nanocrystal formulations on the dissolution rate can also be derived.

The dissolution of a nanoparticulate product such as nanocrystals can be monitored by a number of different techniques. Care must be taken to carefully separate the supernatant from the nanosized particles and no undissolved drug is measured.

In addition to the classical Apparatus II paddle method described in the USP (Niaz et al. 2014), a number of different strategies are established:

- Sampling, separation and determination of supernatant-assay by HPLC or UV spectroscopy after syringe filtration (Juenemann et al. 2011).
- Separation and determination of supernatant-assay by HPLC or UV spectroscopy after dialysis (Bhardwaj et al. 2010).
- In-situ monitoring of drug particle size reduction by turbidity measurement (Jünemann and Dressman 2012)

Other advanced techniques are currently being evaluated for their suitability to describe the dissolution of nanoparticles, which may be suitable to test dissolution rates of nanocrystals, as well:

- Pressure separation by liquid chromatography (Helle et al. 2010) or field-flow fractionation (Jünemann and Dressman 2012) followed by determination of the supernatant drug concentration by HPLC or UV spectroscopy.
- In-situ monitoring of drug particle size reduction by dynamic light scattering (Anhalt et al. 2012).
- In-situ monitoring of drug dissolution by means of an ion-selective electrode (Bohets et al. 2007), or by UV fiber optic spectroscopy (Van Eerdenburgh et al. 2011).

Size reduction of poorly soluble drugs thus leads to an increase in dissolution rate, which can be measured by various approaches. Ultimately, decrease in particle size results in enhanced biopharmaceutical performance of nanocrystals.

Biopharmaceutical Aspects of Nanocrystals

Drug Nanocrystals for Oral Administration

Oral administration is the most preferred route of administration of drug nanocrystal formulations. It has been shown for a number of drugs (Kesisoglou and Wu

2008; Willmann et al. 2010; Quan et al. 2011) that the reduction in size of drug particles translates into higher C_{max} values, as well as an increased AUC after oral administration, and thus an increased bioavailability compared to other formulations of the same drug. Especially drugs with an absorption window in the upper intestinal tract, such as aprepitant (Emed®) or fenofibrate (Tricor®) benefit from fast onset of dissolution after oral administration, as the large percentage of the drug is dissolved at the major absorption site (Wu et al. 2004; Zhu et al. 2010).

Fast dissolution of drug nanocrystals can also lead to an overall faster onset of action (reduction of T_{max}). An example is the naproxen nanocrystal suspension, where patients are benefitting from a faster onset of the pharmacodynamic effect (pain relief) (Merisko-Liversidge et al. 2003).

Poorly soluble drugs administered with food often show increased absorption, which is possibly due to the drug solubilization by bile salts and food components, and by a reduced gastric emptying rate resulting in a longer time period for the solubilization of the drug. By contrast, nanocrystal preparations achieve their highest dissolution rate in the fasted state, which is not increased by the presence of food components (Shono et al. 2010). Thus, for nanocrystal preparations food effects on pharmacokinetic parameters such as AUC, T_{max} and C_{max} are generally not observed.

Drug Nanocrystals for Other Routes of Administration

Nanosizing can serve to enhance the performance of poorly soluble drugs applied by various routes of administration, including the dermal (Mishra et al. 2009), ophthalmic (Tuomela et al. 2014), pulmonary (Steckel et al. 2003), buccal (Rao et al. 2011) and ocular (Kassem et al. 2007) routes.

However, the parenteral route is the most important for non-oral application of nanocrystal drugs (Shi et al. 2009). A challenge in the preparation of nanocrystal suspensions for injection is the physical stability of the drug, i.e. the avoidance of aggregation by addition of stabilizers and Ostwald ripening leading to larger particle sizes (Shi et al. 2009). Larger nanocrystal aggregates may mechanically block microvessels, especially in the lung, leading to embolisms. In addition, larger particles with slower dissolution kinetics will activate the opsonin system and successively be more readily recognized and eliminated by the reticuloendothelial system (RES) (Moghimi et al. 2001). In the cells of the RES, professional phagocytes residing in the liver, spleen, and lung, slower dissolving drug particles form a reservoir, slowly releasing drug into the blood stream. This will profoundly change the drug's pharmacokinetic profile, as was shown, e.g., for itraconazole nanosuspensions (Rabinow et al. 2007). Here, alteration in particle size (from 300 nm to about 580 nm) yielded a reduced C_{max} and a prolonged plasma half-life when compared to itraconazole complexed with HP-β-cyclodextrin solution.

Another challenge is to meet the sterility requirements for parenteral solutions, as terminal γ-irradiation or heat sterilization (autoclavation) may lead to changes in particle size due to physical instability of the nanocrystals. Most commonly used are therefore aseptic processing and sterile filtration to assure formulation sterility (Zheng and Bosch 1997).

Regulatory Status

As stated above, several nanocrystal products, utilizing the manufacturing processes described under 6.1., have entered the market. Thus, regulatory experience as well as a pathway to approval appears to be established in a conventional framework. It is therefore surprising, considering the obvious advantages of this technology, and the number of drug candidates of poor "developability" that not more products are being developed. This has in part been attributed to the lack of interest in IP development by pharmaceutical companies in the past, and the favoring of more "developable" drug candidates. This strategy, no longer valid today, precludes single companies from owning all proprietary knowledge to be used in late-stage manufacturing (Müller and Keck 2012). Companies are therefore lacking experience and capabilities to cover the whole scale-up and (clinical) manufacturing process, which is quite sophisticated for nanocrystal preparations. Nanocrystal technology has therefore often to compete with other enabling techniques to formulate drug candidates of low "developability", which can more readily be performed in-house. On the other hand, recognizing that only very few pharmaceutical "nano" products in the form of liposomes, micelles or nanoparticles have entered the market, nanocrystals appear to be among the most successful nanotechnological enabling technologies to drug delivery. In 2010, the market size for drug nanocrystals was an estimated US$ 596 million, about 44 % of the total nanotechnology drug delivery market (US$ 1.3 billion). It is forecasted that drug nanocrystal technology will achieve a total addressable market (TAM) of about US$ 82 billion by 2021 (Market opportunities in nanotechnology drug delivery and Cientifica Ltd 2012).

The regulatory landscape for drug nanocrystals, however, is about to change due to the ongoing discussion around the evaluation of quality, safety and efficacy of nanomedicines in general. As mentioned above, nanocrystal suspensions are stabilized by adsorption of stabilizers to the particle surface. In addition, in the future functionalization of the nanocrystal surface, e.g., for targeting purposes, may be applied. These modified surfaces may be regarded as an "engineered" surface as described in the EMA reflection paper on the parenteral administration of coated nanomedicine products (EMA 2013). The document reflects on the influence of the modification of nanoparticle surfaces on the variation in opsonization patterns and resulting disposition after parenteral administration as a quality, efficacy and safety aspect of nanomedicines. The relevance of these reflections for coated drug nanocrystal suspensions for parenteral use is evident.

Perspectives for Nanocrystal Technology

True innovations in the nanocrystal area will come from the development of more sophisticated analytical techniques, which will supply more accurate and detailed information on the physico-chemical properties and dissolution kinetics of drug nanocrystals. Successively, increased knowledge of these physicochemical properties must be linked to the in vivo performance (in vivo dissolution and aggregation, opsonization, RES uptake) of nanocrystals to lay the basis for a directed nano-engineering, which may also include a specific alteration of the nanocrystal surface, e.g., by adsorption of targeting moieties to achieve active targeting to specific cells and tissues (Åkerman et al. 2002). Last but not least, manufacturing of nanocrystals may include a bottom-up approach in the form of single-particle resolved printing (Kraus et al. 2007). In any case, formulation design space with regard to routes of administration, variety of dosage forms as well as functionalization are far from being fully exploited.

References

Åkerman ME, Chan WCW, Laakkonen P, Bhatia SN, Ruoslahti E (2002) Nanocrystal targeting in vivo. Proc Nat Acad Sci USA 99:12617–12621.

Al Shaal L, Müller RH, Shegokar R (2010) SmartCrystal combination technology—scale up from lab to pilot scale and long term stability. Pharmazie 12:877–884

Al Shaal L, Shegokar R, Müller RH (2011) Production and characterization of antioxidant apigenin nanocrystals as a novel UV skin protective formulation. Int J Pharm 420:133–140

Ali HSM, York P, Blagden N (2009) Preparation of hydrocortisone nanosuspension through a bottom-up nanoprecipitation technique using microfluidic reactors. Int J Pharm 375:107–113

Ali HSM, York P, Ali AMA, Blagden N (2011) Hydrocortisone nanosuspensions for ophthalmic delivery: a comparative study between microfluidic nanoprecipitation and wet milling. J Control Rel 149:175–181

Anhalt K, Geissler S, Harms M, Weigandt M, Fricker G (2012) Development of a new method to assess nanocrystal dissolution based on light scattering. Pharm Res 29:2887–2901

Auweter H, André V, Horn D, Lüddecke E (1998) The function of gelatin in controlled precipitation processes of nanosize particles. J Disp Sci Technol 19:163–184

Beirowski J, Ingelbrecht S, Arien A, Gieseler H (2011) Freeze-drying of nanosuspensions, 1: freezing rate versus formulation design as critical factors to preserve the original particle size distribution. J Pharm Sci 100:1958–1968

Bhardwaj U, Sura R, Papadimitrakopoulos F, Burgess DJ (2010) PLGA/PVA hydrogel composites for long-term inflammation control following s.c. implantation. Int J Pharm 384:78–86

Bohets H, Vanhoutte K, de Maesschalck R, Cockaerts P, Vissers B, Nagels JL (2007) Development of in situ ion selective sensors for dissolution. Anal Chim Acta 581:181–191

Bruno RP, McIlwrick R (1999) Micofluidizer processor technology for high performance particle size reduction, mixing and dispersion. Eur J Pharm Biopharm 56:29–36

Bushrab NF, Müller RH (2003) Nanocrystals of poorly soluble drugs for oral administration. J New Drugs 5:20–22

Butler JM, Dressman JB (2010) The developability classification system: application of biopharmaceutics concepts to formulation development. J Pharm Sci 99:4940–4954

Chan HK, Kwok PCL (2011) Production methods for nanodrug particles using the bottom-up approach. Adv Drug Deliv Rev 63:406–416

de Waard H, Grasmeijer N, Hinrichs WL, Eissens AC, Pfaffenbach PP, Frijlink HW (2009) Preparation of drug nanocrystals by controlled crystallization: application of a 3-way nozzle to prevent premature crystallization for large scale production. Eur J Pharm Sci 38:224–229

Deshiikan SR, Papadopoulos KD (1998) Modified Booth equation for the calculation of zeta potential. Coll Polym Sci 276:117–124

Di L, Kerns EH, Carter GT (2009) Drug-like property concepts in pharmaceutical design. Curr Pharm Des 15:2184–2194

EMA (2013) Reflection paper on surface coatings: general issues for consideration regarding parenteral administration of coated nanomedicine products. EMA/325027/2013. http://www.ema.europa.eu/docs/en_GB/document_library/Scientific_guideline/2013/08/WC500147874.pdf
Fakes MG, Vakkalagadda BJ, Qian F, Desikan S, Gandhi RB, Lai C, Hsieh A, Franchini MK, Toale H, Brown J (2009) Enhancement of oral bioavailability of an HIV-attachment inhibitor by nanosizing and amorphous formulation approaches. Int J Pharm 370:167–174
Gao L, Zhang D, Chen M (2008) Drug nanocrystals for the formulation of poorly soluble drugs and its application as a potential drug delivery system. J Nanoparticle Res 5:845–862
Gao L, Liu G, Wang X, Liu F, Xu Y, Ma J (2011) Preparation of a chemically stable quercetin formulation using nanosuspension technology. Int J Pharm 404:231–237
Gao L, Liu G, Ma J, Wang X, Zhou L, Li X, Wang F (2013a) Application of drug nanocrystal technologies on oral drug delivery for poorly soluble drugs. Pharm Res 30:307–324.
Gao B, Wang J, Wang D, Zhu Z, Qiao Z, Yang G, Nie F (2013b) A novel preparation method for drug nanocrystals and characterization by ultrasonic spray-assisted electrostatic adsorption. Int J Nanomedicine 8:3927–3936
Gassmann P, Sucker H (1992) Improvements in pharmaceutical compositions. European Patent 0580690
Gassmann P, List M, Schweitzer A, Sucker H (1994) Hydrosols—alternatives for the parenteral application of poorly water soluble drugs. Eur J Pharm Biopharm 40:64–72
Gill P, Moghadam TT, Ranjbar B (2010) Differential scanning calorimetry techniques: applications in Biology and Nanoscience. J Biomol Tech 21:167–193
Hafner A, Lovric J, Lakos GP, Pepic I (2014) Nanotherapeutics in the EU: an overview on current state and future directions. Int J Nanomed 9:1005–1023
Hancock BC, Parks M (2000) What is the true solubility advantage for amorphous pharmaceuticals? Pharm Res 17:397–404
Harrison AJ, Bilgili EA, Beaudoin SP, Taylor LS (2013) Atomic force microscope infrared spectroscopy of griseofulvin nanocrystals. Anal Chem 85:11449–11455
Helle A, Hirsjärvi S, Peltonen L, Hirvonen J, Wiedmer SK, Hyötyläinen T (2010) Novel, dynamic on-line analytical separation system for dissolution of drugs from poly(lactic acid) nanoparticles. J Pharm Biomed Anal 51:125–130
Hunter RJ (2001) Measuring zeta potential in concentrated industrial slurries. Coll Surf A Phyicochem Engin Aspects 195:205–214
Jacobs C, Müller RH (2002) Production and characterization of a budesonide nanosuspension for pulmonary application. Pharm Res 19:189–194
Jiang TY, Han N, Zhao BW, Xie YL, Wang SL (2012) Enhanced dissolution rate and oral bioavailability of simvastatin nanocrystal prepared by sonoprecipitation. Drug Dev Ind Pharm 38:1230–1239
Juenemann D, Jantratid E, Wagner C, Reppas C, Vertzoni M, Dressman JB (2011) Biorelevant in vitro dissolution testing of products containing micronized or nanosized fenofibrate with a view to predicting plasma profiles. Eur J Pharm Biopharm 77:257–265
Juhnke M, John E (2014) Size reduction as integral element for development and manufacturing of engineered drug particles. Chem Engin Technol 37:757–764
Jünemann D, Dressman J (2012) Analytical methods for dissolution testing of nanosized drugs. J Pharm Pharmacol 64:931–943
Junghanns JUAH, Müller RH (2008) Nanocrystal technology, drug delivery and clinical applications. Int J Nanomed 3:295–309
Junyaprasert VB, Morakul B (2014) Nanocrystals for enhancement of oral bioavailability of poorly water-soluble drugs. Asian J Pharm Sci. doi:10.1016/j.ajps.2014.08.005
Kassem MA, Abdel Rahman AA, Ghorab MM, Ahmed MB, Khalil RM (2007) Nanosuspension as an ophthalmic delivery system for certain glucocorticoid drugs. Int J Pharm 340:126–133
Keck CM, Müller RH (2006) Drug nanocrystals of poorly soluble drugs produced by high pressure homogenization. Eur J Pharm Biopharm 62:3–16
Kesisoglou F, Wu Y (2008) Understanding the effect of API properties on bioavailability through absorption modeling. AAPS J 10:516–525

Kesisoglou F, Panmai S, Wu Y (2007) Nanosizing—oral formulation development and biopharmaceutial evaluation. Adv Drug Del Rev 59:631–644

Kraus T, Malaquin L, Schmid H, Riess W, Spencer ND, Wolf H (2007) Nanoparticle printing with single-particle resolution. Nature Nanotechnology 2:570–576.

Lapresta-Fernández A, Salinas-Castillo A, Anderson de la Llana S, Costa-Fernández JM, Domínguez-Meister S, Cecchini R, Capitán-Vallvey LF, Moreno-Bondi MF, Marco MP, Sánchez-López JC, Anderson IS (2014) A general perspective of the characterization and quantification of nanoparticles: imaging, spectroscopic, and separation techniques. Crit Rev Solid State Mater Sci 39:423–458

Lee J, Lee SJ, Choi JY, Yoo JY, Ahn CH (2005) Amphiphilic amino acid copolymers as stabilizers for the preparation of nanocrystal dispersions. Eur J Pharm Sci 24:441–449

Lindfors L, Skantze P, Skantze U, Westergren J, Olsson U (2007) Amorphous drug nanosuspensions. 3. Particle dissolution and crystal growth. Langmuir 23:9866–9874

List M, Sucker H (1988) Pharmaceutical colloidal hydrosols for injection. GB Patent 2200048

Liversidge GG, Cundy KC (1995) Particle size reduction for improvement of oral bioavailability of nanocrystalline danazol in beagle dogs. Int J Pharm 125:91–97

Harper T (2012) Marketing opportunities in nanotechnology drug delivery. Cientifica Ltd. January 2012. http://www.cientifica.com/wp-content/uploads/downloads/2012/04/NDD-White-Paper-Jan-2012.pdf

Merisko-Liversidge E, Liversidge GG (2011) Nanosizing for oral and parenteral drug delivery: a perspective on formulating poorly water-soluble compounds using wet mediamilling technology. Adv Drug Deliv Rev 30:427–440

Merisko-Liversidge E, Liversidge GG, Cooper ER (2003) Nanosizing: a formulation approach for poorly-water-soluble compounds. Eur J Pharm Sci 18:113–120

Mhatre P, Chinchole R, Desai U, Chavan R (2012) Review: nanosuspensions. Int J Pharm Sci Rev Res 13:118–124

Mishra PR, Al Shaal L, Müller RH, Keck CM (2009) Production and characterization of Hesperetin nanosuspensions for dermal delivery. Int J Pharm 371:182–189

Moghimi S, Hunter A, Murray J (2001) Long-circulating and target-specific nanoparticles: theory to practice. Pharmacol Rev 53:283–318

Möschwitzer JP (2013) Drug nanocrystals in the commercial pharmaceutical development process. Int J Pharm 453:142–156.

Mosharraf M, Nystrom C (1995) The effect of particle size and shape on the surface specific dissolution rate of micronized practically insoluble drugs. Int J Pharm 122:35–47

Mulders JJL, Day Ap (2005) Three-dimensional texture analysis. Mater Sci 495–497:237–242

Müller RH, Keck CM (2012) Twenty years of drug nanocrystals: where are we, and where do we go? Eur J Pharm Biopharm 80:1–3

Müller RH, Peters K (1998) Nanosuspensions for the formulation of poorly soluble drugs I. Preparation by a size-reduction technique. Int J Pharm 160:229–237

Müller RH, Böhm BHL, Grau MJ (2000) Nanosuspensions—a formulation approach for poorly soluble and poorly bioavailable drugs. In: Wise DL (ed) Handbook of pharmaceutical controlled release. Marcel Dekker, New York, pp. 345–357

Müller RH, Jacobs C, Kayser O (2003) DissoCubes—a novel formulation for poorly soluble and poorly bioavailable drugs. In: Rathbone MJ, Hadgraft J, Robert MS (eds) Modified-release drug delivery systems. Marcel Dekker, New York, pp. 135–149

Nel AE, Mädler L, Velegol D, Xia T, Hoek EMV, Somasundaran P, Klaessig F, Castranova V, Thompson M (2009) Understanding biophysicochemical interactions at the nano–bio interface. Nat Mater 8:543–557

Niaz MS, Traini D, Young PM, Ghadiri M, Rohanizadeh R (2014) Investigation into physical-chemical variables affecting the manufacture and dissolution ofwet-milled clarithromycon nanoparticles. Pharm Dev Technol 19:911–921

Niwa T, Miura S, Danjo K (2011) Universal wet-milling technique to prepare oral nanosuspension focused on discovery and preclinical animal studies—development of particle design method. Int J Pharm 405:218–227

Noyes A, Whitney W (1897) The rate of solution of solid substances in their own solutions. J Am Chem Soc 19:930–934
O'Brien RW, Cannon DW, Rowlands WN (1995) Electroacoustic determination of particle size and zeta potential. J Coll Interface Sci 173:406–418
Otsuka M, Nishizawa J, Shibata J, Ito M (2010) Quantitative evaluation of mefenamic acid polymorphs by terahertz–chemometrics. J Pharm Sci 99:4048–4053
Pinna N (2005) X-ray diffraction from nanocrystals. Prog Coll Polymer Sci 130:29–32
Quan P, Xia D, Piao H, Shi K, Jia Y, Cui F (2011) Nitrendipine nanocrystals: its preparation, characterization, and in vitro-in vivo evaluation. AAPS PharmSciTech 12:1136–1143
Rabinov BE (2004) Nanosuspensions in drug delivery. Nat Rev Drug Discov 3:785–796
Rabinow B, Kipp J, Papadopoulos P (2007) Itraconazole IV nanosuspension enhances efficacy through altered pharmacokinetics in the rat. Int J Pharm 339:251–260
Raghava Srivalli KM, Mishra B (2014) Drug nanocrystals: a way toward scale-up. Saudi Pharm J. http://dx.doi.org/10.1016/j.jsps.2014.04.007
Rao S, Song Y, Peddie F, Evans AM (2011) Particle size reduction to the nanometer rtange: a promising approach to improve buccal absorption of poorly water-soluble drugs. Int J Nanomedicine 6:1245–1251
Sahoo NG, Kakran M, Shaal LA, Li L, Müller RH, Pal M, Tan LP (2011) Preparation and characterization of quercetin nanocrystals. J Pharm Sci 100:2379–2390
Salazar J, Müller RH, Möschwitzer JP (2014) Combinative particle size reduction technologies for the production of drug nanocrystals. J Pharm. http://dx.doi.org/10.1155/2014/265754
Shi Y, Porter W, Merdan T, Li LC (2009) Recent advances in intravenous delivery of poorly water-soluble compounds. Expert Opin Drug Deliv 6:1261–1282
Shono Y, Jantratid E, Kesisoglou F, Reppas C, Dressman JB (2010) Forecasting in vivo oral absorption and food effect of micronized and nanosized aprepitant formulations in humans. Eur J Pharm Biopharm 76:95–104
Steckel H, Rasenack N, Villax P, Müller BW (2003) In vitro characterization of jet-milled and in situ-micronized fluticasone-17-propionate. Int J Pharm 258:65–75
Tuomela A, Liu P, Puranen J, Rönkkö S, Laaksonen T, Kalesnykas G, Oksala O, Ilkka J, Laru J, Järvinen K, Hirvonen J, Peltonen L (2014) Brinzolamide nanocrystal formulations for ophthalmic delivery: reduction of elevated intraocular pressure in vivo. Int J Pharm 467:34–41
Van Eerdenbrugh B, Froyen L, Martens JA, Blaton N, Augustijns P, Brewster M, Van den Mooter G (2007) Characterization of physico-chemical properties and pharmaceutical performance of sucrose co-freeze-dried solid nanoparticulate powders of the anti-HIV agent loviride prepared by media milling. Int J Pharm 338:198–206
Van Eerdenbrugh B, Van den Mooter G, Augustijns P (2008) Top-down production of drug nanocrystals: nanosuspension stabilization, miniaturization and transformation into solid products. Int J Pharm 364:64–75
Van Eerdenburgh B, Alonzo DE, Taylor LS (2011) Influence of particle size on the ultraviolet spectrum of particulate-containing solutions: implications for in-situ concentration monitoring using UV/Vis fiber-optic probes. Pharm Res 28:1643–1652
Vehring R (2008) Pharmaceutical particle engineering via spray drying. Pharm Res 25:999–1022
Wang GD, Mallet FP, Ricard F, Heng JYY (2012) Pharmaceutical nanocrystals. Curr Opin Chem Engin 1:102–107
Weissig V, Pettinger TK, Murdock N (2014) Nanopharmaceuticals (part I): products on the market. Int J Nanomed 9:4357–4373
Williams HD, Trevaskis NL, Charman SA, Shanker RM, Charman WN, Pouton CW, Porter CJ (2013) Strategies to address low drug solubility in discovery and development. Pharmacol Rev 65:315–499
Willmann S, Thelen K, Becker C, Dressman JB, Lippert J (2010) Mechanism-based prediction of particle size-dependent dissolution and absorption: cilostazol pharmacokinetics in dogs. Eur J Pharm Biopharm 76:83–94

Wirth R (2008) Focused Ion Beam (FIB) combined with SEM and TEM: advanced analytical tools for studies of chemical composition, microstructure and crystal structure in geomaterials on a nanometre scale. Chem Geol 261:217–229

Wu Y, Loper A, Landis E, Hettrick L, Novak L, Lynn K, Chen C, Thompson K, Higgins R, Batra U, Shelukar S, Kwei G, Storey D (2004) The role of biopharmaceutics in the development of a clinical nanoparticle formulation of MK-0869: a Beagle dog model predicts improved bioavailability and diminished food effect on absorption in human. Int J Pharm 285:135–146

Wu L, Zhang J, Watanabe W (2011) Physical and chemical stability of drug nanoparticles. Adv Drug Deliv Rev 63:456–469

Yoshioka M, Hancock BC, Zografi G (1995) Inhibition of indomethacin crystallization in poly(vinylpyrrolidone) coprecipitates. J Pharm Sci 84:983–986

Yu WW, Chang E, Sayes CM, Drezek R, Colvin VL (2006) Aqueous dispersion of monodisperse magnetic iron oxide nanocrystals through phase transfer. Nanotechnol 17:4483–4487

Zheng JY, Bosch HW (1997) Sterile filtration of NanoCrystal drug formulations. Dev Ind Pharm 23:1087–1093

Zhu T, Ansquer JC, Kelly MT, Sleep DJ, Pradhan RS (2010) Comparison of the gastrointestinal absorption and bioavailability of fenofibrate and fenofibric acid in humans. J Clin Pharmacol 50:914–921

Part II
Characterization of NBCDs; Analytical Tools to Consider

Analytical Methods for Determining the Size (Distribution) in Parenteral Dispersions

David F. Driscoll and David F. Nicoli

Contents

Introduction 196
Safety of Pharmaceutical Dosage Forms 197
Active Pharmaceutical Ingredients and Drug Delivery Vehicles: NBCD Examples 198
  Physical Stability 201
  Vehicle Issues: Clinical Implications 202
  API Issues: Clinical Implications 204
Nanoparticle Sizing Techniques 206
  Single-Particle Optical Sizing (SPOS) 207
  SPOS 213
  SPOS: Combination of LE and LS—"LE+LS" 214
  SPOS: Focused Beam Light Extinction (FBLE) Method 217
  SPOS: Focused Beam Light Scattering (FBLS) Method 222
  Single-Particle Techniques: Electrical Sensing Zone (ESZ) Method 226
  Single-Particle Techniques 231
  Single-Particle Techniques 235
  Single-Particle Techniques 240
Synopsis 255
  Summary 255
  Conclusions 255
References 256

D. F. Driscoll (✉)
Stable Solutions LLC, Easton, MA, USA
e-mail: d.driscoll@stablesolns.com

UMASS Medical School, Worcester, MA, USA

D. F. Nicoli
Stable Solutions LLC, Goleta, CA, USA

Particle Sizing Systems LLC, Port Richey, FL, USA

D. J. A. Crommelin, J. S. B. de Vlieger (eds.), *Non-Biological Complex Drugs,* AAPS Advances in the Pharmaceutical Sciences Series 20, DOI 10.1007/978-3-319-16241-6_7

Abstract Analytical methods for nanomedicines are rapidly developing and include several potentially useful techniques. Of utmost importance, the physical methods applied must be robust, able to detect particles as small as 1 nm, and be stability-indicating since the nanostructures are designed as drug carriers. An unstable dispersion prevents the reproducible distribution of the active pharmaceutical ingredient(s), which may have clinically significant consequences. Aggregation, agglomeration and/or coalescence represent abnormal growth in the size of nanoparticles or nano-droplets, and are the hallmarks of an unstable dispersion. Hence, accurate quantification of these populations is essential for their safe use in patients. At the moment, the only official physical methods for pharmaceutical dispersions is for lipid injectable emulsions where dynamic light scattering is applied to determine the mean droplet diameter and light obscuration (or extinction) to monitor the stability of the population of large-diameter fat globules (>5 μm). These two distinctly different size regions of the droplet size distribution have been determined to be clinically important. There are no established pharmacopoeial methods for nanomedicines in the deep submicron range. Stability of the particle or droplet size distribution is particularly important when acutely ill patients in the critical care setting receive injectable dispersions such as propofol for sedation, clevidipine for the treatment of hypertension or amphotericin B to fungal fungal sepsis.

This chapter also discusses the growing number of non-pharmacopoeial analytical approaches that help the formulator to establish well defined and stable formulations for parenteral emulsions and dispersions in the deep submicron (i.e. nanometer) range. Furthermore, it explains the mechanisms behind these different analytical techniques, the size range where they can be applied, and their pros and cons compared with alternative approaches.

Keywords Particle aggregation · Pharmacopoeial methods · Active pharmaceutical ingredient or API · Drug vehicle · Large-diameter tail · Single particle size analysis · Quantitative analysis · Stability-indicating methods · Parenteral dispersions · Infusion safety

Abbreviations

ACF	Autocorrelation function
API	Active Pharmaceutical Ingredient
BSA	Bovine serum albumin
CCD	Charge-coupled device
CDC	Centers for Disease Control and Prevention
CLS	Classical light scattering
CMC	Critical micelle concentration
CV	Coefficient of variation
D	Diffusivity
DLS	Dynamic light scattering

DSD	Droplet size distribution
EDTA	Ethylenediaminetetraacetic acid
EMA	European Medicines Agency
ESZ	Electrical Sensing Zone
FBLE	Focused Beam Light Extinction
FDA	Food Drug Administration
FFF	Field flow fractionation
HNC	Hypernetted chain
I_S	Scattered light intensity
k	Boltzmann's constant
LDT	Large Diameter Tail
LS	Light Scattering
MAC	Monitored anesthesia care
MALS	Multi-angle light scattering
MDD	Mean droplet diameter
MEMS	Micro-Electro-Mechanical Systems
NBCD	Non-Biological Complex Drugs
NIST	National Institute of Standards and Technology
NLT	Not less than
NMT	Not more than
NTA	Nanoparticle Tracking Analysis
PSD	Particle Size Distribution
PBS	Phosphate buffered solution
PC	Phosphatidylcholine
PCS	Photon correlation spectroscopy
PE	Phosphatidylethanolamine
PFAT5	Volume-weighted percentage of fat exceeding 0.05 % (PFAT(5)
Pharm Eur	European Pharmacopoeia
PHD	Pulse height distribution
PI	Polydispersity index
PLGA	Poly(lactide co-glycolide)
PLS	Polystyrene latex
PMD	Particle mass distribution
PSD	Particle size distribution
QELS	Quasi-elastic light scattering
R_H	Hydrodynamic particle radius
RES	Reticulo-endothelial system
RMM	Resonant Mass Measurement
RPS	Resistive pore sensing
S/N	Signal/noise
SDS	Sodium dodecyl sulfate
SLS	Static light scattering
SMR	Suspended microchannel resonator
SNR	Suspended nanoparticle resonator

SPOS	Single-particle optical sizing
T	Temperature
TI	Therapeutic index
TMC	Trimethyl chitosan
TRPS	Tunable Resistive Pulse Sensing
USP	US Pharmacopoeia
η	Viscosity
θ_S	Scattering angle
λ	Wavelength
ρ	Particle density
τ	Characteristic time

Introduction

The main task of federal agencies overseeing drug approvals, such as the United States Food and Drug Administration (FDA) and the European Medicines Agency (EMA), is to ensure the safe use of drug products containing active pharmaceutical ingredients (API) to protect public health. Once an API has been approved for human use, a specific drug *monograph* is developed that outlines the critical physicochemical aspects of the API. In addition, *chapters* are also developed that outline specific methods of analysis appropriate to the API and the formulation. The task of writing monographs and chapters is usually accomplished by the respective pharmacopoeias, e.g., US Pharmacopoeia (USP) and the European Pharmacopoeia (Pharm Eur). There are selected committees within the pharmacopoeial agencies consisting of established experts in pharmacology and medicine, as well as physics and chemistry. Members of these committees often come from industry and academia. The essential physical and chemical data concerning an API are typically provided to the pharmacopoeial agency by the innovator company that developed the drug, reflecting the key information which was originally submitted to the FDA or EMA. From this information, drug monographs are developed that set the standard for the API that will eventually be applicable to future generic applications, which must be met by every manufacturer to ensure pharmaceutical and therapeutic equivalence. Whenever possible, the USP and the Pharm Eur collaborate on both drug monographs and chapters to ensure harmonization between the pharmacopoeias.

This chapter will focus on stability-indicating physical methods for the "carrier" or "vehicle", which is equally important as the API it is to deliver. Much of the historical focus on pharmacopoeial requirements has been on the API and excipients of drug products, and relevant assays to meet labeled requirements. Very little attention has been given to drug carriers, and only recently has there been the introduction of official methods for lipid emulsions applicable for nutritional purposes, as well as for drug delivery vehicles. We shall review the techniques recently employed, as well as others that most accurately characterize the particle/droplet size distributions of these complex dispersions.

Safety of Pharmaceutical Dosage Forms

The clinical significance of these pharmacopoeial tasks is greatly heightened with regard to injectable dosage forms, especially for those introduced directly into the systemic circulation. This includes the *intravascular* dissemination of APIs mostly through the venous circulation, but can also involve arterial routes of administration with nearly instantaneous pharmacological effects. With extravascular administration of drugs (e.g., intradermal, subcutaneous, intramuscular), the onset of pharmacological effects are delayed as the drug must variably pass through several "barriers" en route to the systemic circulation. With intravascular administration however, there are two major concerns that may assume clinical significance. First, introduction of an API directly into the bloodstream is essentially a "point of no return". That is, upon intravenous administration, the pharmacologically active compound, whether administered in error or not, or in the correct or incorrect dose, exerts nearly instantaneous effects that, if toxic, requires immediate intervention to reverse potentially life-threatening effects. In some cases, it is also possible to counter the effects of some extravascular injectable overdoses (subcutaneous, intramuscular) by application of a tourniquet distal to the injection site. It is important to recognize that all therapeutic agents used in medicine are toxic when given in sufficient doses, and that the dose-response relationship is a central tenet in both pharmacology and toxicology. The significance of this relationship is clearly articulated in two premier textbooks in pharmacology and toxicology, as noted below:

> Most pharmaceuticals are threshold poisons; at therapeutic dosing the drug is used to confer a health advantage, but at higher doses the drug may produce a toxic effect. (Osterhoudt and Penning 2011)

> Whatever response is selected for measurement, the relationship between the degree of response of the biological system and the amount of toxicant administered assumes a form that occurs so consistently as to be considered the most fundamental and pervasive concept in toxicology. (Eaton and Gilbert 2013)

Second, the intravascular dosage form must also be carefully prepared to ensure it is safe for administration directly into the systemic circulation. Physical or chemical incompatibilities can produce clinically significant morbidity and mortality. For example, incompatible combinations of intravenously supplied calcium and phosphate salts have resulted in the formation of large-diameter (>5 µm), solid precipitates of dibasic calcium phosphate that have caused occlusion of the microvasculature, producing pulmonary embolism and deaths in patients (Newton and Driscoll 2008). As well, unstable and/or coarse intravenous emulsions extemporaneously prepared in syringes have produced significant hypertriglyceridemia in critically ill, premature infants (Martin et al. 2008).

So clearly, any pharmaceutical dosage form intended for administration directly into the bloodstream poses the most serious health risks, and thus, the focus of this chapter will be on (nano)dispersions intended for intravascular injection. Of course the information in this chapter and the analytical procedures described will be applicable to all dosage forms containing non-biological complex drugs (NBCDs) as dispersions.

Parenteral dispersions intended for intravenous administration, where a drug is carried in a specialized vehicle (e.g., lipid droplets, liposomes, micelles, nanoparticulates), and the implications for safety and efficacy will be discussed. Because of the critical link between dose and effect (i.e., the dose-response relationship), review of the analytical techniques will be concentrated on physical methods of quantitative analysis employing various single-particle sizing techniques, having the highest size resolution, as well as particular, widely used "ensemble" methods, which together span a broad size range (1–1000 nm or larger). Thus, physical methods of analysis that can accurately monitor the droplet/particle size distribution over time are central to this discussion.

Active Pharmaceutical Ingredients and Drug Delivery Vehicles: NBCD Examples

Pharmacopoeial requirements for the APIs are very specific with regard to the concentration range for a particular compound, both for the active ingredient and the dosage form. For example, to meet pharmacopoeial requirements according to the USP for the "Propofol" drug monograph, it states "Propofol contains NLT [not less than] 98.0 % and NMT [not more than] 102.0 % of $C_{12}H_{18}O$", whereas for the Propofol Injectable Emulsion monograph, the 10 % (w/v) formulation defines the concentration requirements as "NLT 90.0 % and NMT 110 % of the labeled amount of propofol ($C_{12}H_{18}O$)" Hence, for every drug monograph it must meet certain *chemical* stability requirements, whereby the methods applied must not only be stability-indicating (i.e., differentiate between parent drug and degradation products), but also ensure that a fixed amount of API is present at all times during the manufacturer-assigned shelf-life of the product (e.g., 24–60 months). The amount of API present at all measurement intervals during its shelf-life must meet a certain percentage of the labeled amount. For many drugs, the pharmacopoeial range for the API in a particular dosage form is ± 10 % of the labeled amount so that, for example, a dosage form containing 100 mg/mL of an API must contain between 90 and 110 mg/mL to meet pharmacopoeial requirements at all times during the defined shelf-life period. But for other drugs, the ranges can be higher or lower, depending primarily on the therapeutic index (TI) of the drug (e.g., ratio of the toxic dose to a therapeutic dose) and/or the limits of the analytical methods for determining its final concentration. For example, the lower the TI, the lower the pharmacopoeial limit, e.g., ± 5 % or less; for a drug with a higher TI, a higher limit is tolerated, e.g., ± 15 % or more.

Pharmacopoeial standards for injectable pharmaceutical *solutions* are mainly concerned with the chemical stability of the API. Of course, concentrations of intrinsically or extrinsically-introduced particulate matter must be minimized, but these particles rarely cause clinically significant harm. This is certainly true since the adoption of specified particle limits. All commercially available aqueous injectables in the US and Europe, for example, must meet certain pharmacopoeial requirements for particulates not detected by unaided visual inspection as described in a *harmonized* chapter on particulate matter in injections from the USP and Pharm Eur. This chapter places limits on the number of particles present based on the volume of an injectable product, as follows:

Solutions for injection supplied in containers with a nominal content of **more** than 100 mL:

- The preparation complies with the test if the average number of particles present in the units tested does not exceed 25 per mL equal to or greater than 10 μm and does not exceed 3 per mL equal to or *greater* than 25 μm.

Solutions for injection supplied in containers with a nominal content of **less** than 100 mL:

- The preparation complies with the test if the average number of particles present in the units tested does not exceed 6000 per container equal to or greater than 10 μm and does not exceed 600 per container equal to or greater than 25 μm.

Intravenous lipid emulsions for nutritional purposes were developed in the late 1950s and the first successful commercially available product was introduced in 1962 under the brand name Intralipid™, followed by several others. It consisted of a 10 %w/v (higher concentrations followed, 20 and 30 %w/v, respectively) soybean oil dispersed in sterile water for injection and stabilized by a phospholipid emulsifier (egg lecithin), with an average droplet diameter approximately between 200 and 400 nm (depending upon the oil concentration). Over time, the oil phase has been modified to include MCT, fish oil, and olive oils in varying ratios. Although the product has been used in critically ill infants and adults since then, it wasn't until approximately 45 years later that an official pharmacopoeial chapter (United States Pharmacopeia, USP37/NF32 2014a) and monograph (United States Pharmacopeia, USP37/NF32 2014b) were adopted. With respect to the pharmacopoeial requirements for the *physical* stability of the emulsion, the USP <729> stipulates that Method I, applying dynamic light scattering, be applied to determine the mean droplet diameter (MDD) and Method II, applying a light obscuration, or light extinction, method that employs a single particle optical sensing technique be applied to determine the extent of the large-diameter tail (PFAT5) and includes the following two globule size limits:

1. "The intensity-weighted mean droplet diameter (MDD) for lipid injectable emulsions must be less than 500 nm or 0.5 μm, irrespective of the concentration of the dispersed phase."
2. "The volume-weighted, large-diameter fat globule limits of the dispersed phase, expressed as the percentage of fat residing in globules larger than 5 μm (PFAT5) for a given lipid injectable emulsion, must not exceed 0.05 %."

In the 1980s, a highly effective anesthetic/sedative known as propofol was reformulated as a 10 % oil-in-water emulsion drug vehicle because of anaphylactoid reactions associated with Cremophor EL™, a nonionic solubilizer/emulsifier used in the original formulation. The current lipid emulsion formulation contains 10 mg/mL of the API with a mean droplet diameter of approximately 200 nm. Due to the extreme water-insolubility of the API (~ 1 g in 10,000 mL), the drug has to be incorporated into the oleaginous phase of an oil-in-water emulsion to be safe for intravenous administration. With an octanol/water partition coefficient of 6761:1, propofol can be readily incorporated into the oil phase of the emulsion. Today, it is one of the most successful injectable anesthetic/sedative agents in the world and has largely enjoyed a successful safety record. In 1995, however, the CDC (Centers for Disease Control and Prevention) reported unusual outbreaks of infections in seven U.S. hospitals involving at least four deaths (Bennett et al. 1995). Consequently, the FDA mandated the formulation to contain a preservative, and EDTA was added. Otherwise, the drug vehicle is essentially identical to the nutritional injectable emulsion formulation, plus the API, and has similar physicochemical requirements. In fact, in 2009 the USP adopted a new monograph for the product entitled "Propofol Injectable Emulsion", stating that it must meet the globule size requirements of USP Chapter <729> (United States Pharmacopeia, USP37/NF32 2014c). Other water-insoluble drugs in lipid injectable emulsions have entered the market, including, for example, diazepam, etomidate and clevidipine, but pharmacopeial monographs for *these have yet to be developed.*

In the 1990's, liposomal drug delivery systems, for drugs such as doxorubicin, daunorubicin, amphotericin B, were developed and introduced into clinical use. Unlike propofol, the API is encapsulated within the aqueous core or within the lipid bilayers of individual vesicle consisting of various phosphatides (PC, PE) and cholesterol. The mean vesicle diameters in these formulations is between 20 and 100 nm, and the API can reside within the lipid bilayer, or inside the aqueous region of the liposome, depending on the drug. *There are no official monographs or chapters for these formulations.*

A number of other NBCDs (e.g., iron-saccharide complexes, polymeric micelles) offer other examples of pharmaceutical dispersions, where the API and the drug vehicle are inextricably linked. That is, both the chemical stability of the API and the physical stability of the carrier must be optimized to ensure a beneficial therapeutic outcome. These NBCDs require a special carrier that determines the safety and efficacy of the API. Not only is the chemical integrity of the API critical, but also the physical stability of the vehicle assumes equal importance. Thus, like all pharmaceutical dispersions, the composition, quality and *in vivo* performance of these NBCDs are highly dependent on the manufacturing processes of the active ingredient, as well as (in most cases) the formulation. *There are no official monographs or chapters for these formulations.*

Thus, there is a great need to develop (harmonized) pharmacopoeial standards for dispersed dosage forms. From a clinical perspective, this is particularly true for those formulations intended for intravenous administration. Generic versions of these products must demonstrate pharmaceutical and therapeutic equivalence in order to ensure a beneficial therapeutic outcome. As an example of the importance of formulation: in the 1990s three different intravenous dispersion formulations of the antifungal agent amphotericin B became available in the U.S., including: (1) Abelcet^{TM}, a "phospholipid" complex with an adult dose of 5 mg/kg/day; (2)

Amphotec™, a "cholesteryl sulfate" complex with an adult dose of 3–4 mg/kg/day; and, (3) Ambisome™, a "phospholipid-cholesterol" complex with an adult dose of 3–5 mg/kg/day. So clearly, changing the drug vehicle changes the effective clinical dose.

Physical Stability

Assuming the chemical stability of the API is maintained throughout the shelf-life of the product, this discussion will focus on the unique physical stability issues of the drug vehicle, and the associated clinical implications. As the API is intended to be uniformly distributed within the stable particle size distribution of the carrier, the ideal dispersion should be one that is a "fine" dispersion and has a narrow particle (or globule) size distribution, and maintains that over the course of the product's shelf-life. If the dispersion becomes unstable, ("coarsens") there are two major consequences that may assume clinical significance: (1) *Vehicle Issues*: agglomeration of particles or globules that comprise the vehicle structure can produce sufficiently large diameters that they may produce an embolism, particularly within the pulmonary microvasculature following intravenous administration; and, (2) *API Issues*: the homogeneous distribution of the API is severely altered, affecting therapeutic efficacy. From a physical stability standpoint, the ideal formulation successfully incorporates the API into the carrier and the vehicle of the final dispersion maintains its physical stability. Factors that may adversely affect the stability of the vehicle (i.e., alterations in temperature, pH, oxygen concentration, light exposure, addition of electrolytes, expansion of the aqueous phase, etc.) must be identified and controlled. Charge-stabilized dispersions, such as triglyceride oil-in-water emulsions may be stabilized by an emulsifier that imparts a net negative surface charge, and zeta potential, upon the droplet surfaces in order to establish mutual repulsion. These effects can be reversed or neutralized by the presence of sufficient concentrations of counterions (i.e., positively-charged cations such H^+, Na^+, Mg^{2+}, Ca^{2+}, Fe^{3+}, Al^{3+}), whereby these ions can adsorb to the negatively-charged surfaces of lipid droplets or can "screen" the repulsive forces between them, leading to agglomeration and coalescence within the dispersed phase. No matter what the mechanism destabilizing the dispersion, growth of nano-sized particles or droplets is consequential and likely influences clinical outcomes. To illustrate the importance of the particle or droplet size distribution of API-containing dispersions, and due to the dearth of pharmacopoeial guidance on these formulations, lipid injectable emulsions as drug vehicles will be used to provide a window into the key issues that are relevant for similar nanodispersions having different vehicle structures, and much smaller mean sizes. With respect to the analytical discussions to follow, USP Chapter <729> entitled: *Globule Size Distribution in Lipid Injectable Emulsions* and the USP Monograph entitled: *Propofol Injectable Emulsion* will serve as reference points for development of future pharmacopoeial articles for consideration regarding nanodispersion vehicles including colloidal vehicle based NBCDs.

Vehicle Issues: Clinical Implications

Drug-containing or drug-free lipid injectable emulsions have a mean droplet diameter (MDD) between 200 and 400 nm, depending upon the formulation. With respect to the physical stability of the dispersions, USP Chapter <729> places an upper MDD limit of 500 nm on vegetable and/or marine-based triglyceride oil-in-water emulsions. It also places a limit on the large-diameter fraction (>5 μm) of the globules found in the over-size tail of the distribution (i.e., the "large-diameter tail"). The main reason for limits on the population of these large-diameter globules is the recognition that they pose a heightened danger of fat embolism by *mechanical* obstruction of the pulmonary microvasculature, since the threshold internal diameter of the capillaries is approximately between 4 and 5 μm. Hence, infusion of sufficient numbers of large-diameter fat globules into the venous system, which provides immediate access to the pulmonary circulation, poses a significant clinical risk to patients.

Pathological accumulation of large-diameter fat globules has been demonstrated in the lungs of rodents, (Driscoll et al. 2005) but fat globule accumulation in the liver also engenders metabolic toxicity. Evidence of heightened oxidative stress in hepatic tissues, enzymatic evidence of liver damage, and a mild systemic inflammatory response via elevated C-reactive protein (CRP) levels in blood have also been shown (Driscoll et al. 2006d, 2008, 2009a). The effects on the liver are not surprising since it plays a major role in the reticuloendothelial system (RES) and phagocytic functions. Of the major organs that participate in the RES (liver, lungs, bone marrow and spleen), the liver and spleen are responsible for between 85 and 95 % of intravascular phagocytic activity (Saba 1970). Accumulation of lipid in the livers of rodents has also been shown to impair Kupffer cell function, thus reducing their capacity to clear bacteria, and causing increased sequestration of microorganisms in the lungs, (Hamawy et al. 1985) and possibly increased risk of pulmonary infections. Impairments in RES function in humans can occur by the intravenous infusion of excessive doses of lipid emulsions (Seidner et al. 1989) and/or infusion of unstable dispersions. The adverse effects can be mitigated by reducing the infusion rate for long-chain triglycerides (LCTs) or by reducing the LCT load by substitution of a portion of the lipid calories with medium-chain triglycerides (MCTs) (Jensen et al. 1990). No such alleviation can occur when the dispersion becomes unstable and forms increasing numbers of large-diameter fat globules that pose embolic risks from mechanical obstruction and also increase the participation of the RES to phagocytize and contain them. As the RES is a key component of the immune system, increased RES activity also triggers a systemic inflammatory response. With respect to pharmaceutical dispersions entering the blood stream, the principal role of the RES is the clearance of foreign colloidal and particulate matter. One can reasonably conclude that the RES plays a critical role for all nanodispersions, and its importance in health and disease is best summed up as follows (Saba 1970):

> The RES [however] does not function as merely a passive scavenger system, but indeed, its functional activity can become activated or depressed by a variety of endogenous and

exogenous factors. For example, the activation of the RES during bacterial infection, neoplastic disease, and diseases of autoimmunity has been suggested to be a physiological host-defense response, while depression of the RES associated with circulatory failure may be a crucial factor in the development and progression of a disease process.

Another issue of clinical importance with regard to the stability of the drug vehicle is the composition of the carrier. In the case of lipid injectable emulsions, the spherical nanodroplets are made from vegetable and/or marine oil triglycerides. They are flexible, which has important toxicological implications when compared to nanoparticulate carriers. To illustrate the differences, one should consider the physiological conditions in which large-diameter fat globules (>5 μm) exceed the internal diameter of blood vessels, such as the pulmonary capillaries. In this case, there is a "temporary" occlusion. The physiological response to this presentation triggers an increase in pulmonary artery pressure. This causes the deformation of the fat globule as it squeezes through the capillary, much in the same way that red blood cells (~7 μm) move through the circulatory system, and the fat globules proceed downstream to vital organs, such as the liver. In contrast, large-diameter aggregates of nanoparticulates are less flexible or even rigid, such that the aforementioned compensatory response by the pulmonary arteries cannot dislodge them. Nanoparticulates can accumulate in sufficient quantities at the microvasculature level, causing pulmonary artery pressures to increase to pathological proportions (e.g., pulmonary hypertension), and eventually induce clinically significant pulmonary embolism. So one can reasonably conclude that the median lethal dose (LD50) would be much lower for rigid (nano)particulate vehicle structures compared to those comprised of flexible, oleaginous ones. The potential toxicological differences between vehicle structures are shown below (Table 1).

Thus, the physical stability of the vehicle is essential to its safety when used in humans. But there is a second component related to the vehicle necessary that must be evaluated before completing the assessment of its toxicological profile, and that is the dose of the toxicant. How many droplets/globules/particulates does it take to cause harm? Is the toxic dose a result of mechanical injury (e.g., embolic phenomena), and/or of the chemical composition/size distribution of the dispersion as these factors affect oxidative stress from organ accumulation (hepatocytes and Kupffer cells), and/or 'saturation' and 'silencing' of the RES? Are there different doses that

Table 1 Parenteral "nanoparticles": physical characteristics and potential toxicity

"Flexible" or "soft" nanoparticles
Example: oil or lipid-based, such as droplet or liposomal structures
Flexible, malleable, deformable
Low embolic risk, higher LD50
"Rigid" or "hard" nanoparticles
Example: solid particulates, such as carbon or graphene structures
Inflexible, rigid, obstructing
High embolic risk, lower LD50

cause mechanical vs. chemical injury? The toxic threshold of these (nano)structures needs to be quantified. The methods of analysis for these droplet/particle-laden dispersions must be robust in terms of being quantifiable so that toxic doses for specific formulation characteristics (e.g. many or few large particles) can be identified.

API Issues: Clinical Implications

Finally, the physical instability of the vehicle greatly influences safety and efficacy of the drug product. Assuming the API is uniformly distributed within the dispersed phase (i.e., based on the droplet/globule or particle size), the concentration is proportional to the sizes in the overall Particle or Droplet Size Distribution (PSD or DSD). In this case, continuing with the example of the innovator product for propofol injectable emulsion (Diprivan™) as the gold standard for all generic versions, we can possibly assess the uniformity of the distribution of the API within the dispersion. But first, we must know how many droplets exist in a stable dispersion. If it were a monodisperse system, assuming for simplicity that all droplets are 200 nm in diameter, and knowing the oil concentration (10 g/dL) and density (soybean oil ~0.91 g/mL), as well as the volume of a 200 nm sphere ($4/3\pi r^3$), the droplet concentration can be estimated to be approximately 2×10^{13} droplets/mL of emulsion. Knowing the octanol/water partition coefficient for propofol is 6761:1, we are confident that most of the propofol is inside the dispersed phase. The free propofol concentration in the aqueous phase of the innovator product has been shown to be approximately 20 μg/mL or approximately 0.2 % of the labeled amount in the formulation (Babi et al. 1995). Of course, the propofol injectable emulsion is not single-sized, but a polydisperse system. For example, we know that the innovator product meets the pharmacopoeial requirements related to globule size limits as outlined in USP Chapter <729>, and that it is mandated in the official propofol injectable emulsion USP monograph. Given the two globule size thresholds (i.e., 500 nm and 5 μm) and physical measurements to meet pharmacopoeial requirements, it is clear that propofol injectable emulsions are polydisperse formulations (e.g., having a droplet size range from approximately 10 nm to 10,000 nm). We also know that the various doses specified in the package insert, which correspond to the FDA approved indications (e.g., induction and maintenance of general anesthesia, initiation and maintenance of MAC sedation), were derived from clinical trials with the propofol injectable emulsion dispersion. Therefore, it can be inferred that the API is uniformly distributed amongst the lipid droplets present, which have a mean droplet diameter of approximately of 200 nm and a PFAT5 of < 0.05 % and are safe and effective.

With regard to nanodispersions having smaller mean droplet or particle diameters, clearly the population (by number) greatly increases. Again, as an example, assuming that such a nanodispersion is monodisperse at 100 nm, the droplet concentration would increase by approximately 1-log higher than at 200 nm, to approximately 2×10^{14}/mL. At 20 nm, recognizing that for a 1-log change in size (200 → 20 nm), there is a corresponding 3-log change in volume, the droplet count would be approximately 2×10^{16} droplets/mL. As will be seen later, it must be appreciated that these magnitudes in droplet concentration, coincident with reduction in size, are critical factors for physical measurement of nanodispersion based NBCDs. Again, the analytical methods must be able to size and quantify the differences, and do so in a manner such that the data sample is statistically relevant or maximized for these highly concentrated nanodispersions, in order to discern a stable (safe) vs. an unstable (unsafe) formulation. For all intents and purposes, these particle/droplet-laden nanodispersions largely render light microscopic techniques (Driscoll et al. 2006c) to be of little applicability from the vanishingly small, and statistically irrelevant sample measurements.

In our current example, the USP requires that the globule size limits for propofol injectable emulsion must meet two criteria: (1) MDD < 500 nm; and, (2) PFAT5 must not exceed 0.05 %. The stability and safety parameter is the PFAT5 value as described, but why such a gap in the size range? The MDD is derived from an ensemble particle sizing method (semi-quantitative), and as such it is a qualitative measure, used to guide the homogenization process, whereas the PFAT5 parameter is *both* a qualitative measure (with respect to the large-diameter tail speaking to the fineness vs. coarseness of the dispersion) and quantitative criterion (regarding the safety/efficacy of the formulation). This would suggest that the mean diameter establishes a "marker" from which the "critical region" in the large-diameter tail can be identified quantitatively in order to optimize the formulation. In fact, it was found that the critical population of droplets in the overall distribution that assumes clinical importance is the population of globules that is approximately 1-log unit higher (5000 nm) than the mean droplet size (500 nm). In fact, the commercial significance of this "mean-tail" relationship has been shown to exist for non-medical applications where we state the following:

> The most devastating effects of agglomeration or coalescence in a commercially prepared dispersion or emulsion often occur in a size range that is approximately ½ to 1 log larger than the mean particle or droplet size. (Nicoli et al. 2006)

We have also tested and observed this critical linkage to performance in a number of non-medical applications such as inorganic colloidal suspensions, (e.g., slurries for chemical-mechanical polishing of silicon wafers used for semiconductor device manufacturing); aqueous polymer suspensions (e.g., paints and coatings); edible oil-in-water emulsions (e.g., beverage concentrates); pigment-based suspensions (e.g., ink-jet printing); dairy products (e.g., homogenized milk); and, asphalt-based oil-in-water emulsions (e.g., roadway surfaces), among others. This "connection" will have to be confirmed for nanodispersion NBCDs based on the analytical methods to be applied. In the case of these dispersions, the potentially clinically significant regions in the tail of the overall droplet/particle distribution, based on the mean diameter of the dispersion, to which critical measurements are applied, are shown below (Table 2).

Therefore, if the emulsion is coarse or becomes unstable, there is a clear shift in the DSD commensurate with the growth of submicron droplets into large-diameter fat globules above 1 μm. When this occurs, there is also a consequential shift in the API distribution when a coarse dispersion is made or if coalescence occurs during

Table 2 Key regions in the large-diameter tail (LDT) based on mean diameter

Mean diameter, nm	Key regions	
	LDT-1 threshold, nm	LDT-2 threshold, nm
10	50	100
25	125	250
50	250	500
100	500	1000
250	1250	2500
500	2500	5000

its shelf-life. The larger droplets or particles will contain substantially more drug and could either cause the NBCD to be sub-therapeutic or toxic. That is, if these large-diameter globules or particles “float” or “sink” in the unstable dispersion, the drug is no longer homogenously dispersed. If this large-diameter population is infused, an inadvertent “bolus” of concentrated medication could occur, potentially producing a toxic response or, alternatively, if they adhere to the intravenous administration set or are otherwise retained in the infusion container and therefore not infused, a sub-therapeutic effect may be observed.

Since the only relevant methods that are pharmacopoeially-approved for nanodispersions at this time exist for lipid injectable emulsions, we shall use USP Chapter <729> as a guide to choose the most appropriate regions of the particle/droplet size distribution and applicable methods (Driscoll 2004) to ensure the safety and efficacy of these much smaller, nanodispersions. Hence, dynamic light scattering will be the principle method used for determining the mean diameters of particles smaller than ≈50–100 nm, while the clinically important “large-diameter tail”, which is approximately ½ to 1-log above the mean size, must be quantified by appropriate single-particle sensing techniques. This “mean-tail” has been established for lipid injectable emulsions (Driscoll 2007, 2009; Driscoll et al. 1995, 2000, 2001a, b, 2003, 2006a, b, 2007a, b, 2009b; Martin et al. 2008) and was the basis for the two-stage procedure (Driscoll 2004) that was ultimately adopted by the USP in 2007 for Chapter <729>. To assess the safety limits of the large-diameter tail of any nanomedicine, i.e., to establish both the therapeutic and toxicological dose-response relationships (especially for injectable products), the method must be able to obtain samples that are accurately sized, quantifiably precise, and statistically relevant.

Nanoparticle Sizing Techniques

There are numerous techniques for particle size analysis that can characterize, with varying degrees of accuracy and resolution, the particle size distribution (PSD) of nanoparticle suspensions. Broadly speaking, the different techniques can be divided into “single-particle” and “ensemble” methods. In ensemble techniques, the detected "signal" to be analyzed consists of a mixture, or "superposition", of the responses generated simultaneously by many particles, possibly representing a large range of sizes (and concentrations). The raw data must therefore be "inverted", or "deconvoluted", by using an appropriate mathematical algorithm to obtain the desired PSD, necessarily resulting in an approximation of the "true" distribution. By contrast, with single-particle techniques the detected "signal" is produced by only one particle at any given instant of time. Hence, obtaining the desired PSD is in theory a simple matter of incrementing the accumulated particle count in the appropriate size bin, or channel, in a multi-channel representation of the PSD.

In general, single-particle techniques are usually preferred over ensemble methods because of the superior size resolution that they inherently are able to deliver, resulting in a PSD that most closely resembles the "true" particle size distribution. This preference is most evident for a broad PSD that possesses substantial "polydispersity" such as a "tail" of over-size particles extending well beyond (above) the main population of particles. A good example is a therapeutic drug product containing a concentrated suspension of protein molecules, in which there is a significant "tail" of aggregated molecules, often extending out to sizes more than 10, or even 100, times larger than the size of the individual, un-aggregated ("native") protein molecules. The latter are small enough to require the use of the well-known ensemble technique of dynamic light scattering (DLS) or another method for size determination, which will be effective at sizing the unaggregated monomer molecules, provided the concentration of much larger aggregate particles is not excessive. PSDs that possess a high degree of polydispersity, typically require a sizing technique that is capable of providing a relatively high degree of resolution, in order to accurately reveal important "details". Hence, this chapter will focus on the principles and characteristics of single-particle techniques for particle counting and sizing, followed by a last section that reviews the underlying principles and capabilities of the DLS ensemble technique.

Single-Particle Optical Sizing (SPOS)

Light Extinction (LE, or LO) Method (Knollenberg and Gallant 1990; Holger 1990; Knapp and Abramson 1990)

Sensors based on single-particle optical sizing (SPOS), historically referred to as optical particle counting (OPC), use a laser light source and specialized optics to generate a thin, ribbon-like beam of light that passes through an optical flow cell, shown schematically in Fig. 1 (left).

The two lateral dimensions of the flow cell are typically in the range of 0.4–1 mm. The cross section of the flow channel and the thickness ($1/e^2$ width) of the incident "ribbon" of laser light define an optical sensing zone, or "view volume", through which particles pass at an appropriate, controlled flow rate. Individual particles are detected by one of two optical means, light extinction or light

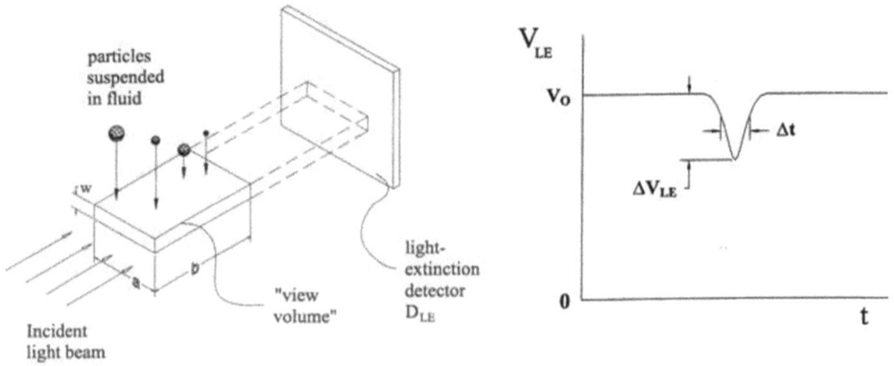

Fig. 1 SPOS: Simplified optical diagram and resulting light extinction (LE) or light obscuration (LO), signal. A ribbon-shaped beam of light passes through an optical flow channel, defining an optical "view volume", and impinges on a distant LE (or LO) detector. In the absence of a particle the signal, V_{LE}, is a steady (maximum) voltage, V_0. Momentary passage of a particle through the view volume results in a negative-going pulse of height ΔV_{LE}, superimposed on V_0

scattering. Appropriate optical elements (focusing and cylinder lenses) are used to produce a light intensity profile that is approximately uniform over the cross section of the flow channel (and "thin" in the direction of particle flow), producing a detector response that is approximately uniform for a particle of a given size, regardless of its trajectory. A sample of suspended particles must typically be diluted to ensure that the particle concentration over the range of detectable sizes lies below the "coincidence limit", so the great majority of particles (>99 %) pass one at a time through the view volume, ensuring that the detected signal pulses and resulting PSD will not be appreciably distorted by more than one particle passing through the view volume at substantially the same time.

The first, and simplest, physical method of particle detection utilized in an SPOS sensor was that of light extinction (LE), historically called light obscuration (LO) in pharmaceutics. Light rays impinging on a particle are scattered, in the case of relatively small particles (i.e., diameter $d \leq \lambda$, the light wavelength), or deflected (i.e., refracted) in the case of relatively large particles (i.e., $d >> \lambda$), away from the axis of the incident light beam, so these rays fail to impinge on a distant detector. Hence, passage of a particle through the view volume results in a momentary reduction in the light flux impinging on the light detector and a corresponding decrease in the signal voltage, in the form of a negative-going pulse relative to the "baseline" voltage V_0 in the absence of a particle, shown schematically in Fig. 1 (right).

A typical LE, or LO, sensor utilizes an optical flow cell of inner dimensions A = 0.4 mm ("width", along the expanded dimension of the incident light beam) and B = 1 mm ("depth", along the axis of the incident beam, in line with the distant LE detector). Assuming a typical flow rate of 60 mL/min for the sample fluid, the velocity of a particle passing through the flow cell is 2.5 m/s. Further assuming a typical focused laser beam width, or "waist" ($1/e^2$), w_0, of 40 um at the center of the flow channel (thus defining the detection view volume, $A \times B \times w_0$), the width Δt of the resulting negative-going signal pulse (assuming $d << w_0$) is ≈ 16 µs.

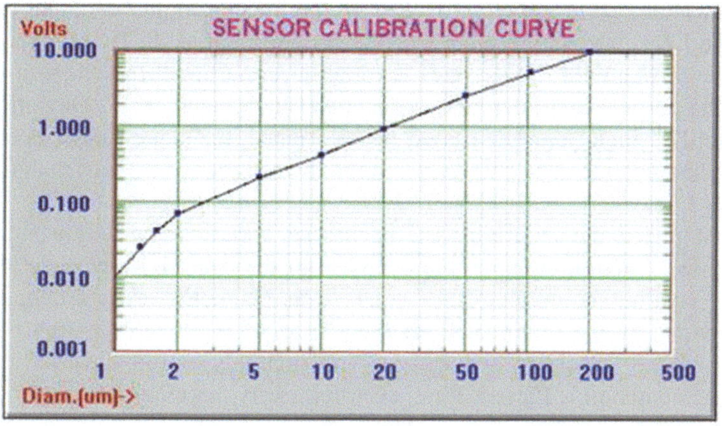

Fig. 2 LE calibration curve: pulse height (V) *vs* particle diameter (μm). Plot of light extinction pulse height (0.025–10 V) vs particle (PSL) diameter (1.33–200 μm) for a typical LE sensor. (Data on File, Particle Sizing Systems LLC)

The particle size is determined from the height, ΔV_{LE}, of the detected LE signal pulse. In general, the larger the particle, the greater ΔV_{LE}. The LE sensor is "calibrated" by measuring its response for certified polystyrene latex (PSL) "standards" of uniform size. The smallest PSL particles that can be reliably detected by this sensor are ≈ 1.3 μm in size, with $\Delta V_{LE} \approx 24$ mV, about 3 × larger than the underlying noise level (≈7 mV, with $V_0 = 10$ V).The pulse height *vs* particle size values comprising the "calibration curve" are shown in Fig. 2 (Data on File, Particle Sizing Systems LLC).

The size of the largest PSL standard particles used for this calibration was 200 um, resulting in a pulse height of 9.78 V, nearly the largest possible value before amplifier "saturation". Over most of the size range (>2 μm) the response curve, after signal conditioning, is approximately linear (log-log), showing that this LE sensor possesses high sizing *resolution.* A given change in particle size, Δd, results in a corresponding, proportional change in measurable pulse height, $\Delta(\Delta V_{LE})$, with similar proportionality over virtually the entire range of measurable particle sizes. A second feature of this calibration curve is that the slope of ΔV_{LE} *vs* d clearly has a point of inflection at about 2 um, where the pulse height response falls more steeply with decreasing size below that diameter. This behavior is the result of a transition in the physical mechanism of light scattering that is responsible for the observed light extinction.

In the size range of 1.3 to about 3 μm the particle size is only moderately larger than the laser wavelength, (785 nm, or 0.785 μm). Hence, as explained by the Mie intra-particle scattering theory, a particle resembles, from a scattering point of view, a thin shell of mass, because much of the scattering originating from the interior of the particle effectively disappears due to destructive interference of the individual scattered light wavelets. The volume of the shell is roughly proportional to the surface area, in turn proportional to the square of the particle diameter, d.

The net scattering intensity, removed from the incident light beam and contributing to the LE signal, is therefore roughly proportional to d^4. By contrast, the LE signal for substantially larger sizes is roughly proportional to the cross sectional area of the particle that intercepts the incident light beam, or d^2. Deviations from this simplified picture are discussed below. Hence, the slope (log-log) of the LE response should indeed be ≈2 × larger at the lowest sizes (<≈2 µm) than for higher sizes (>≈2 µm).

At first glance, the underlying LE mechanism appears to be simple and straightforward. The fraction of light "removed" from the beam by a particle located in the view volume should simply be equal to its geometric cross section, denoted by G, divided by the illuminated area, equal to $A \times w_0$, where $G = \pi r^2$ (r = particle radius), or $\frac{1}{4}\pi d^2$. There are two respects in which this simple, idealized picture of the LE phenomenon is flawed. First, even for a very large particle (>>λ), the fraction of light removed from the incident beam is given by *twice* its cross-sectional area—i.e., $2G/Aw_0$, rather than simply G/Aw_0. This behavior is known as the "Extinction Paradox", simply explained by van de Hulst (van Hulst 1981): "A flower pot in a window prevents only the sunlight falling on it from entering the room, and not twice this amount, but a meteorite of the same size somewhere in interstellar space between a star and one of our big telescopes will screen twice this light." In reality there are *two* contributions to the LE signal. All energy falling on the particle is scattered or absorbed and therefore removed from the incident beam. However, light is also *diffracted* by the particle, and the amount of incident light removed from the beam is equal to the amount of light removed by light scattering. Therefore, the total amount of light removed by the particle is *twice* that removed by light scattering alone.

There is a second way in which the simplified picture of the LE mechanism is flawed. The extinction efficiency, Q, of a particle is defined as the ratio C/G, where C is the effective area of the incident light beam removed by the particle. In the naïve picture suggested above, C = G and therefore Q = 1. In reality, Q depends not only on d, but also on λ, as well as the refractive indices of both the particle and the suspending fluid. The dependence of Q on the particle radius, r, and these other parameters is summarized by the following equation:

$$Q = 2\ (4/\rho)\ \sin(\rho) + (4/\rho^2)\ [1 - \cos(\rho)] \qquad (1)$$

where $\rho = 2x\ (m - 1)$, $x = 2\pi r\ (m_2/\lambda)$ and $m = m_1/m_2$

m_1 refractive index of the particle
m_2 refractive index of the suspending fluid

A graphical representation of the effects of both the particle size and refractive index on the light extinction efficiency, Q, is shown in Fig. 3.

There is a plot of Q *vs* d for particles of PSL (used for sensor calibration) suspended in water ("Q-PSW") and another for droplets of lipid (soybean oil, used in injectable emulsions), also suspended in water ("Q-LIPW"). Several features of the plots shown in Fig. 3 are noteworthy.

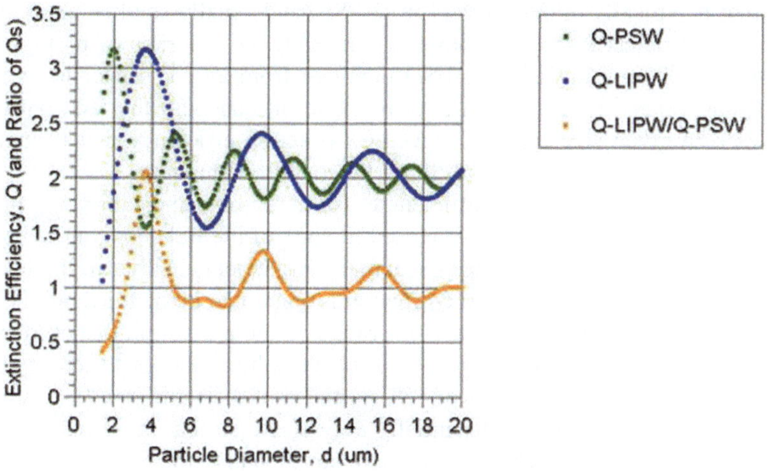

Fig. 3 Extinction efficiency, "Q", and ratio of Qs vs. particle diameter (μm). (Driscoll 2004)

First, Q oscillates as a function of particle size, alternating between values larger and smaller than 2, within a decaying "envelope" that eventually settles to the latter value at large sizes. Q values larger than 2 indicate that the particle blocks *more* light than predicted by G, while Q values smaller than 2 indicate that the particle blocks *less* light than that. This oscillatory behavior is the result of alternating net destructive and constructive interference of scattered light wavelets originating from within the particle with changing particle size, as predicted by the Mie scattering theory.

Second, the "frequency" of oscillation of Q *vs* d—as measured by the difference in particle diameter, Δd, between two adjacent Q = 2 crossing points—depends on the refractive index of the particles. In the case of PSL particles, which have a relatively high refractive index ($n = 1.59$) and therefore a relatively high "contrast" to the surrounding water ($n = 1.33$), the "frequency" of oscillation of Q *vs* d is relatively high. In the case of lipid droplets, which have a significantly lower refractive index (est. $n = 1.43$) and therefore a correspondingly lower contrast to water, the "frequency" of oscillation of Q *vs* d is considerably lower. Hence, the two oscillatory curves have different "frequencies" and therefore cannot remain "in phase" with increasing d. As a result, there are certain size ranges over which the droplets of lipid are more effective than PSL particles in producing light extinction, alternating with other size ranges where the reverse is true. This alternating LE behavior is shown by the third curve in Fig. 3, which plots the *ratio* of the two Q curves as a function of the particle diameter, d. Values of the ratio > 1 indicate that lipid droplets of that size remove more light (and therefore yield a larger LE pulse height) than PSL particles of the same size, and vice-versa for values of the ratio < 1.

The practical importance of this non-linear, oscillating behavior of Q *vs* d should be evident. The sensor is calibrated using "standard" particles of known size, comprised of PSL. However, samples of interest contain particles with a refractive index that is typically quite different from PSL, and usually smaller, as in the case of the lipid droplets discussed above as well as most materials of biological and pharmaceutical significance. The PSD obtained from an LE sensor is, in reality, a "PSL-equivalent" result, which does not take into account the fact that sample particles of a given size typically have an extinction efficiency that is somewhat different—higher or lower—than PSL particles of the same size.

Consequently, the extinction efficiencies for the sample and calibrating particles over the size range of interest can be used to convert the PSL-equivalent PSD to a more accurate PSD appropriate for the actual sample being analyzed. In the case of lipid droplets in water, the needed correction is contained in the third plot in Fig. 3: the ratio (Q-LIPW)/(Q-PSW). In the size range of about 3–5 μm, this ratio exceeds 1, and substantially so in the vicinity of 4 μm, where the extinction efficiency of lipid droplets is 100% larger than that of PSL particles of the same size (i.e., ratio $\approx$2). These lipid droplets will be substantially *over*-sized, so that those particle counts will appear in size bins, or channels, that are larger than 4 μm. Similar behavior occurs again in the vicinity of 9–10 μm, but here the magnitude of over-sizing is considerably smaller, with a maximum of about 30% (i.e., ratio$\approx$1.3). In any case, the percentage deviation in extinction efficiency between the particles of interest and the PSL particles used for calibration decreases with increasing particle size. For particles larger than about 5 μm the LE technique yields PSDs of relatively good absolute size accuracy without making any corrections for differences in extinction efficiency. In the case of smaller particles, in the range of about 1–5 μm, it may be necessary to convert the apparent, "PSL-equivalent" channel sizes comprising the PSD to more accurate channel sizes—both larger and smaller, depending on whether the Q value for sample particles of a given size are significantly enhanced or depressed relative to the Q value for PSL particles.

An example of a successful pharmaceutical application of the SPOS technique based on light extinction is the counting and sizing of over-size fat droplets ("globules") in a fat emulsion made for intravenous administration, mandated for emulsion manufacturers by USP <729> (Method II), "Globule Size Distribution in Lipid Injectable Emulsions". Typical PSD results obtained using the SPOS/LE technique for a variety of injectable fat emulsions of varying stability and quality have been reported extensively (Driscoll 2006).

Of course, the stated success of the SPOS/LE technique in providing accurate and reliable particle size results for this and other applications should be qualified. First, as discussed above and illustrated in Fig. 3, the fact that the refractive index of the particles of interest usually differs from that of the PSL particles used to calibrate the LE sensor will result in errors, both upward and downward, in the reported sizes of the particles. In extreme cases, the particles of interest may be nearly transparent, owing to their refractive index being very close to that of the surrounding fluid. Their reported sizes may then be substantially smaller than their actual physical sizes. However, as discussed above, theoretical corrections can be made to correct for significant reduction in the extinction efficiency, Q, due to the very low "contrast" of these nearly "index-matched" particles. Second, sizing by SPOS/LE takes no account of particle morphology, as the reported diameter is determined from the LE pulse height value that has been averaged over all particle shapes and orientations with respect to the incident light beam. The resulting reported "diameter" is that of an "equivalent sphere", which removes the same fraction of incident light as the actual, non-spherical particles being measured.

SPOS

Light Scattering (LS) Method (Lieberman 1990; LiQuilaz)

The lower size limit that is achievable using the LE technique is typically ≈1 μm, and in practice it is often closer to 1.3–1.5 μm (PSL particles in water), notwithstanding the aforementioned "extinction efficiency" issues. For LE sensors of conventional optical/electronic design, it is not feasible to extend their sensitivity to substantially lower sizes, given signal/noise limitations imposed by the laser light source and electronic processing circuits, among other issues. Instead, a different physical mechanism—light scattering (LS)—is needed to count and size particles smaller than 1 μm.

The design of a typical LS sensor is similar to that of an LE sensor with regard to the shape (intensity profile) of the light beam used to illuminate the flow channel and establish therein a view volume. The difference lies instead with the scheme of detection. Instead of locating the LS detector along the optical axis, it is instead positioned off-axis, so that it "looks" at the same view volume seen by the LE detector, but without being subjected to the intense incident beam. When a particle passes through the view volume a signal pulse is produced, with a height, ΔV_{LS}, that increases monotonically with particle size, provided the range of detection angles is sufficiently small and forward-oriented that there are no "reversals" due to destructive interference within the particle, as described by Mie scattering theory. The pulse width, Δt, is essentially the same as for light extinction pulses (Fig. 1), equal to the time of transit of the particle through the illuminated view volume.

The LS method is able to count and size significantly smaller particles than the LE method. On the one hand, detection of small signal pulses is made easier by the fact that ideally they lie on a relatively "clean", nearly-zero, baseline signal level. In addition, the signal/noise ratio of the pulses, and therefore the lower size detection limit, can often be improved significantly by increasing the intensity of the laser light source. A lower size limit of 0.5 μm can be obtained relatively easily using an LS sensor of simple design with a laser of moderate power (e.g., 35 mW, $\lambda = 785$ nm). As will be discussed, sensors based on a more sophisticated optical design and higher power lasers can reduce the lower limit of detection to ≈0.2 μm.

Along with the great advantage of a lower particle size limit provided by the LS method, there is unfortunately a significant disadvantage, not suffered by the LE method. The LS method suffers from a greatly reduced dynamic size range, because of the strong, ≈6th power dependence of the scattering intensity on particle size when the latter is smaller than, or comparable to, the light wavelength, λ. A simple numerical example is instructive. Assume that the effective lower size limit of an LS sensor is 0.5 μm, producing a signal pulse height, ΔV_{LS}, of 0.025 V (i.e., about three times the average r.m.s., "root mean squared", noise level). Assuming also that Rayleigh's 6th power law applies, then a particle of size 1.35 μm will yield nearly the maximum measurable pulse height of 9.68 V (saturation = 10 V). This particle scatters about 387 times more light than the 0.5 μm particle, and yet it is less than three times larger.

By modifying the optical design and adding appropriate signal-conditioning electronics, it is possible to increase somewhat the dynamic size range of an LS sensor, but not by very much. For example, one optical particle counter claims a working size range of 0.2–2 μm, while another has an indicated size range of 0.3–3 μm (LiQuilaz™). The tenfold dynamic size range offered by these *scattering-only* sensors is adequate for most particle contamination applications of interest. However, the concentration limits imposed by these sensors (10,000/mL) require most samples in non-contamination applications of interest to be diluted to a very large extent—as much as by a factor of 100 million, or more, depending on the lower size limit—in order to avoid significant distortion of the measured PSD due to particle coincidences. It is useful to note that the typical dimensions of the flow cell used to implement an LS sensor, together with the typical fluid flow rate, are such that the resulting shear forces generally remain too low to break up most particle aggregates of interest. Additional concerns may exist regarding break up of aggregates due to dilution of the starting sample. Both concerns can be addressed and alleviated by measuring the PSD using a range of fluid flow rates and pre-dilution factors, to determine whether the measured aggregate concentration and size distribution are affected (i.e., reduced) by these physical effects. Ultimately, it is only by carrying out such experiments on a given particle system of interest that concerns regarding possible breakup of particle aggregates due to flow-induced shear forces produced by fluid entry into the sensor flow cell can be dispelled.

SPOS: Combination of LE and LS—"LE + LS"

The LE and LS methods are complementary with respect to their advantages and disadvantages. The LE method has the advantage of a very large dynamic size range. However, it has the disadvantage of a poor lower size limit, ≈ 1 μm at best. By contrast, the advantage of the LS method is that it can provide a significantly lower particle size limit—at least 0.2 μm, or even 0.1 μm, using special optical, electronics and signal processing methods. But all LS sensors have the significant disadvantage of a poor dynamic size range, typically ≤ 10:1. One way to take advantage of the inherent, complementary strengths of the LE and LS methods is to *combine* them, in effect, into a single, "LE + LS", sensor (AccuSizer FX™, AccuSizer FX Nano™ and Nicomp 380™) so as to retain the advantages of each method. The LE + LS sensor provides *both* a relatively low particle size limit, 0.5 μm, delivered

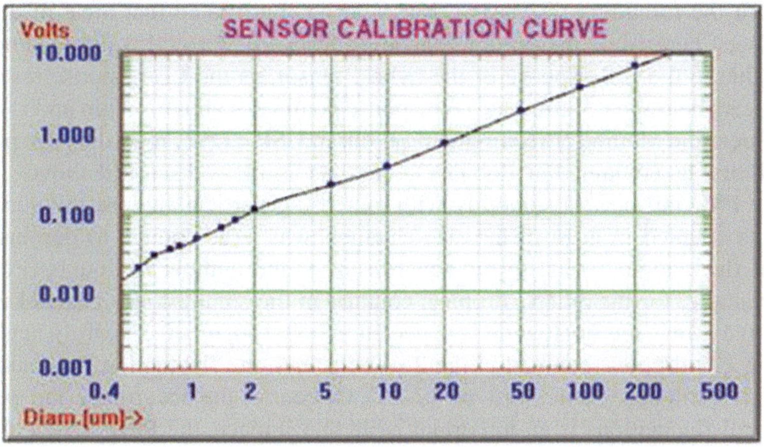

Fig. 4 LE+LS calibration curve: pulse height (V) *vs*. particle diameter (μm). Plot of LE+LS pulse height (0.02–10 V) *vs* particle (PSL) diameter (0.5–200 μm) for a typical LE+LS sensor. (AccuSizer FX™, AccuSizer FX Nano™ and Nicomp 380™)

by LS, and a relatively large upper size limit, 200+ μm, provided by LE. The two signals can be combined after appropriate "conditioning", and a typical calibration curve is shown in Fig. 4.

Figure 4 shows a smooth, approximately linear (log-log scale) response extending down to the new detection limit of 0.5 μm. An example of a successful SPOS analysis result using the "LE+LS" hybrid sensor is seen in Fig. 5 (Data on File, Particle Sizing Systems LLC), showing the "tail" of lipid droplets for a typical injectable fat emulsion discussed earlier, but with the droplet size analysis now extending down to ≈0.5 μm.

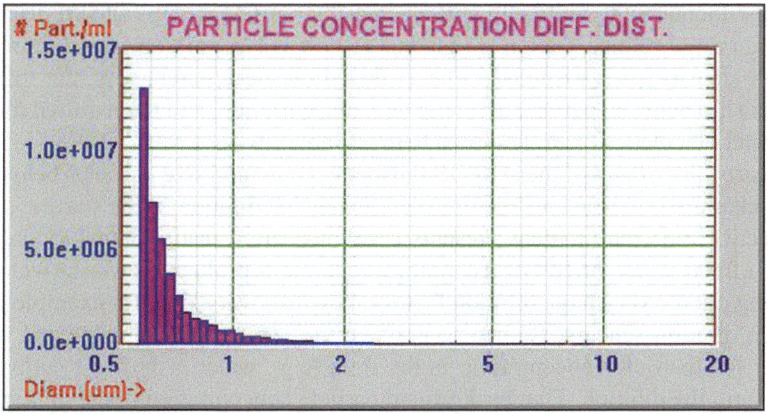

Fig. 5 Injectable fat emulsion PSD from "LE+LS": # droplets/mL vs. particle diameter. (Data on File, Particle Sizing Systems LLC)

From the foregoing discussion, it should be appreciated that most of the PSD "tail" seen here—i.e., the portion extending from ≈0.56 to ≈1.3 μm is almost entirely due to the LS response of the hybrid sensor, as the LE response essentially disappears below ≈1.3 μm. The LE response dominates above ≈2 μm and is used to implement the Method II requirement set out in USP <729>, regarding the percentage of total fat contained in globules larger than 5 μm, as discussed above.

The PSD tail decays smoothly with increasing particle size, as the number of droplets counted in most of the size channels was large enough to dominate statistical fluctuations, equal to the square root of the number of counts (Poisson). For example, for the 34,557 droplets counted in the smallest size channel utilized (0.56–0.59 μm), the fluctuation is expected to be ± 186, or ± 0.5 %, truly negligible. For the 173 droplets counted in the 2-μm channel, the fluctuation is considerably larger, but still acceptable: ± 13, or ± 7.5 %. Of course, the occurrence and extent of statistical fluctuations is an important issue that impacts the PSD results obtained from all particle counting techniques, not limited to SPOS.

Sensors based on the LE and LS methods were originally developed for one, and only one, broad class of applications: contamination analysis. The sole objective was, and largely remains, to measure the concentration and sizes of contaminant particles in various fluids of interest, such as injectable water and saline for medical use, to ensure safety. The particle concentration is invariably below the coincidence limit of the sensors, typically ≈10,000/mL for the smallest particles and significantly less than this value for larger ones, so that the sample to be analyzed requires *no dilution*. By contrast, for most other applications for which the SPOS technique has proven to be a powerful analytical tool, the sample of interest contains, by design, a very large concentration of particles, usually greatly exceeding (by several orders of magnitude) the concentration limit of the sensor. Therapeutic drugs for injection containing a concentrated suspension of proteins which for one reason or another, have become extensively aggregated, possibly resulting in immunogenicity concerns, are an important example. In all of these cases, the starting sample must be diluted, and usually by a very large factor. Furthermore, the extent of dilution required increases dramatically with decreasing particle size, as shown below. The starting particle concentration is assumed to be 1 % (wt/vol), and for simplicity the particle density, ρ, is assumed to be 1. Based on these parameters, the following relationship between particle diameter and concentration, and the required dilutions to avoid coincidence error, is shown below (Table 3).

Clearly, use of any of these SPOS sensors to count and size particles below 1 μm in concentrated suspensions requires very extensive dilution of the starting sample. In practice, this requirement presents a serious, often insurmountable, challenge to the effective use of SPOS technology. The main problem is *not* in technically carrying out the very large dilution factors shown above, using, for example, a 2- or 3-stage dilution scheme. Rather, in practice the difficulty is in achieving a sufficiently low particle concentration in the fluid (e.g., water or buffer solution) used to perform the dilution. That background particle concentration in the diluting fluid must be significantly lower than the final target concentration of the sample *after* dilution. Bringing the background contamination level down to 10–20/mL at 0.5-

Table 3 Particle diameter and concentration: dilution requirements (SPOS sensors)

Particle diameter (μm)	Concentration	Dilution needed
10	1.91×10^{8}/mL	≈2000
5	1.53×10^{9}/mL	≈15,000
2	2.39×10^{10}/mL	≈240,000
1	1.91×10^{11}/mL	≈2,000,000
0.5	1.53×10^{12}/mL	≈15,000,000
0.2	2.39×10^{13}/mL	≈240,000,000

The starting particle concentration is assumed to be 1 % (wt/vol), and for simplicity the particle density, ρ, is assumed to be 1

μm is difficult enough, but achieving this level of cleanliness at 0.2-μm becomes exponentially more difficult and time consuming.

SPOS: Focused Beam Light Extinction (FBLE) Method

The only way in which the maximum concentration limit of an SPOS sensor can be increased greatly—i.e., by a factor of at least 100—over the normal coincidence limit is to make a commensurate major reduction in the view volume of the sensor, and there are only two ways to do this. The first is to reduce the cross-sectional area of the flow channel. However, there are three major disadvantages associated with this approach. First, the maximum particle size must be reduced to avoid frequent clogging. Second, it will be harder, if not impossible, to clean the cell by mechanical means. Third, a reduction in the flow rate of the sample fluid will be needed to avoid an increase in the particle velocity, possibly resulting in a reduced pulse height due to bandwidth limitations of the processing amplifier. But a reduced flow rate will result in reduced particle counts and poorer particle counting statistics, requiring longer data collection times.

The second, and only other, way to increase the concentration limit of an SPOS sensor of conventional design is to reduce the "thin" dimension of the view volume, defined by the thickness of the "ribbon" of laser light impinging on the flow channel. In the LE example used earlier, this illuminated thickness was 40 μm. One might propose to reduce this width by, say, 4X, down to 10 μm, thereby increasing the coincidence limit by 4X. But this would likely be the greatest improvement that could be made in practice, given the physical limits and constraints associated with various alternative optical designs.

The only way in which the coincidence concentration limit can be increased by a large factor is to reduce greatly the other, large dimension of the incident *light beam*, which is currently defined by the physical width of the flow channel ("A" = 0.4 mm in the earlier LE example), through which it passes. Consequently, a new type of SPOS sensor has been developed, involving a radically different optical design, based on a *focused light beam* (AccuSizer FX™, AccuSizer FX Nano™

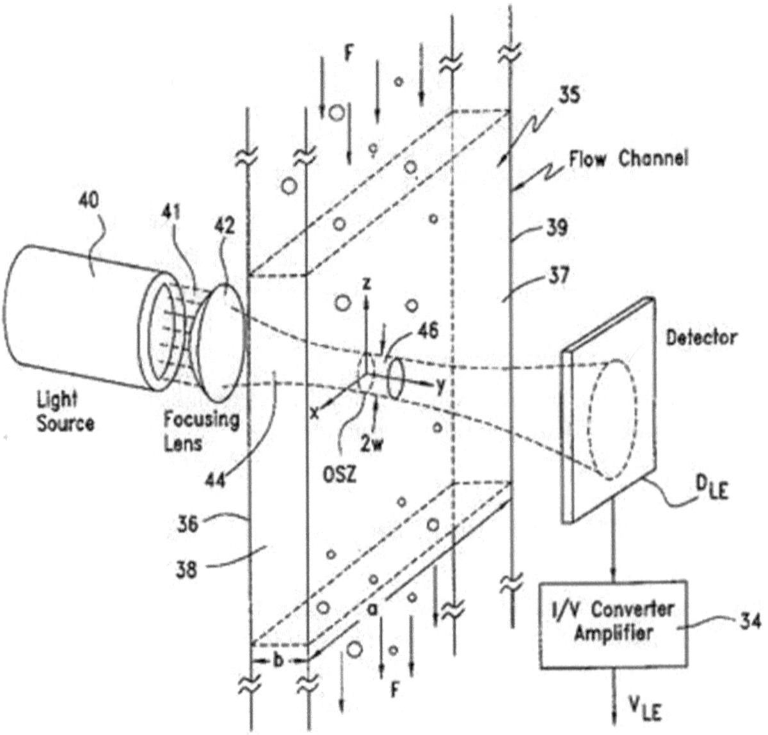

Fig. 6 FBLE: simplified schematic diagram of optical configuration

and Nicomp 380™). A much smaller view volume is then defined by the thin "pencil" of focused light traversing the flow channel. Another major advantage has also resulted—a great increase in sensitivity. A particle passing through the new view volume with greatly reduced cross sectional *area* will effectively remove a much greater fraction of light from the light beam than would be the case for a light extinction sensor of conventional design. The lower size limit will therefore be significantly reduced.

The new "FBLE" sensor, based on focused-beam illumination, is referred to as "focused extinction" and is shown schematically in Fig. 6, (AccuSizer FX™, AccuSizer FX Nano™ and Nicomp 380™) with three noteworthy features.

The light beam from a laser light source is focused within a narrow optical flow cell, after which it impinges on a distant light detector (Driscoll et al. 2007a). Passage of a particle through the "view volume", defined by the focused beam and the opposing walls (narrow dimension) of the flow cell, results in a momentary negative-going pulse, analogous to the response obtained for a conventional LE sensor, shown schematically in Fig. 1.

First, the increased sensitivity due to the much smaller cross-sectional area of illumination reduces the lower detectable size limit down to ≈0.6 μm—a major advance for a light-extinction sensor. Second, the tightly-focused laser beam illu-

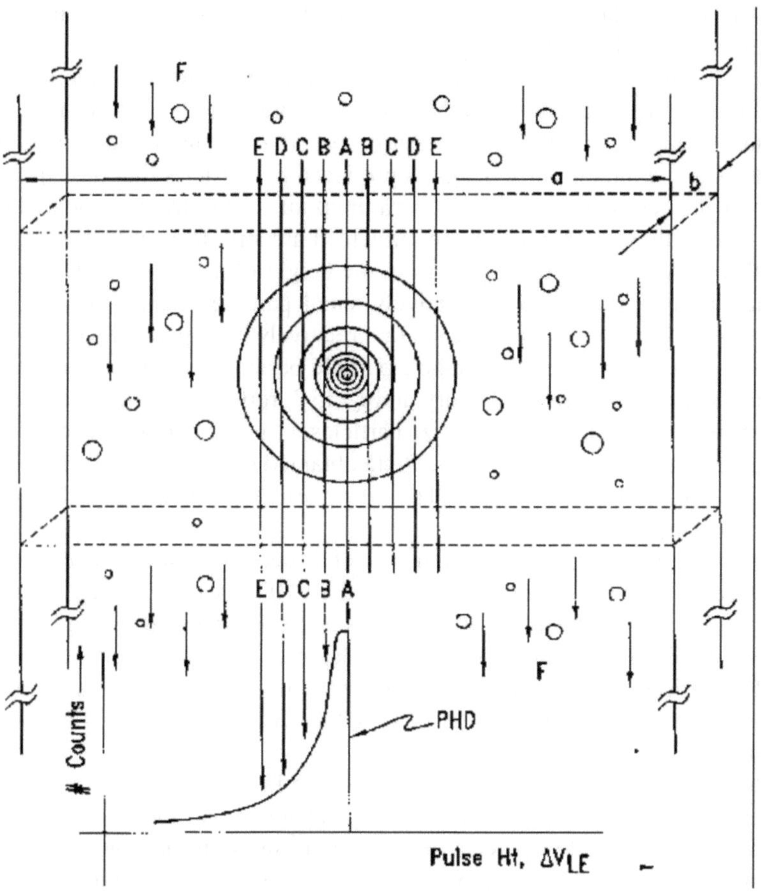

Fig. 7 FBLE: non-uniform illumination and resulting signal response

minates only a small fraction of the particles flowing through the sensor—typically 0.5–1 %. Consequently, it can operate at much higher (≈ 100–200X) particle concentrations without particle coincidence effects. A related advantage is that there is a commensurate reduction in the influence of background contaminant particles. Third, and most importantly, the intensity of the incident light beam is, by definition, no longer approximately uniform across the flow channel. Instead, the beam is highly focused, typically with an approximately Gaussian intensity profile, as shown schematically in Fig. 7 (AccuSizer FX™, AccuSizer FX Nano™ and Nicomp 380™).

The resulting FBLE signal pulse height depends not only on the *size* of the particle, but also on its *trajectory* with respect to the highly non-uniform light intensity profile that defines two of the three dimensions of the view volume. A particle that passes through the central, most intense region of the focused light beam ("A") will remove a larger fraction of the incident light flux than a particle of the *same size* that passes through a less intense region of the beam ("B"—"E"), away from its central axis. The further the particle passes from the beam axis, the lower the resulting light-extinction pulse height, ΔV_{LE}.

The resulting pulse height distribution (PHD) for particles of the *same size* is therefore no longer narrow, as in the case of a conventional LE sensor. Instead, as indicated in Fig. 7, the PHD is necessarily *broad*, covering a wide range of pulse heights. The maximum pulse height is produced by particles that pass through the center of the focused beam, while the minimum results from particles that pass furthest from the center of the beam but which are still detectable. Hence, in principle there is no way to determine from the height of a *single* FBLE signal pulse whether it was produced by a relatively small particle passing through/near the central axis of the focused laser beam, or a relatively large particle passing far from the beam axis, or an intermediate-size particle passing somewhere between these limiting trajectories. This "trajectory ambiguity", reflected in the broadness of the PHD, even in the case of uniform-size particles, is a major consequence of the new focused-beam sensor. Therefore, the PHD obtained during data collection must be deconvoluted to take into account all possible particle trajectories, in order to obtain the final desired PSD.

It is important to appreciate that the use of a "deconvolution" process to remove the influence of the particle trajectory on the PHD does not greatly compromise the resolution of the resulting PSD. An example is provided in Fig. 8, showing the PSD obtained from the FBLE method depicted in Figs. 6 and 7, for a mixture of NIST-certified PSL particles of diameter 0.707-, 0.993-, 1.361- and 2.001-μm, with a total particle concentration of 625,000/mL (Data on File, Particle Sizing Systems LLC).

A total of 58,032 pulses were collected in the PHD from 15 mL of sample volume over 60 s. The four peaks are cleanly separated and their computed (mean) sizes are each within a few percent of the known values. In addition, their individual concentrations (#/mL) are within about 10–15% of the values expected from the

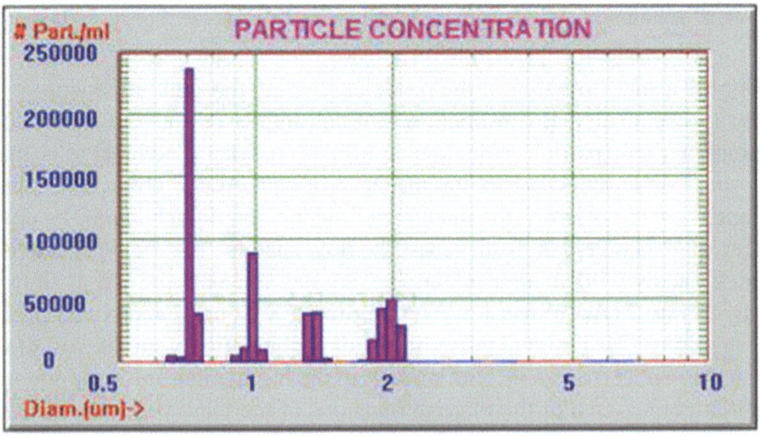

Fig. 8 FBLE PSD for PSL standard particles: #/mL *vs*. particle diameter. Plot of PSD obtained for the "4-modal" mixture of PSL particles (sizes listed above), using the focused-beam light extinction (FBLE) sensor, discussed above

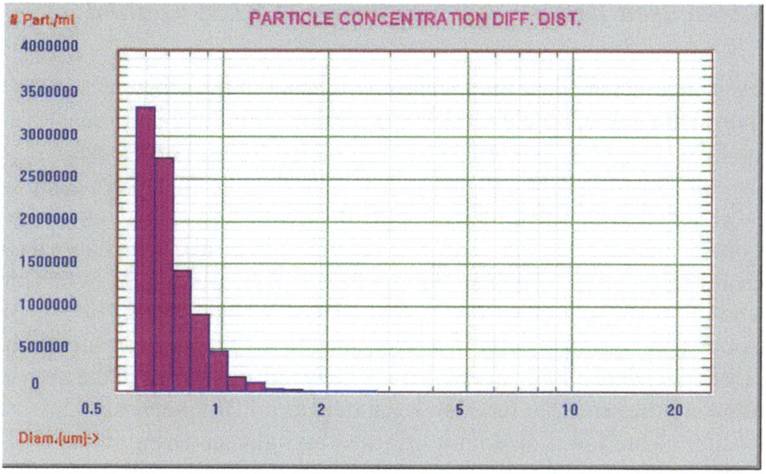

Fig. 9 FBLE PSD for protein IgG (1 % in PBS, pH 7.5): # aggregates/mL *vs*. particle diameter (μm)

starting "cocktail". Results of similar quality can often be obtained at higher particle concentrations, sometimes > 1 million/mL, depending on the complexity of the PSD.

It is also important to appreciate that the new focused-beam light-extinction technique, notwithstanding the necessity of deconvoluting the resulting PHD, still represents an SPOS technique. That is, the detected "signal" consists of individual pulses, each of which is produced by a *single* particle passing through the view volume of the sensor. This behavior is in sharp contrast to the ensemble technique of dynamic light scattering (discussed later), for which particles of *all sizes* contribute *simultaneously* to the measured signal, and deconvolution of the raw data yields PSDs that necessarily have a relatively poor size resolution.

Results obtained for an aggregated protein (IgG, M.W. 150 kD, 10 mg/mL, in PBS), analyzed *without dilution* using a FBLE sensor, are shown in Fig. 9 (AccuSizer FX™, AccuSizer FX Nano™ and Nicomp 380™).

A total sample volume of 7.5 mL was analyzed, with 419,424 particles detected (60 s). The resulting concentration of particles larger than 0.57 μm was computed to be 9.34×10^6/mL, decreasing to below 10^3/mL at 15 μm. The actual starting sample volume was 1 mL. The analysis was made by automatically adding together the individual segments of PHD data collected during ten successive pulls (using a high-resolution syringe pump) through the FBLE sensor, with 0.75 mL/pull. The pulled sample volume was pushed back into the sample vial after each sample pull. This tenfold increase in the effective sample volume and corresponding particle counting statistics is made possible by the fact that only a small fraction (≈0.5–2 %, depending on the particle size) of the particles passing through the FBLE sensor are detected. Following each sample pull the suspended particles will have become so randomized within the sample fluid that essentially a completely different set of particles will contribute to the next segment of PHD data during the subsequent sample pull—certainly for at least the first ten pulls.

SPOS: Focused Beam Light Scattering (FBLS) Method

The view volume created by the focused light beam in the sensor described above, also serves, with appropriate optics, as the starting point for a more sensitive sensor based on light scattering. Scattered light originating from a particle passing through the view volume, defined by the focused incident light beam and opposing walls of the flow channel, is captured over a range of forward angles by suitable light collection optics and directed onto an LS detector. A simplified schematic diagram of the optical design, used to implement the FBLS sensor, based on light scattering, is shown in Fig. 10 (AccuSizer FX™, AccuSizer FX Nano™ and Nicomp 380™).

The optical design also includes a provision for the forward transmitted beam to be detected independently by a separate LE detector (not shown in the above depiction), allowing the sensor to function separately as a FBLE sensor.

Extension of the lower size limit afforded by light scattering in the FBLS sensor is again influenced by the 6th power dependence of the scattering intensity on particle diameter. First, as the particle size drops, the scattered intensity will eventually fall to a level that is too low to be detected reliably. As before, if the problem

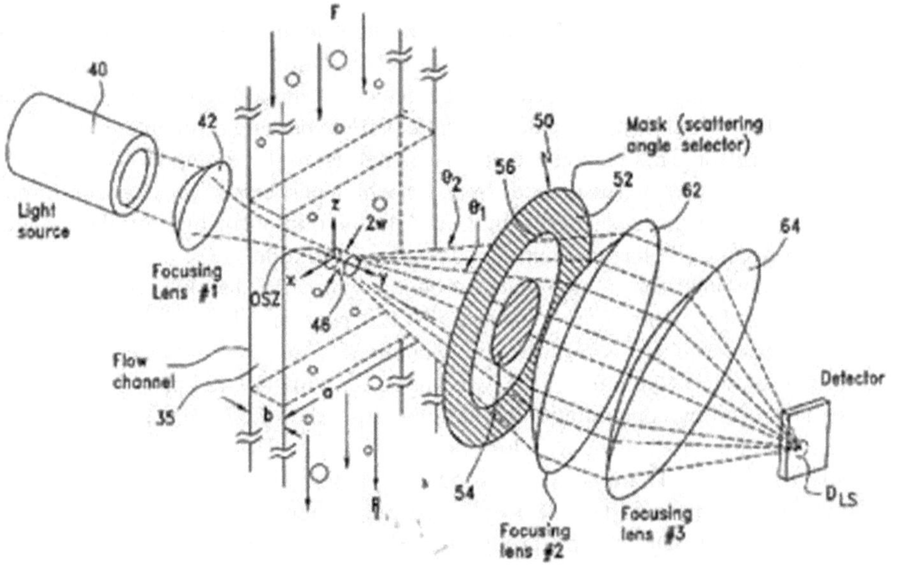

Fig. 10 FBLS: simplified schematic diagram of optical configuration. As in the FBLE optical design, the light beam from a laser light source is focused within a narrow optical flow cell. However, in the FBLS sensor the beam that emerges from the cell is blocked, so that it cannot impinge on a distant light detector (Driscoll et al. 2007a). Instead, scattered light originating from a particle passing through the "view volume" is collected by a first lens over a range of relatively small forward angles defined by an appropriate "mask", and focused by a second lens onto a light detector. The resulting positive-going signal pulse is analogous to that produced by a conventional LS sensor

consists simply of inadequate detection sensitivity, then the answer is to increase the intensity of the incident light (laser) beam. However, this will be effective only if background noise isn't the limiting factor; otherwise, the prevailing noise level will simply scale up with increasing laser power. Of considerable advantage is the fact that the power density of the focused light beam is much higher than that found in a traditional LS sensor. This factor effectively increases the signal/noise (S/N) ratio, making detection and signal conditioning easier at the lowest limits of detection (≈0.15 μm).

Second, the detectable size range, ≈0.15 to 0.6 μm, of the FBLS sensor implies a more than 4000-fold range of LS pulse heights, imposing a great demand on the signal processing system. If one assumes that a pulse height of only 0.020 V for a 0.15-μm particle would provide an adequate S/N ratio, then the expected pulse height for a 0.6-μm particle is ≈82 V—more than eight times the maximum 10 V saturation value. Hence, without the use of nonlinear compression techniques, a single amplification gain cannot be used to cover the desired size range. Instead, two different amplifier gains are needed to span the range of pulse heights associated with the seemingly modest (less than fourfold) particle size range that the sensor is designed to cover in the light scattering mode. Consequently, the FBLS sensor includes a first-stage amplifier that can be automatically switched from "low gain" (LG) to "high gain" (HG), which together cover the overall size range.

Evaluation of the accuracy and resolution achievable by the FBLS sensor is best done using a "multimodal" sample. A typical PSD obtained from a mixture of five PSL particle "standards" (0.3-, 0.34-, 0.4-, 0.5- and 0.6-μm) using the low gain (LG) mode of detection is shown in Fig. 11 (Data on File, Particle Sizing Systems LLC).

The resulting peaks are narrow, cleanly separated and close to their expected sizes (within ≈5 %) and concentration values (within ≈15–20 %). The total number of particles counted during the 60-s analysis time (15 mL of sample fluid) was 69,308, with a concentration of ≈669,000/mL. A PSD obtained from a mix-

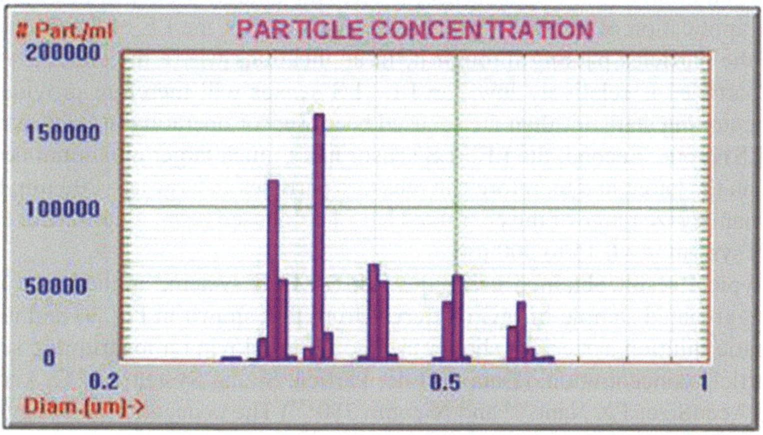

Fig. 11 FBLS (LG) PSD for PSL standard particles: #/mL *vs.* particle diameter (μm). Plot of PSD obtained for a "5-modal" mixture of submicron PSL particles (sizes listed above), using the focused-beam light scattering (FBLS) sensor, discussed above, operating in low-gain (LG) mode

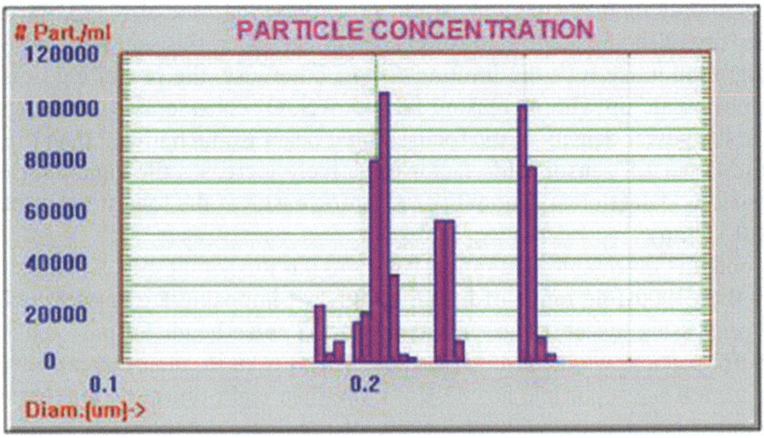

Fig. 12 FBLS (HG) PSD for PSL standard particles: #/mL *vs.* particle diameter. Plot of PSD obtained for a "3-modal" mixture of submicron PSL particles (sizes listed above), using the focused-beam light scattering (FBLS) sensor, discussed above, operating in high-gain (HG) mode

ture of three smaller PSL "standards" (0.2-, 0.24- and 0.3-µm) using the high gain (HG) mode of detection is shown in Fig. 12 (Data on File, Particle Sizing Systems LLC).

The peaks are again cleanly separated, with their sizes and concentrations close to the expected values. The total number of particles counted in 60 s (15 mL of sample fluid) was 48,303, with a concentration of ≈597,000/mL.

An important variation of the system design has been developed, allowing the user to "piggyback" the LE+LS sensor onto the FBLS sensor, using the former to obtain the large-size portion of the PSD, starting at ≈0.56 µm. There are two reasons why the addition of this second sensor is often advantageous, depending on the particular application of interest. First, as discussed earlier, the LE+LS sensor detects *all* of the particles passing through it. In applications where the concentration of large particles is relatively low, the LE+LS sensor will therefore provide much better counting statistics than the focused beam sensor operating in light-extinction (FBLES) mode. Second, the LE+LS sensor has a much larger maximum detection size than the latter, and therefore can measure particles as large as ≈200 µm. Hence, the dynamic size range of the combined LE+LS/focused beam, dual-sensor (FBLS-LELS) system is ≈0.15 to 200 µm.

A typical result obtained using the FBLS-LELS system, applied to a moderately aggregated sample of IgG (different from that shown in Fig. 9) and using an automatic dilution system, is shown in Fig. 13, employing a logarithmic scale for the particle concentration: (Data on File, Particle Sizing Systems LLC; AccuSizer FX™, AccuSizer FX Nano™ and Nicomp 380™) The concentration of aggregates measured by the dual-sensor system ranges from ≈10^7/mL at 0.2 µm to ≈10/mL at 20 µm—a 1,000,000-fold range of concentrations.

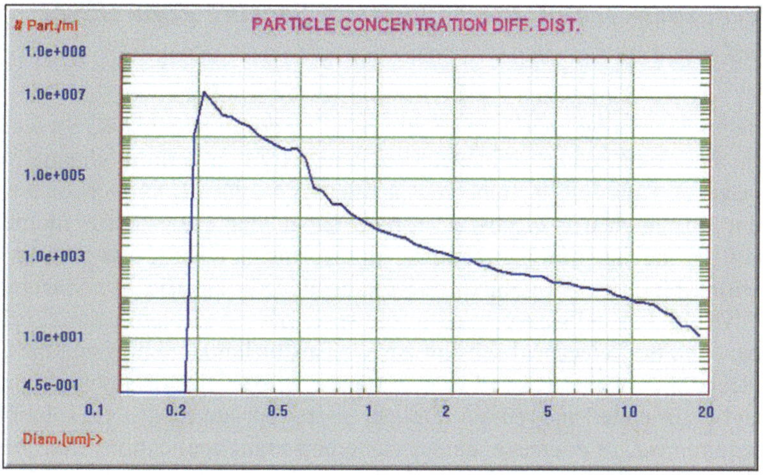

Fig. 13 FBLS-LELS PSD for protein IgG (1 % in PBS, pH 7.5): # aggregates/mL *vs*. particle diameter (μm)

The PSD segment contributed by the LE+LS sensor in Fig. 13, with an overall particle concentration of 2.32×10^5/mL above ≈0.56 μm, is shown in Fig. 14 (Driscoll et al. 2001a).

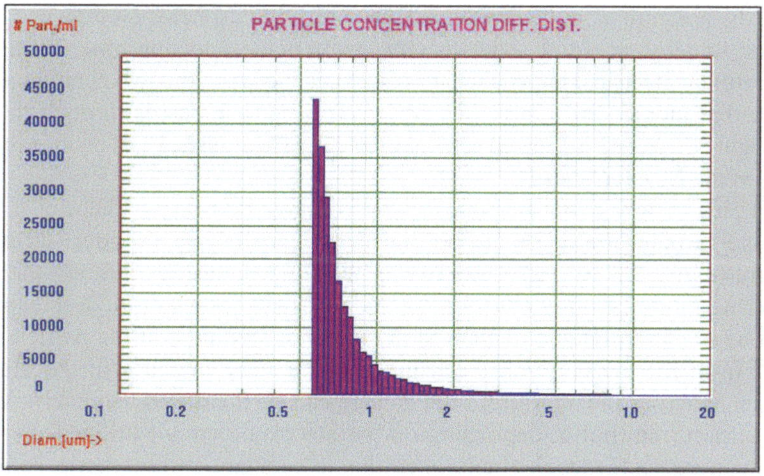

Fig. 14 LE+LS PSD for protein IgG (1 % in PBS, pH 7.5): # aggregates/mL *vs* particle diameter (μm). Its smoothness is due to superior particle counting statistics, because all of the detectable particles passing through this sensor are counted—needed, because of the diminishing concentration of large aggregate particles with increasing size, particularly above 1 μm

Single-Particle Techniques: Electrical Sensing Zone (ESZ) Method

The "electrical sensing zone" (ESZ), or "resistive pore sensing" (RPS), technique, based on a transient reduction in electrical conductivity of a fluid "channel" caused by passage of a particle is somewhat analogous to the LE version of the SPOS technique, in which it is electrical current, rather than light, that is momentarily "blocked" by passage of a particle through the sensing zone It was first developed for counting blood cells by measuring the changes in electrical conductance of a conductive fluid as cells suspended in the latter passed through a small orifice or aperture—i.e., the "electrical sensing zone". This technique, known as the Coulter Principle, in theory can be used to count and size particles of any type, provided they can be suspended and remain stable in an appropriate electrolyte solution. This latter requirement, of course, excludes some important applications from being effectively addressed by the ESZ technique, because of incompatibility with specific buffer formulation requirements.

A tubular-shaped vessel ("aperture tube") containing an electrolyte solution, with a precision aperture of a fixed, known size mounted in its wall, is immersed into a beaker containing particles suspended in a similar electrolyte solution, such as sodium chloride in water or lithium chloride in isopropyl alcohol. Two platinum electrodes, one immersed in the fluid inside the aperture tube and the other immersed in the sample fluid outside the tube, inside the beaker, are connected to an electronic circuit that applies a voltage between the two fluid bodies, connected electrically only by the aperture. Lines of equal voltage potential are established within the aperture, and the lines also form active hemispheres on both ends of the aperture, thereby extending somewhat the length of the detection zone. The aperture constitutes an electrical impedance, or resistance, between the two conducting fluid bodies, that momentarily increases whenever a particle passes through the aperture. Figure 15 shows a stylized schematic picture (Multisizer™ 4) of an instrument based on the ESZ technique.

Particles in low concentration can be counted by passing them at a steady flow rate through the aperture, facilitated by establishing a modest differential pressure (e.g., $\approx 2 \times 10^4$ Pa) between the two fluid bodies, or through the use of a metering pump, as shown in Fig. 15. As a particle passes through the aperture, a volume of electrolyte equivalent to the immersed volume of the particle is displaced from the sensing zone, causing a momentary increase in the electrical impedance across the aperture. This change can be measured as either a voltage pulse or a current pulse, depending on the electronic circuit utilized. Typically, the system uses a feedback circuit operating in constant-current mode, so that the momentary increase in aperture resistance causes a proportional increase in the voltage across it—i.e., a voltage pulse, where the pulse height, ΔV, relative to the background voltage, V_o, in the absence of a particle is proportional to the volume of the sensed particle. A volume-weighted particle size distribution (PSD) is thus generated from the succession of voltage pulses. If constant particle density is assumed, ΔV is also proportional to the particle mass, and a mass-weighted PSD is thereby produced.

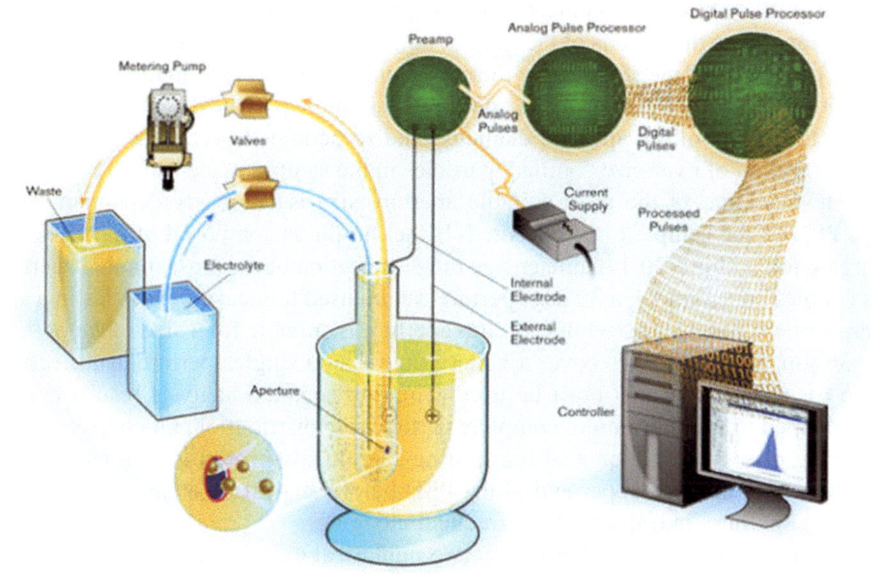

Fig. 15 Schematic diagram of an instrument based on the ESZ technique. (From www.beckman-coultier.com—Reprinted with permission)

If the volume of liquid passing through the aperture is precisely controlled and known, the concentration of the sample (i.e., the number of particles/mL) can also be determined. The flow rate of fluid/particles passing through the aperture is very small—e.g., typically <0.1 mL/min for a 20-μm aperture—and a sensitive function of the pressure differential across the aperture, or metering pump flow rate. The voltage pulses are digitized using a high-resolution, high-speed multi-channel electronic analyzer. The resulting pulse height value can therefore be determined accurately, as well as the pulse width and shape, allowing the instrument in principle to discriminate between normal, "legitimate" pulses and those that are distorted by particle coincidences and other non-ideal effects (discussed further below). Such discrimination capability in principle allows the instrument to operate effectively at significantly higher count rates than would normally be permitted to ensure low (e.g., 1 %) coincidence levels.

The aperture sizes used most often in applications that can be addressed by ESZ range from 20 to 1000 μm. In practice, the particle sizes that can be measured for a given aperture size range from approximately 2–60 % of the aperture diameter, resulting in a dynamic size range of 30:1. The lower detection limit of ≈2 % of the aperture size is determined by hydraulic and electrical ("Johnson") noise, which appears as small voltage pulses that are indistinguishable from pulses generated by actual detectable particles, appearing as an excessive number of "fines" at the lowest end of the size scale. The upper size limit of ≈60 % of the aperture size is determined by two factors. First, there is a maximum size at which the response remains linear, dictated by the diameter of the particle relative to the aperture diameter. The pulse height starts to become nonlinear when the particle diameter exceeds about 45 % of the latter, due to distortions of the lines of equal electrical potential that form around large particles. Second, in practice the maximum effective size is limited by the need to avoid frequent clogging of the aperture. This limit is typically chosen to be no more than 60 % of the aperture diameter, and sometimes lower, depending on the concentration of over-size "outlier" particles in the sample suspension.

The selection of the most suitable aperture size is obviously dependent upon the PSD of the sample to be analyzed. If the sample is composed of particles that largely fall within a 30:1 diameter size range, selection of the most suitable aperture is simple. For example, a 30 μm aperture can be used to measure particles in a size range of ≈0.6 to 18 μm, while a 60 μm aperture is suitable for a size range of ≈1.2 to 36 μm. If the particles cover a wider range than a single aperture can measure, two (or more) apertures must be used. The two separate analysis results can be overlapped to provide a more complete particle size distribution. Of course, careful filtering of a second aliquot of the starting sample suspension is then essential in order to "cut off" the upper end of the PSD before analysis is attempted using the smaller aperture, in order to avoid frequent clogging.

It is instructive to consider briefly two numerical examples, using different orifice sizes. First, an aperture diameter of 19 μm in theory allows particles as small as about 0.4 μm to be detected and sized, with a maximum size in practice of ≈11 μm. The nominal maximum particle concentration resulting in 1 % particle coincidence is estimated to be 1,100,000/mL. Assuming an aperture fluid flow rate of 0.064 mL/min, the resulting particle count rate is about 70,000/min (Karuhn 1998). Lower flow rates are often utilized, resulting in correspondingly lower particle count rates. An example of a PSD result of relatively low size range obtained by ESZ for a "trimodal" mixture of submicron particles, extending down to 0.4 μm, is shown in Fig. 16 (MultisizerTM 4).

Higher count rates can be achieved, but at the expense of increased particle coincidence and associated distortion of the resulting PSD (and somewhat reduced particle counts, discussed below). Of course, successful measurement of particles as small as 0.4 μm requires an operating environment with a very low electrical noise level. It also requires very effective pre-filtering of the starting solution, in order to avoid frequent clogging of the aperture. In a second example, assuming a 48-um aperture, the minimum detectable particle size in theory is ≈1 μm, and the maximum is ≈29 um. The nominal maximum particle concentration for < 1 % coincidence is now much lower: 70,000/mL. Assuming a flow rate of 0.44 mL/min, the resulting particle count rate is ≈31,000/min (Karuhn 1998).

It is instructive to look more closely at the non-ideal effects of particle coincidences on the behavior and resulting performance of particle size analysis results by ESZ (Karuhn 1998). As discussed above, during the passage of a particle through the aperture, the resistance, R, of the sensing zone is momentarily increased. The magnitude of the resistance increase, ΔR, relative to its value, R, in the absence of a particle, is a function of several variables,

$$\Delta R/R = \left(V/V_1\right)\left[\left(P/(P-P_o\right)-\left(a/A\right)\right]^{-1} \tag{2}$$

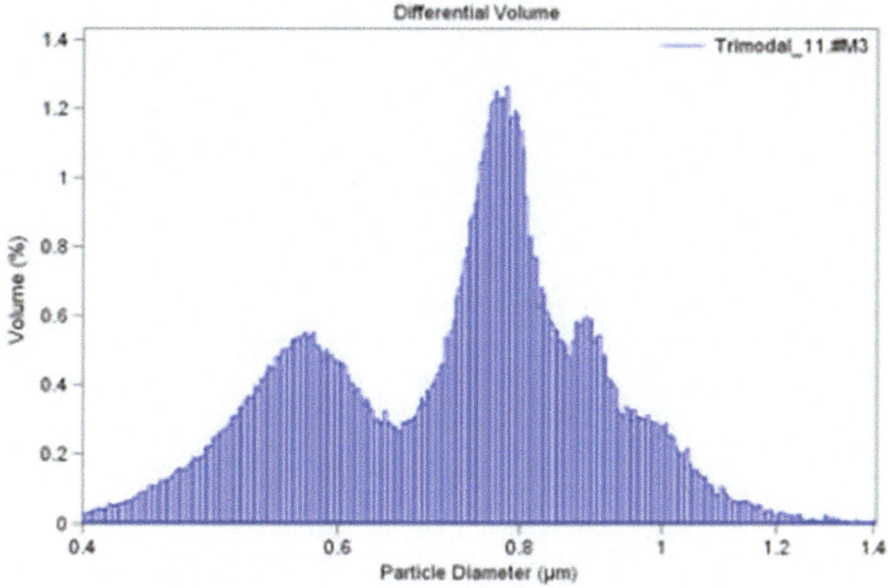

Fig. 16 PSD obtained by ESZ: particle volume (%) *vs*. particle diameter (0.4–1.4 μm). (From www.beckman-coultier.com—Reprinted with permission)

where,

V is the volume of the particle;
V_1 is the volume of the sensing zone;
P is the resistivity of the particle;
P_o is the resistivity of the electrolyte;
a is the cross-sectional area of the particle (normal to the axis of the aperture);
A is the area of the aperture normal to its axis.
In the case of the smallest detectable particles, where a<<A, Eq. 2 above becomes,

$$\Delta R/R \approx (V/V_0)\ \left[(P/(P-P_o)/P\ \right] \tag{3}$$

In typical cases where the particles are insulators, having a much higher resistivity than that of the electrolyte, i.e., $P >> P_o$, Eq. 3 becomes even simpler,

$$\Delta R/R \approx V/V_1 \tag{4}$$

Figure 17 shows how particles can re-circulate and interact with the sensing zone downstream of the aperture (Karuhn 1998).

Recirculation can result in secondary coincidence if the particles are not flushed away from the detection zone following their initial transit through the latter. This issue may become important for the smallest apertures, for which the fluid flow rate is particularly low—i.e., <0.1 mL/min. When a particle traverses the aperture

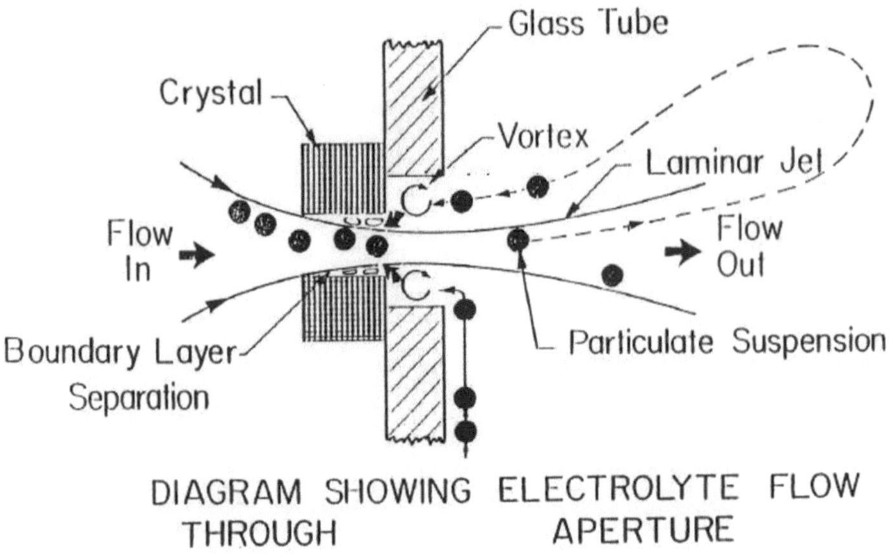

Fig. 17 Schematic representation of the effects of particle recirculation in ESZ

without coincidence with another particle, flow distortion or aperture edge effects, the signal pulse shape is smooth and symmetric. If a particle enters the aperture at a high entrance angle, there will be an electrical artifact that adds to the otherwise ideal, symmetric pulse shape, resulting in an increase in the measured pulse height. When two particles are in some form of secondary coincidence and/or transit near the aperture wall, where the current density is highest, there will be additional changes in the shape and height of the resulting signal pulses.

Coincidence results in a "count loss" and over-sizing, since the response of the detector is linear with the volume of the particle. Regarding over-sizing, in the simplest case in which two particles are in the orifice at the same time, the system will report the passage of a single particle, having a volume (deduced from the pulse height) equal to the combined volumes of the two particles. In the case in which the two particles are not fully within the detection zone at the same time, the resulting pulse will indicate a volume that represents some integrated average of the two individual particle volumes. Regarding under-counting, the magnitude of the count loss is related to the sample concentration, the volume of sample fluid passed through the aperture and the aperture diameter (Karuhn 1998). Because the required parameter values are known, the apparent count can be easily corrected to obtain a true particle count:

$$N = n_1 + n_2 \tag{5}$$

where,

N the true particle count
n_1 the apparent particle count

n_2 the additional particle counts that are "missing" due to coincidences,

$$n_2 = p\ (n_1/1000)^2 \tag{6}$$

where,

$$p = \text{the coincidence factor} = 2.5(D/100)^3 \times 500/V \tag{7}$$

And

D the aperture diameter (um)
V the fluid/particle sample volume (uL)

Equation 7 then becomes,

$$N = n_1 + \left[1.25\times10^{-9} D^3 n_1^2\right]/V \tag{8}$$

If the contribution due to particle coincidence, n_2, is less than 5 % of the apparent count, n_1, then usually no correction needs to be made to the latter, because the resulting error in the true particle count is small enough to be ignored.

Single-Particle Techniques

Tunable Resistive Pulse Sensing (TRPS) Method (qNano™; Vogel et al. 2011; Kozak et al. 2011; van der Pol et al. 2010; Anderson et al. 2013; Roberts et al. 2010; Willmott et al. 2010)

The Tunable Resistive Pulse Sensing (TRPS) method represents a novel variation and improvement on the traditional ESZ method. As shown schematically in Fig. 18 (qNano™), the pore, or aperture, through which particles pass individually, is "tuned" in size by stretching or relaxing, to an adjustable degree, an elastomer membrane that contains it.

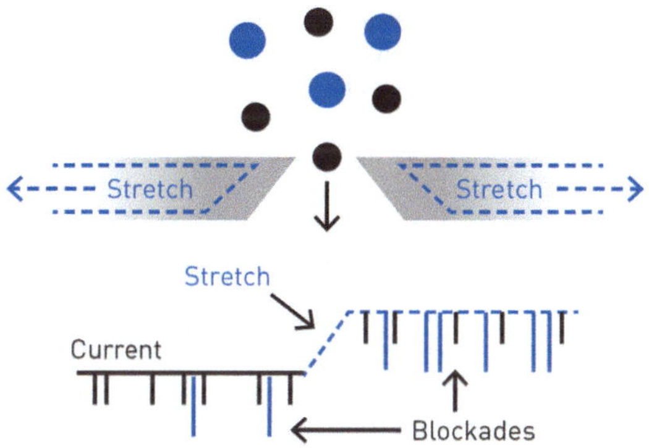

Fig. 18 Schematic representation of the stretchable pore upon which TRPS is based. (From http://www.izon.com—Reprinted with permission)

This capability allows the resulting device to measure particles of significantly lower size, and over a substantially wider size range, than can be accomplished using an aperture of fixed size, as utilized in a conventional ESZ instrument. The size range claimed for the TRPS system is 50 nm to 10 μm, resulting in a dynamic size range of 200:1, compared to the 30:1 range claimed for instruments based on conventional ESZ. The usable concentration range for this system is quite wide, from 10^5 to 10^{12} particles/mL. The typical sample volume is 40 μL, with an average analysis (data collection) time of 10–15 min, resulting in the counting and sizing of some thousands of particles, depending on the range of particle sizes and concentration.

The pore size in the TRPS system is reduced to its minimum size in order to obtain an acceptable S/N ratio for the smallest measurable particles, i.e., 50 nm. If one assumes that the 2 % pore-diameter criterion for the minimum size applies here, as for conventional, fixed-aperture ESZ devices, the minimum working pore size would be ≈2.5 μm, or perhaps somewhat smaller to enhance the S/N ratio. In order to accommodate the largest particles, 10 μm, the pore would in theory need to be stretched to ≈20–25 μm, or about 8–10 times its minimum size. All adjustments of pore size are carried out automatically and quickly by the system, in order to maintain reasonable particle throughput. One potential advantage of TRPS over conventional ESZ is that it should be able to "clear" a clogged particle by stretching the membrane and widening the pore size after a clog has occurred. Over-size particles that block the pore are "cleared" by stretching the membrane and opening the pore to its maximum size, without requiring the fluid flow across the pore to be reversed, as practiced by conventional ESZ devices.

PSD results of high resolution are obtained over a wide range of particle concentrations. Figure 19 (qNano™) shows the PSD obtained for a trimodal mixture of PSL particles of sizes ≈220-, 400- and 780-nm, displayed on a concentration scale of 2.8×10^7 particles/mL.

Figure 20 (qNano™) shows the PSD result obtained for liposomes of smaller size, ≈70–150 nm, and considerably higher total concentration, 5.8×10^{11} particles/mL.

The linearity of response is demonstrated in Fig. 21 (qNano™), where the count rate of the device is plotted as a function of liposome concentration, obtained using

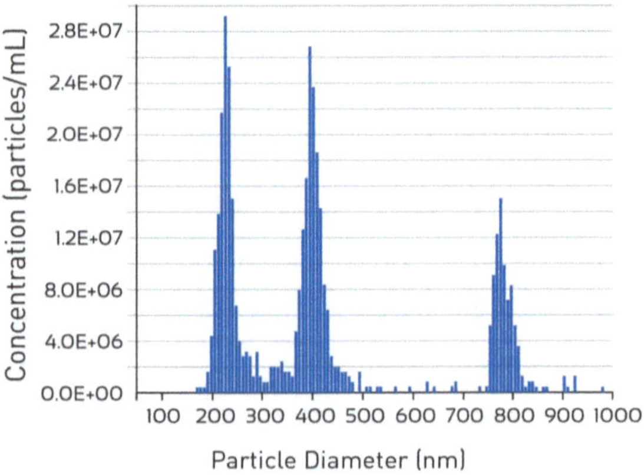

Fig. 19 PSD obtained from TRPS for a PSL trimodal: #/mL *vs*. particle diameter (nm). (From http://www.izon.com—Reprinted with permission)

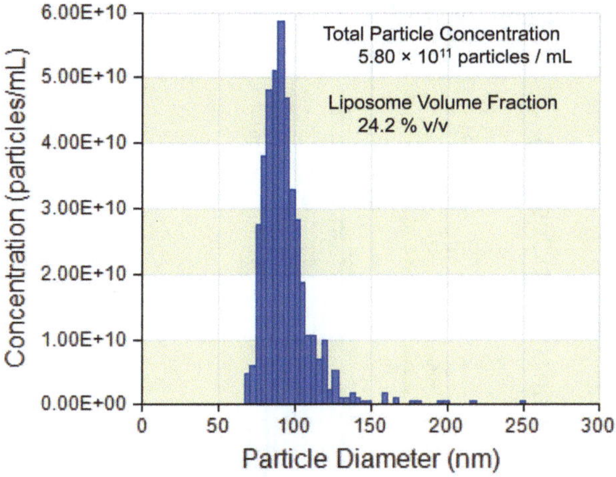

Fig. 20 PSD obtained from TRPS for liposomes: #/mL *vs*. particle diameter (nm). (From http://www.izon.com—Reprinted with permission)

dilution factors ranging from 1:10 to 1:100. The count rate, ranging from ≈ 150/min to 1600/min, is seen to increase in proportion to the liposome concentration, ranging from 5.28 to 57.5 × 10^9/mL.

Finally, Figs. 22 and 23 (qNano™), show the PSDs obtained for a significantly aggregated virus samples of lentivirus and a much less aggregated (but still significantly polydisperse) sample of adenovirus, respectively, showing the

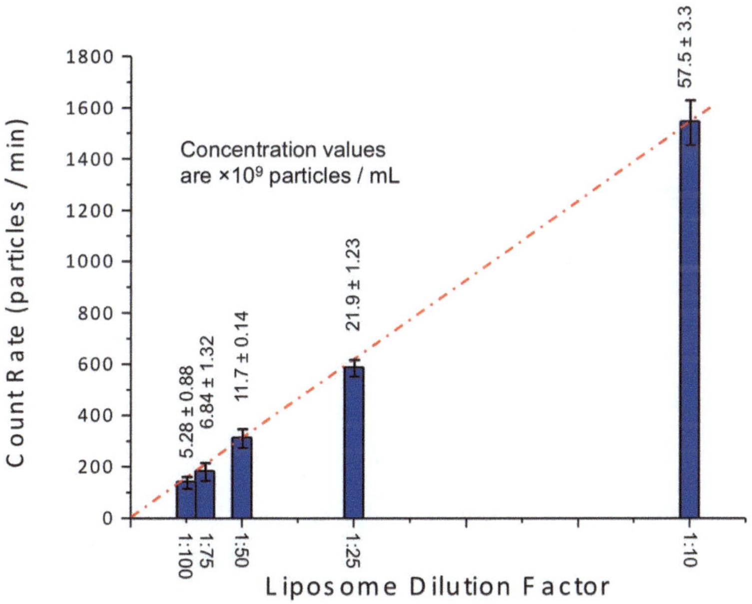

Fig. 21 TRPS linearity study: count rate (#/min) *vs*. liposome concentration. (From http://www.izon.com—Reprinted with permission)

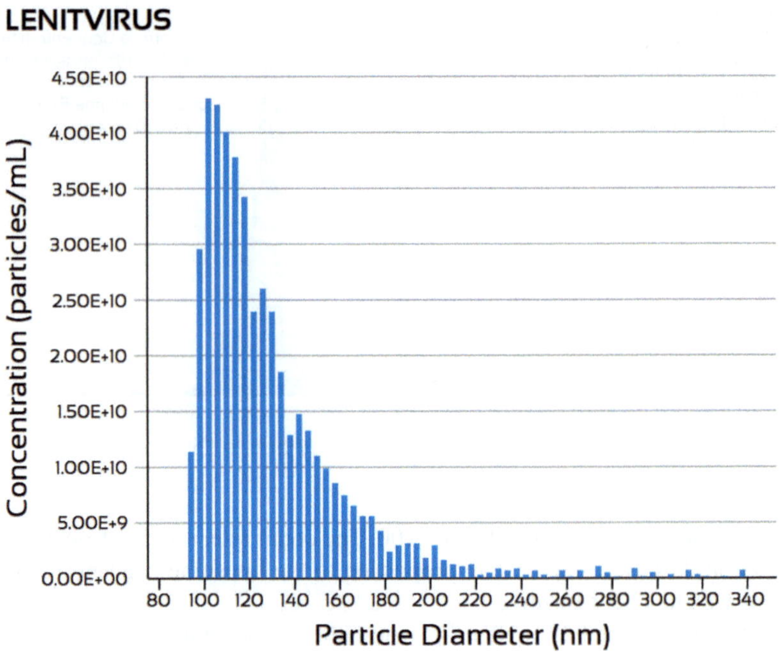

Fig. 22 PSD obtained from TRPS for Lentivirus particles: #/mL *vs*. particle diameter (nm). (From http://www.izon.com—Reprinted with permission)

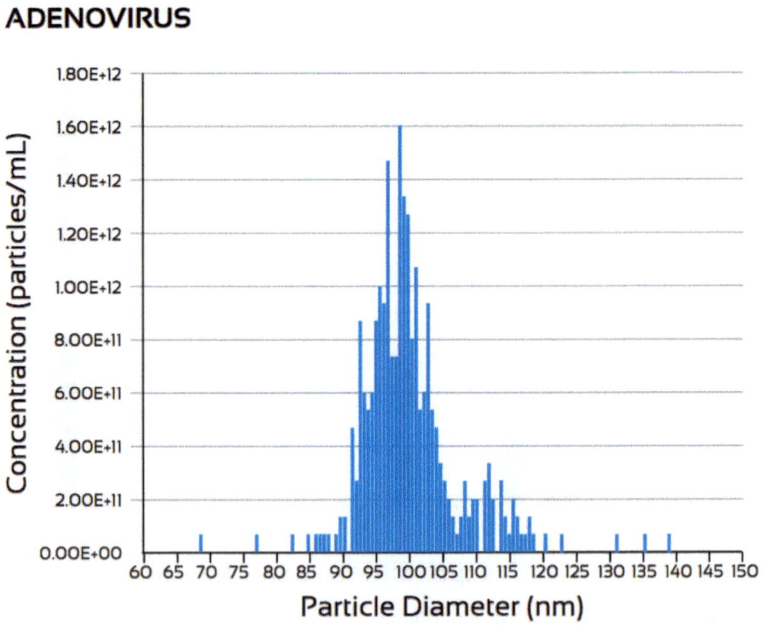

Fig. 23 PSD obtained from TRPS for Adenovirus particles: #/mL *vs*. particle diameter (nm). (From http://www.izon.com—Reprinted with permission)

"tails" of primary and aggregated larger virus particles: #/mL vs particle diameter (>≈90 nm).

In the former case of lentivirus particles, one can clearly distinguish dimer and trimer aggregates from the primary virus particles, in addition to a smoothly decaying tail of higher oligomers.

Single-Particle Techniques

Resonant Mass Measurement (RMM) Method (Archimedes™; Burg et al. 2007; Godin et al. 2007; Chunara et al. 2007; Olcum et al. 2014)

The principle of resonant mass measurement (RMM) can be used to determine the buoyant mass of individual sub-micron particles with high resolution and absolute accuracy. A mechanically resonant structure, such as a beam suspended at one end, resonates at a specific frequency, which depends on its physical dimensions, mass and stiffness. The resonant frequency decreases when a mass is added to the free end of the beam. By accurately measuring the shift in resonant frequency, the mass added to the beam can be determined with very high precision.

Recently, the RMM principle has been extended in several ways, using MEMS (Micro-Electro-Mechanical Systems) fabrication methods to produce very small resonators, sensitive to extremely small amounts of added mass, including an embedded fluidic microchannel, shown schematically in Fig. 24: (Archimedes™)

The resulting device, known as a suspended microchannel resonator (SMR), has a size scale of ≈100 μm with a total suspended mass of only a few nanograms. The microchannel allows particles suspended in fluid to pass through the resonator and be measured one at a time, provided the flow rate of the particle-bearing fluid is adjusted to an appropriate, extremely low value.

Second, these SMRs (and smaller, newer generation, more sensitive suspended nanoparticle resonators, or SNRs) are contained in a vacuum environment, even while the particles are being measured in liquid suspension. This critical feature allows the SMRs to vibrate without dissipating energy unnecessarily, resulting in a very high Q, or quality factor, typically exceeding 10,000, where Q is defined as the ratio of the resonant frequency to the width of the resonant frequency response curve. A high Q value, in turn, allows the resonant frequency to be measured to very high precision. Also, advanced MEMS fabrication methods allow the fluid inside the resonator to remain at ambient pressure without reducing the Q.

Third, the resonant frequency can be measured relatively quickly and with very high resolution. Specialized detection optics, electronics and signal processing techniques, combined with the inherently high Q, allow measurement of the resonant frequency to within a few parts per billion over a 1 kHz bandwidth (i.e., at a rate of 1000 times per second). These techniques applied to current SMR devices allow the mass of individual particles to be measured with a resolution below 1 fg (10^{-15} g) at relatively high particle throughput. For perspective, this mass sensitivity is approximately 1 million times better than that provided by a high-end quartz-crystal microbalance.

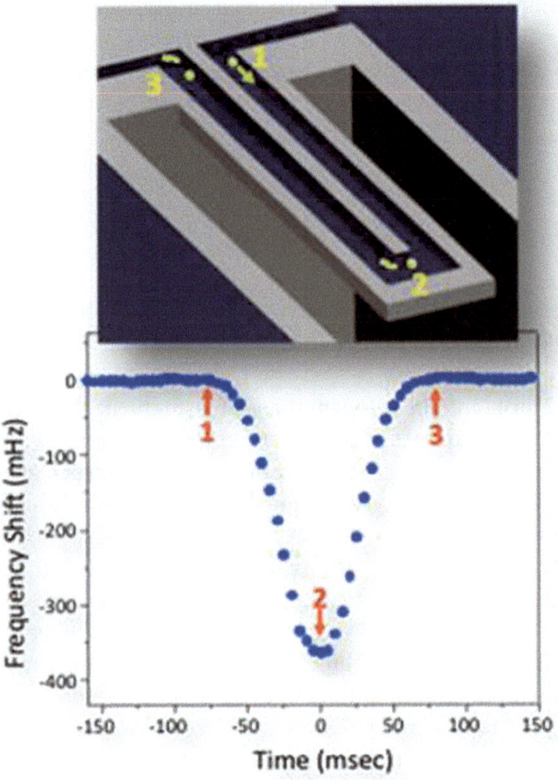

Fig. 24 Stylized picture of the fluidics microchannel and frequency shift for RMM. (From http://www.malvern.com—Reprinted with permission)

To measure a particle, the resonant frequency is monitored as the particle passes through the microchannel in the resonator, as indicated in Fig. 24. The maximum shift, or excursion, in the frequency occurs when the particle is at the tip of the resonator and is proportional to the particle's buoyant mass—i.e., the mass of the particle over that of the fluid it displaces. The frequency excursions for a large number of particles are measured one at a time, in rapid succession (typically at a rate of $\approx$ 5–10/s, depending on the application). From this frequency-shift information, Archimedes' Principle is used to build up a distribution of mass, and therefore size (given the particle density), for the population of particles analyzed. During a measurement the instrument "feels" the buoyant mass of a particle as it passes through the microchannel resonator, and clearly the buoyant mass increases not only with particle size but also with the density mismatch between the particle and fluid.

The mass and size of a particle can be determined from some simple relationships. The first step relates the frequency excursion obtained for a particle to its buoyant mass, using the microchannel resonator's "sensitivity" $\boldsymbol{S}$, which relates the frequency excursion to buoyant mass with a simple constant, with units [mHz/fg]. The sensitivity $\boldsymbol{S}$ is a fixed value for each resonator, reflecting a simple linear relationship over the entire size range of measurable particles, determined by a simple calibration procedure. The dry mass, volume and "equivalent sphere" diameter can then be calculated for each particle individually. The resulting values are cast as histograms that form either a particle mass distribution (PMD) or a PSD.

The RMM technique as currently practiced commercially yields particle mass and size results with an absolute accuracy typically at, or better than, the 1 % level. For example, using a mixture of four different NIST-traceable polystyrene latex (PSL) standards, the mean sizes of each of the sub-populations were determined to be 518-, 796-, 904- and 994-nm, within 0.1–0.5 % of the manufacturer's specified sizes of 519-, 799-, 903- and 994-nm, respectively. Comparable absolute accuracy and resolution have also been achieved for mixtures of much smaller, but also much heavier, gold nanoparticles (density 19.3 g/cm^3). A trimodal distribution of 57-, 79- and 100-nm gold particles yielded mean sizes of 57-, 81- and 99-nm.

A notable advantage of the RMM technique is its high resolution—i.e., the ability to distinguish between two populations of particles that differ only slightly in mass or size. Figure 25 shows the PSD obtained for a mixture of NIST-traceable PSL particles of 0.994 and 1.034 μm, differing in size by only 4 % (van Hulst 1981).

The resulting populations are cleanly separated. The measured standard deviations of the two populations are 4.5 and 5.2 nm, showing that these particles are considerably "better" (i.e., more uniform) than indicated by the 10 nm standard deviation value provided by the manufacturer.

A useful consequence of the high resolution provided by the RMM technique is its ability to characterize the extent of aggregation of particles. Because it is responsive to mass, its output is directly proportional to the order of aggregation. A good example is provided by 1-μm PSL beads exposed to concentrated IgG pro-

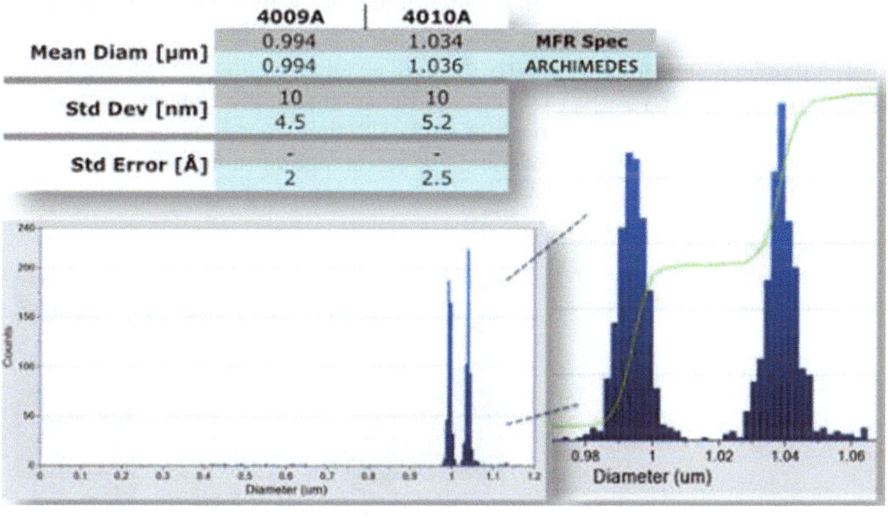

Fig. 25 RMM: response to two populations of closely-spaced PSL particles. (From http://www.malvern.com—Reprinted with permission)

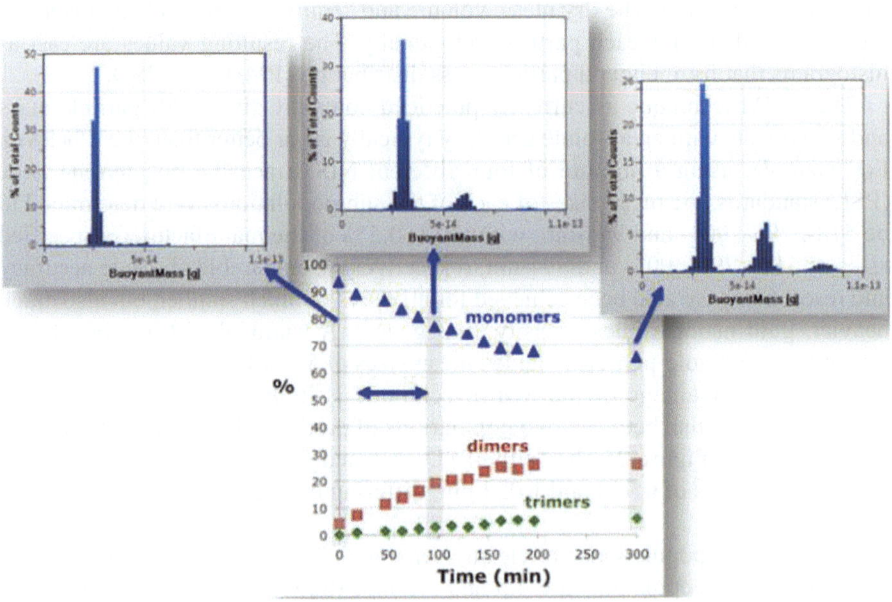

Fig. 27 RMM: PSDs obtained for PSL particles plus aggregates over time. (From http://www.malvern.com—Reprinted with permission)

tein (1 mg/mL). The protein molecules coat the beads, reducing their net surface charge and inter-particle repulsions, thereby causing them to aggregate relatively quickly over time. Common applications include immunoassays and selective separation of macromolecules. Figure 26 shows three representative frequency responses for a PSL monomer, dimer and trimer, along with their photographic images (Archimedes^{TM}).

The dimers and trimers are clearly distinguishable from the monomers, because each of their frequency shifts (indicated by the heights of the "pulses" seen above) is an integral multiple of the monomer shift. The time course of the agglutination process is seen in Fig. 27, and of note, the percentage of monomer particles falls

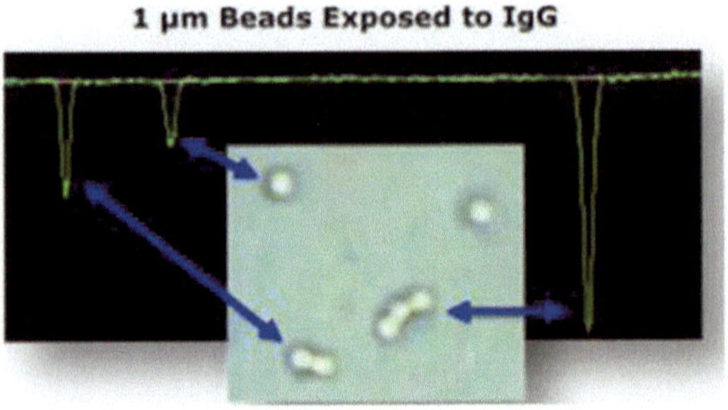

Fig. 26 RMM: signals/images obtained for IgG-coated PSL monomers, dimers, trimers. (From http://www.malvern.com—Reprinted with permission)

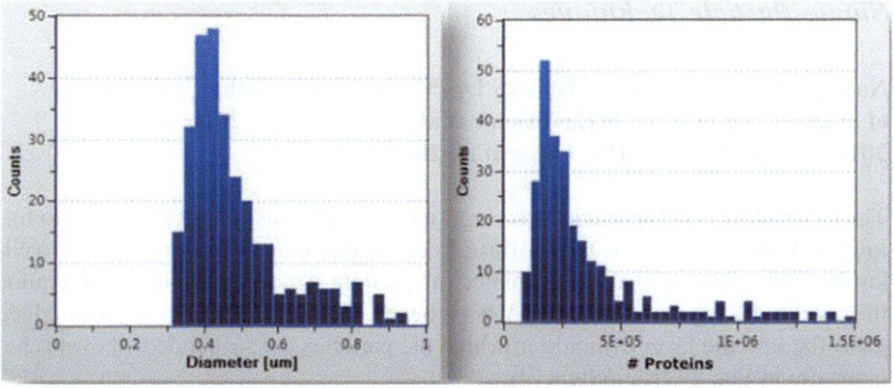

Fig. 28 RMM: IgG aggregates *vs*. diameter (*left*) and vs. # protein molecules per aggregate. The histogram of particle counts vs size (*left-hand plot*) shows the size distribution of IgG protein aggregates obtained from 4 μL of native sample suspension. The measured concentration of aggregates above 300 nm was 4 × 106/mL. The instrument reportedly can accept aggregate particle concentrations up to about 5 × 108/mL, so that many formulations, typically of high concentration, should be measurable without dilution. The viscosity limit provided by current SMR devices is ≈100 mPa•s. Because the underlying measurement is of particle mass, the distribution can also be expressed as the number of IgG molecules making up each aggregate, also shown (*right-hand plot*). (From http://www.malvern.com—Reprinted with permission)

monotonically with elapsed time, while the percentages of dimers and trimers, formed from the monomers, increase in the process, as expected (Archimedes™).

A second example is the aggregation of protein molecules used in therapeutic drug products. The current-generation SMR technology practiced commercially can reportedly count and size protein aggregates as small as ≈150 nm, with an equivalent lower size limit of ≈40–50 nm for nanoparticles of gold suspended in water. Future commercial application of smaller, more sensitive, "first-generation" SNR devices is expected to reduce the lower size limit for proteins in typical aqueous preparations to ≈35–40 nm. A representative measurement of an aggregate "tail" for IgG protein (M.W. 150 kDa) using the currently available instrument is shown as # counts vs particle diameter in Fig. 28 (left-hand plot) (Archimedes™).

The results of a significant academic research effort, aimed at improving substantially the performance of RMM devices, have been reported recently (Olcum et al. 2014). The performance achieved is significantly improved over that of existing commercial devices, approaching the thermomechanical noise limit and thus permitting measurement of 10-nm gold particles and exosomes, ranging in size from 30 to 100 nm. Improving the design of SNRs in order to reach attogram-scale mass resolution requires both an increase in mass sensitivity and a reduction in frequency noise. The former is proportional to the resonant frequency of the cantilever and inversely proportional to its mass, while the latter decreases with increasing oscillation amplitude.

Single-Particle Techniques

Nanoparticle Tracking Analysis (NTA) Method (NanoSight™; Carpenter et al. 2012; Zhou et al. 2012; Jiskoot et al. 2012; Li et al. 2011; Zölls et al. 2011; Mickisch et al. 2010; Filipe et al. 2010, 2012)

The technique of Nanoparticle Tracking Analysis (NTA) uses light scattering to follow individual particles as they diffuse freely in liquid and then employs the well-known Stokes Einstein equation connecting particle size and diffusion to determine their individual mean diameters. A laser beam is directed through a prism-edged glass flat into the layer of liquid in which the particles are suspended, shown schematically in Fig. 29 (NanoSight™).

The angle of incidence of the laser beam and the refractive indices of the glass and water are such that the beam refracts downward, toward the glass-water interface, resulting in a "compressed" beam with a reduced profile and high intensity that passes through the relatively thin layer of sample liquid. The particles illuminated by this beam scatter light and are detected by a microscope with 20X magnification, outfitted with a high-sensitivity CCD (charge-coupled device) camera. The latter, operating at 30 frames/s, creates a digitized video file of the diffusing particles within a field of view of approximate dimensions 100um × 80um × 10um.

The particles moving randomly under Brownian motion can be viewed directly by eye using the microscope, but, more importantly, they can be analyzed automatically from the digital camera frame images. Proprietary software is used to identify and track virtually simultaneously the centers of each of many particles from one captured digitized frame picture to the next, as they diffuse randomly within the layer of liquid. Of course, Brownian motion takes place in all three dimensions, while NTA tracks particles in only two dimensions—i.e., in the x-y plane, paral-

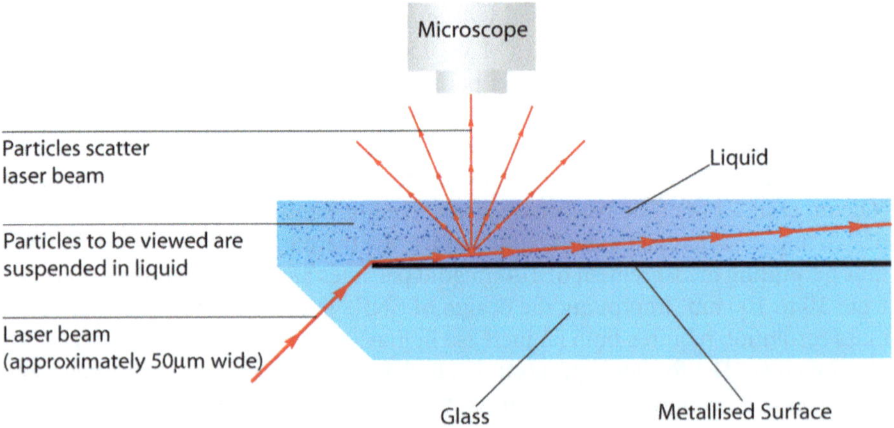

Fig. 29 Schematic diagram of the optical configuration used to implement NTA. (From http://www.nanosight.com—Reprinted with permission)

lel to the glass/liquid interface. The image analysis software determines the mean squared displacement (per sec) in two dimensions, $(x,y)^2_{avg}$, of each tracked particle, proportional to its diffusion coefficient, D, and the time, t, over which the particle has randomly diffused,

$$Dt = (x, y)^2_{avg} / 4 \tag{9}$$

The spherical-equivalent hydrodynamic diameter, d, of the particle is then obtained from D using the well-known Stokes-Einstein equation,

$$D = kT / 3\pi\eta\, d \tag{10}$$

where k is Boltzmann's constant, T the temperature (deg-K) and η the viscosity (Pa • s) of the liquid in which the particles are suspended.

The equation above, of course, forms the basis of the well-known DLS technique (discussed in the next Section), in which the mean particle diameter is calculated from the mean diffusivity, obtained from the time behavior of the fluctuations in the scattered light intensity produced by *many* particles at any given instant of time. In contrast, each of the diffusivity values obtained by NTA refers to a *single* particle. Hence, the lowest size limit of NTA is determined by the lowest level of scattered light that can be reliably detected, used only for the purpose of tracking a given particle. Of course, the intensity of light scattered by a particle depends on many variables, especially its size. For submicron particles, the scattered intensity follows Rayleigh's 6th power law. Other variables include the laser output power, wavelength and beam size, angle of polarization, refractive index (real and imaginary) and shape of the particle, refractive index of the suspending solvent, and the efficiency of the light-collecting optics and sensitivity of the digital camera.

In practice, given that the laser light source and detection system are usually fixed, the variable that effectively determines the lower size limit of NTA is the "contrast" of the particles—i.e., the difference in refractive index between the particles and the liquid in which they are suspended. For particles of high refractive index (high contrast), such as colloidal gold or silver, the lower size limit for detection is ≈10–15 nm. For particles of lower refractive index, such as those of biological or pharmaceutical interest, the lower size limit is closer to 25–35 nm. For very weakly scattering particles, such as liposomes, exosomes and proteins, the smallest visible particles may be ≈40 nm. The upper sizing limit, ≈1–2 μm, is the point at which the Brownian motion becomes so slow that the particle movement from one camera frame to the next is so small as to be comparable to the small centering errors inherent in the tracking software.

As for all other single-particle techniques, a sufficient number of particles must be analyzed to ensure that the results are statistically meaningful. A particle concentration in the range of 10^7 to 10^{10}/mL is said to yield reliable PSDs within 30–60 s. Of course, the importance of statistical counting fluctuations will obviously depend on the size range (degree of polydispersity) and the shape of the distribution (e.g., decaying tail vs a single peak). Regarding absolute concentration determination, the scattering volume can be estimated from the fixed optical field of view and depth of the illuminating beam. The number of particles counted in a given size class in the scattering volume can be extrapolated to the number per mL of the sample fluid. However, the effective scattering volume in which particles are detected and subsequently tracked and sized is not a "fixed" quantity. Rather, it is a function of particle size and refractive index, and also the power, wavelength and beam size of the laser, as well as camera sensitivity. Hence, calibrations should be carried out using actual samples of interest, of known concentrations, and not just PSL standard particles.

If the particle concentration is less than 10^7/mL, an extended analysis time (2–3 min) is needed to obtain statistically meaningful PSDs. Conversely, if the concentration is too high, $> 10^{10}$/mL, the Brownian motion trajectories of neighboring particles will cross each other relatively frequently, resulting in confused identities before a good estimate of their size can be made, thus degrading the quality of the resulting PSD. The accuracy of estimates of the number and concentration of particles in a particular size class will depend on how much light they scatter, which will determine their effective scattering volume and therefore estimates of their concentration. The polydispersity of the underlying distribution is very important. A lower number of larger particles (e.g., aggregates) may not be detected and counted/sized with the same accuracy as a higher number of smaller particles.

The basic principles of particle diffusion under which NTA operates are the same as those on which DLS is based. The Stokes-Einstein relationship (Eq. 10) assumes that the particles are effectively spherical and non-interacting, diffusing freely in the infinite dilution limit. However, inter-particle repulsive forces between charged particles may become sufficiently strong at high particle and/or low electrolyte concentrations to increase significantly the diffusivity and therefore decrease correspondingly the computed particle size. Assuming the temperature and viscosity of the sample fluid are well known, the absolute accuracy achievable by NTA is effectively the same as DLS—ideally within 2%. Also, since both methods operate in the time domain, they are "absolute" methods, requiring no calibration. The absolute accuracy of NTA was found to agree very well with the results produced by two commercial DLS instruments, using PSL calibration standards of 50-, 100-, 200- and 400-nm diameter (Filipe et al. 2010).

An important property of NTA is the fact that it does not suffer from intensity weighting, unlike the DLS method, in which a small number of large particles contribute disproportionately to the scattered intensity signal and resulting intensity autocorrelation function, from which the final PSD is obtained. Therefore, in theory NTA has inherently higher size resolution than DLS, in addition to possessing the fundamental advantage that it does not require mathematical "inversion" of the autocorrelation data using a fitting procedure that is ill-conditioned to one extent or another—the ultimate factor that limits size resolution in DLS. The superior size resolution provided by NTA compared to DLS for several closely-spaced bimodal mixtures of PSL standard particles (60- and 100-nm; 100- and 200-nm; 200- and 400-nm; and 400- and 1000-nm, displayed from top to bottom) is shown in Fig. 30 (Filipe et al. 2010).

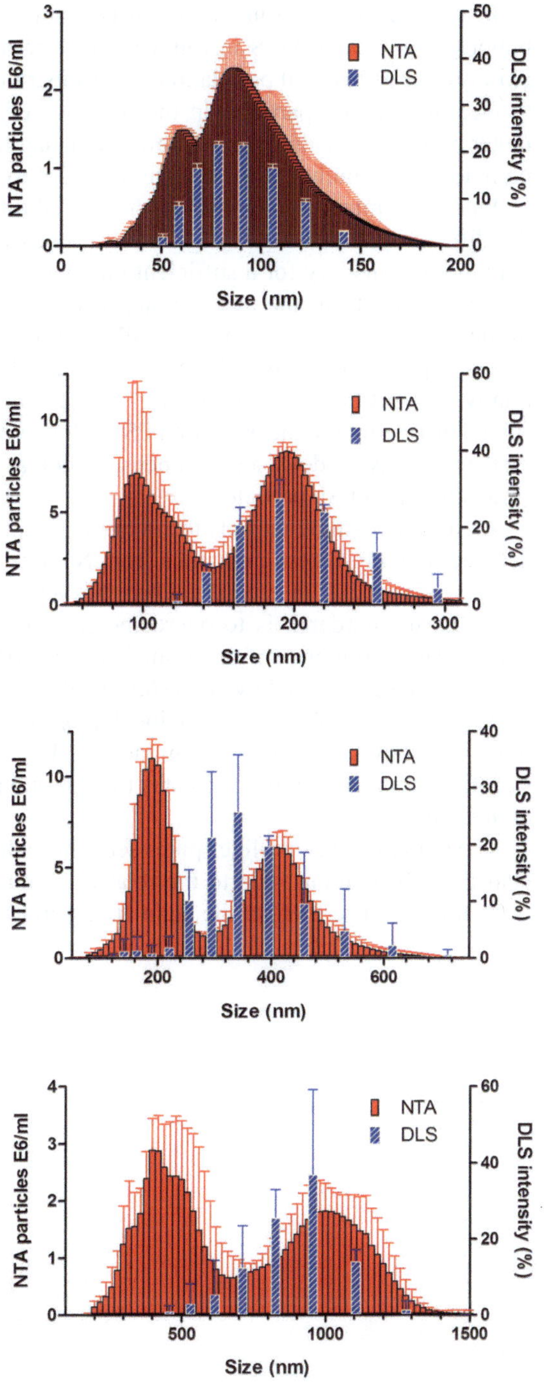

Fig. 30 Nanoparticle tracking analysis versus dynamic light scattering: particle size distributions obtained for four different closely-spaced polystyrene latex standards. (From Filipe et al. 2010—Reprinted with permission)

In all cases this DLS instrument was unable to perceive the two separate distributions. Some other DLS instruments may be more successful in resolving such closely-spaced bimodal populations. The superior ability of NTA compared to DLS to measure smaller particles in the presence of relatively high concentrations of larger particles has also been established, using mixtures of 100- and 400-nm PSL particles, with number ratios ranging from 3:1 to 300:1 (Filipe et al. 2010).

Finally, it must be appreciated that the ability of NTA to estimate accurately the diffusivity, D, of a given particle depends on its ability to track the Brownian motion trajectory for a sufficient number of "random" steps so as to obtain an accurate estimate of the average step-length—i.e., $(x,y)^2_{avg}$. However, the depth of the scattering volume is very small, ≈10 μm. Therefore, small particles, which diffuse rapidly, may remain in the scattering volume (and be detectable) for only a relatively short time, thus producing "limited duration trajectories", resulting in artificial broadening of the measured PSD. However, the reduction in accuracy can be mathematically modeled and compensated for, at least in the simple case of uniform (monodisperse) PSL particles.

As for other single-particle techniques, the quality (reproducibility and absolute accuracy) of a PSD result obtained from NTA depend greatly on statistics—i.e., the number of particles in each size class that have been tracked. Clearly, the broader the PSD, the harder it is to overcome statistical noise and achieve reproducible results. The importance of this issue is illustrated in Fig. 31 (NanoSight™), which shows the PSD obtained by NTA (60 s analysis) from a sample of human IgG protein (1 mg/mL) before and after incubation at 70 °C for 24 min. The plot on the left (t=0) shows a relatively "compact" peak, centered above 100 nm, indicating a significant population of aggregates (as IgG monomers are only ≈8–9 nm in size) at the outset.

The numerous needle-like spikes seen throughout the PSD are the result of large statistical fluctuations, due to the relatively small number of IgG aggregates that were tracked. The plot on the right (t=24 min) indeed shows significant growth

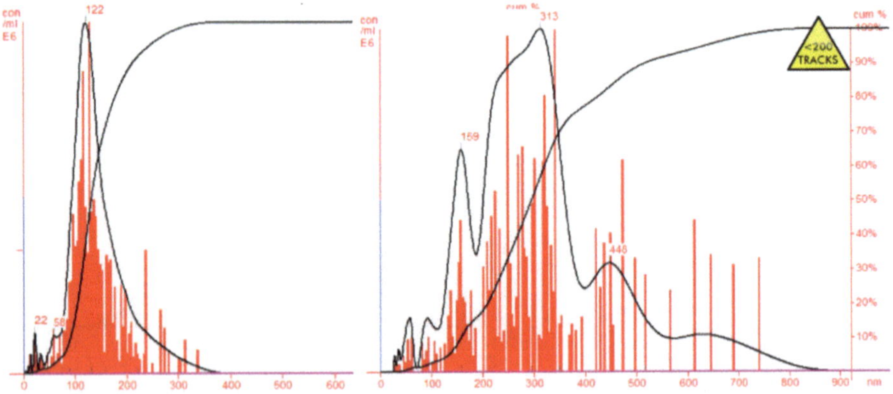

Fig. 31 PSDs obtained by NTA for aggregates of IgG, before/after thermal incubation. (From http://www.nanosight.com—Reprinted with permission)

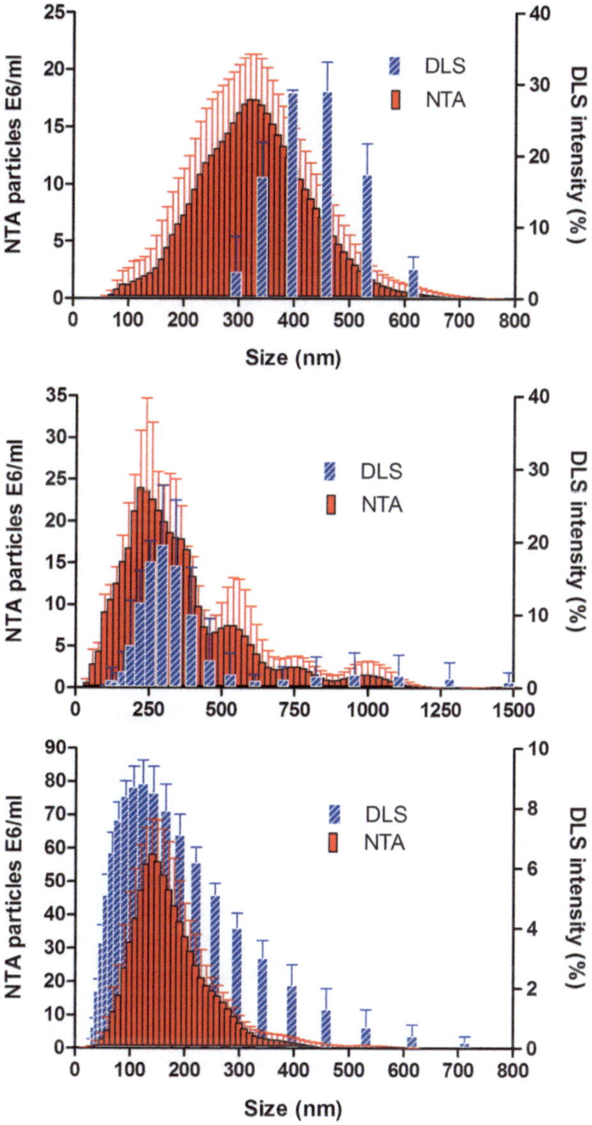

Fig. 32 Particle size distributions obtained by nanoparticle tracking analysis versus dynamic light scattering for TMS, PLGA nanoparticles and liposomes. (From Filipe et al. 2010—Reprinted with permission)

in the size and number of aggregates. However, here the effects of limited particle counting statistics are even more evident.

Figure 32 (Filipe et al. 2010) shows PSD results obtained from NTA, compared to DLS, for three different kinds of nanoparticles used for drug delivery applications; TMC (trimethyl chitosan) (top plot), PLGA (poly(lactide-co-glycolide)) (middle plot) and liposomes (bottom plot).

In the case of TMC particles, the mean diameter obtained by NTA was 320 nm, 90 nm smaller than the mean size obtained from DLS. This shift is explained by the fact that the PSDs produced by DLS are intensity-weighted, whereas NTA produces number-weighted PSDs, and the extent of the shift to higher sizes by DLS is significant for distributions which are broad. In the case of PLGA particles, the mean size obtained by NTA was 322 nm, actually slightly higher than the intensity-weighted mean diameter given by DLS. This different shift behavior is ascribed to the difficulty DLS has in analyzing polydisperse samples, resulting from the use of an ill-conditioned inversion algorithm. Finally, in the case of liposomes, the mean size obtained by NTA was 154 nm, again larger than the intensity-weighted value given by DLS. A possible explanation is that DLS has a lower detection size limit than NTA. Therefore, smaller particles (<30 nm) may decrease the intensity-weighted mean size in DLS.

Finally, it has been pointed out by independent evaluation that successful use of NTA requires several, potentially time-consuming, optimization steps by a skilled operator, including arriving at an optimal sample concentration. It has further been observed that the operator can easily choose settings that either ignore or emphasize the presence of certain particles, which can make the results dependent, to a degree, on individual judgment and experience (Filipe et al. 2010).

Dynamic Light Scattering (DLS) Method (Chu 1991; Berne and Pecora 2000; Charles et al. 1981; Brown 1996)

The technique of dynamic light scattering (DLS), also known as photon correlation spectroscopy (PCS) and quasi-elastic light scattering (QELS), has proven to be a powerful tool for characterizing the particle size distribution (PSD) of nanoparticles from ≈1 to 1000 nm (1 μm). Notwithstanding its inherent inability to count individual particles, DLS technique has nevertheless proven to be effective in the "deep submicron" size range, <50 nm, where most other techniques are either unreliable or lack any capability whatsoever.

DLS determines the diffusion coefficient, or diffusivity, D, of particles suspended in liquid by analyzing the temporal fluctuations in scattered light intensity, I_S, caused by Brownian motion of the particles. Laser light (wavelength λ) is focused into a sample cell containing particles suspended in liquid. The particles scatter light, provided their refractive index differs from that of the liquid. For very small particles ($d \ll \lambda$) in the "Rayleigh region" the scattered intensity, I_S, is approximately independent of the scattering angle, θ_S (after correction for the effective scattering volume, $\sim 1/\sin\theta_S$), varying as the square of the particle volume, V, or the 6th power of the particle diameter, d, for spherical particles,

$$I_S = g^2\left(d, n_p, n_s\right)V^2 I_o \tag{11a}$$

or,

$$I_S = g^2\left(d, n_p, n_s\right)d^6\, I_o \tag{11b}$$

where g^2 is the "intraparticle scattering factor", accounting for interference of the scattered wavelets originating from different points within a given particle, described by the Mie theory. This quantity is a function of the particle diameter, the laser wavelength and also the refractive indices n_p and n_s of the particle and suspending liquid, respectively.

The physical principle underlying DLS may be understood by considering the simplified case in which there are only *two* very small ("point") particles in suspension. The illuminated particles re-radiate ("scatter") light, and the resulting scattered waves "interfere" everywhere in space. The net detected scattering intensity, I_S, depends on the *relative phase* of the two waves when they mix "coherently" at the detector, depending only on the difference in their optical path lengths—from the laser "wave front" to each respective particle and then to the detector. Intensity I_S ranges from a *maximum* value, when there is total constructive interference between the two waves (when the two optical path lengths differ by an integral multiple of λ, or $N\lambda$), to *zero*, for total destructive interference (when the path lengths differ by $[N \pm 1/2]\lambda$).

The suspended particles constantly undergo Brownian motion, caused by random collisions of the surrounding solvent molecules, resulting in corresponding temporal fluctuations in the difference of the optical path lengths, causing the scattering intensity, I_S, to fluctuate between a *maximum* value and zero. In real situations involving a large number of particles, there are a large number of individual scattered waves that mix at the detector, so that I_S fluctuates by a smaller relative magnitude (described by the root mean square, or RMS, amplitude) with respect to the mean intensity value, and hence statistically unlikely to reach either the extreme maximum or zero values. But the principle of fluctuations caused by variations in the phase differences of the interfering waves remains the same.

The quantity of interest, D, is effectively "buried" in the fluctuating, "noisy" I_S signal, but it can be easily extracted from the noise using the mathematical operation of autocorrelation. The intensity autocorrelation function (ACF), denoted by $C(\Delta t)$, is defined by

$$C(\Delta t) = < I_S(t) \times I_S(t - \Delta t) > \tag{12}$$

where I_S at the instantaneous, current time, t, is multiplied by I_S at the earlier time, $t - \Delta t$, for a given value of Δt. The symbol $< >$ denotes an ensemble (running) average over evolving time, t. Quantity $C(\Delta t)$ is computed simultaneously for each of many different values of Δt, ranging from "small" to "large" relative to the *characteristic time*, τ, of the fluctuations in I_S, where τ depends on D. The smaller the particles, the larger is D, and the smaller is τ.

In the simplest case of uniform particles, autocorrelation of I_S yields a very simple result: a single decaying exponential function,

$$C(\Delta t) = A \exp(-\Delta t/\tau) + B \tag{13}$$

where τ is related to D by

$$\tau = 1/2DK^2 \tag{14}$$

and K is the "scattering wave vector", connecting time and distance travelled (by diffusion),

$$K = (4\pi n/\lambda)\sin(\theta/2) \tag{15}$$

where n is the refractive index of the suspending liquid.

All that remains (for uniform particles) is to obtain the hydrodynamic particle radius, R_H, or diameter, d, from D, using the well-known Stokes-Einstein relation, given earlier by Eq. 10.

One of the principal advantages of the DLS technique, apart from its ability to measure the average size of particles as small as 1 nm, is that it is an "absolute" technique. That is, the diffusivity of uniform particles depends only on their *size*, and not their *composition*—i.e., their physical properties, especially refractive index.

Of course, applications of interest usually involve particles that are *not* uniform in size, but instead are spread across a range of sizes, defining the PSD. A single diffusivity value is then no longer sufficient to describe the Brownian motion of the particles; instead, a *distribution* of D values is required. Consequently, $C(\Delta t)$ *vs* Δt no longer behaves like a single decaying exponential function, but instead is comprised of a *mixture* of decaying exponential functions, each corresponding to a different particle size. The extent to which the ACF departs from single-exponential behavior depends on the "polydispersity" of the sample.

A mathematical algorithm is needed to "invert" the raw ACF data, $C(\Delta t)$, to obtain the best estimate for the PSD. In cases where the distribution of diffusivities is approximated well by a log-normal shape, whether narrow or moderately broad, the simple method of cumulants can be used to invert $C(\Delta t)$ to obtain a reasonable estimate of the PSD (Kopell 1972). The first step in the analysis is to obtain the "reduced" ACF, $C^R(\Delta t)$, by subtracting the measured baseline ("B" in Eq. 13 from the contents of each of the "channels" comprising $C(\Delta t)$ and then taking the logarithm of each of the resulting values,

$$C^R(\Delta t) = \ln[C(\Delta t) - B] \tag{16}$$

The second step consists of performing a least-squares fit of a polynomial (usually a quadratic) function of Δt to the reduced ACF,

$$C^R(\Delta t) \approx a_o + a_1\,\Delta t + a_2(\Delta t)^2 + \ldots \tag{17}$$

The most basic, essential parameters of the PSD—mean size and width, or standard deviation—are obtained from the coefficients of the linear and quadratic terms, a_1 and a_2, respectively. In the idealized case of uniform particles discussed above, $C^R(\Delta t)$ is simply a straight line with a negative slope, a_1, proportional to D, and the curvature term, a_2, equals zero. In the case of an actual distribution of particle sizes, coefficient a_1 equals the initial ($\Delta t = 0$) slope of $C^R(\Delta t)$, proportional to the mean diffusivity, from which the mean diameter is obtained (Eq. 10). The remaining quadratic "curvature" term is given by coefficient a_2, the square root of which

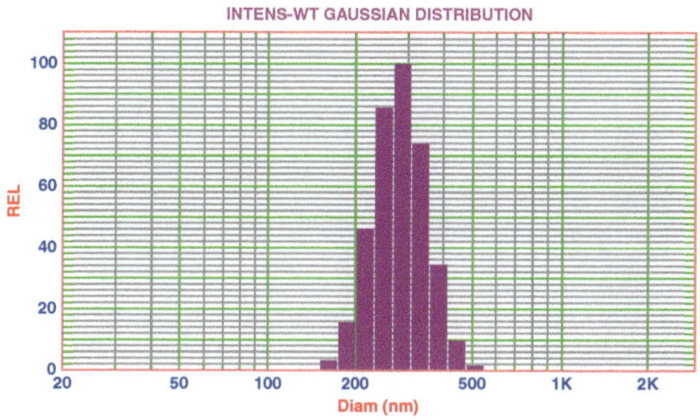

Solid Particle Fat-Emulsion.1	Detector: PMT Printout ID: Injectable IV fat emulsion	
Run Time hr:min:sec 00:09:46		Mean Diam. (nm) 289.4
Chi Squared 0.79		Coeff. Of Var'n. 0.214
Auto B.adj. 0.10%		Stnd. Dev. (nm) 61,929
Ch.1. Data x 1000 2967		Norm. Stnd. Dev. 0.214
Z-Average Diff. Coeff. cm2/sec 1.61e-008		Variance (P.I.) 0.046

Fig. 33 PSD obtained by DLS for an injectable fat emulsion, using cumulants analysis

is proportional to the half-width of the distribution of diffusivities, which in turn is related to the half-width, or standard deviation (or "coefficient of variation", CV) of the PSD, estimated to be log-normal in shape.

A good example of a system having a moderately broad PSD, well characterized using cumulants analysis, is an injectable fat emulsion. The PSD obtained for this submicron oil-in-water emulsion is shown in Fig. 33 (Data on File, Particle Sizing Systems LLC).

The intensity-wt mean diameter, 289 nm, is obtained from the mean diffusivity, 1.61×10^{-8} cm^2/s, in turn obtained from the least-squares fit coefficient a_1. The standard deviation, 61.9 nm, is obtained from coefficient a_2. The corresponding CV (mean diameter divided by the standard deviation) is 0.214, and the square of this quantity, the "variance", also called the "polydispersity index" (P.I.), is 0.046. The "chi-squared" value describes how well the quadratic function of Δt fits the ACF data. Its low value in this case (0.79) indicates a very good fit. Therefore, given the fact that the computed PSD is only moderately broad, as measured by the CV (≈ 0.21), the computed values of intensity-wt mean diameter and standard deviation should be considered to be quantitatively reliable.

A second example, involving much smaller particles, is a suspension of anionic micelles: 0.1 M sodium dodecyl sulfate (SDS), plus added 0.3 M NaCl as shown in Fig. 34 (Data on File, Particle Sizing Systems LLC).

The addition of salt, 0.3 M NaCl, to the surfactant preparation was done for a very important reason. The SDS micelles carry a relatively large (negative) electrical charge, because of significant dissociation of their Na+ counter-ions. In the absence of added electrolyte, there are strong electrostatic repulsive forces between neighboring charged micelles, speeding up their Brownian motion. As a result, the computed micellar size will be significantly *smaller* than it really is. The extent of enhancement of D and corresponding repression of the computed particle size can easily approach, and will often exceed, 50 %, depending on the magnitude of the electrical charge on the particles (influenced by pH and other factors) and the surfactant concentration. The addition of salt ions serves to "screen", to a lesser or greater extent, the electric field that emanates from each charged particle, thereby reducing the electrical potential at any given distance and the corresponding width

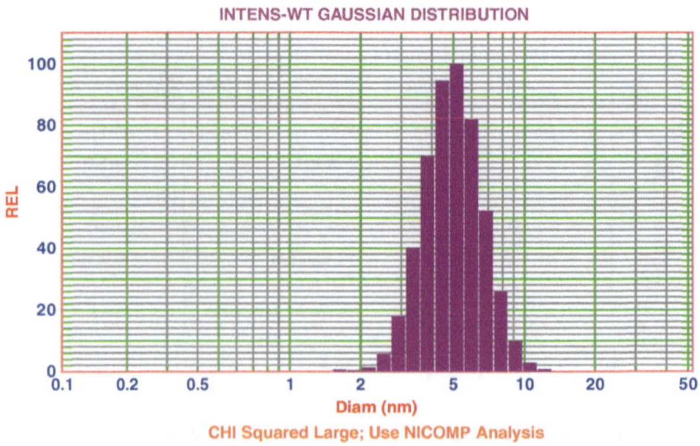

Solid Particle	Detector:	PMT
SDS-NaCl-Micelles.1	Printout ID:	SDS Micelles:0.1M SDS+0.3M NaCL
Run Time hr:min:sec		Mean Diam. (nm)
00:01:32		5.2
Chi Squared		Coeff. Of Var'n.
5.38		0.289
Auto B.adj.		Stnd. Dev. (nm)
0.00%		1,496
Ch.1. Data x 1000		Norm. Stnd. Dev.
3		0.289
Z-Average Diff. Coeff. cm2/sec		Variance (P.I.)
9.48e-007		0.084

Fig. 34 PSD obtained by DLS for anionic SDS micelles, using cumulants analysis. The channel width automatically chosen for the ACF was only 0.5 μs, much smaller than the value of 35 μs used for the fat emulsion, because the diffusivity of the micelles is much larger and hence the intensity fluctuations much faster. The resulting intensity-wt mean diameter is only 5.2 nm, obtained from the computed mean diffusivity, 9.48×10^{-7} cm^2/s. The standard deviation is 1.5 nm, with a corresponding CV of 0.29 and variance, or polydispersity index, of 0.084. Hence, this PSD is somewhat broader than that obtained for the IV fat emulsion, albeit with a mean diameter that is almost 60 times smaller than that of the fat emulsion droplets. The resulting chi-squared value of 5.38 is larger than the nominal threshold value for rejection (Newton and Driscoll 2008), but it is still low enough to allow this PSD result to be considered reliable

of the electrical "double layer" surrounding each particle. This action in turn reduces the strength of the repulsive inter-micellar forces that otherwise increase the diffusivity of the particles. Of course, the addition of salt to a colloidal suspension has a potential "downside" as well, in that it can also promote particle aggregation.

An example of the effect of inter-particle repulsive forces on particle diffusivity, in this case SDS micelles, is shown in Fig. 35 (Data on File, Particle Sizing Systems LLC).

Values of D are plotted for two different surfactant concentrations—[SDS] = 0.05 M and 0.1 M—and three different added salt concentrations—[NaCl] = 0.05, 0.1 and 0.3 M. As expected, for a given concentration of micelles, proportional to [SDS], D increases with decreasing added salt concentration, [NaCl]. As well, for a given added salt concentration, D increases with increasing [SDS]. The higher the micellar concentration, the shorter the average distance between neighboring micelles and therefore the greater the influence of inter-micellar repulsive forces on the diffusivity. The magnitude of this effect will clearly be greater for the smallest values of added [NaCl], for which the degree of screening of the electrical fields between neighboring micelles is the least. Similar results for a variety of cationic micellar systems have been reported extensively (Dorshow et al. 1983; Dorshow and Nicoli 1985; Ortega et al. 1990.

Consequently, it should be clear from Fig. 35 that the "size" of 5.2 nm for micelles of 0.1 M [SDS] with 0.3 M [NaCl] calculated from the measured diffusivity is somewhat smaller than the actual, physical size, due to incomplete screening of the inter-micellar repulsive forces. The true micellar size can be easily determined by effectively removing the influence of repulsive interactions, which have

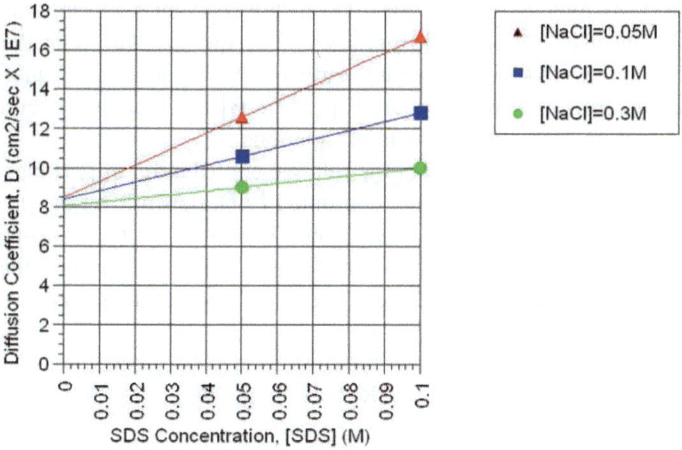

Fig. 35 DLS: effect of [SDS] and [NaCl] on micellar diffusivity

been shown, using both hypernetted chain (HNC) analysis (Zhou et al. 2012) and linear interaction theory (Dorshow et al. 1983; Batchelor 1976; Felderhof 1978), to result in linear dependence of D on [surfactant]. The measured diffusivity values are extrapolated to the hypothetical value, D_o, that would be obtained at the critical micelle concentration (CMC), where the surfactant concentration, [SDS], is nearly zero. From Fig. 35, this extrapolated diffusivity, D_o, is approximately 8×10^{-7} cm²/s—about 20 % smaller than the value obtained for [SDS] = 0.1 M. The resulting micellar size is therefore estimated to be ≈6.2 nm, almost 20 % larger than the value obtained without correcting for the influence of repulsions. At the lowest added salt concentration, [NaCl] = 0.05 M, the measured diffusivity is $\approx 1.65 \times 10^{-6}$ cm²/s, fully twice as large as D_o. Hence, the reported micellar size for [SDS] = 0.1 M and [NaCl] = 0.05 M is 50 % smaller than the true value, in the absence of repulsive forces. It should be clear that careful attention must be paid to PSD results obtained by DLS in applications involving electrically charged nanoparticles, such as concentrated proteins, in aqueous suspension.

There are numerous important applications, involving greater polydispersity, for which the simple 2-parameter cumulants method for analysis of the ACF data is ineffective. Many can be handled successfully by using a more generalized and powerful method of "inverting" the starting ACF data, the most effective of which is based on a mathematical algorithm known as the inverse Laplace transform (ILT) (Chu 1991; Berne and Pecora 2000; Charles et al. 1981 Brown 1996). Unlike the simple method of cumulants, the ILT approach makes no assumptions regarding the shape of the PSD and is especially well suited to analyzing distributions that can be approximated as a bimodal. In general, the ACF consists of a mixture of decaying exponential functions, each of which represents a different diffusivity, D_I (and hence, particle radius, R_I),

$$C(\Delta t) = A\left[\sum\nolimits_I g(D_I)\exp(-D_I K^2 \Delta t)\right]^2 + B \tag{18}$$

where sum $\sum_I$ is over all diffusivities, D_I, in the relevant size range and $g(D_I)$ is the intensity-wt distribution of diffusivities—i.e., the "answer" that the inversion algorithm produces, from which an intensity-wt PSD is obtained. Equation 18 can be rewritten in terms of a "reduced" ACF, $[C(\Delta t)-B]^{1/2}$, which then becomes simply a weighted sum of decaying exponential functions.

A central property of DLS is that the relative contribution of particles of a given size to the ACF, and therefore to the computed PSD, depends on the proportion of scattered light intensity that they contribute. Therefore, in the nano-particle size range, the "unknown" distribution of diffusivities, $g(D_I)$, is heavily "tilted" toward the largest particles because of the very strong, 6th-power dependence of intensity on particle size:

$$g(D_I)=f(R_I)\times V_I^2\times g_I^2(R_I) \tag{19a}$$

$$\text{or,}\quad g(D_I)=f(R_I)\times R_I^6\times g_I^2(R_I) \tag{19b}$$

where $g_I^2(R_I)$ is the "intra-particle scattering factor", accounting for a reduction (if any) of the net scattering intensity produced by particles of a given size due to partial destructive interference of scattered light wavelets originating from different points within the particle. In the case of very small particles in the Rayleigh region, where $R_I \ll \lambda$, all of the wavelets originating from a given particle are essentially in phase, so that g_I^2 becomes ≈ 1 and therefore can effectively be ignored. The largest particles in the distribution will then dominate the ACF and therefore the resulting computed PSD. In some cases this behavior is advantageous, as, for example, in achieving a high sensitivity to small populations of aggregated particles that scatter much more light than un-aggregated primary particles. This behavior then allows a small population of aggregates to be characterized in a semi-quantitative way, despite the fact that they cannot be counted individually by DLS.

An example is provided in Fig. 36 (Data on File, Particle Sizing Systems LLC), which shows the intensity-wt PSD obtained from a protein suspension—0.05 % (0.5 mg/mL) BSA (bovine serum albumin, M.W. 66 kD.).

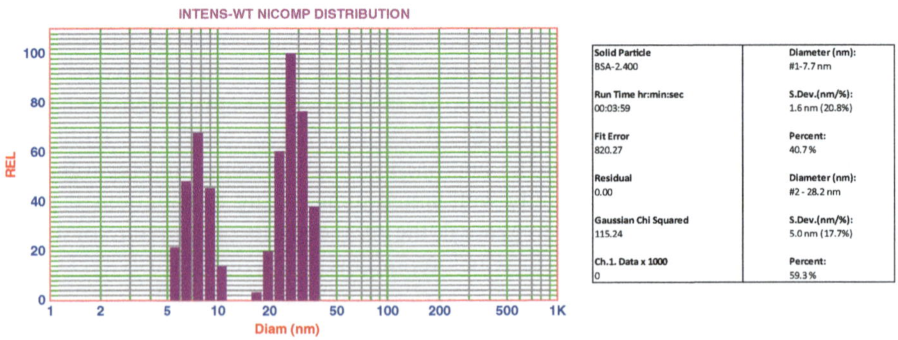

Fig. 36 Intensity-wt PSD obtained by dynamic light scattering for BSA, using multimodal ILT analysis

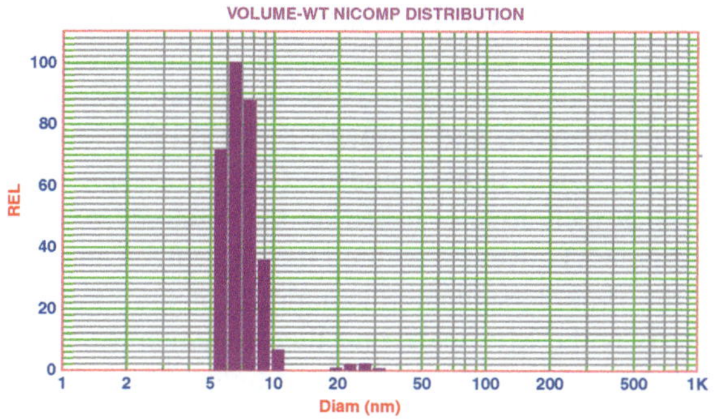

Solid Particle BSA-2.400	Diameter (nm): #1-7.0 nm
Run Time hr:min:sec 00:03:59	S.Dev.(nm/%): 1.2 nm (17.6 %)
Fit Error 820.27	Percent: 97.1 %
Residual 0.00	Diameter (nm): #2 - 25.8 nm
Gaussian Chi Squared 115.24	S.Dev.(nm/%): 4.1 nm (16.0 %)
Ch.1. Data x 1000 0	Percent: 2.9 %

Fig. 37 Volume-wt PSD for BSA, obtained from the intensity-wt PSD

This bimodal PSD provides valuable information. The first peak, centered at 7.7 nm, accounts for about 41 % of the intensity-wt ACF, according to the inversion analysis. This represents the "native", or primary, BSA molecules. The second peak, centered at 28.2 nm, represents, in effect, the "center of scattering" of the aggregate population, which in reality should resemble a smoothly decaying "tail". The concentration (#/mL) of aggregated particles is normally expected to fall monotonically with increasing size, while the scattered intensity produced by them increases as the 6th power of their diameter. The size at which these decaying and rising curves cross is the location where the analysis algorithm will place the second peak—i.e., at the "center of scattering". It serves to provide qualitative measure of the average size and concentration of the aggregate "tail" population.

The corresponding volume-wt PSD is shown in Fig. 37 (Data on File, Particle Sizing Systems LLC), obtained from the intensity-wt PSD by dividing the contents of each size bin, R_I, by the particle volume, or R_I^3, and also by the intra-particle scattering factor, $g_I^2(R_I)$, where the latter is ≈ 1:

In volume-wt terms, the PSD heavily favors the primary, un-aggregated BSA molecules, as expected. According to the analysis calculations, fully 97 % of the total volume of the particles is accounted for by the native molecules, represented by Peak #1, with an average size of 7.0 nm. The expected size of BSA molecules is about 6.5 nm. The moderately higher value found by DLS analysis is attributed to the population of small aggregates—mainly dimers—which cannot be discriminated by the inversion algorithm as a separate size peak. Instead, their presence serves to "pull up" slightly the average diameter of the primary peak, mostly consisting of un-aggregated molecules. The remaining 3 % of the particle volume, or mass, is ascribed to the entire aggregate population, represented by Peak #2. What is invariably striking to those not familiar with DLS analysis is the powerful influence of intensity weighting, taught earlier, on the computed PSD. One sees from Figs. 36 and 37 that aggregates representing only about 3 % of the total volume, or mass, of the particles represent almost 60 % of the intensity-wt ACF. In cases where the extent of aggregation is higher, and/or the average size of the aggregates is larger, the second, aggregate peak in the intensity-wt PSD will overshadow to an even greater extent the first peak belonging to un-aggregated primary particles. This behavior helps to make the DLS technique a very effective "aggregate detector", albeit on only a semi-quantitative basis.

In closing this section, it is useful to acknowledge briefly the usefulness of another ensemble technique—the "static" counterpart, and historically the predecessor, of dynamic light scattering, known as classical light scattering (CLS), or static light scattering (SLS). Like DLS, the SLS technique is based on the intensity of light scattered from a population of particles. However, unlike DLS, it utilizes only the time-average value, rather than the time-dependent fluctuations, of the scattered intensity. The basis of SLS analysis is the Mie scattering theory, which describes the dependence of the scattered light intensity, for a given particle size, d, on the scattering angle, θ, using laser light of a well-defined wavelength, λ, as the illuminating light source. Hence, this technique is often referred to as multi-angle light scattering, or MALS.

For a particle of diameter d sufficiently large compared to λ (i.e., d>λ/20) the scattered light intensity varies in a measurable way with θ, due to the interference of individual scattering "wavelets" of different phase originating from different points within the particle. The angular variation of the net scattered intensity is a function of d and the refractive indices of the particle and the surrounding fluid, n_p and n_s, respectively, as described by the "intra-particle scattering factor", g_I^2, discussed briefly above (See Eqs. 19a, b). In many applications of biological and pharmaceutical interest, n_p is relatively close to n_s, in which case the Rayleigh-Gans approximation (Chu 1991; Berne and Pecora 2000) provides a good estimate of g_I^2,

$$g_I^2 = \left\{3[\sin(Kd/2) - (Kd/2)\cos(Kd/2)] / (Kd/2)^3\right\}^2 \tag{20}$$

where K, the scattering wave vector, is a function of θ, as given by Eq. 15, and the scattered intensity is proportional to g_I^2, as shown in Eqs. 19a and b.

In applications of practical interest, the PSD is usually significantly "polydisperse"—e.g., "tails" of aggregated protein molecules used in therapeutic drugs. In these cases, the scattering intensity *vs* angle data consist of a superposition of plots for each constituent particle size, each behaving with scattering angle like Eq. 20, and it is therefore necessary to "invert" the measured intensity *vs* angle data to obtain the desired PSD. In practice it is usually difficult, if not impossible, to use the MALS technique to obtain an accurate approximation of the underlying PSD. First, there are significant limitations inherent in the mathematical algorithm needed to deconvolute the angular dependent intensity data. Analogous to the deconvolution of the autocorrelation function data required in DLS analysis, "inversion" of the raw data in MALS often results in relatively poor accuracy and, especially, resolution of the resulting computed PSD, because of the "ill-conditioned" nature of the problem. Second, there is a diminishing dependence of the intensity on the scattering angle with decreasing particle size, as described by g_I^2 in Eq. 20. For example, in the case of 100 nm particles, with λ = 650 nm and n = 1.33 (water), g_I^2 is almost 30 % larger at θ = 10 than at 170°. However, this "angular dissymmetry" drops to 8 % for 50 nm particles and is only about 1 % for 20 nm particles. Hence, in practice MALS is virtually unusable in identifying particles smaller than about 15–20 nm, where there is almost no measurable angular dependence. In contrast,

DLS has been shown to produce accurate measurements of the average particle size all the way down to a couple of nanometers.

In practice, therefore, MALS has limited usefulness for determining the PSD of nano-particle suspensions, especially in the deep-submicron part of the size scale. The limitations inherent in the deconvolution, or inversion, process can be mitigated by combining MALS with a separation technique, such as size-exclusion chromatography or field flow fractionation (FFF) (DAWN™). In this case, deconvolution of the intensity *vs* angle data becomes significantly easier, given that the range of particle sizes contributing to the measured data is greatly reduced for each time/data point by the separation, or fractionating, process. However, the limitation imposed by the diminishing sensitivity of the angular dissymmetry with decreasing size remains a serious obstacle for the effective use of the MALS technique for particle size analysis at smallest sizes. Instead, DLS is considerably more useful, especially when used in conjunction with a separation technique.

Synopsis

Accurate physical characterization of NBCD nanostructures, particularly when used as delivery vehicles for active pharmaceutical ingredients, is paramount to their safe use. This includes accurate particle sizing as well as an appropriate sample size to ensure the safe and efficacious delivery of nanomedicines. This requirement is especially important when these nanomedicines are given intravenously, and in critically ill patients, for whom such therapy is often life-saving. At present, the only pharmacopoeial guidance with respect to the physical characterization of parenteral dispersions is for lipid injectable emulsions (intravenous fat emulsions) used for nutritional purposes or as a drug vehicle for poorly water-soluble drugs. This dosage form, however, is on the higher side of what is considered a nanodispersion (e.g., mean droplet size: ~200 to 500 nm), but nonetheless it provides a useful framework for constructing pharmacopoeial limits that ensure their efficacy, therapeutic equivalence and ultimately, safety.

Summary

Of the critical single-particle techniques reviewed in this chapter, the table below summarizes the approximate particle size range and count rate for each (Table 4).

Conclusions

Accurate particle sizing can be achieved by many methods, including "ensemble" and "single-particle counting" techniques. Ensemble techniques, such as dynamic light scattering, are important to establish the mean particle size, from which the

Table 4 Sample sizes of single-particle techniques

Method	Size range (μm)	Particle count rate
LE	≈1–400	5000–10,000/s
LS	≈0.2–2	5000–10,000/s
LE+LS	0.5–400	5000–10,000/s
FBLE	0.6–20	1500–3000/s
FBLS	0.15–0.6	1500–3000/s
FBLS-LELS	0.15–400	1500–3000/s
ESZ	≈0.4–1000	≈10–1000/s[a]
TRPS	≈0.05–10	≈2–100/sa[a]
RMM	≈0.15–10	≈5–10/s
NTA	≈0.04–1	≈10–100/min[b]

[a] Depends on pore size and particle polydispersity
[b] Depends on particle size and polydispersity

critical measurements in the large-diameter tail of dispersions can be directed and optimized. For the latter, given the extraordinary number of particles, droplets, liposomes, micelles, etc. in a nanodispersion (see Table 3), the chosen counting method must provide results based on a statistically relevant sample in order to represent accurately the quality and stability of the formulation.

We have tried to inform the reader of what we believe are likely to be the most useful particle size analysis techniques for nanodispersions. The field of analytics continues to develop. Each dispersion will have its own features/characteristics that may be more amenable to one method over another, even though the selected method(s) may be less robust. We believe each "category" of NBCDs will likely have its own set of standards. How the categories are determined (e.g., drug class, vehicle class, particle size, particle shape, particle/droplet concentration, route of administration, etc.) will likely come from continued study and application of various methods. Ultimately this experience will be brought together and evaluated by individual experts who will be assigned to specific panels at the various pharmacopoeial agencies and who will develop the pharmacopoeial methods that ensure the safety and efficacy of the nanodispersion.

References

AccuSizer FX™, AccuSizer FX Nano™ and Nicomp 380™ www.pssnicomp.com

Anderson W, Kozak D, Coleman VA et al (2013) A comparative study of submicron particle sizing platforms: accuracy, precision and resolution analysis of polydisperse particle size distributions. J Colloid Interface Sci 405:322–333p

Archimedes™ www.affinitybio.com and www.malvern.com

Babi J, Doenicke A, Mönch V (1995) New formulation of propofol in an LCT/MCT emulsion. EHP 1:15–21

Batchelor GK (1976) Transport properties of two-phase materials with random structure. J Fluid Mech 74:1–29

Bennett SN, McNeil MM, Bland LA et al (1995) Postoperative infections traced to contamination of an intravenous anesthetic, propofol. N Engl J Med 333:147–154

Berne BJ, Pecora R (2000) Dynamic light scattering. Dover Publications, Inc., New York

Brown W (1996) Light scattering. Clarendon Press, Oxford

Burg TP, Godin M, Knudsen SM et al (2007) Weighing of biomolecules, single cells and single nanoparticles in fluid. Nature 446:1066–1069

Carpenter JF, Cherney B, Rosenberg AS (2012) The critical need for robust assays for quantitation and characterization of aggregates of therapeutic proteins (Chapter 1). In: Mahler H-C, Jiskoot W (eds) Analysis of aggregates and particles in protein pharmaceuticals.Wiley, West Sussex

Centers for Disease Control (CDC) and Prevention: Follow-up on septicemias associated with contaminated intravenous fluids. Morb Mortal Wkly Rep 1971;20:91–92

Charles S, Johnson Jr CS, Gabriel DG (1981) Laser light scattering. Dover Publications, Inc., New York

Chu R (1991) Laser light scattering, 2nd edn. Dover Publications, Inc., New York

Chunara R, Godin M, Knudsen SM et al (2007) Mass-based readout for agglutination assays. Appl Phys Lett 91:193902

Data on File, Particle Sizing Systems LLC, Port Richey, FL, USA

DAWN™ www.wyatt.com

Dorshow RB, Nicoli DF (1985) The role of intermicellar interactions in interpretations of micellar diffusion coefficients. Physics of Amphiphiles: Micelles, Vesicles and Microemulsions. Soc. Italiana di Fisica, Bologna, pp 429–447

Dorshow RB, Bunton CA, Nicoli DF (1983) Comparative study of intermicellar interactions using dynamic light scattering. J Phys Chem 87:1409–1416

Driscoll DF (2004) Examination of selection of light-scattering and light-obscuration acceptance criteria for lipid injectable emulsions. Pharm Forum 30:2244–2253

Driscoll DF (2006) Lipid injectable emulsions: pharmacopeial and safety issues. Pharm Res 23:1959–1969

Driscoll DF (2007) Globule size distribution of 20 % lipid injectable emulsions: compliance with USP proposed requirements. Am J Health-Syst Pharm 64:2032–2036

Driscoll DF (2009) The pharmacopeial evolution of Intralipid injectable emulsions in plastic containers: from a coarse to a fine dispersion. Int J Pharm 368:193–198

Driscoll DF, Bhargava HN, Li L et al (1995) The physicochemical stability of complex intravenous lipid dispersions as total nutrient admixtures. Am J Hosp Pharm 52:623–634

Driscoll DF, Bacon MN, Bistrian BR (2000) Physicochemical stability of two different types of intravenous lipid emulsion as total nutrient admixtures. J Parenter Ent Nutr 24:15–22

Driscoll DF, Giampietro K, Wichelhaus DP et al (2001a) Physicochemical stability assessments of lipid emulsions of varying oil composition. Clin Nutr 20:151–157

Driscoll DF, Etzler F, Barber TA et al (2001b) Physicochemical assessments of parenteral lipid emulsions: light obscuration versus laser diffraction. Int J Pharm 219:21–37

Driscoll DF, Nehne J, Peterss H et al (2003) Physicochemical stability of intravenous lipid emulsions as all-in-one admixtures for the very young. Clin Nutr 22:489–495

Driscoll DF, Ling PR, Quist WC, Bistrian BR (2005) Pathological consequences from the infusion of unstable lipid emulsion admixtures in guinea pigs. Clin Nutr 24:105–113

Driscoll DF, Silvestri AP, Nehne J et al (2006a) The physicochemical stability of highly concentrated total nutrient admixtures (TNAs) intended for fluid-restricted patients. Am J Health-Syst Pharm 63:79–85

Driscoll DF, Parikh M, Silvestri AP et al (2006b) Establishing a stability window for MCT-LCT lipid based parenteral nutrition: applying pharmacopeial standards. Am J Health-Syst Pharm 63:2135–2143

Driscoll DF, Nehne J, Peterss H et al (2006c) Physical assessments of injectable lipid emulsions via microscopy supported by light extinction and light scattering techniques. Int J Pharm Comp 10:309–315

Driscoll DF, Ling PR, Bistrian BR (2006d) Pathological consequences to reticuloendothelial system organs following infusion of unstable all-in-one mixtures in rats. Clin Nutr 25:842–850

Driscoll DF, Silvestri AP, Mikrut BA, Bistrian BR (2007a) Stability of adult-based total nutrient admixtures with soybean oil-based lipid injectable emulsions: the effect of glass versus plastic packaging. Am J Health-Syst Pharm 64:396–403

Driscoll DF, Ling PR, Bistrian BR (2007b) Physical stability of 20% lipid injectable emulsions via simulated syringe infusion: effects of glass vs. plastic product packaging. J Parenter Ent Nutr 31:148–153

Driscoll DF, Ling PR, Silvestri AP, Bistrian BR (2008) Fine vs. coarse complete all-in-one admixture infusions over 96 hours in rats: fat globule size and hepatic function. Clin Nutr 27:889–894

Driscoll DF, Ling PR, Andersson C, Bistrian BR (2009a) Hepatic indicators of oxidative stress and tissue damage accompanied by systemic inflammation in rats following a 24-hour infusion of an unstable lipid emulsion admixture. JPEN 33:327–335

Driscoll DF, Ling PR, Bistrian BR (2009b) Pharmacopeial compliance of fish oil-containing parenteral lipid emulsion mixtures: globule size distribution and fatty acid analyses. Int J Pharm 379:125–130

Eaton DL, Gilbert SG. Principles of toxicology. In: Klassen CD (ed) Casarett & Doull's toicology: the basic science of poisons, 8th edn. McGraw-Hill Medical, New York, p 22

Felderhof BU (1978) Diffusion of interacting particles. J Phys A: Math Gen 11:929–944

Filipe V, Hawe A, Jiskoot W (2010) Critical evaluation of nanoparticle tracking analysis (NTA) by NanoSight for the measurement of nanoparticles and protein aggregates. Pharm Res 27:796–810

Filipe V, Poole R, Olubukayo O et al (2012) Detection and characterization of subvisible aggregates of monoclonal IgG in serum. Pharm Res 29:2202–2212

Godin M, Bryan AK, Burg TP et al (2007) Measuring the mass, density and size of particles and cells using a suspended microchannel resonator. Appl Phys Lett 91:123121

Hamawy KJ, Moldawer LL, Georgieff M et al (1985) The effect of lipid emulsions on reticuloendothelial system function in the injured animal. JPEN 9:559–565

Holger TS (1990) Performance of monochromatic and white light extinction particle counters. Proceedings International Conference on Particle Detection, Metrology and Control, Institute of Environmental Sciences, Parenteral Drug Association, Inc., Arlington, VA (Feb. 5–7, 1990), p 269

Jensen GL, Mascioli EA, Seidner DL et al (1990) Parenteral infusion of long and medium-chain triglycerides and reticuloendothelial system function in man. JPEN 14:467–471

Jiskoot W, Randolph TW, Volkin DB (2012) Protein instability and immunogenicity: roadblocks to clinical application of injecable protein delivery systems for sustained release. J Pharm Sci 101:946–954

Karuhn R (1998) Theory, practice, and applications of electrozone sensing technology. Modern Methods of Particle Size Distribution: Assessment and Characterization, American Chemical Society, New Orleans, LA (Feb. 28, March 1, 1998)

Knapp JZ, Abramson LR (1990) A systems analysis of light extinction particle detection systems. Proceedings International Conference on Particle Detection, Metrology and Control, Institute of Environmental Sciences, Parenteral Drug Association, Inc., Arlington, VA (Feb. 5–7, 1990), p 283

Knollenberg RG, Gallant RC (1990) Refractive index effects on particle size measurement by optical extinction. Proceedings International Conference on Particle Detection, Metrology and Control, Institute of Environmental Sciences, Parenteral Drug Association, Inc., Arlington, VA (Feb. 5–7, 1990), p 154

Kopell DE (1972) Analysis of macromolecular polydispersity in intensity correlation spectroscopy. The method of cumulants. J Chem Phys 57:4814–4820

Kozak D, Anderson W, Vogel R et al (2011) Advances in resistive pulse sensors: devices bridging the void between molecular and microscopic detection. Nano Today 6:531–545

Li Y, Lubchenko V, Vekilor PG (2011) The use of dynamic light scattering and Brownian microscopy to characterize protein aggregation. Rev Sci Instrum 82:053196

Lieberman A (1990) Light scattering systems for particle counting and sizing. Proceedings International Conference on Particle Detection, Metrology and Control, Institute of Environmental Sciences, Parenteral Drug Association, Inc., Arlington, VA (Feb. 5–7, 1990), p 353

LiQuilaz™, Model S02 and Model S03, www.pmeasuring.com

Martin CR, Dumas GJ, Shoae C et al (2008) Incidence of hypertriglyceridemia in critically ill neonates receiving lipid injectable emulsions in glass versus plastic containers: a retrospective analysis. J Pediatr 152:232–236

Mickisch S, Tantipolphan R, Wiggenhorn M et al (2010) Subvisible particles in a monoclonal antibody formulation analyzed by nanoparticle tracking analysis and microflow imaging. 2010 AAPS National Biotechnology Conference, San Francisco, CA (May 16–19, 2010)

Multisizer™ 4 www.beckmancoulter.com

NanoSight™ www.nanosight.com and www.malvern.com

Newton DW, Driscoll DF (2008) Calcium and phosphate compatibility: revisited again. Am J Health-Syst Pharm 65:73–80

Nicoli DF, Driscoll DF, Bistrian BR (2006) Stability assessments of dispersions and emulsions. US Patent No. 7,150,996 B2, December 19, 2006

Olcum S, Cermak N, Wasserman SC et al (2014) Weighing nanoparticles in solution at the attogram scale. Proc Natl Acad Sci U S A 111:1310–1315

Ortega F, Bacaloglu R, McKenzie DC et al (1990) Static and dynamic light scattering study of strongly interacting micelles: hypernetted chain vs dilute gas approximation. J Phys Chem 94:501–504

Osterhoudt KC, Penning TM (2011) Drug toxicity and poisoning. In: Brunton L, Chabner B, Knollman B (eds) Goodman & Gilman's the pharmacological basis of therapeutics, 12th edn. McGraw Hill Medical, New York, p 73

qNano™ www.izon.com

Roberts GS, Kozak D, Anderson W et al (2010) Tunable nano/micropores for particle detection and discrimination: scanning ion occclusion spectroscopy. Small 6:2653–2658

Saba TM (1970) Physiology and physiopathology of the reticuloendothelial system. Arch Intern Med 126:1031–1052

Seidner DL, Mascioli EA, Istfan NW et al (1989) Effects of long-chain triglyceride emulsions on reticuloendothelial system in function in humans. JPEN 13:614–619

United States Pharmacopeia, USP37/NF32 (2014a) Chapter <729>. Globule size distribution in lipid injectable emulsions. United States Pharmacopeia, Rockville, pp 360–363

United States Pharmacopeia, USP37/NF32 (2014b). Lipid injectable emulsion monograph. United States Pharmacopeia, Rockville, pp 3565–3566

United States Pharmacopeia, USP37/NF32 (2014c). Propofol injectable emulsion monograph. United States Pharmacopeia, Rockville, pp 4452–4453

van der Pol E, Hoekstra AG, Sturk A et al (2010) Optical and non-optical methods for detection and characterization of microparticles and exosomes. J Throm Haemo 8:2596–2607

van Hulst HC (1981) Light scattering by small particles. Dover Publications, Inc., New York

Vogel R, Willmott GR, kozak D et al (2011) Quantitative sizing of nano/microparticles with a tunable elastomeric pore sensor. Anal Chem 83:3499–3506

Willmott GR, Vogel R, Yu SS et al (2010) Use of tunable nanopore blockade rates to investigate colloidal dispersions. J Phys: Condens Matter 22:454116

Zhou C, barnard JG, Qi W et al (2012) Application of nanoparticle tracking analysis in characterization of protein aggregates, 2012 Workshop on Protein Aggregation and Immunogenicity, July 2012, Beaver Run Resort and Conference Center, Breckenridge, CO, USA

Zölls S, Tantipolphan R, Wiggenhorn M et al (2011) Particles in therapeutic protein formulations, part 1: overview of analytical methods. J Pharm Sci 101:914–935

NBCD Pharmacokinetics and Bioanalytical Methods to Measure Drug Release

Vishakha V. Ambardekar and Stephan T. Stern

Contents

General 262
Drug Release as a Fundamental NBCD Property for Bioequivalence Assessment 264
Bioanalytical Methods for Measurement of NBCD Drug Release 267
  Chromatographic Methods 267
  Liquid-Liquid Extraction 272
  Equilibrium Methods 273
Indirect Methods for Estimation of NBCD Drug Release 276
  Evaluation of Total Drug and NBCD Platform Pharmacokinetic Profiles 276
  Modeling and Simulation 277
  Analysis of Metabolite Pharmacokinetics 279
Conclusions and Future Directions 281
References 282

Abstract A primary regulatory challenge for generics of non-biological complex drugs (NBCDs) (i.e., NBCD similars or nanosimilars) is evaluation of bioequivalence (i.e., pharmacokinetic equivalence). NBCD pharmacokinetics are highly dependent upon drug release kinetics and dynamic tissue distribution, with the simultaneous existence of both NBCD encapsulated (e.g., bound) and unencapsulated (e.g. free) forms of the active pharmaceutical ingredient (API). For this reason, regulators have focused on the importance of evaluation of NBCD drug release, and the pharmacokinetics of both the encapsulated and unencapsulated forms of the API, which is challenging from a bioanalytical perspective. While many separation methods are currently available to evaluate drug release from NBCD formulations and measure encapsulated/unencapsulated forms of the API, including both direct and indirect methods, the most appropriate method for any particular NBCD type ultimately depends upon a combination of existing experience, scientific intuition, and trial and error. Presently, the available separation techniques have arisen from repurposing of small molecule preparatory and protein binding methods, none of

S. T. Stern (✉) · V. V. Ambardekar
Nanotechnology Characterization Laboratory, Cancer Research Technology Program, Leidos Biomedical Research, Inc., Frederick National Laboratory for Cancer Research, Frederick, MD, USA
e-mail: sternstephan@mail.nih.gov

D. J. A. Crommelin, J. S. B. de Vlieger (eds.), *Non-Biological Complex Drugs,* AAPS Advances in the Pharmaceutical Sciences Series 20, DOI 10.1007/978-3-319-16241-6_8

which adequately address all concerns, such as the problems of process-induced drug release and accurate determination of unbound drug. This chapter will review the existing regulatory guidance for NBCD generic bioequivalence, as well as methods to evaluate NBCD drug release and pharmacokinetics.

Keywords Nanomedicine · Non-biological complex drugs · Pharmacokinetics · Bioanalytical methods · Drug release methods · Equilibrium dialysis · Liquid-liquid extraction · Modeling and simulation · Metabolite pharmacokinetics · Size exclusion chromatography · Solid phase extraction · Ultracentrifugation · Ultrafiltration

Abbreviations

API	Active pharmaceutical ingredient
EMA	European Medical Agency
EPR	Enhanced permeation and retention
FDA	Food and Drug Administration
LLE	Liquid-liquid extraction
MALS	Multi angle light scattering
mTHPC; temoporphin	meta-tetra (hydroxyphenyl) chlorine
mPEG-PLA	methoxy polyethylene glycol polylactide
MPS	Mononuclear phagocytic system
NBCD	Non-biological complex drugs
PD	Pharmacodynamic
PK	pharmacokinetic
PEG	Polyethylene glycol
PEG-PLA	Polyethylene glycol polylactide
PLA-POLOX	Poloxamer adsorbed polylactide
SEC	Size exclusion chromatography
SPE	Solid phase extraction
TNF	Tumor necrosis factor
Au-TNF	Tumor necrosis factor-colloidal gold nanoparticle
USP	United States Pharmacopeia
UV	Ultra-violet
Vd	Apparent volume of distribution

General

Most suspensions in the nanoscale size range (e.g., liposomes, emulsions, micelles, and polymeric, metallic, and solid lipid nanoparticles) are non-biological complex drugs (NBCDs). In contrast to the NBCDs described in Chapters "Glatiramoids", "Iron Carbohydrate Complexes", and "Low Molecular Weight Heparins", where the active pharmaceutical ingredient (API) mainly is the formulation, these colloidal suspension types of NBCDs clearly consist of two parts; a complex formulation and a well characterizable often small molecule API. These complex nanoscale drug formulations of the API often provide superior therapeutic efficacy and/or safety in comparison to the free API or legacy drug counterparts. Nanoscale NBCD formulations can improve therapy by altering one or more of the encapsulated API's pharmacokinetic (PK) properties, such as tissue distribution or clearance (Onoue et al. 2014). Indeed, nanoscale NBCD formulations have been reported to improve absorption upon extravascular administration (Mei et al. 2013), reduce metabolism and clearance (Jiang et al. 2007; Zhang et al. 2014), and enhance distribution to target tissues, while decreasing distribution to off-target tissues (Duncan and Gaspar 2011; Jiang et al. 2007). These PK advantages of nanoscale NBCD formulations stem from a unique set of tunable physicochemical properties, such as composition, geometry and surface characteristics that influence in vivo stability, release of active component(s), movement across physiological barriers and interactions at the molecular level (Albanese et al. 2012; Blanco et al. 2011; Desai 2012).

Due to their size (>7 nm), nanoscale NBCDs are generally excluded from glomerular filtration, as well as tissues with intact vascular endothelium. This vascular confinement of nanoscale NBCDs decreases the volume of distribution of the encapsulated API (Stern et al. 2010). Subsequent release of the encapsulated API from the nanoscale NBCD typically occurs in one or more of the following compartments: (1) Large NBCDs (100's nm-micron), or those lacking a steric hydrophilic surface coating (e.g., polyethylene glycol (PEG)), are rapidly opsonized and accumulate in organs of the mononuclear phagocytic system (MPS), releasing API mainly in Kupffer cells of the liver and splenic macrophages (He et al. 2010; Owens and Peppas 2006; Prantner and Scholler 2014); (2) Nanoscale NBCDs which evade the MPS show extended circulation, and accumulate and release API in tissues with compromised vascular endothelium (e.g., inflammatory sites, tumor) through the enhanced permeation and retention (EPR) effect (Maeda 2012; Torchilin 2011); (3) For the majority of nanoscale NBCDs that demonstrate extended circulation, or conversely are unstable (e.g., "burst" releasing or simple solubilizing agents) (Stern et al. 2013; Zou et al. 2013), the systemic vascular space is also a primary site of API release. NBCD encapsulated and unencapsulated API concentrations in the vascular space reflect exposure to less accessible peripheral target tissues, in that tissue accumulation removes NBCD encapsulated API from the circulation and unencapsulated API in the circulation is in dynamic equilibrium with peripheral tissues. Therefore, both encapsulated and unencapsulated API profiles are critical measures of the overall NBCD PK and determinants of safety and efficacy.

Typically only the total (encapsulated and unencapsulated) API concentration-time profile is measured in preclinical and clinical NBCD PK studies (Stern et al. 2010). This total API PK profile is often dominated by the encapsulated fraction of the API which is confined to the vascular space, with minimal contribution from the released, unencapsulated fraction (Fig. 1) (Stern et al. 2010). For example, clinical plasma API fractions of a liposomal formulation of amphotericin B, separated by the ultrafiltration method (see below), were found to have the rank order: total drug > liposome encapsulated>>liposome unencapsulated (Bekersky et al. 2002). Given that the unencapsulated API is active and is ultimately responsible for the observed efficacy and toxicity of the NBCD, measuring the total (encapsulated and

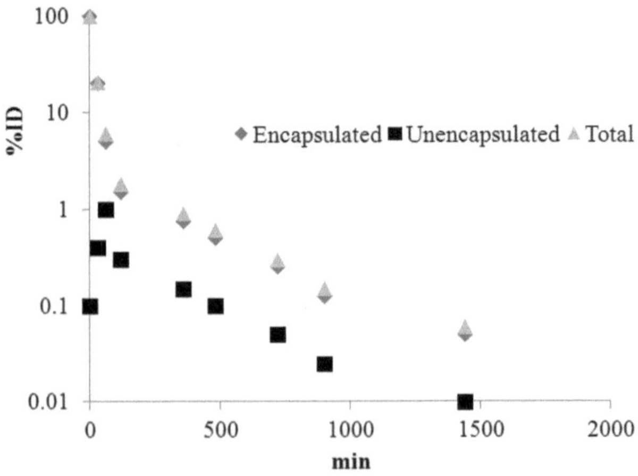

Fig. 1 Typical NBCD PK profile. The total API PK profile is often dominated by the NBCD encapsulated fraction which is confined to the vascular space, with minimal contribution from the released, unencapsulated fraction. Plasma time profiles are expressed as the percentage of injected dose (%ID). ♦ Encapsulated API, ■ Unencapsulated API, ▲ Total API

unencapsulated) API concentration-time profile may be insufficient and often misleading when attempting to establish dose linearity, determine PK/pharmacodynamic (PD) relationships, evaluate exposure across species, estimate first in man doses, or compare NBCD formulations for bioequivalence (Stern et al. 2010).

The current chapter provides a critical analysis of the regulatory implications of NBCD PK, and the commonly used methods to estimate NBCD drug release and measure unencapsulated fraction. We will focus exclusively on intravenously administered NBCD, as this class dominates the regulatory and scientific literature. We will not cover NBCDs administered by the oral, intramuscular, subcutaneous, intraperitoneal, intradermal or intrathecal routes, including depot or implant systems, which have their own unique challenges.

Drug Release as a Fundamental NBCD Property for Bioequivalence Assessment

Establishment of therapeutic equivalence is a statutory requirement for regulatory approval of small molecule generic drugs (CFR 2013). Therapeutic equivalence of small molecule generics is assessed by establishing both pharmaceutical (i.e., physicochemical) equivalence and bioequivalence (i.e., PK equivalence) (CFR 2013). Bioequivalence of a small molecule generic is typically evaluated by clinical crossover PK studies with well-established acceptance criteria (FDA 2001).

Although the term “generic” has been applied to NBCD follow-on products, in the past this has meant identical to a reference product. Since NBCDs are complex formulations, there will always be some degree of polydispersity and batch-to-batch variation, such that no batch will be absolutely identical to a reference batch. As with complex biological products, the regulator may ask the originator for comparability studies in cases in which they have made significant changes to their manufacturing process. Since follow-on versions of these types of complex products will never be identical to the originator product, the term similar (i.e. NBCD similar, nanosimilar) may be more appropriate than the term generic (Ehmann et al. 2013), as is used for follow-on biological products, i.e., biosimilars (PHS). For NBCDs, the challenge is to identify meaningful differences between the follow-on and reference products that will affect therapeutic equivalence. With regard to NBCDs therapeutic equivalence, NBCD drug release is a highly relevant property to examine.

Many parenteral NBCDs can be considered systemic controlled release formulations; in addition to potentially altering the tissue distribution of the API, parenteral NBCDs may also modulate release of the free, pharmacologically active form of the drug. Consistent with traditional oral controlled release medications, drug release kinetics is considered a fundamental property of the NBCDs discussed in this chapter. Similar to *in vitro* drug release experiments performed for oral controlled release formulations using a simulated gastric environment (e.g., USP apparatus (USP 2011)), drug release from parenteral NBCD formulations can also be simulated *in vitro* in a suitable physiological matrix (e.g., blood or plasma). The *in vitro* drug release characteristics of the NBCD can then be used as a quality attribute of the formulation for purposes of lot release (critical quality attribute) or drug comparability (EMA 2011b, c, 2013a, b; FDA 2012a, 2013b, c). It is important that drug release for parenteral NBCDs be performed in an appropriate physiological matrix (e.g., plasma, blood, cellular fraction, etc.), as evaluation of drug release *in vitro* in simple buffered media may not necessarily correlate with drug release *in vivo* (Yasui et al. 1995), unlike traditional oral controlled release formulations. In some cases, bioequivalence can be assessed exclusively based on *in vitro* drug release data and physicochemical attributes, an example being nanoparticle-based solubilizing agents that behave as parenteral drug solutions (FDA 2014b). This waiver of clinical PK studies for assessment of parenteral drug solution bioequivalence is based on the same scientific rationale as waivers of clinical bioavailability and bioequivalence studies for highly soluble/highly permeable oral drugs (BCS class I), which also only require drug release (i.e., dissolution rate) testing to establish bioequivalence as the drug is considered immediately and highly bioavailable (FDA 2000). These are exceptions, however, and in most cases establishment of therapeutic equivalence of a follow-on NBCD product requires clinical bioequivalence studies, as well as additional non-clinical (e.g. toxicology, efficacy, PK, tissue distribution) and clinical (e.g., efficacy and safety) studies in certain instances.

Non-clinical evaluation of PK and tissue distribution, as well as efficacy and toxicity, can be informative for evaluation of therapeutic equivalence. Obviously, it is often not feasible to determine clinical tissue distribution of a NBCD, apart from imaging agents. Thus, comparison of the preclinical tissue distribution of follow-on and reference product may be meaningful, especially in cases where the NBCD tissue distribution has important pharmacological or toxicological significance (e.g., actively targeted therapy, known off-target deposition related to side effect). Likewise, toxicity and efficacy studies in clinically relevant animal models may hold value and have been used in assessment of NBCD follow-on products in the past, specifically a Caeylx/Doxil nanosimilar as described in detail below (EMA 2011a). Clinical safety and efficacy studies would be the highest regulatory bar for therapeutic equivalence assessment, but also represents a logistical and financial burden on the competitor product. For NBCD follow-on products, like biosimilars, clinical safety and efficacy studies may not be justified if therapeutic equivalence can be inferred from prior product experience regarding the therapeutic relevance of physicochemical attributes, preclinical efficacy and toxicity, and clinical PK. The addition of PD biomarkers to preclinical or clinical PK studies allows evaluation of PK-PD relationships, and can further support therapeutic equivalence of the follow-on and reference products. Evaluation of clinical immunogenicity as part of PK studies may also be informative in some cases, without adding substantial cost.

Parenteral NBCDs are a special case with regard to bioequivalence assessment, because of their unique PK attributes. As stated earlier, the total NBCD (encapsulated and unencapsulated) PK profile for a parenterally administered drug is often dominated by the encapsulated form of the API (Fig. 1) (Bekersky et al. 2002). Furthermore, it is the tissue distribution (e.g., mononuclear phagocytic system) of the encapsulated API that often dictates total PK profile (Gregoriadis 1991), while drug release kinetics and inherent API PK underlie the unencapsulated profile. Thus, total NBCD PK may be very misleading if used exclusively as a measure of bioequivalence, as differences in drug release kinetics may not be reflected in the total NBCD profile. Since unencapsulated API exposure is often the main contributor to systemic toxicity, the unencapsulated concentration-time profile should be a primary safety metric for regulatory consideration. Indeed, drug release rates have been shown to dramatically alter the preclinical toxicity of NBCD formulations (Mayer et al. 1994). Only evaluation of both the total and unencapsulated API PK in preclinical or clinical studies can establish meaningful bioequivalence of a follow-on NBCD in comparison to that of the innovator product.

Measurement of drug release from its NBCD encapsulated form is a requirement included in all NBCD guidance documents and reflection papers published by the Food and Drug Administration (FDA) and European Medical Agency (EMA), respectively, to date (EMA 2011b, c, 2013a, b; FDA 2012a, 2013b, c). The EMA and FDA have developed regulatory guidance for liposome (EMA 2011b; FDA 2013c, 2014a, c), iron colloid (EMA 2011c, 2013b; FDA 2012a, c, 2013b) and nanoparticulate albumin (FDA 2012b) follow-on products, that include the recommendation for in vitro drug release studies conducted in physiologically relevant media (e.g., plasma, blood), and bioequivalence determinations based on both unencapsulated and encapsulated API PK. EMA guidance for preclinical and clinical development of block copolymer micelles also includes these recommendations (EMA 2013a). A relevant and timely example of the regulatory significance of NBCD PK complexity comes from the recent European and US regulatory experience with follow-on liposomal doxorubicin formulations.

Doxil® (US) and Caelyx® (Europe), the innovator doxorubicin liposome formulation, experienced a worldwide shortage in 2013 spurring development of follow-on forms of the drug (FDA 2013a). Comparability studies for a Caelyx follow-on product in Europe included non-clinical efficacy, toxicity, PK and tissue distribution, as well as clinical bioequivalence (EMA 2011a). As part of the pivotal clinical bioequivalence studies, the EMA required assessment of the unencapsulated API PK profile in addition to total API profile. Unfortunately, while the total API PK of the innovator (reference) and follow-on (test) product were deemed bioequivalent, this was not the case for the unencapsulated PK prompting the EMA to reject market authorization of the Caelyx follow-on product on the grounds that the product did not meet the established bioequivalence standards (EMA 2011a). Conversely, in the United States, the FDA granted market approval to the same follow-on product, although the details of the requisite bioequivalence studies are not available (FDA 2013a). It is noteworthy that a Doxil crossover study included in the original European Caelyx bioequivalence studies found Doxil to be bioequivalent to the Caelyx follow-on product based both on encapsulated and unencapsulated PK (EMA 2011a). This discordance between the bioequivalency of the follow-on product to Caelyx and Doxil may have to do with the fact that the Doxil crossover study included a larger patient population with greater statistical power than the Caelyx arm of the study. Importantly, this is an example of how analysis of total API PK in the absence of unencapsulated API PK can give different regulatory conclusions with regard to bioequivalence and product comparability.

Bioanalytical Methods for Measurement of NBCD Drug Release

The measurement of drug release from intravenously administered NBCD formulations is limited by the availability of analytical methods that can accurately and precisely quantify the encapsulated and unencapsulated fractions of the NBCD in physiologically relevant media, such as blood and plasma. The majority of the analytical methods currently available for the measurement of NBCD drug release are adapted from techniques conventionally used for bioanalytical clean-up of small molecule biological samples or small molecule protein binding assessment. As such, the challenges for the use of these techniques for measuring drug release of NBCD in biological matrices is the selection of the most appropriate method, based on a thorough understanding of the underlying separation mechanisms, and optimization of this method for the NBCD (Thies et al. 1990). The most commonly used methods to measure drug release from NBCD formulations include chromatographic methods (size exclusion (SEC), solid phase extraction (SPE)), liquid-liquid extraction and equilibrium methods (e.g., dialysis, ultrafiltration, and ultracentrifugation) (Table 1, Fig. 2) (Druckmann et al. 1989; Mayer and St-Onge 1995; Thies et al. 1990; Zhao et al. 2009).

Chromatographic Methods

Size Exclusion Chromatography

SEC, also known as gel-filtration or gel-permeation chromatography, is a separation technique based on the principle of size fractionation utilizing a stationary phase

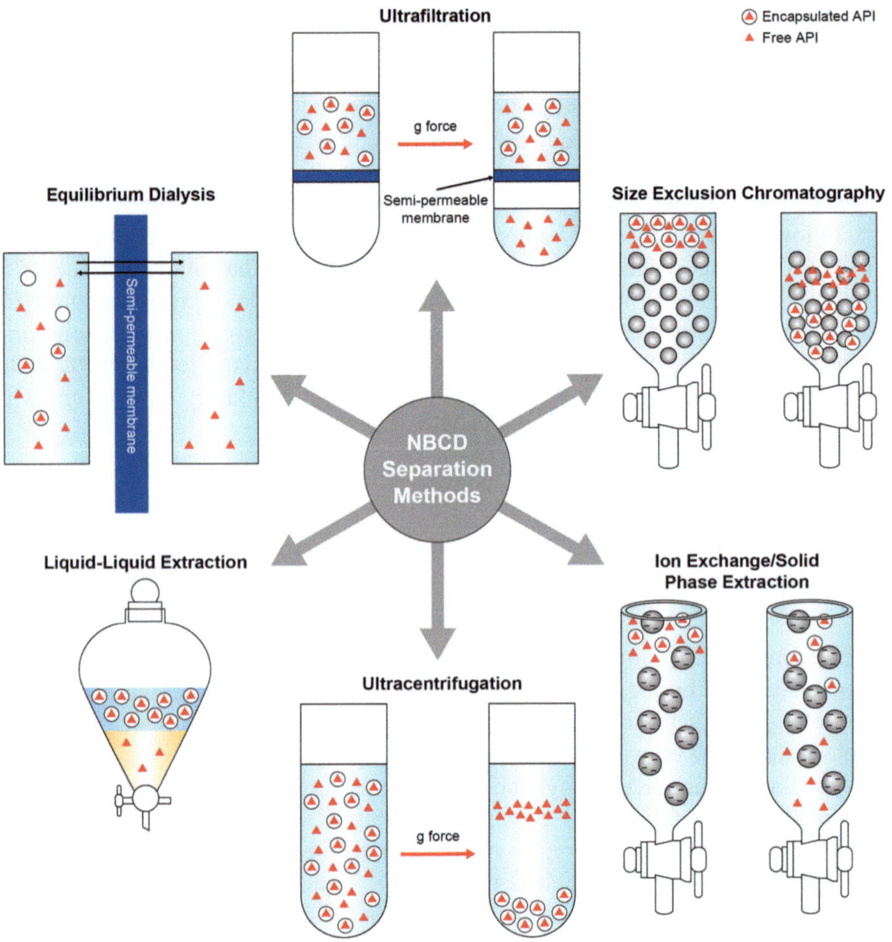

Fig. 2 NBCD separation methods

of porous polymeric particles (Hagel 2001). Most nanoscale NBCDs being larger than the pore size of the packing material are not retained by the column and elute early, while the unencapsulated analyte molecules being much smaller are trapped within the pores of the stationary phase and have greater elution volumes and subsequently longer elution times. The major advantages of SEC include rapid and robust separation of large NBCDs from smaller unencapsulated API, relatively low elution volumes, and less potential for stationary phase-NBCD interaction (Table 1) (Boyd et al. 2006; Kaiser et al. 2003). As such, SEC is a popular and commonly used technique for determining encapsulation efficiencies of NBCD formulations, and has also been applied to the separation of NBCD encapsulated and unencapsulated API fractions in biological matrices, such as plasma. Commonly used detection techniques for both SEC and SPE (see below), dependent upon the physicochemi-

Table 1 Summary of general methods to separate NBCD encapsulated and unencapsulated fractions

Separation method	Advantages	Disadvantages	References
Membrane filtration			
Equilibrium dialysis	Free drug in equilibrium w/protein	Long equilibration time can lag release kinetics	(Wallace et al. 2012)
	Low encapsulated drug contamination	API instability	
	Low non-specific binding of API	Difficult to accurately differentiate protein bound and encapsulated API	
	Less process-induced drug release?	Sample dilution	
Ultrafiltration	Free drug in equilibrium w/protein	Non-specific binding of API to filter membrane	(Bekersky et al. 2002; Gardner et al. 2008a; Magin and Chan 1987; Mayer and St-Onge 1995; Wallace et al. 2012)
	Low encapsulated drug contamination	Dissociation of the bound drug	
	Fast separation	Difficult to accurately differentiate protein bound and encapsulated API	
	No sample dilution		
	Less process-induced drug release?		
Chromatography			
Size exclusion/ Ion exchange/ Solid phase extraction	Fast separation	Potential encapsulated drug contamination	(Druckmann et al. 1989; Magin and Chan 1987; Mayer and St-Onge 1995; Thies et al. 1990)
		Labor intensive	
		Free drug not in equilibrium w/protein	
		Process-induced drug release?	
		sample dilution	
Liquid-liquid extraction			
	Fast separation	Potential encapsulated drug contamination	(Zhao et al. 2009)
		Free drug not in equilibrium w/protein	
		Large sample volume	
		Matrix effects due to the organic solvents	
		Process-induced drug release?	

Table 1 (continued)

Separation method	Advantages	Disadvantages	References
Ultracentrifugation			
	Free drug in equilibrium w/protein	Long equilibration time can lag release kinetics	(Wallace et al. 2012)
	Low non-specific binding of API	API instability	
	No sample dilution	Encapsulated drug contamination, back diffusion of encapsulated drug	
		Process-induced drug release?	

cal properties of the NBCD and API, include multi angle light scattering (MALS), refractometry, ultra-violet (UV)/fluorescent spectroscopy and scintillation counting. For example, the release of a hydrophobic photosensitizer, meta-tetra (hydroxyphenyl) chlorin (mTHPC; temoporphin), from PEGylated and non-PEGylated liposomal formulations in human plasma was measured using dextran and agarose-based SEC columns (Decker et al. 2013; Reshetov et al. 2012). The amount of mTHPC in the liposomal encapsulated (early eluting) and unencapsulated (late eluting) SEC fractions were compared between formulations. A more rapid release of mTHPC was observed for Pegylated liposomes (42 % in 30 min, 74 % over 24 h) in comparison to the non-Pegylated liposomes (22 % in 30 min and 91 % over 24 h), which followed a slower release pattern. Differences in release characteristics between the Pegylated and non-Pegylated liposome were attributed to the fact that PEG can decrease retention of bilayer-incorporated hydrophobic drugs (Decker et al. 2013). In contrast to rapidly releasing, bilayer incorporated hydrophobic drugs, hydrophilic drugs, such as doxorubicin, are encapsulated within the aqueous liposomal core and can be retained for a relatively longer time period. Very low unencapsulated doxorubicin fraction (< 10 % of the encapsulated drug) was observed by SEC analysis of plasma samples from a clinical PK study of liposomal doxorubicin, suggesting very slow release in vivo (Druckmann et al. 1989). Besides liposomes, SEC has also been used to examine the stability of other NBCD formulations, such as ferritin-based nanoparticles (Vannucci et al. 2012) and poly (L-lysine) (PLL)-based dendrimers (Boyd et al. 2006; Kaminskas et al. 2008) following intravenous administration in rodents. These studies suggest that the SEC drug release technique can be applied to a wide variety of NBCD types. However, while SEC is an easy, rapid and relatively inexpensive technique, its application for measuring drug release from NBCDs is limited by several concerns, such as on-column drug release resulting from sample dilution, and lack of high throughput (Druckmann et al. 1989).

Solid Phase Extraction

A common method used for separation of NBCD encapsulated and unencapsulated fractions in biological matrices is SPE, which has the advantages of fast sample processing speeds and low sample volume requirements (Table 1) (Thies et al. 1990). SPE utilizes a solid phase sorbent (e.g., silica-, polymeric-based, etc.) functionalized with either hydrophobic (C18, phenyl, etc.) or ionic (NH_4, COOH, etc.) groups, to effect a separation of the encapsulated and unencapsulated analyte based on differences in hydrophobicity or charge, respectively. Mixed-mode SEC/SPE phases that incorporate the size exclusion mode of separation are also available. Although potentially applicable to many different NBCD types, SPE has predominantly been used for measuring drug release from liposomal NBCDs which have inherent stability, "non-equilibrium" drug release characteristics, and surface properties that allow for separation from unencapsulated API using selective chromatography (e.g., reverse phase, ion exchange) (Druckmann et al. 1989; Gómez-Hens and Fernández-Romero 2006; Jiang et al. 2011; Thies et al. 1990). Silica based hydrophobically modified SPE cartridges are most commonly employed for separation of liposome encapsulated and unencapsulated forms of the analyte through reverse-phase interactions (Ahmad et al. 2005; Bellott et al. 2001; Deshpande et al. 2010; Thies et al. 1990; Zamboni et al. 2007, 2009a, b, 2011a, b). Similar to other chromatographic separation techniques, including SEC, the unencapsulated analyte usually has a longer retention time on the SPE column than the corresponding NBCD encapsulated form, which elutes earlier (Deshpande et al. 2010; Gómez-Hens and Fernández-Romero 2006).

A major problem with SPE, and potentially all separation methods to some degree, is that it is unknown as to how the separation method itself may influence drug release. NBCD surface properties would be expected to change over time upon exposure to biological media, potentially influencing interaction with the solid phase. Interaction with the solid phase may disrupt the NBCD leading to artificially high drug release estimates. Successive extraction of the same sample can give insight into the potential for process induced drug release, but there is currently no control that can adequately address this concern. Interactions between the solid phase sorbent and NBCD can be minimized through changes in solid phase chemistries and mobile phase, and passivation of the solid phase (e.g. by albumin treatment) (Bellott et al. 2001; Thies et al. 1990; Yamamoto et al. 2011). Clearly, it is important to use a separation method that is accurate and preserves the NBCD release characteristics.

Unlike other methods discussed in this review, SPE does have the advantage of being amenable to automated sample handling. For example, SPE has been combined with SEC in a semi-automated method (Bellott et al. 2001). Similarly, an online SPE-SPE-HPLC system has been developed and validated to improve throughput, extraction efficiency and accuracy of estimation of liposomal encapsulated and unencapsulated doxorubicin fractions from plasma samples in vitro and in vivo (Yamamoto et al. 2011). While SPE has been used successfully for measuring encapsulated and unencapsulated fractions of liposomal NBCDs, it is unclear whether this technique can be applied to other NBCD types. For example, amine-terminated

PAMAM dendrimers interact extensively with polar SPE sorbents through adsorption, resulting in instability, poor and variable recovery of the encapsulated analyte, and inaccurate estimation of the unencapsulated analyte after SPE processing (van Haandel and Stobaugh 2010). This, together with the inability of SPE to separate protein-bound and unbound fractions of the unencapsulated API, have favored the use of equilibrium techniques over SPE for measuring NBCD drug release, as discussed in the following sections.

Liquid-Liquid Extraction

Another common technique that has been applied to estimate drug release from NBCD formulations is extraction of the desired drug fraction(s) by liquid phases. Similar to SPE discussed above, in which partitioning of the free drug occurs between the liquid matrix and a solid sorbent, liquid-liquid extraction (LLE) involves partitioning of the analyte between two immiscible liquid phases (Kole et al. 2011; Żwir-Ferenc and Biziuk 2006). In contrast to SPE discussed above, LLE tends to be less selective, have lower analyte recoveries and utilizes greater sample volumes. It is also less amenable to automation.

LLE is one of the oldest methods used for sample clean-up prior to chromatographic and other bioanalytical analysis. The use of LLE for measuring drug release is based on the differential solubility of the free drug and the NBCD encapsulated drug in a water immiscible organic solvent. A variety of organic solvents, including ethyl acetate, dichloromethane, chloroform, n-heptane, etc., can be used for LLE; their selection being dependent on the nature of the analyte and the corresponding NBCD. For example, Zhao et al., used n-heptane to measure the free and nanoparticle-bound fraction of a model compound, pyrene, in rat plasma in vitro and in vivo (Zhao et al. 2009). In this example, the pyrene analyte was highly soluble in n-heptane, while the polymer used for NBCD preparation, methoxy polyethylene glycol polylactide (mPEG-PLA), was practically insoluble in the same solvent. Thus, released pyrene could be extracted by n-heptane, while encapsulated pyrene remained in the aqueous phase. Variations in the PK profiles of total and unencapsulated pyrene were observed after intravenous administration, indicating release of the API from the polymer platform (Zhao et al. 2009). While this method for measurement of pyrene release from mPEG-PLA was validated in vitro (Zhao et al. 2009), and holds promise for the measurement of drug release from some NBCD formulations, LLE may be unsuitable for measuring drug release from lipid-based NBCDs such as liposomes and solid-lipid nanoparticles, as it could interfere with the integrity of the NBCD due to lipid solubilization (Yang et al. 2013). Additional challenges limiting the widespread use of LLE methods for measurement of drug release include the need for relatively large sample volumes, matrix effects due to the use of organic solvents, emulsification of the immiscible organic phase by amphiphilic drugs and/or NBCD components, lack of automation and concerns regarding the use of hazardous organic solvents (Table 1) (Kole et al. 2011). As such, SPE is a preferred method over LLE for measuring drug release (Żwir-Ferenc and Biziuk 2006).

Equilibrium Methods

Unencapsulated drug in the circulation can exist in different forms that are in equilibrium with each other: blood cell-bound, plasma protein-bound and free/unbound drug. Only the free/unbound drug is biologically active. Thus, in addition to measuring encapsulated and unencapsulated drug fractions, measuring the free drug, blood cell-bound, and protein-bound fractions of the unencapsulated drug can be pivotal to understanding and comparing NBCD behavior in vivo (Sparreboom et al. 1999). While SPE can separate the encapsulated and unencapsulated analyte fractions in plasma, it is unable to distinguish between the free and protein-bound components of the unencapsulated fraction (Bellott et al. 2001; Druckmann et al. 1989; Thies et al. 1990). Indeed, SPE methods are a poor measure of free/unbound drug as they have been shown to extract unencapsulated drug from plasma proteins, inflating the "free" drug estimate (Druckmann et al. 1989; Mayer and St-Onge 1995; Thies et al. 1990). As this extraction of the free drug from protein is not necessarily complete (Mayer and St-Onge 1995), it is confusing as to exactly what fraction of unencapsulated drug is being measured in plasma by SPE methods, free and plasma protein-bound, or some fraction of the latter. It is expected that this lack of unbound drug specificity may be a problem for LLE and SEC methods as well, although this has not been specifically addressed.

Several simple and accurate equilibrium techniques have been developed to measure free/unbound drug, and used traditionally to assess small molecule protein binding (Vuignier et al. 2010). Many of these equilibrium methods, including ultrafiltration, equilibrium dialysis, and ultracentrifugation, have also been adapted to assess drug release from NBCD formulations (Table 1). These methods are referred to as equilibrium techniques, because unlike the methods listed above, the measured unbound drug is in "equilibrium" with plasma protein and formulation components. Both ultrafiltration and equilibrium dialysis involve use of a low molecular weight cutoff membrane to generate a plasma filtrate or dialysate, respectively, containing the unbound drug fraction (Bekersky et al. 2002; Mayer and St-Onge 1995; Wallace et al. 2012). In the case of ultracentrifugation, the unbound drug fraction is separated from the encapsulated and bound fractions by high speed centrifugation of the plasma sample (Wallace et al. 2012). While equilibrium dialysis dilutes the unbound fraction in the process of equilibrium with the buffer reservoir, this is not a problem in the case of ultrafiltration or ultracentrifugation in which the unbound drug concentration is undiluted by the separation technique. The ultrafiltration and equilibrium dialysis methods have the advantage of having low potential for process-induced drug release, since they involve mild physical modes of separation that are least likely to disrupt the NBCD complex. An additional advantage of the ultrafiltration and equilibrium dialysis methods is low encapsulated drug contamination of the free drug isolate (Mayer and St-Onge 1995). By contrast, process induced drug release and encapsulated drug contamination of the free drug fraction are still a concern for the ultracentrifugation technique.

Ultrafiltration

For ultrafiltration, separation of NBCD fractions is achieved through the use of low speed centrifugation (centrifugal ultrafiltration) or application of hydrostatic pressure (pressure ultrafiltration). While both centrifugal and pressure ultrafiltration have been used to measure drug release from NBCD formulations in buffered media (Wallace et al. 2012), centrifugal ultrafiltration is a more commonly used technique for separation of NBCD fractions in biological media such as plasma. For example, the release of ^{99m}Tc-fluconazole from mPEG-PLA and poloxamer adsorbed PLA (PLA-POLOX) nanocapsules were compared in the presence of 70 % mouse plasma. Free and protein-bound ^{99m}Tc-fluconazole were separated from their corresponding encapsulated fraction using ultrafiltration devices with an average membrane pore size of 0.1 μm (de Assis et al. 2008). Slower drug release kinetics were observed for PLA-PEG nanocapsules than for PLA-POLOX nanocapsules, possibly due to the presence of covalently bound PEG, which may decrease drug extraction by plasma proteins (de Assis et al. 2008). Centrifugal ultrafiltration has also been validated for determining total and unbound paclitaxel concentrations in plasma from patients receiving intravenous administration of Abraxane® and Taxol® (Gardner et al. 2008b). In this case, unbound paclitaxel was separated from the corresponding formulation-encapsulated and protein-bound fractions using ultrafiltration devices with a molecular weight cut-off of 30 kDa. Pharmacokinetic studies revealed higher unbound paclitaxel exposure of Abraxane relative to Taxol, which was thought to contribute to a greater therapeutic response by the former. Besides polymeric and albumin-based nanoparticles, ultrafiltration has also been used for evaluating drug release from liposomal formulations in plasma after intravenous administration (Bekersky et al. 2002; Krishna et al. 2001; Mayer and St-Onge 1995). Low free drug concentrations in the liposome ultrafiltrate were measured using sensitive radiolabeling (Krishna et al. 2001; Mayer and St-Onge 1995) or liquid chromatography coupled tandem mass spectrometry analysis (Bekersky et al. 2002).

The ultrafiltration method has also been used to differentially measure protein bound and NBCD encapsulated drug fractions, by first establishing the relationship between the ultrafilterable (free) drug concentration and % protein binding in the absence of formulation components (Bekersky et al. 2002). Using this technique, the protein-bound drug fraction of a liposomal formulation was estimated indirectly by interpolating the protein-bound fraction from the established ultrafilterable drug concentration-%protein binding correlation. The liposomal encapsulated drug fraction was then measured by subtracting the ultrafilterable concentration and protein bound concentration from the total drug concentration. One caveat for this method of measuring protein bound drug is that formulation-induced alterations in protein binding, or equilibrium binding of the free drug to the formulation itself, is not reflected in the established binding relationship in the absence of the formulation components. This is an important point, as formulations have been shown to alter protein binding (ten Tije et al. 2003), as well as reversibly bind free drug (Sparreboom et al. 1999), and this could result in inaccurate estimation of protein bound and encapsulated drug fractions (Bekersky et al. 2002). In addition to the above concerns, ultrafiltration is also prone to membrane clogging by samples with high particle content, as well as non-specific binding of free drug to the membrane.

Equilibrium Dialysis

Another commonly used method to estimate protein binding and drug release from NBCD is equilibrium dialysis. Similar to ultrafiltration, equilibrium dialysis employs a semi-permeable membrane between a donor chamber, containing the sample in plasma, and an acceptor chamber, containing physiological buffer. Molecules smaller than the membrane pore size (e.g. free drug) pass from the donor to the acceptor chamber by simple diffusion while larger molecules (e.g. NBCD encapsulated drug and protein-bound drug) are retained in the donor compartment. Despite being regarded as a "gold-standard" for estimating plasma protein-binding (Singh et al. 2012), equilibrium dialysis can often produce confusing results when used as a primary method for evaluating drug release. Given that equilibrium dialysis is predominantly a diffusion-controlled process, the appearance of the free drug molecules in the acceptor chamber is dependent on the rate of diffusion of free drug molecules across the membrane. The rate of diffusion across the membrane is dependent upon the concentration gradient of the unencapsulated drug, which is likely to be low for a controlled release NBCD formulation. As such, the rate of appearance of free drug molecules in the acceptor chamber may not always be indicative of the rate of drug release, as it may lag drug release, especially for slowly diffusing drug molecules with fast NBCD release rates (Wallace et al. 2012).

Ultracentrifugation

Besides ultrafiltration and equilibrium dialysis methods, ultracentrifugation has also been used for separating free drug from protein-bound and NBCD-encapsulated fractions (Ricci et al. 2006). Major advantages of the ultracentrifugation technique are the lack of sample dilution and low non-specific binding (Table 1). Ultracentrifugation is a membrane-free separation technique that relies on high centrifugal forces and long centrifugation times to sediment the NBCD encapsulated and protein-bound fractions, leaving the free, unbound drug fraction in the supernatant. These long processing times can lag drug release similar to equilibrium diffusion described above. The use of ultracentrifugation for measuring drug release is further limited by contamination of the supernatant with the encapsulated drug fraction, as a result of back diffusion or the inability to completely sediment the entire nanoparticle population (Wallace et al. 2012). Obviously nanoparticles within a population that have a density similar to that of plasma water will not sediment. There is also the possibility of process-induced drug release due to the high centrifugal forces, and the lack of high throughput that is a concern for processing large numbers of pharmacokinetic samples. For these reasons, there are few examples of use of the ultracentrifugation method for processing of NBCD biological samples.

Indirect Methods for Estimation of NBCD Drug Release

Several indirect methods for estimation of drug release from NBCDs in the absence of actual measurement of the unencapsulated fraction exist. These indirect methods can be complimentary to bioanalytical methods that can separate encapsulated and unencapsulated drug fractions, or alternatively may be the only methods available in cases where the NBCD is not amenable to established separation techniques. Indirect methods for estimation of NBCD drug release include: (1) evaluation of total drug and NBCD platform pharmacokinetic profiles; (2) modeling and simulation; and (3) analysis of metabolite pharmacokinetics.

Evaluation of Total Drug and NBCD Platform Pharmacokinetic Profiles

The total drug (encapsulated and unencapsulated) and NBCD platform PK, and concentration ratio of total drug to NBCD platform, can be used to estimate drug release from NBCD. The concentration ratio of the total drug to NBCD platform (e.g., API:lipid ratio for liposomal formulation) in the circulation should be identical to the initial loading and remain constant if (1) the formulation is stable and drug is not released, and (2) the unencapsulated drug accounts for an insignificant fraction of the total drug concentration. As discussed above in the introduction to NBCD pharmacokinetics, the encapsulated drug generally dominates the total drug concentration profile, and unencapsulated drug generally represents a small fraction of this total. This dominance of the encapsulated fraction in the total drug concentration profile is dependent upon a faster clearance of the unencapsulated drug, which is not always the case. For example, early non-stealth liposomal formulations of doxorubicin were cleared faster by the MPS than free doxorubicin, and so in this case, free doxorubicin dominated the pharmacokinetic profile (Gabizon and Martin 1997). As drug is released from the NBCD formulation, the PK profile of the total drug diverges from that of the NBCD platform, and the concentration ratio of total drug to NBCD platform decreases. Thus, the divergence in total drug and NBCD platform PK, and resulting decrease in API: NBCD platform concentration ratio, can be useful for estimation of drug release.

Concentrations of total drug (unencapsulated and encapsulated) and NBCD platform in biological matrix can be measured by use of established bioanalytical methods, or alternatively use of dual radiotracers. Obviously, the use of radiotracers would be more applicable to preclinical studies, and it is important to ensure that the radiolabeled NBCD has the same physicochemical and biological properties as the unlabeled NBCDs under development. As an example of radiolabeling, the stability of a PEGylated liposomal C6-ceramide formulation was evaluated in rats by the use of dual radiotracers (Zolnik et al. 2008). In this study, the ^{14}C-C6-ceramide and ^{3}H-DSPC lipid radiotracer plasma profiles, for the ceramide and liposome, respectively, were observed to diverge rapidly in vivo upon administration of the C6-ceramide

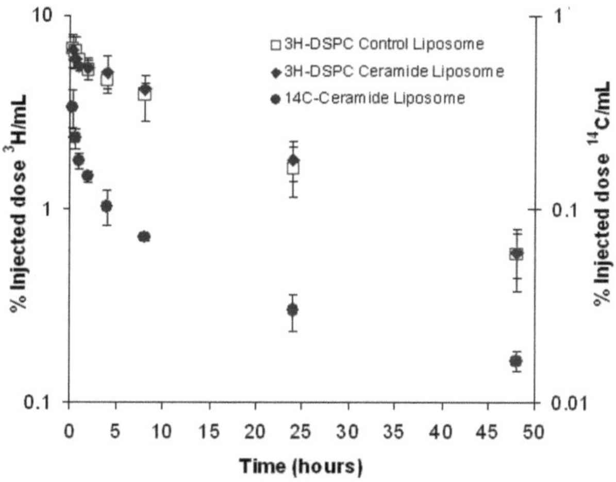

Fig. 3 [^{14}C]-C6-ceramide and [^{3}H]-DSPC plasma profiles. Plasma time profiles are expressed as the percentage of injected dose per milliliter. ♦ [^{3}H]-DSPC of ceramide liposomes, □ [^{3}H]-DSPC of control liposomes on the *y*1-axis, ● [^{14}C]-C6-ceramide on the *y*2-axis. Each point represents the mean ± S.D. from *n*=4 to 5 rats. (Reproduced from (Zolnik et al. 2008))

liposome, supporting instability of the formulation (Fig. 3). Interestingly, the apparent volumes of distribution (Vd) were strikingly different for the ceramide and liposome components. While ceramide had a very large Vd, 1020 mL/kg, the liposome Vd was only 63 mL/kg, comparable to the rat plasma volume of 40 mL/kg (Gillen et al. 1994). NBCD platforms that do not undergo rapid MPS accumulation, such as those with steric hydrophilic surface coating (e.g., PEG coating), are often confined to the plasma space (Stern et al. 2010). The larger apparent Vd for ceramide in relation to the liposome, as well as larger clearance (165 vs. 3.6 mL/h*kg, for ceramide and liposome, respectively) suggests rapid drug release immediately upon injection. These data agreed with in vitro blood partitioning studies and in vivo tissue distribution studies, which suggested that the mechanism of ceramide release from the liposome was through a bilayer exchange mechanism in which ceramide "flipped" from liposome bilayer to blood cell and endothelial bilayers (Zolnik et al. 2008). This bilayer exchange mechanism is also in agreement with the theoretical bilayer loading mechanism for ceramide in the liposome, and with the fact that ceramide release apparently did not result in liposome rupture and rapid liposome clearance by the MPS.

Modeling and Simulation

Another indirect method for estimation of drug release kinetics, as well as unencapsulated and encapsulated drug concentration-time profiles, is the use of modeling and simulation. Drug release and associated pharmacokinetic parameters can be estimated by fitting of pharmacokinetic models to total drug and NBCD platform PK profiles. The unencapsulated and encapsulated drug concentration-time profiles can then be simulated using the established model and fitted parameter estimates. Thus, modeling and simulation can be a powerful tool for qualification of bioanalytical

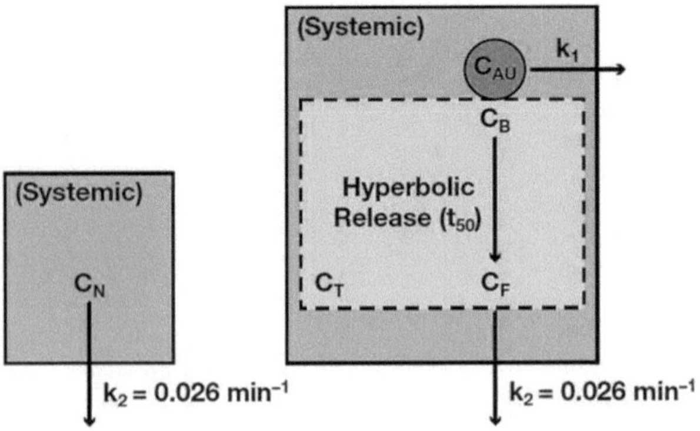

Fig. 4 Pharmacokinetic model for Au-TNF release. The first-order elimination rate for gold (k_1), hyperbolic time to 50 % TNF release (t_{50}), and time zero concentrations for gold ($C_{Au}(0)$) and Au–TNF ($C_T(0)$) were obtained by fitting the TNF release model to the blood concentration-time data for gold (C_{Au}) and Au–TNF (C_T). The first-order elimination constant (k_2) and concentration at time zero ($C_N(0)$) for native TNF (C_N) were fixed in the TNF release model, based on a single compartment fitting of the native TNF blood concentration-time data. The released, unencapsulated (C_F) and bound, encapsulated (C_B) TNF concentrations were then simulated using the hyperbolic stability equation, $C_{Rel} = C_T(0) * \exp(-k1*t)*(t/(t+t50))$. (Reproduced from (Stern et al. 2010))

separation methods, and in vitro NBCD drug release assays. There are even instances in which modeling and simulation have been used to improve upon analytical methods of NBCD drug fraction separation by subtracting in-process drug release from the encapsulated and unencapsulated fraction measurements (Bulitta et al. 2009). Alternatively, modeling and simulation may be the only tool available to estimate drug release if bioanalytical separation techniques are unavailable for the NBCD.

Drug release studies performed by Stern et al. for a tumor necrosis factor (TNF)-colloidal gold nanoparticle (Au-TNF) formulation is an example of a case in which modeling and simulation were used in the absence of available bioanalytical separation techniques (Stern et al. 2010). TNF drug release from Au-TNF was estimated by modeling of rat total blood TNF and gold pharmacokinetic data. Total blood TNF was measured using a standard ELISA, whereas total blood Au concentrations were measured by inductively coupled plasma mass spectrometry. A compartmental model was then fitted to the total TNF and gold blood profiles (Fig. 4 and 5a), and this model was then used to estimate drug release kinetics, and simulate the unencapsulated TNF profile in comparison to the unformulated, native TNF profile (Fig. 5b). In comparison to the unformulated, native TNF, the simulated unencapsulated TNF maximal peak concentration was 25 % of the native TNF and occurred 34 min later. This lower Cmax for the colloidal gold TNF formulation agreed with the reduced toxicity observed in the clinic for the Au-TNF formulation in comparison to native TNF (Libutti et al. 2010).

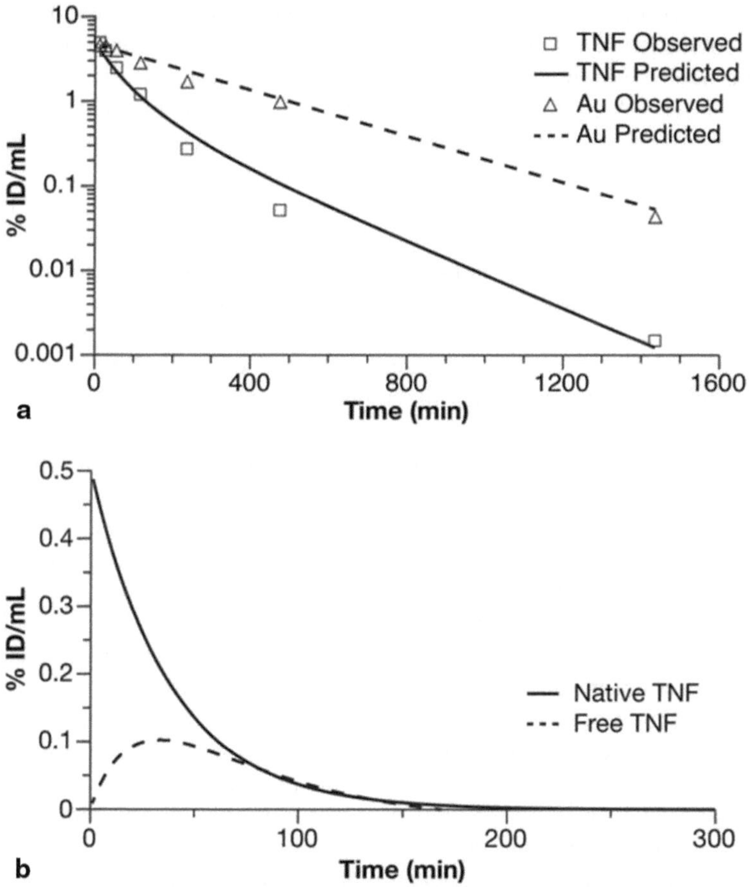

Fig. 5 **a** Au–TNF release model fit. The Au–TNF release model was fit to the total TNF and Gold (Au) blood concentration-time data, expressed as %ID/mL. Lines represent the fit of the Au–TNF release model to the blood concentration-time data. Data are presented as mean blood concentrations (n=4 to 5 rats). **b** Au–TNF release model simulation. The native TNF and released, unencapsulated TNF blood concentration data were simulated using the Au–TNF release model and estimated pharmacokinetic parameters. Lines represent the simulated blood concentration-time data for the unencapsulated and native TNF. (Reproduced from (Stern et al. 2010))

Analysis of Metabolite Pharmacokinetics

Since only the free API is generally metabolized, metabolite data can be very useful for estimation of free drug exposure. Both the EMA and FDA recommend the use of metabolite pharmacokinetic data as a supportive metric for bioequivalence assessment, and as a primary metric in cases where the active drug cannot be accurately measured (EMA 2008; FDA 2002). For liposomal products, the EMA has also recommended, “Quantification of at least one metabolite regardless of its pharmacological activity may facilitate to assess and compare a release rate, since metabolism of the active substance takes place only after release from the liposomes”

(EMA 2011b). In the Caelyx generic bioequivalence study discussed above, the pharmacokinetics of the major doxorubicin metabolite, doxorubicinol, was compared between the follow-on and innovator products as a secondary measure of bioequivalence (EMA 2011a). As per the EMA Caelyx generic bioequivalence report, doxorubicinol PK, "...could be considered a surrogate for free doxorubicin." and "...may also reflect rate and extent of liposomal release."

In a study by Zou et al., curcumin metabolite data was used to assess free curcumin exposure of a polymeric nanoscale curcumin formulation in comparison to solvent formulated curcumin in rats (Zou et al. 2013). Curcumin metabolite data was analyzed, as opposed to curcumin itself, as curcumin is metabolically unstable and cannot be accurately measured. The polymeric curcumin had greater cumulative biliary and urinary curcumin metabolite clearance over an 8 h collection period than the solvent curcumin, which precipitated upon injection and accumulated in lung capillaries. The greater biliary and urinary clearance suggested rapid release of curcumin from the polymeric nanoparticle, and greater free curcumin exposure than the solvent curcumin control. Polymer encapsulated curcumin concentrations in the plasma, which could be measured as the encapsulated curcumin was metabolically stable, only accounted for 10 % of the curcumin dose at the earliest time point, suggesting a burst release mechanism that agreed with the metabolite data.

In addition to estimation of total unencapsulated drug exposure, metabolite data can also be used for modeling and simulation of unencapsulated drug concentration-time profiles, since metabolite clearance is proportional to free drug concentrations. For example, modeling of rat biliary hydroxydocetaxel, the primary metabolite of docetaxel, was used to estimate conversion of a polymeric docetaxel prodrug and simulate released docetaxel concentration-time profiles (Stern et al. 2013). A compartmental model was used to simultaneously fit rat plasma PK data for both the polymeric docetaxel prodrug, Procet 8, and the commercial docetaxel formulation, Taxotere, (Fig. 6). This compartmental model was then used to simulate the converted (released) docetaxel profile (Fig. 7). Model simulation predicted formation rate-limited kinetics for the docetaxel released from prodrug, and a peak plasma docetaxel concentration for the prodrug 70 times lower than that of Taxotere. Supporting the predictive nature of this model, the model estimated hydroxydocetaxel clearance was identical to scaled in vitro rat hydroxydocetaxel intrinsic clearance estimates. Additionally, the model estimated prodrug conversion rate was identical to the prodrug hydrolysis rate in rat plasma in vitro, qualifying the in vitro prodrug hydrolysis assay as predictive of in vivo hydrolysis. Qualifying the predictive nature of the in vitro prodrug hydrolysis assay further supported its use in assessing batch-to-batch reproducibility of the formulation, and the setting of in vitro hydrolysis rate as a critical quality attribute of the polymeric docetaxel prodrug formulation for lot release purposes (personal communication, Lawrence Mayer, Celator Pharmaceuticals, July 22, 2014).

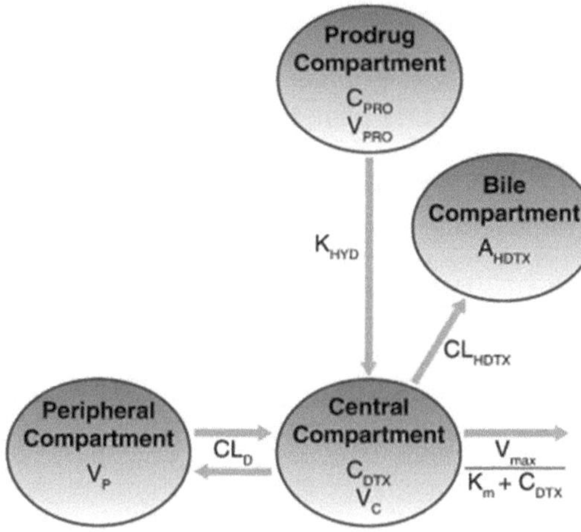

Fig. 6 Compartmental biliary clearance model. This is a two compartment model with linear distribution clearance (CL_D), linear clearance for docetaxel (*DTX*) oxidation to hydroxydocetaxel (*HDTX*) (CL_{HDTX}) and nonlinear clearance ($V_{max}/(K_m + C_{DTX})$ representing all remaining clearance routes (e.g., urine, other biliary metabolites). The parameters C_{PRO} and V_{PRO} represent the polymeric docetaxel prodrug (Procet 8) plasma concentration and volume of the polymeric docetaxel prodrug compartment, respectively. Prodrug hydrolysis to DTX is described by a first order rate (K_{HYD}). The parameters C_{DTX} and V_C represent the DTX plasma concentration and volume of the central compartment, respectively. The parameters V_P and A_{HDTX} represent the volume of the peripheral compartment and the cumulative amount of HDTX excreted in the bile, respectively. (Reproduced from (Stern et al. 2013))

Conclusions and Future Directions

In summary, NBCD pharmacokinetics and bioanalytical methods development are an exciting and timely area of drug delivery research. With greater regulatory scrutiny of follow-on NBCD products, regulators have apparently come to consensus regarding the requirement for evaluation of unencapsulated and encapsulated drug PK for bioequivalence determination. While many methods are currently available to evaluate drug release from NBCD formulations, including both direct and indirect methods, the most appropriate method for any particular NBCD type ultimately depends upon a combination of existing experience, scientific intuition and trial and error. Presently, the available separation techniques have arisen from repurposing of small molecule preparatory and protein binding methods, none of which adequately address all concerns, such as the problems of process-induced release and accurate determination of unbound drug. Clearly there is a need for new bioanalytical methods that can accurately quantify NBCD drug fractions and drug release. Modeling and simulation of NBCD drug release and unencapsulated drug have great potential to support bioanalytical method development, and in certain instances possibly even stand alone as a primary drug development tool for regulatory consideration. Further development in these areas have the potential to not only inform drug developers of important aspects of NBCD formulations that govern therapy, but also streamline the regulatory process at the same time.

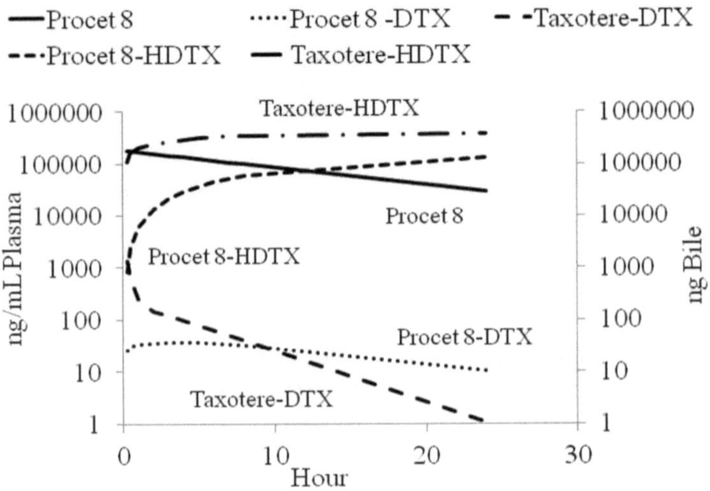

Fig. 7 Biliary clearance model simulation. The simulated plasma docetaxel (*DTX*), plasma polymeric docetaxel prodrug (Procet 8), and cumulative biliary hydroxydocetaxel (*HDTX*) profiles are displayed for prodrug and commercial Taxotere. Cumulative biliary hydroxydocetaxel metabolites are expressed in ng, and plasma prodrug and docetaxel concentrations are expressed in ng/mL. Note: prodrug concentrations are displayed as docetaxel equivalents. (Reproduced from (Stern et al. 2013))

Acknowledgements Acknowledgement: This project has been funded in whole or in part with Federal funds from the Frederick National Laboratory for Cancer Research, National Institutes of Health, under contract HHSN261200800001E. The content of this publication does not necessarily reflect the views or policies of the Department of Health and Human Services, nor does mention of trade names, commercial products or organizations imply endorsement by the US Government. The authors thank Dr. Rachael M. Crist for assistance with the preparation of the manuscript.

References

Ahmad A, Wang YF, Ahmad I (2005) Separation of liposome-entrapped mitoxantrone from nonliposomal mitoxantrone in plasma: pharmacokinetics in mice. Methods Enzymol 391:176–185

Albanese A, Tang PS, Chan WC (2012) The effect of nanoparticle size, shape, and surface chemistry on biological systems. Annu Rev Biomed Eng 14:1–16

Bekersky I, Fielding RM, Dressler DE, Lee JW, Buell DN, Walsh TJ (2002) Plasma protein binding of amphotericin B and pharmacokinetics of bound versus unbound amphotericin B after administration of intravenous liposomal amphotericin B (AmBisome) and amphotericin B deoxycholate. Antimicrob Agents Chemother 46:834–840

Bellott R, Pouna P, Robert J (2001) Separation and determination of liposomal and non-liposomal daunorubicin from the plasma of patients treated with Daunoxome. J Chromatogr B Biomed Sci Appl 757:257–267

Blanco E, Hsiao A, Mann AP, Landry MG, Meric-Bernstam F, Ferrari M (2011) Nanomedicine in cancer therapy: innovative trends and prospects. Cancer Sci 102:1247–1252

Boyd BJ, Kaminskas LM, Karellas P, Krippner G, Lessene R, Porter CJ (2006) Cationic poly-L-lysine dendrimers: pharmacokinetics, biodistribution, and evidence for metabolism and bioresorption after intravenous administration to rats. Mol Pharm 3:614–627

Bulitta JB, Zhao P, Arnold RD, Kessler DR, Daifuku R, Pratt J, Luciano G, Hanauske AR, Gelderblom H, Awada A, Jusko WJ (2009) Mechanistic population pharmacokinetics of total and unbound paclitaxel for a new nanodroplet formulation versus Taxol in cancer patients. Cancer Chemother Pharmacol 63:1049–1063

Code of Federal Regulations (CFR) (2013, April) Content and format of an abbreviated application, 21 CFR 314.94. http://www.accessdata.fda.gov/scripts/cdrh/cfdocs/cfcfr/cfrsearch.cfm?fr=314.94. Accessed 24 April 2015

de Assis DN, Mosqueira VC, Vilela JM, Andrade MS, Cardoso VN (2008) Release profiles and morphological characterization by atomic force microscopy and photon correlation spectroscopy of 99mTechnetium-fluconazole nanocapsules. Int J Pharm 349:152–160

Decker C, Steiniger F, Fahr A (2013) Transfer of a lipophilic drug (temoporfin) between small unilamellar liposomes and human plasma proteins: influence of membrane composition on vesicle integrity and release characteristics. J Liposome Res 23:154–165

Desai N (2012) Challenges in development of nanoparticle-based therapeutics. AAPS J 14:282–295

Deshpande NM, Gangrade MG, Kekare MB, Vaidya VV (2010) Determination of free and liposomal amphotericin B in human plasma by liquid chromatography-mass spectroscopy with solid phase extraction and protein precipitation techniques. J Chromatogr B Analyt Technol Biomed Life Sci 878:315–326

Druckmann S, Gabizon A, Barenholz Y (1989) Separation of liposome-associated doxorubicin from non-liposome-associated doxorubicin in human plasma: implications for pharmacokinetic studies. Biochim Biophys Acta 980:381–384

Duncan R, Gaspar R (2011) Nanomedicine(s) under the microscope. Mol Pharm 8:2101–2141

Ehmann F, Sakai-Kato K, Duncan R, Hernan Perez de la Ossa D, Pita R, Vidal JM, Kohli A, Tothfalusi L, Sanh A, Tinton S, Robert JL, Silva Lima B, Amati MP (2013) Next-generation nanomedicines and nanosimilars: EU regulators' initiatives relating to the development and evaluation of nanomedicines. Nanomedicine (Lond) 8:849–856

European Medicines Agency (EMA), Committee for Medicinal Products for Human Use (CHMP) (2008) Draft, Guideline on the investigation of bioequivalence. http://www.ema.europa.eu/docs/en_GB/document_library/Scientific_guideline/2009/09/WC500003011.pdf. Accessed 24 April 2015

European Medicines Agency (EMA), Committee for Medicinal Products for Human Use (CHMP) (2011a) CHMP assessment report: doxorubicin sun. http://www.ema.europa.eu/docs/en_GB/document_library/Application_withdrawal_assessment_report/human/002049/WC500112957.pdf. Accessed 24 April 2015

European Medicines Agency (EMA), Committee for Medicinal Products for Human Use (CHMP) (2011b) Draft, Reflection paper on the data requirements for intravenous liposomal products developed with reference to an innovator liposomal product. http://www.ema.europa.eu/docs/en_GB/document_library/Scientific_guideline/2011/07/WC500109479.pdf. Accessed 24 April 2015

European Medicines Agency (EMA), Committee for Medicinal Products for Human Use (CHMP) (2011c) Reflection paper on non-clinical studies for generic nanoparticle iron medicinal product applications. http://www.ema.europa.eu/docs/en_GB/document_library/Scientific_guideline/2011/04/WC500105048.pdf. Accessed 24 April 2015

European Medicines Agency (EMA), Committee for Medicinal Products for Human Use (CHMP) (2013a) Joint MHLW/EMA reflection paper on the development of block copolymer micelle medicinal products. http://www.ema.europa.eu/docs/en_GB/document_library/Scientific_guideline/2013/02/WC500138390.pdf. Accessed 24 April 2015

European Medicines Agency (EMA), Committee for Medicinal Products for Human Use (CHMP) (2013b) Reflection paper on the data requirements for intravenous iron-based nano-colloidal products developed with reference to an innovator medicinal product. http://www.ema.europa.eu/docs/en_GB/document_library/Scientific_guideline/2013/09/WC500149496.pdf. Accessed 24 April 2015

Food and Drug Administration (FDA), Center for Drug Evaluation and Research (CDER) (2000, August) Guidance for industry: waiver of in vivo bioavailability and bioequivalence studies for immediate-release solid oral dosage forms based on a biopharmaceutics classification system. http://www.fda.gov/downloads/Drugs/Guidances/ucm070246.pdf. Accessed 24 April 2015

Food and Drug Administration (FDA), Center for Drug Evaluation and Research (CDER) (2001, January) Statistical approaches to establishing bioeuivalence. http://www.fda.gov/downloads/Drugs/Guidances/ucm070244.pdf. Accessed 24 April 2015

Food and Drug Administration (FDA), Center for Drug Evaluation and Research (CDER) (2002, July) Guidance for industry: bioavailability and bioequivalence studies for orally administered drug products—general considerations. http://www.fda.gov/downloads/drugs/developmentapprovalprocess/howdrugsaredevelopedandapproved/approvalapplications/abbreviatednewdrugapplicationandagenerics/ucm154838.pdf. Accessed 24 April 2015

Food and Drug Administration (FDA), Office of Generic Drugs (2012a, March) Draft guidance on iron sucrose bioequivalence. http://www.fda.gov/downloads/Drugs/GuidanceCompliance-RegulatoryInformation/Guidances/UCM297630.pdf. Accessed 24 April 2015

Food and Drug Administration (FDA), Office of Generic Drugs (2012b, September) Draft guidance on paclitaxel. http://www.fda.gov/downloads/drugs/guidancecomplianceregulatoryinformation/guidances/ucm320015.pdf. Accessed 24 April 2015

Food and Drug Administration (FDA) (2012c, December) Draft guidance on ferumoxytol. http://www.fda.gov/downloads/Drugs/GuidanceComplianceRegulatoryInformation/Guidances/UCM333051.pdf. Accessed 24 April 2015

Food and Drug Administration (FDA) (2013a, February) FDA approval of generic version of cancer drug Doxil is expected to help resolve shortage. http://www.fda.gov/newsevents/newsroom/pressannouncements/ucm337872.htm. Accessed 24 April 2015

Food and Drug Administration (FDA) (2013b, April) Therapeutic equivalence of generic iron complex products. (Solicitation for a study on Na ferric gluconate; Nulecit (followon upon 505j approval) vs Ferrlecit (originator)). www.fbo.gov/index?s=opportunity&mode=form&id=592788989854da145c8e7b6d103c898d&tab=core&tabmode=list&. Accessed 24 April 2015

Food and Drug Administration (FDA), Office of Generic Drugs (2013c, November) Draft guidance on doxorubicin hydrochloride. http://www.fda.gov/downloads/Drugs/…/Guidances/UCM199635.pdf. Accessed 24 April 2015

Food and Drug Administration (FDA), Office of Generic Drugs (2014a, April) Draft guidance on amphotericin B. http://www.fda.gov/downloads/Drugs/GuidanceComplianceRegulatoryInformation/Guidances/UCM384094.pdf. Accessed 24 April 2015

Food and Drug Administration (FDA), Office of Generic Drugs (2014b, April) Draft guidance on verteporfin. http://www.fda.gov/downloads/Drugs/GuidanceComplianceRegulatoryInformation/Guidances/UCM384173.pdf. Accessed 24 April 2015

Food and Drug Administration (FDA) (2014c, July) Draft guidance on daunorubicin citrate. http://www.fda.gov/downloads/Drugs/GuidanceComplianceRegulatoryInformation/Guidances/UCM406256.pdf. Accessed 24 April 2015

Gabizon A, Martin F (1997) Polyethylene glycol-coated (pegylated) liposomal doxorubicin. Rationale for use in solid tumours. Drugs 54(Suppl 4):15–21

Gardner ER, Dahut W, Figg WD (2008a) Quantitative determination of total and unbound paclitaxel in human plasma following Abraxane treatment. J Chromatogr B Analyt Technol Biomed Life Sci 862:213–218

Gardner ER, Dahut WL, Scripture CD, Jones J, Aragon-Ching JB, Desai N, Hawkins MJ, Sparreboom A, Figg WD (2008b) Randomized crossover pharmacokinetic study of solvent-based paclitaxel and nab-paclitaxel. Clin Cancer Res 14:4200–4205

Gillen CM, Takamata A, Mack GW, Nadel ER (1994) Measurement of plasma volume in rats with use of fluorescent-labeled albumin molecules. J Appl Physiol (1985) 76:485–489

Gómez-Hens A, Fernández-Romero JM (2006) Analytical methods for the control of liposomal delivery systems. TrAC Trends Anal Chem 25:167–178

Gregoriadis G (1991) Overview of liposomes. J Antimicrob Chemother 28(Suppl B):39–48

Hagel L (2001) Gel-filtration chromatography. Curr Protoc Protein Sci Chapter 8:Unit 8.3

He C, Hu Y, Yin L, Tang C, Yin C (2010) Effects of particle size and surface charge on cellular uptake and biodistribution of polymeric nanoparticles. Biomaterials 31:3657–3666

Jiang W, Kim BY, Rutka JT, Chan WC (2007) Advances and challenges of nanotechnology-based drug delivery systems. Expert Opin Drug Deliv 4:621–633

Jiang W, Lionberger R, Yu LX (2011) In vitro and in vivo characterizations of PEGylated liposomal doxorubicin. Bioanalysis 3:333–344

Kaiser N, Kimpfler A, Massing U, Burger AM, Fiebig HH, Brandl M, Schubert R (2003) 5-Fluorouracil in vesicular phospholipid gels for anticancer treatment: entrapment and release properties. Int J Pharm 256:123–131

Kaminskas LM, Boyd BJ, Karellas P, Krippner GY, Lessene R, Kelly B, Porter CJ (2008) The impact of molecular weight and PEG chain length on the systemic pharmacokinetics of PEGylated poly l-lysine dendrimers. Mol Pharm 5:449–463

Kole PL, Venkatesh G, Kotecha J, Sheshala R (2011) Recent advances in sample preparation techniques for effective bioanalytical methods. Biomed Chromatogr 25:199–217

Krishna R, Webb MS, St Onge G, Mayer LD (2001) Liposomal and nonliposomal drug pharmacokinetics after administration of liposome-encapsulated vincristine and their contribution to drug tissue distribution properties. J Pharmacol Exp Ther 298:1206–1212

Libutti SK, Paciotti GF, Byrnes AA, Alexander HR Jr, Gannon WE, Walker M, Seidel GD, Yuldasheva N, Tamarkin L (2010) Phase I and pharmacokinetic studies of CYT-6091, a novel PEGylated colloidal gold-rhTNF nanomedicine. Clin Cancer Res 16:6139–6149

Maeda H (2012) Vascular permeability in cancer and infection as related to macromolecular drug delivery, with emphasis on the EPR effect for tumor-selective drug targeting. Proc Jpn Acad Ser B Phys Biol Sci 88:53–71

Magin RL, Chan HC (1987) Rapid separation of liposomes using ultrafiltration. Biotechnol Tech 1:185–188

Mayer LD, St-Onge G (1995) Determination of free and liposome-associated doxorubicin and vincristine levels in plasma under equilibrium conditions employing ultrafiltration techniques. Anal Biochem 232:149–157

Mayer LD, Cullis PR, Bally MB (1994) The use of transmembrane pH gradient-driven drug encapsulation in the pharmacodynamic evaluation of liposomal doxorubicin. J Liposome Res 4:529–553

Mei L, Zhang Z, Zhao L, Huang L, Yang XL, Tang J, Feng SS (2013) Pharmaceutical nanotechnology for oral delivery of anticancer drugs. Adv Drug Deliv Rev 65:880–890

Onoue S, Yamada S, Chan HK (2014) Nanodrugs: pharmacokinetics and safety. Int J Nanomedicine 9:1025–1037

Owens DE 3rd, Peppas NA (2006) Opsonization, biodistribution, and pharmacokinetics of polymeric nanoparticles. Int J Pharm 307:93–102

Prantner AM, Scholler N (2014) Biological barriers and current strategies for modifying nanoparticle bioavailability. J Nanosci Nanotechnol 14:115–125

Public Health Service (PHS) Act Sect. 351(k). http://www.fda.gov/regulatoryinformation/legislation/ucm148717.htm. Accessed 24 April 2015

Reshetov V, Zorin V, Siupa A, D'Hallewin MA, Guillemin F, Bezdetnaya L (2012) Interaction of liposomal formulations of meta-tetra(hydroxyphenyl)chlorin (temoporfin) with serum proteins: protein binding and liposome destruction. Photochem Photobiol 88:1256–1264

Ricci M, Giovagnoli S, Blasi P, Schoubben A, Perioli L, Rossi C (2006) Development of liposomal capreomycin sulfate formulations: effects of formulation variables on peptide encapsulation. Int J Pharm 311:172–181

Singh JK, Solanki A, Maniyar RC, Banerjee D, Shirsath VS (2012) Rapid Equilibrium Dialysis (RED): an in-vitro high-throughput screening technique for plasma protein binding using human and rat plasma. J Bioequiv Bioavailab S14:1–4

Sparreboom A, van Zuylen L, Brouwer E, Loos WJ, de Bruijn P, Gelderblom H, Pillay M, Nooter K, Stoter G, Verweij J (1999) Cremophor EL-mediated alteration of paclitaxel distribution in human blood: clinical pharmacokinetic implications. Cancer Res 59:1454–1457

Stern ST, Hall JB, Yu LL, Wood LJ, Paciotti GF, Tamarkin L, Long SE, McNeil SE (2010) Translational considerations for cancer nanomedicine. J Control Release 146:164–174

Stern ST, Zou P, Skoczen S, Xie S, Liboiron B, Harasym T, Tardi P, Mayer LD, McNeil SE (2013) Prediction of nanoparticle prodrug metabolism by pharmacokinetic modeling of biliary excretion. J Control Release 172:558–567

ten Tije AJ, Verweij J, Loos WJ, Sparreboom A (2003) Pharmacological effects of formulation vehicles: implications for cancer chemotherapy. Clin Pharmacokinet 42:665–685

Thies RL, Cowens DW, Cullis PR, Bally MB, Mayer LD (1990) Method for rapid separation of liposome-associated doxorubicin from free doxorubicin in plasma. Anal Biochem 188:65–71

Torchilin V (2011) Tumor delivery of macromolecular drugs based on the EPR effect. Adv Drug Deliv Rev 63:131–135

United States Pharmacopeia (USP) (2011, December) Chapter <711>, Dissolution. http://www.usp.org/sites/default/files/usp_pdf/EN/USPNF/2011-02-25711DISSOLUTION.pdf. Accessed 24 April 2015

van Haandel L, Stobaugh JF (2010) Bioanalytical method development for a generation 5 polyamidoamine folic acid methotrexate conjugated nanoparticle. Anal Bioanal Chem 397:1841–1852

Vannucci L, Falvo E, Fornara M, Di Micco P, Benada O, Krizan J, Svoboda J, Hulikova-Capkova K, Morea V, Boffi A, Ceci P (2012) Selective targeting of melanoma by PEG-masked protein-based multifunctional nanoparticles. Int J Nanomedicine 7:1489–1509

Vuignier K, Schappler J, Veuthey JL, Carrupt PA, Martel S (2010) Drug-protein binding: a critical review of analytical tools. Anal Bioanal Chem 398:53–66

Wallace SJ, Li J, Nation RL, Boyd BJ (2012) Drug release from nanomedicines: selection of appropriate encapsulation and release methodology. Drug Deliv Transl Res 2:284–292

Yamamoto E, Hyodo K, Ohnishi N, Suzuki T, Ishihara H, Kikuchi H, Asakawa N (2011) Direct, simultaneous measurement of liposome-encapsulated and released drugs in plasma by on-line SPE-SPE-HPLC. J Chromatogr B Analyt Technol Biomed Life Sci 879:3620–3625

Yang F, Wang H, Liu M, Hu P, Jiang J (2013) Determination of free and total vincristine in human plasma after intravenous administration of vincristine sulfate liposome injection using ultra-high performance liquid chromatography tandem mass spectrometry. J Chromatogr A 1275:61–69

Yasui K, Fujioka H, Nakamura Y (1995) Controlled release by Ca(2+)-sensitive recombinant human tumor necrosis factor-alpha liposomes. Chem Pharm Bull (Tokyo) 43:508–511

Zamboni WC, Strychor S, Joseph E, Walsh DR, Zamboni BA, Parise RA, Tonda ME, Yu NY, Engbers C, Eiseman JL (2007) Plasma, tumor, and tissue disposition of STEALTH liposomal CKD-602 (S-CKD602) and nonliposomal CKD-602 in mice bearing A375 human melanoma xenografts. Clin Cancer Res 13:7217–7223

Zamboni WC, Ramalingam S, Friedland DM, Edwards RP, Stoller RG, Strychor S, Maruca L, Zamboni BA, Belani CP, Ramanathan RK (2009a) Phase I and pharmacokinetic study of pegylated liposomal CKD-602 in patients with advanced malignancies. Clin Cancer Res 15:1466–1472

Zamboni WC, Strychor S, Maruca L, Ramalingam S, Zamboni BA, Wu H, Friedland DM, Edwards RP, Stoller RG, Belani CP, Ramanathan RK (2009b) Pharmacokinetic study of pegylated liposomal CKD-602 (S-CKD602) in patients with advanced malignancies. Clin Pharmacol Ther 86:519–526

Zamboni WC, Eiseman JL, Strychor S, Rice PM, Joseph E, Zamboni BA, Donnelly MK, Shurer J, Parise RA, Tonda ME, Yu NY, Basse PH (2011a) Tumor disposition of pegylated liposomal CKD-602 and the reticuloendothelial system in preclinical tumor models. J Liposome Res 21:70–80

Zamboni WC, Maruca LJ, Strychor S, Zamboni BA, Ramalingam S, Edwards RP, Kim J, Bang Y, Lee H, Friedland DM, Stoller RG, Belani CP, Ramanathan RK (2011b) Bidirectional pharmacodynamic interaction between pegylated liposomal CKD-602 (S-CKD602) and monocytes in patients with refractory solid tumors. J Liposome Res 21:158–165

Zhang J, Li X, Huang L (2014) Non-viral nanocarriers for siRNA delivery in breast cancer. J Control Release 190:440–50

Zhao X, Mou D, Wan J, Xu H, Yang X (2009) A novel method for the separation and determination of non-encapsulated pyrene in plasma and its application in pharmacokinetic studies of pyrene-loaded MPEG-PLA based nanoparticles. Nanotechnology 20:125701

Zolnik BS, Stern ST, Kaiser JM, Heakal Y, Clogston JD, Kester M, McNeil SE (2008) Rapid distribution of liposomal short-chain ceramide in vitro and in vivo. Drug Metab Dispos 36:1709–1715

Zou P, Helson L, Maitra A, Stern ST, McNeil SE (2013) Polymeric curcumin nanoparticle pharmacokinetics and metabolism in bile duct cannulated rats. Mol Pharm 10:1977–1987

Żwir-Ferenc A, Biziuk M (2006) Solid phase extraction technique—trends, opportunities and applications. Polish J Environ Stud 15:677–690

Part III
Closely related Complex Drugs

Low Molecular Weight Heparins, Biological Drugs close to Non-Biological Complex Drugs

Isabel Rodrigo, Sofía Caruncho, Concepción Alonso, Antonio Gómez-Outes and Barbara Mulloy

Contents

Introduction 294
Chemistry and Manufacturing 296
  Chemical Structure 296
  Anticoagulant Action and Relationship of Molecular Weight to Activity 298
  Manufacturing Process 302
  Aspects to Consider for the Quality and Safety of LMWHs During Manufacturing 304
Control 305
  The Need for Proper Control in All Steps of LMWHs Production 305
  Quality Regulatory Requirements in the EU and in US 306
  Potency 307
  Measurement of Molecular Weight 307
  Impurities 310
  New Quality Requirements: Species Identification 311
  Challenges 312
Pharmacology 313
  Introduction 313
  Mechanism of Action 313
  Pharmacokinetics 315
  Indications 315
  Posology and Monitoring 318
  Adverse Effects 318
Regulatory Status 319
Prospects, Innovations, Breakthroughs: Innovation and Future of Heparin and LMWH Like Molecules 321
  Synthetic Analogs of Heparin Pentasaccharide 322
  Understanding in Vivo Synthesis to Translate to Chemical Synthesis 322
  Chemo-enzymatic Synthesis of Heparins 323
  Bioengineered Heparin in Mammalian Cells 324
References 325

I. Rodrigo (✉) · S. Caruncho · C. Alonso
Biological Products and Biotechnology Division Medicines for Human Use Department, Spanish Agency for Medicines and Medical Devices (AEMPS), C/Campezo 1, Edificio 8, 28022 Madrid, Spain
e-mail: irodrigo@aemps.es

D. J. A. Crommelin, J. S. B. de Vlieger (eds.), *Non-Biological Complex Drugs*, AAPS Advances in the Pharmaceutical Sciences Series 20, DOI 10.1007/978-3-319-16241-6_9

Abstract Low molecular weight heparins (LMWHs) are drug substances widely used as anticoagulants after parenteral administration. They are biological substances obtained from mammalian tissues and they are closely related to non-biological complex drugs because of their heterogeneity and their complex characterization. LMWHs are highly sulfated glycosaminoglycans obtained by partial chemical or enzymatic depolymerization of unfractionated heparin, which is prepared usually by purification from porcine mucosa. Their heterogeneity and polydispersity is the result of the biosynthetic route and of the manufacturing process. Regulatory requirements and control methods are in place in Europe and in the United States to guarantee the quality and safety of these products. LMWHs interfere with the coagulation cascade mainly by interacting through a specific pentasaccharide sequence with antithrombin (AT), and accelerating the inhibition of factor Xa and, to a lesser extent, thrombin (factor IIa). Attempts to produce chemically or enzymatically synthesised LMWHs have not succeeded to displace the currently used LMWH of natural origin.

Keywords Heparin · Anticoagulant · Low molecular weight heparin · Glycosaminoglycan · Biosimilar · Unfractionated heparin

Abbreviations

ACS	Acute Coronary Syndromes
ACT	Activated Clotting Time
AEMPS	Spanish Agency for Medicines and Medical Devices
aPTT	Activated Partial Thromboplastin Time
ASMF	Active Substance Master File
AT/AT III	Antithrombin/Antithrombin III
BID	Twice Daily
CE	Capillary Electrophoresis
CEP(s)	Certificate(s) of Suitability
CHO	Chinese Hamster Ovary
CMDh	Coordination Group for Mutual Recognition and Decentralised Procedures, human
CS	Chondroitin Sulfate
DS	Dermatan Sulfate
DVT	Deep Venous Thrombosis
EDQM	European Directorate for the Quality of Medicines and Health Care

A. Gómez-Outes
Pharmacology and Clinical Evaluation Division, Medicines for Human Use Department, Spanish Agency for Medicines and Medical Devices (AEMPS), C/Campezo 1, Edificio 8, 28022 Madrid, Spain

B. Mulloy
Institute of Pharmaceutical Science, King's College London, Franklin Wilkins Building, 150 Stamford Street, Waterloo Campus, SE1 9NH London, UK

ELISA	Enzyme-Linked ImmunoSorbent Assay
EMA	European Medicines Agency
EMEA	European Agency for the Evaluation of Medicinal Products
EU	European Union
FDA	Food and Drugs Administration
GAG	Glycosaminoglycan
GMP	Good Manufacturing Practice
h	hours
HIT	Heparin Induced Thrombocytopenia
HPIC	High Performance Ion Chromatography
HPLC	High Performance Liquid Chromatography
HPMCP	Hydroxypropyl methylcellulose phtalate
HPSEC	High Performance Size Exclusion Chromatography
HS3st1	Heparan Sulfate 3-O-sulfotransferase 1
IU	International Units
IUBMB	International Union of Biochemistry and Molecular Biology
IV	Intravenously
LMWH(s)	Low Molecular Weight Heparin(s)
MI	Miocardial Infarction
min	Minutes
Mn	Number average molecular weight
Mw	Weight average molecular weight
NCBD(s)	Non-Biological Complex Drug(s)
NDST	N-deacetylase/N-sulfotransferase
NIBSC	National Institute for Biological Standards and Control
NMR	Nuclear Magnetic Resonance
NSTEACS	Non-ST-Elevation-ACS
OD	Once-Daily
OMCL	Official Medicines Control Laboratory
OSCS	Oversulfated Chondroitin Sulfate
OSTs	2-, 6-, and 3-O-sulfotransferases
PCI	Percutaneous Coronary Intervention
PCR	Polymerase Chain Reaction
PD	Pharmacodynamic
PF4	Platelet factor 4
Ph. Eur.	European Pharmacopoeia
RI	Refractive index
SAX-HPLC	Strong Anion Exchange-High Perfomance Liquid Chromatography
SC	Subcutaneously
STEMI	ST-Elevation-MI
TFPI	Tissue Factor Pathway Inhibitor
THR	Total Hip Replacement
TID	Thrice Daily
TKR	Total Knee Replacement
Tmax	Time to maximum concentration

UFH	Unfractionated Heparin
US	United States
USP	United States Pharmacopeial Convention
USP-NF	United States Pharmacopeia Convention and National Formulary
UV	Ultraviolet
VTE	Venous Thromboembolism
WHO	World Health Organization

Introduction

Low molecular weight heparins (LMWHs) are widely used in therapy as anticoagulants. They are obtained from mammalian tissues, at the present time mainly from porcine mucosa, and they are active substances in biological medicinal products. Then, why is there a chapter about LMWHs in a book devoted to non-biological drugs? Complexity and borderline are the keys.

LMWHs are obtained by depolymerisation of heparin, which is a natural substance present in mammalian tissues. Both unfractionated heparin (UFH) and LMWHs are used as active substances in medicinal products due to their anticoagulant properties. Their main use is the prevention and treatment of thromboembolic disorders. Authorised LMWH products in the EU contain bemiparin sodium, certoparin sodium, dalteparin sodium, enoxaparin sodium, nadroparin calcium, parnaparin sodium, reviparin sodium and tinzaparin sodium, whereas in the US, products containing tinzaparin sodium, dalteparin sodium and enoxaparin sodium are authorized, but only the two last products are marketed (Jongen and de Kaste 2011; Gray et al. 2008).

The term "heparin" was coined by Howell and Holt in 1918 (from the Greek "hepar", liver), on the basis of the cuorin and heparphosphatides with anticoagulant activity described by McLean in 1916, who is recognised as the discoverer of heparin (Gómez-Outes et al 2012).

Heparin and LMWHs are carbohydrates. In nature heparin is synthesised by mast cells, which are abundant in gut mucosa and lungs. Heparin in mast cell granules is covalently linked to a protein core as the proteoglycan serglycin. During the manufacturing process proteolysis is performed to separate the proteinic part and then the carbohydrate backbone is purified. This purified sugar, the heparin, is a highly sulfated glycosaminoglycan (GAG), composed of various repeating disaccharide units. Then, heparin is partially depolymerised to obtain the LMWHs. Chemical heterogeneity and molecular weight polydispersity make up the structural complexity of this group of biological drugs and contribute to the fact that full characterization is not easily performed, as for Non-Biological Complex Drugs (NBCDs).

Why do we talk about borderline? Directive 2001/83/CE gives the definition of biological substance: A biological substance is a substance that is produced by or extracted from a biological source and that needs for its characterisation and the determination of its quality a combination of physico-chemical-biological testing, together with the production process and its control. In addition, Directive 2001/83/CE specifically enumerates a series of products which should be considered biological medicinal products. Although UFH and LMWHs are not included in this short list, they are biological substances because they meet the definition of biological substance given in the Directive (biological origin and "complex" characterisation),.

There are other GAGs in a "grey-zone" between chemical and biological that are employed as active substances in medicinal products (CMDh 2007; CMDh 2012). They are obtained from animal tissues and are complex to characterise mainly due to their heterogeneity (polydispersity and/or chemical heterogeneity). These are substances like hyaluronic acid/sodium hyaluronate (glucosaminoglycan), chondroitin sulfate (CS) sodium (galactosaminoglycan) or danaparoid sodium and sulodexide (mixtures of molecules of the galactosaminoglycan dermatan sulfate and the glucosaminoglycan heparan sulfate).

LMWHs share with NBCDs the complexity of their characterisation. They undergo a multistep and extensive process of purification and chemical restructuring. Therefore, they share regulatory challenges. In addition, specific aspects to consider for LMWHs are the adventitious agents (mainly viral) safety and the necessity to establish a biological (anticoagulant) activity in International Units (WHO 2014).

In this chapter, the chemical structure of LMWHs will be discussed, which is characterised, as mentioned, by heterogeneity and polydispersity. Of special importance is a specific pentasaccharide sequence, which is responsible for the biological anticoagulant activity. The manufacturing process will also be described, starting from the porcine mucosa, which is purified to UFH, which itself is an active substance, but also the intermediate to produce LMWH by partial depolymerization.

After the problems encountered in 2008 of adulteration of heparin products with oversulfated chondroitin sulfate (OSCS), new methods of control have been introduced both by the FDA and in pharmacopeial compendia including the United States Pharmacopeia Convention and National Formulary (USP-NF) and European Pharmacopoeia (Ph. Eur.). Control methods for LMWHs will be described. These include methods to analyse the biological anticoagulant activity and molecular size distribution, as well as methods to determine impurities or even adulteration.

We will describe how heparins and LMWHs interfere with the coagulation cascade by interacting mainly with antithrombin (AT) and accelerate the inhibition of factor Xa and thrombin (factor IIa), as well as other pharmacological aspects. Other aspects will also be discussed, such as pharmacokinetics, clinical uses, and specific adverse reactions such as heparin induced thrombocytopenia (HIT) or its posology.

Then, the regulatory status of LMWHs as biological medicinal products will be discussed, highlighting the different conceptions in Europe and in the US. Within the current regulatory framework, a generic version of LMWH cannot be approved in Europe, but biosimilar products may be considered for authorisation. In the US, however, a different paradigm for marketing approval, allowing generic versions for LMWHs, has been developed.

Finally, mention will be made of synthetic or semisynthetic heparin like substances with anticoagulant activity, like the already authorised synthetic fondaparinux, or the new chemo-enzymatically synthesized substances under development. These approaches are conceived with the ambition to substitute for animal-derived heparins and LMWHs in the future in order to overcome the theoretical viral risk and possible supply problems.

Chemistry and Manufacturing

Chemical Structure

LMWHs are obtained by depolymerisation of UFH. Both, LMWHs and UFH, are highly sulfated, linear and heterogeneous members of the GAG family of carbohydrates. Heparin is biosynthesized as the polysaccharide chains of a proteoglycan in granules of mast cells (Carlsson and Kjellén 2012). It is abundantly found in mucosal tissues such as the lungs and intestines. Heparin is the glycosidic part of the proteoglycan serglycin, obtained after proteolytic digestion, and is comprised of disaccharide repeat units of N-sulfamido/N-acetylglucosamine linked 1→4 to an uronic acid, either α-L-iduronic acid or β-D-glucuronic acid, whereby each dissacharide unit may contain up to 3 sulfates. In the biosynthetic process, an unsulfated polysaccharide is initially formed consisting of a repeating disaccharide shown in Fig. 1a, in which N-acetylglucosamine alternates with glucuronic acid. Post-polymerization sulfotransferase and epimerase enzymes then act to introduce N-sulfamidoglucosamine and iduronic acid. Differential O-sulfonation of each disaccharide, leading to structural heterogeneity, is possible at the 2-O of the uronic acid as well as the 6-O position of N-sulfo/N-acetylglucosamine (Carlsson and Kjellén 2012). The most common disaccharide unit, which is present in more than 70 % of heparin, is a trisulfated disaccharide, composed of a 2-O-sulfated α-L-iduronic acid and 6-O-sulfated, N-sulfated α-D-glucosamine, [IdoA(2S)-GlcNS(6S)] (Jongen and de Kaste 2011) (Fig. 1b). A number of additional minor disaccharides structures are also present, which contribute to its heterogeneity. Most heparin chains have an intermediate level of sulfation, approximately 2.5 sulfo groups/disaccharide, and are formed of long segments of fully sulfated sequences with interspersed undersulfated domains (Liu et al. 2009). Due to its sulfation heparin has the highest negative charge density of any known biological macromolecule, which is critical to its biological activity.

Heparin is a polydisperse substance composed of different molecules with various chain lengths. Unfractionated heparins consist of a polymer mixture with an average molecular weight of 15–19 kDa (Mulloy et al. 2014). After depolymerization of unfractionated heparin, LMWHs present an average molecular weight of less than 8000 Da. Each different LMWH shows a characteristic molecular weight (Table 1). This property is related to their ability to act as anticoagulant agent by interacting with AT, thereby mediating the inhibition of factor Xa and factor IIa (thrombin), as will be explained later.

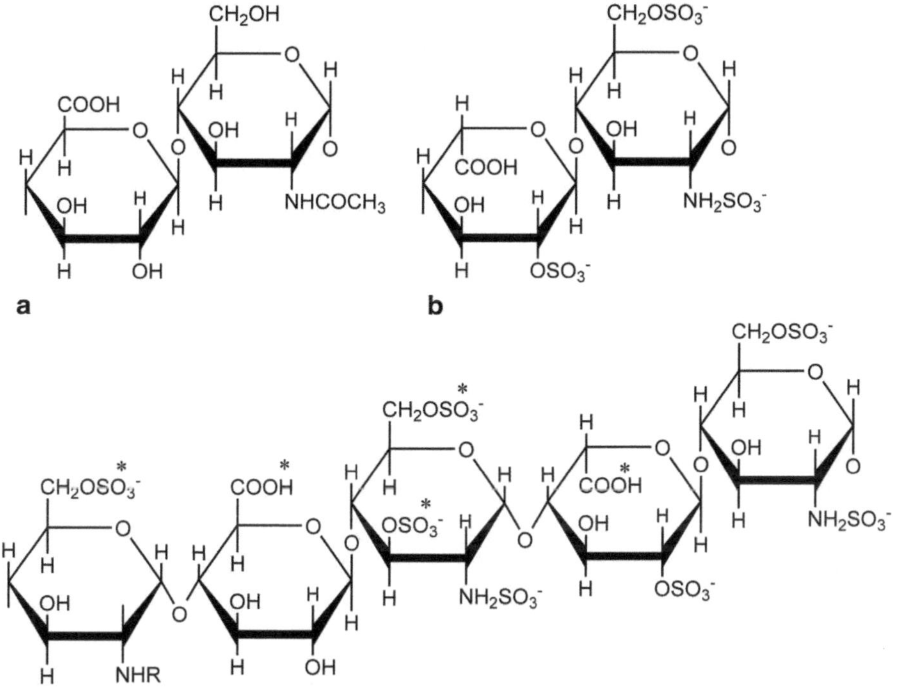

a b c

$R = COCH_3$ or SO_3^-

Fig. 1 Innate complexity of heparin structure as a result of the biosynthetic process: **a** the unsulfated disaccharide that forms the initial polysaccharide chain of heparin, and eventually forms less than 30 % of heparin drug substance. **b** the trisulfated disaccharide that is the predominant structure in heparin, making up more than 70 % of heparin drug substance. Intermediate structures, lacking the 2-O-sulfate or 6-O-sulfate are also found in heparin, as also is 6-O-sulfated, N-acetylated glucosamine. **c** the pentasaccharide structure that forms the minimal core of the antithrombin-binding sequence in heparin. Substituents that are essential for high-affinity antithrombin binding are indicated with an asterisk. Reproduced with permission from Gray et al. 2008

Heterogeneity in heparin and LMWHs is the result of the biosynthetic process of these products, with the possibility of differential O-sulfonation (different grade, different positions), differential substitution possibilities (N-sulfonation or N-acetylation) and different acid sugars (iduronic or glucuronic acid) (Fig. 1a, b), together with polydispersity of the chain length. In addition to this, further complexity is introduced during the manufacturing process of heparin (Fig. 2a, b) and LMWHs (Fig. 2c–e). Depending on the chemical reagents used during the manufacturing process, different chemical structures arise.

Table 1 LMW Heparins. (Adapted with permission from Gray et al. 2008)

LMWH (INN)	Ph. Eur. monograph	Product-specific requirements	Weight average molecular weight (Mw) (with EP limits)	Ratio anti-Xa/anti-IIa activity[a] (IU/mg) (with EP limits)
Enoxaparin sodium	1097	1,6 anhydro structure at reducing end of 15–25% of molecules	4500[b] (3800–5000)	3.9 (3.3–5.3)
Dalteparin sodium	1196	Nitrite, NMT 5 ppm Boron, NMT 1 ppm	6000[b] (5600–6400)	2.5 (1.9–3.2)
Tinzaparin sodium	1271		6500[b] (5500–7500)	1.6 (1.5–2.5)
Parnaparin sodium	1252		5000[b] (4000–6000)	2.3 (1.5–2.5)
Nadroparin calcium	1134	N-NO groups NMT 0.25 ppm Free sulfates NMT 0.5%	4300[b] (3600–5000)	3.3 (2.5–4.0)
Certoparin	none		5400[c]	2.4
Bemiparin	none		3400[d]	9.7
Reviparin	none		4400[c]	4.2

[a] Values measured at NIBSC
[b] Characteristic value of Mw from monograph in European Pharmacopoeia
[c] Value measured at NIBSC (method described in Mulloy et al. 1997)
[d] Value of Mw measured at AEMPS

Anticoagulant Action and Relationship of Molecular Weight to Activity

Structure and activity relationship studies (Mourier and Viskov 2004) have been particularly successful to explain the anticoagulant properties of heparin and LMWHs.

An important feature in the structure of heparin and LMWHs is the existence of a pentasaccharide sequence which specifically binds antithrombin III (AT III), contributing in this way to their overall mechanism of action as anticoagulant drugs (Choay et al. 1983; Lindahl et al. 1984). The sequence is GlcNAc/NS (6S)-GlcA-GlcNS(3S,6S)-IdoA(2S)-GlcNS(6S) (Thunberg et al. 1982) (Fig. 1c). This sequence, containing the unusual 3,6 di-O-sulfated, 2-N-sulfated glucosamine residue, does not occur in every heparin molecule (Gray et al. 2008).

Following the identification of the specific sequence in heparin that confers high affinity for antithrombin, shown in Fig. 1c (Choay et al. 1983; Lindahl et al. 1984), it became clear that, while potentiation of antithrombin-mediated inhibition of factor Xa requires only the interaction between heparin and antithrombin, potentiation

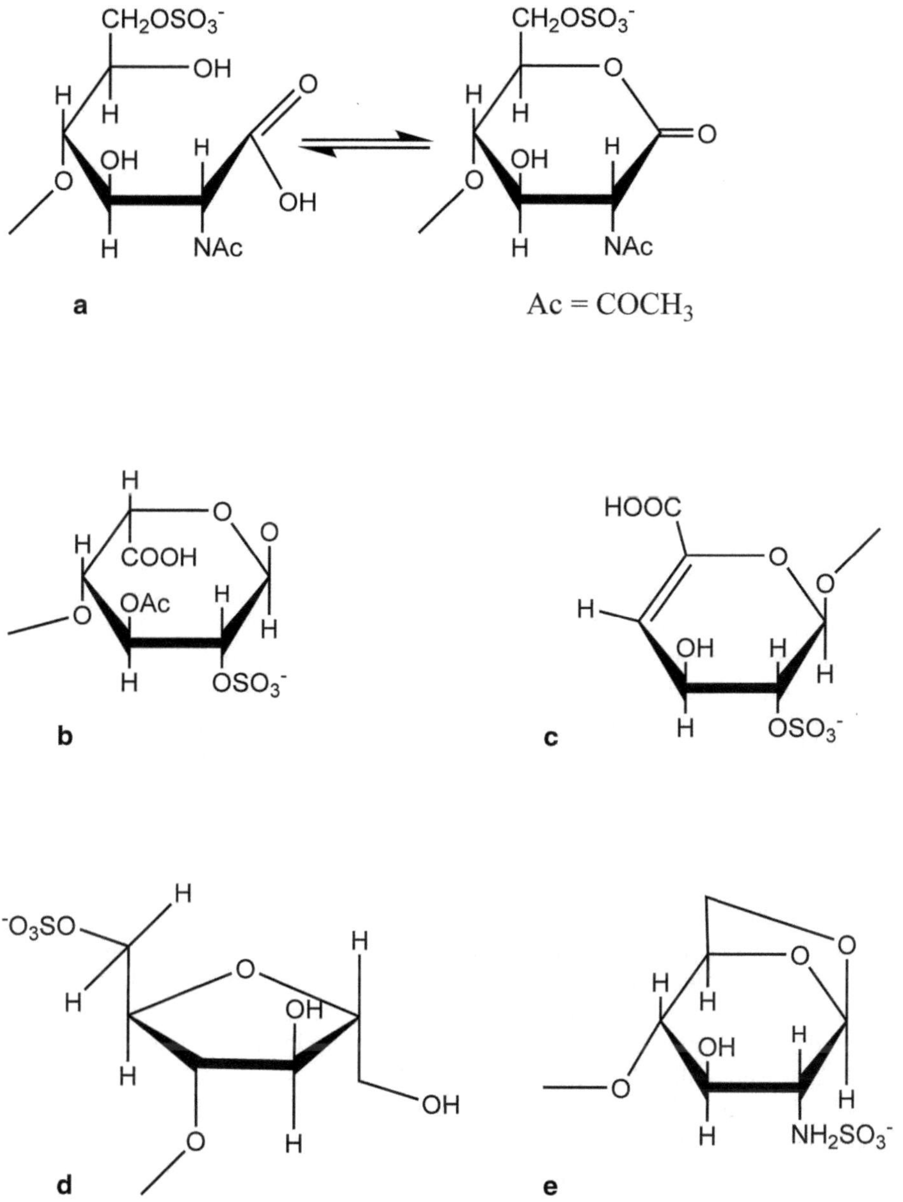

Fig. 2 Elements of complexity introduced by the manufacturing process of heparin and LMWHs: **a** formation of a 2-N-acetyl-2-deoxy-glucono-1,5-lactone structure as a result of oxidation of the anomeric hydroxyl of N-acetyl glucosamine at the reducing terminus of heparin (Kellenbach et al. 2011) (Mourier et al. 2011). **b** 3-O-acetylated iduronic acid resulting from a minor side-reaction of the manufacturing process (Mourier et al. 2012). **c** the unsaturated uronic acid found at the non-reducing terminus of LMWH products depolymerized by a beta-elimination mechanism, either chemically or enzymatically. **d** the 6-sulfated 2,5 anhydromannitol structure at the reducing terminus of nitrous acid depolymerized LMWH products. **e** the 1,6 anhydro, N-sulfated glucosamine at the reducing terminus of a proportion of molecules in enoxaparin (Guerrini et al. 2010)

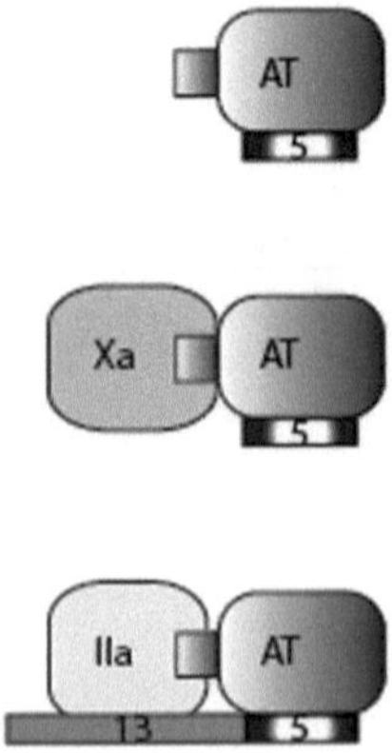

Fig. 3 Molecular weight and antithrombin-mediated inhibition of factor Xa and factor IIa. From *top* to *bottom*: Antithrombin (AT) binds to the high-affinity heparin pentasaccharide (5) and is activated. Activated antithrombin binds to and inhibits factor Xa. Activated antithrombin and an extended heparin molecule (5 + 13) both interact with factor IIa for inhibition

of thrombin (factor IIa) inhibition requires the formation of a ternary complex of antithrombin, thrombin and heparin (Fig. 3; Petitou et al. 1999). The antithrombin-heparin complex necessary to inhibit factor Xa may be as short as the five monosaccharide units constituting the minimum antithrombin-binding motif, but the ternary complex requires, in addition, an extension of about thirteen further monosaccharide residues, making up a heparin molecule with a molecular weight of roughly 5400 Da; this motif has been termed the 'C domain' (Al Dieri et al. 2003). As LMWH products contain a substantial amount of material with a molecular weight lower than 5000 Da, they have higher anti-Xa than anti-IIa activity; the ratio of these two activities in IU must be not less than 1.5 to meet the specification of the EP (Anon 2014a).

For LMWH preparations developed in later years, the molecular weight distribution is a defining characteristic of each individual product. Table 1, adapted from Gray et al (Gray et al. 2008) lists the weight average molecular weight (Mw), method of depolymerisation, and anti-Xa/anti-IIa activity ratio. Figure 4a shows the molecular weight distributions of four LMWH products.

The differences between the molecular weight profiles of LMWH products affect not only their anticoagulant activity, but also their capacity to be neutralised by protamine. For example, tinzaparin has a higher proportion of material in the higher molecular weight range than enoxaparin (Fig. 4). In addition, the enzyme that is used to manufacture tinzaparin can cleave the antithrombin-binding sequence, so that the smallest molecules in tinzaparin are depleted in anti-Xa activity (Shriver et al. 2000). Protamine has been found to incompletely neutralise the anti-Xa activity of heparin molecules below an octasaccharide in size, even when present in considerable excess (Schroeder et al. 2011). Therefore, as the anti-Xa activity of enoxaparin has a much higher proportion of its anti-Xa activity expressed by very small molecules, it cannot so readily be neutralised by protamine (Schroeder et al. 2011).

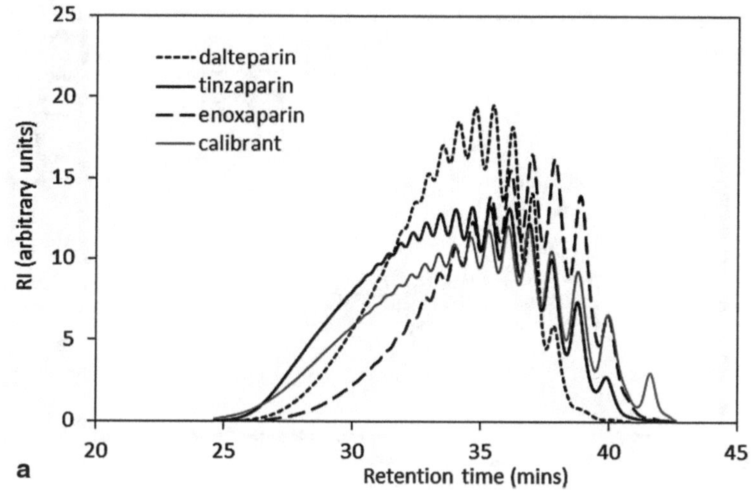

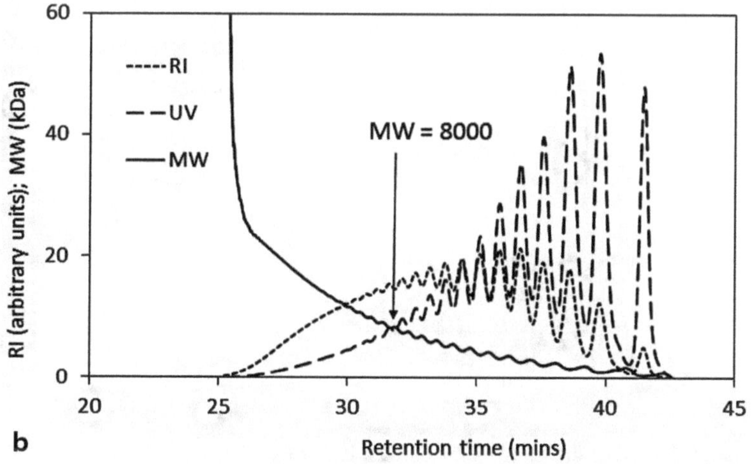

Fig. 4 **a** Molecular weight chromatograms for three LMWH products and a chromatography calibrant, the 2nd International Standard LMWH for molecular weight Calibration. HPSEC conditions are as described in Mulloy et al. 1997. **b** Overlaid chromatographic traces with detection by RI (*short dashes*) and UV at 234 nm (*long dashes*) of a calibrant material for HPSEC of LMWH. The RI trace is proportional to the concentration by weight, and the UV trace is proportional to the molar concentration of sample in the detector. The ratio of RI/UV is therefore proportional to the molecular weight of sample in the detector. From these calculations, a mathematical function (usually a third-degree polynomial) can be fitted to describe the relationship of molecular weight to retention time for the chromatographic system. A plot of MW (*solid line*) corresponding to retention times shown is superimposed, with the MW = 8 kDa point marked. The Ph. Eur. criterion for all LMWH products is that the proportion of material with MW < 8 kDa must not be less than 60 %. Note that the calculated molecular weight for short retention times rises very fast: extrapolation of the calibration curve outside the directly calibrated range is likely to give inaccurate and imprecise results

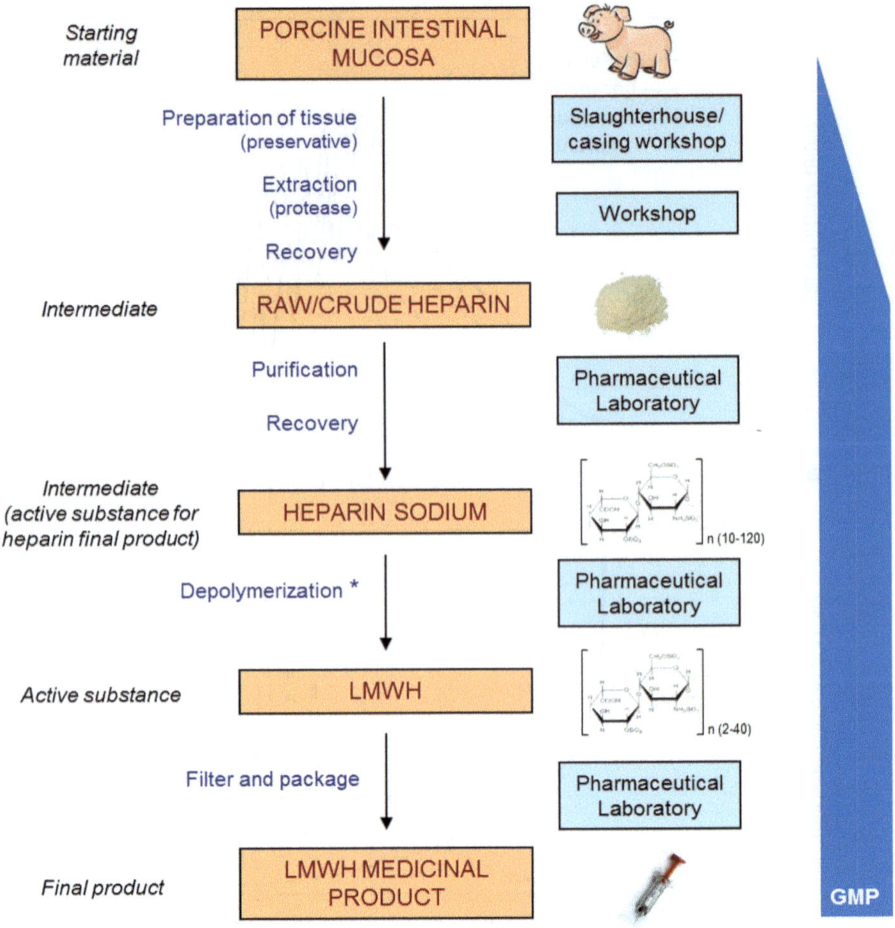

Fig. 5 General scheme of the LMWHs manufacturing process. See text for details. (* Different depolymerization methods originate different LMWHs)

Manufacturing Process

The starting materials of LMWHs are animal tissues, usually bovine lungs or bovine, porcine or ovine intestinal mucosa. Presently, in the EU and the US, marketed LMWHs are only obtained from porcine intestinal mucosa. The manufacturing process of LMWH involves two phases: extraction and purification of the intermediate, unfractionated heparin sodium (or heparin calcium) and depolymerisation or fragmentation of the UFH to obtain the LMWH (Fig. 5).

Production of Unfractionated Heparin

The production of UFH involves two main phases: the extraction and isolation of crude heparin from porcine intestinal mucosa to obtain crude heparin and further purification of the heparin. The process is illustrated in Fig. 5.

First of all, porcine intestinal mucosa is collected from healthy pigs in slaughterhouses. This part of the manufacturing process is not always performed under 'good manufacturing practice' (GMP) conditions. Special care should be taken to avoid contamination, especially contamination with material from other animal species (see control section below) which could coexist in the abattoirs. Typically, the next steps to produce what is frequently called raw heparin or crude heparin are performed in the slaughterhouse or in workshops near slaughterhouses (Liu et al. 2009; Bhaskar et al. 2012). The porcine mucosa is then treated with sodium metabisulfite or sodium chloride to preserve the tissues. Next, proteases are added to solubilise and separate the proteinic part of the heparin molecules, which is usually followed by heat deactivation of the protease. Then, heparin is recovered either by precipitation with a quaternary ammonium salt or by a chromatographic step using an anion exchange resin. After resolubilizing, the heparin can be precipitated again with ethanol or methanol (Bhaskar et al. 2012). In this way the crude heparin is produced, and can be shipped to other facilities for further purification.

The next steps in heparin manufacturing should be performed under GMP conditions. They are mainly devoted to further purification of the crude heparin. The impurities to remove may have been introduced during crude heparin extraction or may come from the starting material. They are other GAGs, such as dermatan sulfate (DS) or CS, residual protein or nucleotides, extraneous cations, heavy metals, solvents or salts other than heparin, as well as possible microbiological residues (Liu et al. 2009). Different procedures may be used to purify the heparin. Usually the crude heparin is dissolved in water and filtered at low pH to remove residual proteins, followed by alkaline bleaching with peroxide, permanganate, ozone or hypochlorite, for product sanitization and depyrogenation. Cation exchange can then be performed to remove extraneous cations, and ethanol precipitation can be used to reduce residual nucleic acids. Remaining salts can be removed by membrane filtration (Liu et al. 2009). In this way, and after drying and proper analyses, unfractionated heparin would be ready to use as active substance or to be used as intermediate in the manufacturing of LMWHs.

Depolymerisation of Heparin to Produce LMWHs

After obtaining the unfractionated heparin, LMWHs are produced by chemical or enzymatic partial depolymerization. This part of the process is performed in a pharmaceutical laboratory under GMP conditions.

Depending on the depolymerization method used, different LMWHs are obtained (Table 1) with different chemical and biological characteristics (see Fig. 2). The most common method used is deaminative cleavage with nitrous acid or an organic nitrite. The mechanism of the method is to break the heparin chain at N-sulfated glucosamine residues. This action leaves a characteristic anhydromannose reducing-end residue which is usually reduced to anhydromannitol (Fig. 2d, Gray et al. 2008). This process is used in the manufacturing of dalteparin, nadroparin, reviparin and certoparin.

The second most common depolymerisation method is the chemical or enzymatic beta-elimination. This leaves an unsaturated uronic acid residue at the non-reducing end (Fig. 2c) (Gray et al. 2008). The chemical beta-elimination is performed with alkali on the heparin benzyl ester in the case of enoxaparin or with alkali on the quaternary ammonium salt in the case of bemiparin. Enzymatic cleavage (also involving a beta-elimination mechanism) is used to produce tinzaparin. In this case, heparinase from the bacterium Pedobacter heparinus (Flavobacterium heparinum) is used. The enzyme belongs to the family of lyases, 4.2.2.7 in the IUBMB nomenclature (NC-IUBMB 2014).

A third method used by the pharmaceutical industry to obtain LMWH is the oxidative depolymerisation which does not leave a specific fingerprint structure in the product (Gray et al. 2008). This method is carried out in the manufacturing of parnaparin, in which hydrogen peroxide and a cupric salt are used as oxidising reagents.

In these ways each specific LMWH active substance is produced. Manufacturing of the final product is then a relatively simple process, which normally involves dissolving of the active substance, pH adjustment, sterilizing filtration and aseptic filling in vials or syringes. Sometimes preservatives (benzyl alcohol) or stabilizers (sodium metabisulfite) are used, but the most frequently used finished products containing LMWH are preservative-free, single-dose LMWH solutions in prefilled syringes.

Aspects to Consider for the Quality and Safety of LMWHs During Manufacturing

Due to the biological origin of LMWHs some specific aspects should be considered during their manufacturing. First of all, the starting material should be properly controlled. The source material must be derived from animals fit for human consumption following ante- and post-mortem veterinary inspections. Information on the species and country of origin should be included in the dossier for the authorization of the medicinal product and traceability from the final medicinal product up to slaughterhouse level should be guaranteed. These aspects are important in order to be able to assess the potential risk of transmission of adventitious agents as well as to have the possibility to trace back in the case a problem arises. Manufacturers should also take steps to ensure that the heparin supply chain is not contaminated with any material of other animal species.

The risk of adventitious agents is always a concern for animal derived products. The putative risk of prion transmission has been overcome in EU and US by marketing products of porcine origin only. Concerning viral transmission risk, as indicated in Ph. Eur. monograph 5.1.7 (Viral Safety), a risk assessment should be carried out. Complementary measures could be used to assure viral safety: selection of source materials and testing for viral contamination, testing the capability of the production process to remove and/or inactivate viruses and testing for viral contamination at appropriate stages of production. In the case of heparins, many of the manufacturing steps that are used to purify the product or to depolymerise the heparin contribute to the safety of the product, because they are very aggressive processes which can inactivate or remove viruses. Virus validation studies should be performed, according to CPMP/BWP/268/95 (Virus Validation Studies, EMEA 1996), to demonstrate that there are steps in the manufacturing process able to inactivate or remove viruses. Although other problems, related with bad practices, have occured with heparins, until now there are no known cases of viral or prion transmission due to heparins or LMWHs.

Control

The Need for Proper Control in All Steps of LMWHs Production

In early 2008 there was an international health crisis related with contamination of several heparin batches. A marked increase was reported in serious adverse events associated with heparin therapy, affecting thousands of patients. These events included development of symptoms such as rash, fainting, racing heart, extremely low blood pressure and serious allergic reactions (Liu et al. 2009). This resulted in nearly 100 deaths alone in the US. It also affected the EU and Asia, although no fatal events were reported in Europe (EMA 2008).

These adverse side-effects resulted in the withdrawal of a number of heparin batches from the US and European markets, followed by an investigation of the presence of contaminants in suspected batches. A number of laboratories participated in the investigation, including laboratories from the pharmaceutical industry and FDA and the European Official Medicines Control Laboratories (OMCLs). As a result of this research, and the rapid response of the FDA, the presence of oversulfated chondroitin sulfate (OSCS) was found as the cause of the adverse effects. Work by all the other organisations confirmed this initial finding.

OSCS is a semisynthetic polysaccharide which is obtained by the chemical sulfonation of natural chondroitin sulfate, a much cheaper substance than heparin. OSCS has a molecular weight of 18 kDa, comparable to heparin, a slightly higher charge density and also shows anticoagulant activity. The structure of the contaminant was confirmed and reported (Guerrini et al. 2008). Not only unfractionated heparin, but also LMWH was found to be contaminated (EMA 2008; Zhang et al. 2008; Viskov et al. 2009a). The acute response caused by OSCS was associated with an anaphylactoid reaction generated by the formation of bradykinin through activation of the kinin-kallikrein pathway in human plasma and by complement activation. Bradykinin caused strong vasodilation and a sometimes extreme hypotension (Kishimoto et al. 2008). The contamination had been traced to heparin originated from Chinese manufacturers (Laurencin and Nair 2008; Jongen and de Kaste 2011; Mintz and Liu 2013).

OSCS-contaminated heparins were first distinguished from control heparins by optical rotation, capillary electrophoresis (CE), and ^{1}H-NMR (Liu et al. 2009). Nuclear magnetic resonance (NMR) represents the most reliable method to detect OSCS and other contaminants and impurities. Both OSCS and DS can be quantified using ^{1}H-NMR (Beyer et al. 2008). The methyl group signals of heparin, DS and OSCS resonate at certain specific chemical shifts and are easily quantified based on integration of the NMR signals. As little as 0.1 % DS impurity and OSCS contaminant is detectable by ^{1}H-NMR.

New methods were introduced to control the absence of the contaminant OSCS. It should be noted, that even highly contaminated batches of heparin would have passed all the controls required by the pharmacopoeias at that time. Thus, in March 2008 the FDA published the CE and NMR detection methods for OSCS and in August 2008 the Ph. Eur. introduced in the Production section of the Heparin calcium and Heparin sodium monographs the need to determine the presence of this contaminant by NMR spectroscopy and CE. These safety measures applied to the UFH also protect LMWHs, as all LMWHs must be prepared from UFH that meets the new criteria. In a second phase in 2010 (implementation date January 2011) these tests for Heparin sodium/calcium were replaced in the Ph. Eur. by new requirements in the Identification and Tests sections using ^{1}H-NMR (McEwen et al. 2008) and strong anion-exchange liquid chromatography (SAX-HPLC).

Quality Regulatory Requirements in the EU and in US

Heparins and LMWHs in Europe and in the US should comply with the quality regulatory requirements established in the Ph. Eur. or in the USP-NF.

The Ph. Eur. contains a general LMWH monograph (0828, "Heparins, low-molecular-mass") (Anon 2014a), outlining those properties that all LMWH products have in common, with methods for the determination of potency, molecular weight distribution and others. Five separate monographs (1097, "Enoxaparin sodium"; 1196, "Dalteparin sodium"; 1134, "Nadroparin calcium"; 1252, "Parnaparin sodium"; 1271, "Tinzaparin sodium") deal with individual LMWH products (see Table 1), containing acceptance criteria and any product-specific methods. Specifically, the molecular weight and potency ranges are defined for each particular LMWH. In addition, the LMWH should be obtained from unfractionated heparin that complies with the monograph on Heparin sodium (0333) (Anon 2014b) or Heparin calcium (0332) (Anon 2014c). The method to assay the potency of unfractionated heparin is described in the general chapter 2.7.5.

In the USP-NF, there is no general LMWH chapter, but general chapter <208> describes anti-Xa and anti-IIa potency assays for both unfractionated and LMWH (Anon 2014d), and <209> (Anon 2014e) describes a method for molecular weight determinations for LMWH products. There are also two individual product monographs, for enoxaparin and dalteparin. The monograph on heparin sodium deals with the quality of the unfractionated heparin from which the LMWH should derive.

Potency

Establishment of potency, the quantitative measure of biological activity, is of crucial importance for any biological product. For LMWHs it consists of the measurement of the anticoagulant activity. The preferred method to analyse the anticoagulant activity of heparin and LMWHs in vitro is a chromogenic assay, which is more specific for antithrombin-mediated anti-Xa and anti-IIa anticoagulant activities than plasma-based coagulation based assays. It is important to consider that the establishment of potency requires the measurement of activity relative to a relevant preparation or standard traceable to the current WHO International Standard.

The prescribed method to assess the potency of LMWH in the Ph. Eur. and USP-NF is a chromogenic assay to measure anti-factor Xa and anti-factor IIa activities. This is an in vitro method based on the ability of the LMWH to accelerate the inhibition of the coagulation factor (Xa or thrombin) by ATIII. Commercial reagents are used (ATIII, bovine factor Xa or human thrombin and the chromogenic substrates) and a reference preparation is needed. The absorbance of sample and reference preparations is measured and analysed by the parallel-line statistical method (Ph. Eur. and USP-NF) or the slope ratio assay (USP-NF). The established limits in the general LMWH Ph. Eur. monograph are that the anti-Xa activity should not be less than 70 IU/mg and that the ratio of anti-Xa/anti-IIa should not be less than 1.5. More restricted limits may be indicated in each specific LMWH monograph.

In the case of the UFH used to produce the LMWHs, the USP-NF recommended method is also the chromogenic assay, based on the same principle as for LMWH. Ph. Eur. will change from the clotting assay to the chromogenic assay on January 2015 (Ph. Eur. Supplement 8.3). Anti-Xa and anti-IIa activity are measured and potency is established based on anti-IIa activity. The minimum potency established is 180 IU/mg (Ph. Eur.)/180 USP Heparin Units/mg (USP-NF). In addition, the ratio of anti-factor Xa activity to anti-factor IIa activity should be between 0.9–1.1.

This 180 IU/mg minimum potency limit, raised in 2010 in the Ph. Eur., is an appropriate value for heparin obtained from porcine mucosa. However, it is difficult to reach for heparin of other origins. In this sense, Ph. Eur. has restricted the source of the heparin sodium and heparin calcium in its monographs from the lungs of cattle or from the intestinal mucosae of pigs, cattle or sheep, to heparin of porcine origin only (Ph. Eur. Supplement 8.3) per 1/1/15, to reflect the fact that all heparin and LMWH medicinal products on the European market are of porcine origin.

Measurement of Molecular Weight

The polydispersity and heterogeneity of heparin makes it effectively impossible to determine its molecular weight with complete accuracy. Mass spectrometry can be used to characterize oligosaccharides separated by HPLC, but even current state of the art instruments don't have sufficient resolution to separate every molecular species in a sample of LMWH. In addition, there are technical difficulties with mass spectrometric analysis of heparin such as low and variable ionization efficiency and facile sulfate loss (Li et al. 2012). The molecular weight distributions of polydisperse compounds such as heparin are usually described using average molecular weights such as the number average molecular weight Mn and the weight average molecular weight Mw. The ratio between these two quantities, Mw/Mn is referred to as the polydispersity (Scheme 1).

Scheme 1: Average molecular weights for a polydisperse polymer such as unfractionated or LMWH

The number average molecular weight, $\overline{M_n}$, is defined:

$$\bar{M}_n = \frac{\sum_i N_i M_i}{\sum_i N_i}$$

where N_i is the number of molecules at molecular weight M_i

The weight average molecular weight, $\overline{M_w}$, is defined:

$$\bar{M}_w = \frac{\sum_i g_i M_i}{\sum_i g_i}$$

where g_i is the weight of the sample at molecular weight M_i

This may be rewritten as:

$$\bar{M}_n = \frac{\sum_i N_i M_i^2}{\sum_i N_i M_i}$$

The number average and weight average of a polydisperse sample are different, and the ratio between them, the polydispersity PD, is a measure of the breadth of the molecular weight distribution.

$$PD = \frac{\bar{M}w}{\bar{M}n}$$

PD has a value of 1 for a monodisperse sample, for which $\overline{M_n} = \overline{M_w}$.

For polydisperse polymers PD will be greater than 1.

Methods for molecular weight analysis of LMWHs have used a number of techniques, including electrophoresis (Buzzega et al. 2008; Edens et al. 1992) and most commonly, high performance size exclusion chromatography (HPSEC). HPSEC of heparin using high resolution silica columns was rapidly developed in the 1980s, and the importance of using heparin itself to prepare molecular weight markers was recognised (Harenberg et al. 1983). Although individual manufacturers have developed their own molecular weight determination methods for LMWH, the importance of a compendial method that was robust to inter-laboratory variation was recognised.

Chromatographic methods for the determination of heparin molecular weights are not complex, and require only isocratic chromatography on a suitable column. The method of calibration has been a more interesting challenge. Some groups have used light scattering detection techniques that do not require a calibrant, in spite of the technical challenges involved with such relatively low molecular weight molecules (Komatsu et al. 1993). Low-angle laser light scattering (Komatsu et al. 1993), multiangle laser light scattering (Knobloch and Shaklee 1997) and the triple detector system of right-angle light scattering, refractive index and viscometry (Bertini et al. 2005) have all been successfully applied. Laser light scattering techniques have also been used to characterize heparin molecular weight markers for use as chromatography calibrants (Kristensen et al. 1991; Mulloy et al. 2000).

Another method of molecular weight characterization using a calibrant is possible if the heparin calibrant material is prepared by beta-eliminative depolymerisation of an unfractionated heparin sample. LMWHs of this type have, in theory, an unsaturated uronic acid residue at the non-reducing end of each molecule (the enzyme used, heparinase I, acts both exolytically (Ernst et al. 1998) and endolytically (Jandik et al. 1994), so that the original non-reducing ends are mostly reduced to disaccharides and discarded). In practice, only a proportion of the molecules in the sample will contain this chromophore, and the assumption needs to be made that this proportion is the same at all molecular weights (Nielsen 1992). The unsaturated uronic acid absorbs UV light at around 234 nm, so absorbance at this wavelength may be assumed to be proportional to molar concentration; refractive index (RI) increment is proportional to the concentration by mass of the sample (Tumolo et al. 2004). The ratio between UV and RI is therefore proportional to its molecular weight. This property of heparinase digested LMWH was used by Nielsen (Nielsen 1992) to propose a convenient calibration method for HPSEC of LMWH, as illustrated in Fig. 4b. Suitable calibrants were developed by both the WHO and the Ph. Eur., and the Ph. Eur. adopted a version of the Nielsen method for the LMWH monograph (van Dedem and Nielsen 1991). On the basis of results from the Nielsen method, a Broad Standard calibration table was developed for use with the WHO calibrant for those circumstances where only RI detection, rather than simultaneous acquisition of RI and UV detected chromatograms, is possible (Mulloy et al. 1997).

Once the first batch of each of these calibrants was exhausted, replacement for the EP standard proved difficult, as the batch of material specially produced for the purpose was not ideally suited to characterization by the Nielsen method, and some methodological difficulties were encountered by participants in the study (Mulloy et al. 2007).

The USP has also developed a general chapter <209> for molecular weight determinations for LMWH, with calibration on the basis of a Broad Standard table. However, the USP-NF monograph for enoxaparin includes a specific molecular weight method for this product, using HPSEC with a specific set of narrow heparin standards. The USP-NF is also the first pharmacopoeia to introduce a molecular weight method for unfractionated heparin and new acceptance criteria into its heparin sodium monograph (Mulloy et al. 2014).

Impurities

The analysis of impurities is of crucial interest for any substance for pharmaceutical use. In the case of substances of animal origin, in addition to the process related impurities, special efforts are made in controlling impurities which might be in the starting material. Therefore, control of impurities of LMWHs is prescribed both in LMWH and unfractionated heparin monographs. As for other active substances and medicinal products, impurities such as residual solvents, heavy metals and others, should be limited. We will mention here some of the impurities that are specific for these types of products.

Because LMWHs are derived from mammalian tissues, impurities such as nucleic acids and proteins have to be controlled. This is done by testing the UFH from which they are derived. For nucleotidic impurities, methods based on different principles are prescribed in USP-NF and Ph. Eur. monographs. The Ph. Eur. method is based on measurement of the absorbance in solution, whereas the USP-NF is a more sophisticated method based on quantitative chromatography of the free nucleosides obtained after enzymatic digestion with benzonase, phosphodiesterase I and alkaline phosphatase.

Due to the above mentioned contamination problem with OSCS in 2008, new methods as ^{1}H-NMR and SAX-HPLC were included to be able to detect this and other contaminants.

SAX-HPLC allows the differentiation of natural contaminants linked to the production process, such as DS and CS, from chemically synthesized contaminants, such as OSCS. In the Ph. Eur. monograph of UFH a limit of 2 % has been set for the sum of DS and CS, which co-elute in this method. No peaks corresponding with OSCS should be detected in any case.

^{1}H-NMR is a high resolution technique introduced in the heparin monographs to be able to detect the presence of OSCS in case of intentional contamination. ^{1}H-NMR can be used to identify the specific signals of heparin as well as to detect the presence of other contaminants, such as DS and OSCS. No signals corresponding with OSCS should be detected in any case.

In the case of the USP-NF monograph, there are also limits for galactosamine in total hexosamine, which is a measure of other possible natural impurities containing galactosamine, such as DS or CS. The sample is hydrolysed to hexosamines, which are then separated by High Performance Ion Chromatography (HPIC) coupled to a pulsed amperometric detector. The limit for galactosamine in total hexosamine is 1 %.

New Quality Requirements: Species Identification

Although in some countries it is possible to produce heparin from other mammals than pig, the medicinal products marketed in the EU and in US are all obtained from porcine mucosa. Both Ph. Eur. and USP-NF unfractionated heparin monographs indicate the need to state in the label the animal species from which it is derived. However, no special requirements were established to confirm the species of origin. With the deliberate contamination episode with OSCS the regulatory authorities thought of the possibility of other putative contamination risks, including the risk of adulteration of porcine heparin with heparin from other species.

In this sense, Ph. Eur. has reacted by modifying the UFH monographs to include the need for a reliable quality management system throughout production and the need to verify the source species and the absence of material from the other species (cattle or sheep) (Anon 2014b, c). From January 2015, when these monographs will apply only to porcine heparin, the method chosen, which can be polymerase chain reaction (PCR), should be able to detect the presence of heparin of other species at 0.1 % (m/m heparin) (Ph. Eur. Supplement 8.3).

On the other hand, the FDA has elaborated a document, "Guidance for Industry, Heparin for Drug and Medical Device Use: Monitoring Crude Heparin for Quality" (FDA 2013) to alert the stakeholders participating in the whole manufacturing process of heparin of the potential risk of crude heparin contamination with OSCS or with non-porcine ruminant material. The FDA recommends that, in addition to the USP-NF monograph tests for heparin sodium to detect OSCS, other testing to detect contamination of crude heparin should be performed, including the identification of the animal origin of heparin. Thus, FDA recommends that drug manufacturers receiving crude heparin perform on each lot before use a test to confirm the species origin as well as a test for OSCS. The test to confirm the species origin should be able to detect ruminant material. The FDA has posted a real time PCR method ("heparin Multiplex Real-Time Assay"), with a sensitivity of 0.5 % (w/w) ruminant (bovine, ovine, caprine) material in porcine crude heparin to verify the porcine origin of the crude heparin (Peters 2014), but alternative methods can be used instead. The document gives other recommendations about the quality systems and the need of the manufacturers to audit and qualify their crude heparin suppliers.

Other PCR based methods have been published and are being used to test the marketed products (Concannon et al. 2011; Auguste et al. 2012; Huang et al. 2012). In all cases the PCR testing should be performed prior to any treatment that could eliminate DNA in the sample and prior to DNA purification which is required during the manufacturing process.

There are other methods able to discriminate between the species of origin, like immunological ones, such as ELISA or radial immunodiffusion assay (Levieux et al. 2002; Rivera et al. 2002), or methods based on the analysis of disaccharides (Houiste et al. 2009). They have the advantage of offering the possibility of performing the analysis at later stages of the manufacturing process. However, their sensitivity might not be enough for the intended purpose. Therefore, at the present moment, quantitative PCR is the method of choice.

Challenges

Some challenges remain for the future in the analysis and control of LMWHs. One of them is to satisfy the need for international reference substances which are essential to standardize results in most of the mentioned techniques to control LMWHs.

Without doubt, one of the main challenges in the medicinal product area is the difficulty to detect and to fight against counterfeiting and adulteration. This has been evidenced for heparins with OSCS contamination and it is a concern at the moment, not only for the regulatory authorities, but also for the industry, which is not always able to control their raw material suppliers. The more controls are imposed, the more difficult it will be for deceivers to remain undetected. However, unless the test is made at the final stage of the manufacturing process, it will be very hard to find foolproof methods to detect intended contamination.

Finally, another important challenge for the very near future is to define when an active substance is considered to be "similar" or "the same" to another existing innovator, to be considered for approval as biosimilar (EU) or generic (US). This is a common issue for other non-biological complex drugs (NBCD) unlike conventional small molecules drugs. To demonstrate similarity/sameness, an extensive analytical characterization program applying state of the art techniques will be necessary, in addition to the tests defined in the respective pharmacopoeial monographs,

The FDA has developed an approach to determine the sameness of generic and innovator LMWH products when evaluating the dossier of a generic version of enoxaparin (Lee et al. 2013). Several criteria are used to evaluate the chemical and biological characteristics of the active substances, in addition to assays to determine that there is no increase in the immunogenicity risk. The approach consists of the demonstration of equivalence based on what the FDA calls the 'integrated analytical technologies', which means demonstration of equivalence based on an array of appropriate analytical methods. These criteria include the demonstration of equivalence of physicochemical properties, equivalence of heparin source material (porcine intestinal mucosa) and way of depolymerization, equivalence in disaccharide building blocks, fragment mapping and sequence of oligosaccharide species, equivalence in biological and biochemical assays, and equivalence of the in vivo pharmacodynamic profile. Therefore, in addition to the monographs tests, other sensitive analytical techniques are required to compare innovator and copy product, including chain mapping methods, analysis of oligosaccharides or disaccharides after partial or total digestion, determination of sequence of oligosaccharides, and/or NMR analysis.

Independently of the different status that copies of the innovator product would have in the US or in the EU, as generic or as biosimilar, respectively, and of the necessity or not of clinical comparative studies, a quality comparability exercise is essential, based on comprehensive characterization using state of the art techniques. Also, to better understand the mechanism by which heparin and LMWHs mediate their function, analytical methods providing molecular level structural characterization are critical to assure their quality and safety (Jones et al 2011; Mulloy 2012). With the advance of novel technologies and the increase in the sensitivity of existing techniques this is expected to be more feasible in the future.

Pharmacology

Introduction

A number of key studies performed during the late sixties and the seventies resulted in the identification and later development of LMWHs. The collection of low-molecular-weight heparin fractions, which was possible due to the development of the heparin-sepharose column (Iverius 1971), allowed understanding the mechanism of action of heparin. In 1973, Rosenberg and Damus had shown that UFH binds to antithrombin and induces a conformational change that converts antithrombin into a much more efficient inhibitor of factor Xa and thrombin (factor IIa) of the coagulation cascade (Rosenberg and Damus 1973). In 1976, Johnson and colleagues reported that low-molecular-weight fractions prepared from UFH had progressively less effect on the activated partial thromboplastin time (aPTT), reflecting a less marked inhibition of thrombin, as they were reduced in molecular size, while still inhibiting activated factor Xa (Johnson et al. 1976). In addition, for an equivalent antithrombotic effect, low-molecular-weight fractions produced less bleeding in experimental models than UFH (Carter et al. 1982; Bergqvist et al. 1985).

These studies showed the antithrombotic potential and higher selectivity of low-molecular-heparin fractions on factor Xa inhibition as compared to UFH, which resulted in several LMWHs manufactured following different patented methods of depolymerisation (Table 1) that conferred to each LMWH distinctive biochemical, pharmacokinetic and pharmacodynamic profiles. After clinical development during the 1980s, several LMWHs reached the market. Dalteparin (Kabi 2165; Fragmin®) was first launched in Germany for anticoagulation during haemodialysis and haemofiltration in 1985 and is currently launched in more than 80 countries. Nadroparin (CY 216; Fraxiparine®), was launched in France in 1986 and is currently marketed in over 100 countries. Enoxaparin (PK 10169; Lovenox®/Clexane®) was marketed in Europe in 1987 and in the US and Canada in 1993, being the best-selling LMWH worldwide. Other LMWHs include bemiparin, certoparin, parnaparin, reviparin and tinzaparin (Gómez-Outes et al. 2012). The various LMWHs differ to some extent in their pharmacokinetic properties, anticoagulant profiles, approved indications and recommended dosing regimens. Therefore, these drugs are not necessarily interchangeable.

Mechanism of Action

Like heparin, LMWHs produce their major anticoagulant effect by catalyzing AT-mediated inhibition of coagulation factor Xa and, to a lesser extent, thrombin (Fig. 6). All pentasaccharide-containing heparin chains are able to bridge and inhibit factor Xa, but only pentasaccharide-containing heparin chains composed of at least 18 saccharide units are of sufficient length to bridge AT to thrombin (Garcia et al. 2012; Gray et al. 2012). Virtually all molecules of UFH contain at least 18

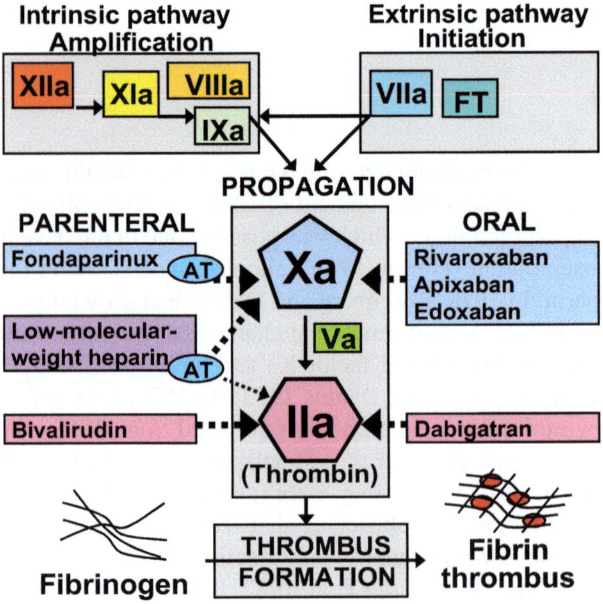

Fig. 6 Current anticoagulants and their targets in the coagulation cascade. (Note: A *continuous line* indicates activation; *dashed line* indicates inhibition. A *thin (dashed) line* indicates a low binding affinity (e.g.: LMWH-AT has a lower binding affinity to FIIa than to FXa))

saccharide units, and heparin has, by definition, an anti-Xa to anti-IIa ratio of 1:1. In contrast, 50–75 % of LMWH chains are too short to catalyze thrombin inhibition (Garcia et al. 2012) and therefore commercial LMWHs have anti-Xa to anti-IIa ratios between 2:1 and 8:1 depending on their molecular size distribution. There is not sufficient evidence to conclude that differences in anti-Xa to anti-IIa ratio among the LMWHs could result in differences in efficacy or safety.

Other mechanisms that may also contribute to LMWH anticoagulant activity are their ability to induce the release of tissue factor pathway inhibitor (TFPI), which inhibits the factor VIIa–tissue factor complex, and to a lesser extent, the inhibition of factor IXa and the release of heparin co-factor II that inhibits thrombin formation (Gray et al. 2012). In addition, LMWH attenuate the release of von Willebrand factor, which is a predictor of outcome in acute coronary syndromes (de Caterina et al. 2013). However, the clinical relevance of these properties is uncertain.

Although the use of heparin has been as an anticoagulant for decades, attention has also been drawn to its non-anticoagulant activities. At the molecular level, heparin inhibits the function, expression and/or synthesis of adhesion molecules, cytokines, angiogenic factors and complement. These properties could be useful in inflammatory diseases like asthma, or in the treatment of cancer. The therapeutic potential of heparin-derived oligosaccharides with anti-inflammatory or anti-angiogenic properties must be explored using well-designed clinical studies (Ludwig 2009; Shastri 2014; Alam et al. 2014).

Pharmacokinetics

Depolymerization of heparin yields low-molecular weight fragments that exhibit reduced binding to proteins and cells. Compared to UFH, LMWHs provide a more predictable dose-response relationship, due to reduced binding to plasma proteins other than AT and a longer half-life due to decreased binding to macrophages and endothelial cells (Garcia et al. 2012). After subcutaneous injection, the bioavailability of LMWHs is over 90 % (Table 2). Anti-Xa levels peak 2–3 h after dosing and the elimination half-life of LMWHs is 3–6 h after subcutaneous injection. As LMWHs are mainly cleared by the kidneys, their elimination half-life may be prolonged in patients with renal insufficiency. Different LMWH with different median chain lengths have different half-lives: LMWH with longer chain lengths are generally endowed with shorter half-lives than LMWH with shorter chain lengths, and therefore are less prone to accumulation (de Caterina et al. 2013).

Indications

LMWHs are effective and safe for the prevention and treatment of venous thromboembolism (VTE), for the prevention of clotting during haemodialysis and for the treatment of acute coronary syndromes (ACS). Indications may differ among distinct LMWH. Therefore, special compliance with the approved indications to each proprietary medicinal product is required. Hereafter, we briefly summarize the available evidence in each indication.

LMWHs are first-line agents for prevention of VTE after major orthopaedic surgery (Falck-Ytter et al. 2012). On the basis of moderate-quality evidence, the use of LMWH for the initial prophylaxis period (10–14 days) is expected to prevent 13 VTE event per 1,000 patients undergoing major orthopedic surgery, assuming a baseline risk (with no prophylaxis) of 1 % for pulmonary embolism and 1.8 % for symptomatic deep venous thrombosis (DVT). Extending thromboprophylaxis up to 35 days postoperation (compared with 10–14 days) in total hip replacement (THR) will result in 20 fewer symptomatic VTE per 1,000 without an appreciable increase in major bleeding, but extended-duration prophylaxis is associated with excess minor bleeding (Eikelboom et al. 2001). Extending thromboprophylaxis up to 35 days postoperation compared with 10–14 days has been poorly studied in total knee replacement (TKR). The baseline risk of VTE during post-discharge period after TKR is lower than after THR (1.4 vs 4.3 %) (Eikelboom et al. 2001). Therefore, there is a limited potential benefit of extending thromboprophylaxis after total knee replacement.

In acutely ill patients at risk for VTE, LMWH prophylaxis for 10–14 days reduces the risk of fatal and non-fatal pulmonary embolism as compared to placebo, and is not associated to a significant increase in major bleeding (Dentali et al. 2007). Extended-duration for additional 4 weeks reduces VTE rates compared with placebo but is associated to increased major bleeding events (Hull et al. 2010).

Table 2 Clinical characteristics of heparin and LMWH[a]

Characteristic	Heparin sodium	LMWH[b]
Pharmacodynamics		
Target	Factor Xa-factor IIa (1:1 ratio)	Factor Xa > factor IIa
Type of inhibition	Indirect (bind to AT)	Indirect (bind to AT)
Antidote	Protamine	Protamine (partial neutralisation)
Pharmakinetics		
Bioavailability (%)	100 (IV), 60 % (SC)	90–98 (SC)
Tmax	Few min (IV), 4–6 h (SC)	2–3 h
Binding to proteins and cells	High	Low
Half-life	0.5–1 h	3–6 h
Metabolism	Depolymerisation	Desulfation, depolymerisation (< 10 %)
Renal excretion (%)	Medium	High
Biliary excretion (%)	Medium	Low
Drug interactions	PD interactions with other antithrombotics	PD interactions with other antithrombotics
Date of first approval		
EU	1939 (Sweden)	1985 (dalteparin)
US	1939	1993 (enoxaparin)
Clinical particulars		
Indications[c]	Prevention and treatment of deep vein thrombosis and pulmonary embolism	Prevention and treatment of deep vein thrombosis and pulmonary embolism
	Prevention of clotting during extracorporeal circulation and haemodialysis	Prevention of clotting during haemodialysis
	Treatment of acute coronary syndromes and acute peripheral arterial occlusion	Treatment of acute coronary syndromes
Route of administration[c]	IV (bolus followed by infusion), into the arterial line in haemodialysis (generally hourly doses), or SC	SC, OD or BID, or into the arterial line in haemodialysis (generally single doses)

Table 2 (continued)

Characteristic	Heparin sodium	LMWH[b]
Dose range[c]	Prophylaxis: 5,000 IU BID or TID, SC; Treatment: option a) 80 units/kg IV bolus + 18 units/kg/h IV infusion, dose-adjusted (see monitoring test); option b) 333 IU/kg SC followed by 250 units/kg BID	Prophylaxis: 2,000–5,000 IU, OD, SC Treatment: 115–200 IU/kg OD or 100 IU/kg BID, SC
Monitoring test[c]	aPTT (1.5–2.5 times the control value) and/or ACT (≈250–500 s)	Anti-Xa chromogenic assay (≈0.6–1.3 anti-Xa units/ml, 4 h after dosing)[d]
Adverse effects	Common: different types of bleedings, transaminase increase, HIT type I (transient); Unfrequent: HIT type II (severe); Rare: allergic reactions, cutaneous necrosis, alopecia, hyperkaliemia; Frequency unknown: osteoporosis	Common: injection site hematoma and other types of bleedings, transaminase increase, HIT type I (transient); Rare: HIT type II (severe), allergic reactions, cutaneous necrosis, alopecia, hyperkaliemia; Frequency unknown: osteoporosis
Common contraindications	Hypersensitivity to heparin or its derivatives;	
	Heparin-induced thrombocytopenia, immunological (type II)	
	Active major bleeding and conditions with a high risk of uncontrolled haemorrhage;	
	In patients receiving heparin for treatment rather than prophylaxis, locoregional anaesthesia in elective surgical procedures is contra-indicated	

[a] *ACT* activated clotting time, *BID* twice daily, *aPTT* activated partial thromboplastin time, *AT* antithrombin, *EU* European Union, *h* hours, *HIT* heparin-induced thrombocytopenia, *IV* intravenously, *LMWH* low-molecular-weight heparin, *min* minutes, *OD* once-daily, *PD* pharmacodynamic, *SC* subcutaneously, *TID* thrice daily, *Tmax* time to maximum concentration, *US* United States of America;

[b] Ardeparin, bemiparin, certoparin, dalteparin, enoxaparin, nadroparin, parnaparin, reviparin, tinzaparin

[c] LMWH are not necessarily interchangeable. Special attention and compliance with the approved indications, posology and instructions for use specific to each proprietary medicinal product are therefore required

[d] Not generally necessary. Anti-Xa monitoring may be of interest in obese patients, renal insufficiency or pregnancy

LMWHs are first-line options for the initial treatment of acute VTE (deep vein thrombosis or pulmonary embolism) (Kearon et al. 2012). LMWHs are also drugs of first choice for extended treatment (beyond 3 months) in patients with VTE associated with active cancer, as they have been shown to be superior to vitamin K antagonists in this setting (Akl et al. 2008).

LMWHs are as safe as UFH in terms of bleeding complications and as effective as UFH in preventing extracorporeal circuit thrombosis. A meta-analysis of 11 clinical trials found that LMWH did not significantly affect the number of bleeding events, bleeding assessed by vascular access compression time or extracorporeal circuit thrombosis as compared with UFH (Lim et al. 2004).

In acute coronary syndromes (ACS), the net clinical benefit [composite of death, myocardial infarction (MI) or major bleeding by 30 days] was in favour of enoxaparin versus UFH among the ST-elevation-MI (STEMI) population undergoing percutaneous coronary intervention (PCI) (Navarese et al. 2011). Therefore, enoxaparin may provide additional benefit over UFH in STEMI patients undergoing PCI. Enoxaparin may also provide some benefit over UFH in STEMI patients receiving fibrinolytic therapy in those patients <75 years of age and with an estimated creatinine clearance >30 ml/min (de Caterina et al. 2013). The net clinical benefit of enoxaparin versus UFH seems neutral among the non-ST-elevation-ACS (NSTEACS) population (Murphy et al. 2007). Limited data are available for dalteparin in patients with NSTEACS (FRISC study group 1996). Data are insufficient to recommend other LMWHs, apart from enoxaparin and dalteparin, in ACS.

Posology and Monitoring

LMWHs are usually given in fixed doses for thromboprophylaxis and in weight-adjusted doses for treatment of thromboembolism (Table 2). Posologies differ among LMWHs. Therefore, special attention and compliance with the approved posology to each proprietary medicinal product are required. Anticoagulant monitoring is not needed in most patients, but it may be of interest in some subpopulations (e.g., extreme body weights, renal insufficiency or during pregnancy), mainly when LMWHs are used at treatment doses. The anti-Xa level measured by a chromogenic assay is the recommended test when monitoring is performed (Garcia et al. 2012). The target range for peak anti-Xa levels (measured 4 h after dosing) will depend on the type of LMWH, dosing interval and indication (prophylaxis or treatment). At treatment doses, it usually ranges from 0.6 to 1.00 units/ml for twice-daily administration and from 0.85 to 1.3 units/ml for once-daily administration (Garcia et al. 2012).

Adverse Effects

Different types of bleedings are the main adverse effects of LMWH treatment. Protamine partially neutralises the anticoagulant effect of LMWH, which may be needed in case of serious bleeding. A dose of 1 mg protamine sulfate is needed per 100 anti-Xa units of LMWH up to a maximum single dose of 50 mg (Garcia et al. 2012).

Serious heparin-induced thrombocytopenia (HIT) type II (as opposed to mild transient and benign HIT type I) is a life-threatening complication of exposure to heparin that results from an autoantibody directed against platelet factor 4 (PF4) in complex with heparin. LMWH offers an improved safety profile compared to UFH as shown by a three-fold lower incidence of serious HIT due to reduced binding to platelets and PF4 (Table 2). However, once HIT antibodies are formed, there is 100 % cross reactivity with LMWH. Therefore LMWH cannot avoid HIT sensitization nor can they be used for HIT therapy. LMWH also shows a lower risk of osteoporosis than UFH, due to decreased binding to bone cells (Garcia et al. 2012). Transaminase elevations >3 ULN have been reported to occur in 5 % of patients receiving UFH and in 4.3–13 % of patients receiving LMWH (Arora and Goldhaber 2006). The hepatotoxic effects remain confined to transaminase elevations, and are not associated with cholestasis or jaundice. Other non-hemorrhagic side effects of LMWH and UFH are uncommon and may include skin adverse reactions that can progress to necrosis, alopecia, and hypersensitivity reactions (Garcia et al. 2012).

Regulatory Status

Heparins are considered in Europe as biological medicinal products. The definition for a product to be considered a biological medicinal product is given in the Directive 2001/83/CE:

A biological medicinal product is a product, the active substance of which is a biological substance. A biological substance is a substance that is produced by or extracted from a biological source and that needs for its characterisation and the determination of its quality a combination of physico-chemical-biological testing, together with the production process and its control. In addition, Directive 2001/83/CE specifically mentions a series of products which should be considered, with no doubt, biological medicinal products (immunological, products derived from human blood and plasma, recombinant products, monoclonal antibodies and advanced therapy medicinal products). Heparins and other substances of animal origin are not included in this list. However, they fit in the definition of a biological product, because of their origin and because they need for their characterization a combination of physico-chemical and biological testing, together with a robust production process and its control.

For this reason, the Coordination Group for Mutual Recognition and Decentralized Procedures–Human (CMDh) in Europe issued a document explaining how the definition of biological product should be applied. On scientific grounds, a number of products should be considered biological medicinal products, because they meet the legal criteria of biological origin and complexity (CMDh 2012). The CMDh provides a non-exhaustive list of substances of non-recombinant origin which should be considered biological active substances, where heparins and LMWHs are included (CMDh 2007).

There are regulatory consequences for this classification of LMWHs as biological substances in Europe. The main one is that generic products can not be authorised, but the "biosimilar" approach applies. This requires demonstration of comparability between the original product and the biosimilar one. In addition, bibliographic applications are normally not applicable. The Active Substance Master File (ASMF) procedure is not applicable to biological medicinal products because the marketing authorisation holder should take full responsibility of the product. This is not possible without having full access to the quality data, which would allow the applicant the full knowledge of the manufacturing process and its control including the starting material, viral safety or aspects to guarantee traceability. Also for this reason the European Directorate of Quality of Medicines (EDQM 2009) excluded these biological products from the Certification Procedure in 2009 and will not issue new Certificates of Suitability (CEPs) for heparins or LMWHs. Therefore, the CEP should not substitute for the full quality information in Module 3 of the dossier of the medicinal product.

Recently the EMA issued a guideline on the use of starting materials and intermediates collected from different sources in the manufacturing of non-recombinant biological medicinal products (EMA 2013a). In this guideline it is clarified that the starting material for heparins and LMWHs is pooled porcine intestinal mucosae. Different intermediates may exist and be qualified for use in the manufacture of LMWHs, such as resin bound heparin, partly purified crude heparin or heparin sodium/calcium. However, these intermediates shall not be considered as starting materials. This means that all information on the whole manufacturing process starting from porcine mucosa should be detailed in the dossier of the medicinal product.

In addition the guideline also clarifies that certain variability in the early manufacturing steps of these products is acceptable. As manufacturers of LMWHs and heparins often need to have several suppliers, triggered by the high demand for the starting material to ensure product supply and the increasing difficulty in finding starting materials suppliers, variability in sourcing and/or initial manufacturing steps may be needed. Thus, if a manufacturer decides to use starting materials or intermediates from different sources and/or a different manufacturing process for the early production steps it should be shown that comparable active substances are consistently obtained in terms of relevant quality attributes irrespective of the process applied.

Taking all these aspects into account for the authorization of a medicinal product containing LMWH, there is critical information that should be included in the dossier. This includes information to guarantee traceability from the slaughterhouse to the final product lot, the country of origin of the animals and confirmation that animals are healthy and inspected by veterinaries. A viral risk assessment should be provided with the information of the studies of validation of the capacity of the manufacturing process to inactivate or remove viruses.

LMWHs in Europe and in the US

LMWH products were invented and developed in Europe, and they came into use in the US some years later. The regulatory approach to low molecular weight heparin has been notably different in these two areas.

In particular, the regulatory authorities in the US (US FDA) and Europe (EMA) have each developed guidelines for potential producers of 'follow-on' LMWH products (see the section on Challenges above), and the contrast between these two documents is interesting, in that it illustrates the practical consequences of regarding LMWH on the one hand as a biological, for which follow-on products are termed 'biosimilars' and on the other hand as a complex non-biological, for which the follow-on products are regarded as more akin to the 'generic' small molecule drugs (Gray and Mulloy 2009). Both authorities recognise that absolute structural description of a LMWH product at a molecular level is not achievable, but draw different conclusions.

For the EMA (EMA 2013b), currently the lack of a complete structural description means that the clinical efficacy and side-effects of a 'biosimilar' LMWH must be investigated in clinical trials. But, at the time of writing of this chapter, this guidance is under review (Senior 2013).

In the guidance from the FDA CDER (FDA 2014), it is proposed that a copy of a LMWH product, indistinguishable from it in structural terms, will have –predictably-similar properties in the patient, so that pre-licensing tests in patients are not necessary. Generic enoxaparin is now available in the US, and it will in time become clear whether the FDA's structural approach has any consequences for patient well-being.

Prospects, Innovations, Breakthroughs: Innovation and Future of Heparin and LMWH Like Molecules

The use of unfractionated heparin as anticoagulant started in the 1930s and it has been and it is still widely used in prevention and treatment of venous thromboembolism. The development of LMWHs in the 1980s improved some important aspects like subcutaneous bioavailability and longer half-life, allowing for self-administration (Hirsh and Levine 1992). The most recently approved LMWH and the shortest one (Mw of 3600), is bemiparin sodium, which was authorised in Europe in 2000 (Martínez-González and Rodríguez 2010). Other even shorter LMWHs have been tested in clinical trials (Gómez-Outes et al. 2011), such as semuloparin/AVE5026 (Mw 2400), which results from selective depolymerisation of heparin by the phosphazene base which preserves the AT binding sequences (Viskov et al. 2009b; Lassen at el. 2009; Lawson et al. 2012; Fisher et al. 2013) or RO-14, obtained by selective chemical depolymerization (Liu et al. 2014; Rico et al. 2011; Vignoli et al. 2011). Attempts have been made to search for new methods of depolymerisation of heparin, like partial photolysis using titanium dioxide (Higashi et al. 2012) or ultrasonic-assisted radical depolymerization of heparin (Achour et al. 2013).

As already mentioned, in 2010 the first generic LMWH was approved in US, a generic version of enoxaparin (Editorial Nature Biotechnology 2010). After that, other enoxaparin generics have been approved in the US, whereas in Europe applications for biosimilars of enoxaparin are presently in progress. It is expected that other successful applications for generic/biosimilar versions of LMWH products will broaden the choice of medicines used for anticoagulation.

Despite the development and marketing of new oral anticoagulants, the use of heparin and LMWHs of animal origin has not diminished, due to their good safety profile and pharmacological properties. Therefore, the development of new LMWHs or LMWH-like substances obtained by other means than from animal tissues that could overcome the theoretical risk of transmitting adventitious agents, known or still unknown, and the possible supply problems due to hypothetical lack of the starting porcine or bovine material, is very attractive. Several attempts are being developed in this direction that could be an alternative to heparins and LMWHs from animal origin for anticoagulation in the future.

Synthetic Analogs of Heparin Pentasaccharide

Because the active motif of heparins and LMWHs has been identified as the pentasaccharide sequence, it is logic to deduct that a GAG as short as this sequence could be chemically synthesized and that this molecule could be used as active substance to substitute LMWH as anticoagulant.

In 2001 a new medicinal product was authorized containing fondaparinux sodium as active substance (Petitou and van Boeckel 2004). Fondaparinux sodium is a synthetic heparin pentasaccharide that contains the minimum ATIII binding site in heparin. Fondaparinux is chemically synthesized in a process involving about 50 steps with an overall yield of about 0.1 % (Petitou et al. 1989; Xu et al. 2011). Fondaparinux sodium shows particular properties, like having specific anti-Xa activity, a long half-life and the excess of anticoagulant activity cannot be reversed by protamine. In 2011 a generic version of fondaparinux was approved by the FDA.

Idraparinux is a new pegylated fondaparinux analog. However, it exhibited a high-risk bleeding effect in clinical trials due to its very long half-life (van Gogh Investigators et al. 2007). Clinical development of idraparinux, which has no antidote, was stopped in favour of idrabiotaparinux (Gómez-Outes et al. 2011), a biotinylated form of idraparinux that can be neutralized with avidin (Savi et al. 2008; Paty et al. 2010). However, clinical development of idrabiotaparinux was stopped in 2009.

Understanding in Vivo Synthesis to Translate to Chemical Synthesis

The in vivo synthesis of heparin occurs in mast cells, which are mostly found in the intestine, lung, liver and skin of higher animals. Heparin is synthesized in a multistep process that involves multiple enzymes in the endoplasmic reticulum and the Golgi apparatus of cells. In vivo, all GAG except hyaluronic acid are covalently linked to proteoglycan protein cores and modified by a large series of enzymes present in the Golgi apparatus. A core protein is synthesized in the endoplasmic reticulum. The GAG chains are linked to the core protein through a linkage tetrasaccharide. In the case of heparin, the polysaccharide chains are linked to serine residues in the Ser-Gly repeat regions of the small protein serglycin. Heparin and heparan sulfate chains are elongated as they transit the Golgi, via condensation polymerization reactions catalyzed by the exostosin enzymes. In this way, a repeating 1->4-glycosidically-linked copolymer of D-glucuronic acid and N-acetyl-D-glucosamine is extended, forming a linear homocopolymer (Carlsson and Kjellén 2012).

Sulfate groups are added to the polysaccharide chain in a specific order by the enzymes N-deacetylase/N-sulfotransferase (NDST) and specific sulfotransferases (2-, 6-, and 3-O-sulfotransferases, OSTs). The epimerization of glucuronic acid residues to iduronic acid is performed via C5 epimerases. Specifically, the 3-O-sulfotransferase family of enzymes is responsible for the addition of the sulfate group to the 3 position of the disulfated monosaccharide GlcNS (6S) which is important for the binding of ATIII and the overall anti-coagulant activity of heparin. In this way heparin and heparan sulfate are synthesised in vivo by mammalian cells, heparin being rich in N- and O-sulfate and L-iduronic acid, whereas heparan sulfate is rich in N-acetyl-D-glucosamine and D-glucuronic acid, so that heparin carries more sulfo groups (Lord and Whitelock 2014; Liu et al. 2009).

Chemo-enzymatic Synthesis of Heparins

Understanding how heparin and heparan sulfate are synthesised in vivo has allowed researchers to search for ways to synthesize these molecules in the laboratory.

There have been some attempts to produce heparin like molecules by chemo-enzymatic synthesis starting from E. coli K5 as source of the polysaccharide backbone. Bacterial fermentation is first employed to synthesize heparosan, a bacterial capsular polysaccharide composed of repeating units of glucoronic acid 1->4 N-acetyl glucosamine, which is treated with the necessary isolated enzymes.

In this way, Lindahl and colleagues produced what was called “neoheparin”, from heparosan, which was chemically de-N-acetylated and N-sulfonated. Then it was treated with C5Epi followed by chemical per-O-sulfation and selective O-desulfation. This approach, although producing high yields on the gram scale of anticoagulant and anti-thrombin-binding heparin, produced non-natural sequences, not present in mammalian heparin, which could be a potential risk (Lindahl et al. 2005).

Similarly, Liu and colleagues have developed methods to produce chemo-enzymatically synthesized heparin (Linhardt et al. 2007, Chen et al. 2005). All the enzymes required for the biosynthesis of heparin and heparan sulfate have been cloned and expressed. They have started from E. coli K5 N-acetyl heparosan and have used two chemical steps followed by four enzymatic steps. In this way they produced milligram quantities of bioengineered heparin which is able to bind ATIII and with an anticoagulant activity of about 180 U/mg. Others have also used similar approaches to prepare other heparin-like polysaccharides and oligosaccharides (Kuberan et al. 2003a, b; Kane et al. 2006).

More recently, the chemoenzymatic synthesis, in milligram quantities, of two pentasaccharide LMWHs designed to contain the AT binding domain of porcine and bovine heparin, respectively was published (Xu et al. 2011). They are similar in structure to the chemically synthesized fondaparinux, but they are obtained chemo-enzymatically by a 10- and 12- steps and with a yield of 45 and 37%, respectively. The starting material is a disaccharide prepared from heparosan obtained by fermentation. This disaccharide is elongated using different enzymes, including glycosiltransferases, sulfotransferases and C5-epimerase. These molecules showed affinity to AT, displayed in vitro anticoagulant activity (anti Xa) and showed comparable pharmacokinetics to fondaparinux in a rabbit model. Chemo-enzymatic methods have recently been used to develop homogeneous low molecular weight heparins of up to twelve saccharides that could be neutralized with protamine (Xu et al. 2014), thus potentially improving patient safety.

Although promising, optimization and economic evaluation of these chemo-enzymatic processes will be necessary before these products can be used in therapy as large amounts of material are needed, first for the clinical trials and then, if a marketing authorization is granted, for therapeutic use.

Bioengineered Heparin in Mammalian Cells

As heparins are obtained from natural proteoglycans present in mammalian cells, and with the increasing application of molecular biology in the production of medicines, the idea of producing recombinant heparin comes to mind at first thought. However, unlike proteins, GAG synthesis does not rely on a DNA/RNA template, but requires multiple Golgi and endoplasmic reticulum localized enzymes to be expressed by mammalian cells, including those involved in glycosaminoglycan chain elongation, epimerisation and modification by sulfate. Most cell types, and particularly those used for recombinant protein expression, do not express enzymes involved in heparin biosynthesis, such as 3-O sulfotransferase, at a level that gives complete heparin chains (Lord and Whitelock 2014). Therefore, other approaches different than the "typical" recombinant protein approach using bacteria/yeast/mammalian cells, which were successfully used for the synthesis of therapeutic (glyco)proteins, should be developed.

Because Chinese hamster ovary (CHO) cells are capable of producing heparan sulfate, which shares the biosynthetic pathway with heparin, Linhardt and colleagues hypothesized that heparin could be produced in metabolically engineered CHO cells. They developed stable human NDST2 and mouse heparan sulfate 3-O-sulfotransferase 1 (HS3st1) expressing cell lines. While heparan sulfate biosynthesis was increased, the level of anti-coagulant activity of the GAG produced was not as large as that of heparin isolated from natural sources (Baik et al. 2012). When Golgi-targeted HS3st1 was used in the same cells to localize the enzyme in the Golgi apparatus, they produced the AT-binding site with anti-Xa activity, however still not at the same level as observed in heparin (Datta et al. 2013). A further step in this direction is that recent transfection of murine mastocytoma cells with HS3st1 has produced a heparan sulfate/heparin substance with anticoagulant activity (Gasimli et al. 2014).

Another area of research which would improve the quality of life of patients is the development of LMWHs with acceptable and reproducible oral bioavailability. This seems to be quite a challenging task, but some attempts are being made in that direction (Martínez-González and Rodríguez 2010). A recent study in rats has shown promising results with the use of nanoparticles prepared with thiolated chitosan and the pH-sensitive polymer hydroxypropyl methylcellulose phthalate (HPMCP) by an ionic cross-linking method (Fan et al. 2014). Oral formulations are also being developed for low molecular weight heparin derivatives with anti-angiogenic properties potentially applicable to cancer patients, by chemical conjugation with tetrameric deoxycholic acid and physical complexation with deoxycholylethylamine (Alam et al. 2014).

In conclusion, a cost-effective method for preparing new synthetic or semi-synthetic LMWH is desirable. Due to the complexity and peculiarities of these types of molecules, an interdisciplinary approach is required, including areas of expertise of chemical synthesis, biochemistry, molecular biology, cellular biology and metabolomics. Basic research should evolve to applied research in order to be able to reach industrial production scales. However, at this moment the likeliness of synthetic heparins replacing heparins and LMWHs of animal origin in the near future is only very small. Therefore, close control of LMWHs, their intermediates, starting material and their manufacturing processes should be maintained to guarantee that this family of the most widely used anticoagulants continues to be safe and efficacious for our patients.

References

Achour O, Bridiau N, Godhbani A et al (2013) Ultrasonic-assisted preparation of a low molecular weight heparin (LMWH) with anticoagulant activity. Carbohydr Polym 97:684–689

Akl EA, Barba M, Rohilla S et al (2008) Low-molecular-weight heparins are superior to vitamin K antagonists for the long term treatment of venous thromboembolism in patients with cancer: a Cochrane systematic review. J Exp Clin Cancer Res 27:21

Al Dieri R., Wagenvoord R, van Dedem GW et al (2003). The inhibition of blood coagulation by heparins of different molecular weight is caused by a common functional motif–the C-domain. J Thromb Haemost 1:907–914

Alam F, Al-Hilal TA, Chung SW et al (2014) Oral delivery of a potent anti-angiogenic heparin conjugate by chemical conjugation and physical complexation using deoxycholic acid. Biomaterials 35:6543–6552

Anon (2014a) Heparins, low-molecular-mass. In: European Pharmacopoeia, 8.0 (ed), European Directorate For The Quality of Medicines and Healthcare, Strasbourg, 2392–2394

Anon (2014b) Heparin sodium. In: European Pharmacopoeia, 8.0 (ed), European Directorate for the Quality of Medicines and HealthCare, Strasbourg, 2390–2391

Anon (2014c) Heparin calcium. In: European Pharmacopoeia, 8.0 (ed), European Directorate for the Quality of Medicines and HealthCare, Strasbourg, 2388–2389

Anon (2014d) USP 37 <208> Anti-factor Xa and anti-factor IIa assays for unfractionated and low molecular weight heparins. In: United States Pharmacopeia and National Formulary (USP 37-NF 32), The United States Pharmacopeial Convention, Rockville (MD), 152–155

Anon (2014e) USP 38 <209> Low molecular weight heparin molecular weight determinations. In: United States Pharmacopeia and National Formulary (USP 38), The United States Pharmacopeial Convention, Rockville (MD), p. -in press
Arora N, Goldhaber SZ (2006) Anticoagulants and transaminase elevation. Circulation 113:e698–e702
Auguste C, Dereux S, Rousset M, Anger P (2012) Validation of quantitative polymerase chain reaction methodology for monitoring DNA as a surrogate marker for species material contamination in porcine heparin. Anal Bioanal Chem 404:43–50
Baik JY, Wang CL, Yang B et al (2012) Toward a bioengineered heparin: challenges and strategies for metabolic engineering of mammalian cells. Bioengineered 3:227–231
Bergqvist D, Nilsson B, Hedner U et al (1985) The effects of heparin fragments of different molecular weight in experimental thrombosis and haemostasis. Thromb Res 38:589–601
Bertini S, Bisio A, Torri G et al (2005). Molecular weight determination of heparin and dermatan sulfate by size exclusion chromatography with a triple detector array. Biomacromolecules 6:168–173
Beyer T, Diehl B, Randel G et al (2008) Quality assessment of unfractionated heparin using ^{1}H nuclear magnetic resonance spectroscopy. J Pharm Biomed Anal 48:13–19
Bhaskar U, Sterner E, Hickey AM et al (2012) Engineering of routes to heparin and related polysaccharides. Appl Microbiol Biotechnol 93:1–16
Buzzega D, Maccari F, Volpi N (2008) Fluorophore-assisted carbohydrate electrophoresis for the determination of molecular mass of heparins and low-molecular-weight (LMW) heparins. Electrophoresis 29:4192–4202
Carlsson P, Kjellén L (2012) Heparin biosynthesis. Handb Exp Pharmacol 207:23–41
Carter C, Kelton J, Hirsh J et al (1982) The relationship between the hemorrhagic and antithrombotic properties of low molecular weight heparins in rabbits. Blood 59:1239–1245
Chen J, Avci FY, Muñoz EM et al (2005) Enzymatic redesigning of biologically active heparan sulfate. J Biol Chem 280:42817–42825
Choay J, Petitou M, Lormeau JC et al (1983) Structure-activity relationship in heparin: a synthetic pentasaccharide with high affinity for antithrombin III and eliciting high anti-factor Xa activity. Biochem Biophys Res Commun 116:492–499
CMDh (2007) Overview of Biological Active Substances of Non-Recombinant Origin, CMDh, June 2007. http://www.hma.eu/fileadmin/dateien/Human_Medicines/CMD_h_/procedural_guidance/Compilation_Biological_Active_Substance_non-recombinant_origin.pdf. Accessed 12 Sept 2014
CMDh (2012) CMDh questions and answers biologicals. CMDh/269/2012, Rev0. http://www.hma.eu/20.html. Accessed 11 July 2014
Concannon SP, Wimberley PB, Workman WE (2011) A quantitative PCR method to quantify ruminant DNA in porcine crude heparin. Anal Bioanal Chem 399(2):757–62
Datta P, Li G, Yang B et al (2013) Bioengineered Chinese hamster ovary cells with Golgi-targeted 3-O-sulfotransferase-1 biosynthesize heparan sulfate with an antithrombin-binding site. J Biol Chem 288:37308–37318
De Caterina R, Husted S, Wallentin L et al (2013) Parenteral anticoagulants in heart disease: current status and perspectives (Section II). Position paper of the ESC working group on thrombosis-task force on anticoagulants in heart disease. Thromb Haemost 109:769–786
Dentali F, Douketis JD, Gianni M et al (2007) Meta-analysis: anticoagulant prophylaxis to prevent symptomatic venous thromboembolism in hospitalized medical patients. Ann Intern Med 146:278–288
Edens RE, Al-Hakim A, Weiler JM et al (1992) Gradient polyacrylamide gel electrophoresis for determination of molecular weights of heparin preparations and low-molecular-weight heparin derivatives. J Pharm Sci 81:823–827
Editorial Nature Biotechnology (2010) The identity problem. Nat Biotech 28:877
EDQM (2009) The EDQM position on CEP Applications for Biological Substances. PA/PH/CEP (09) 152 rev 01 (EN). https://www.edqm.eu/site/cep_the_edqm_position_on_cep_applications_for_biolpdf-en-18486-2.html. Accessed 12 Sept 2014

Eikelboom JW, Quinlan DJ, Douketis JD (2001) Extended-duration prophylaxis against venous thromboembolism after total hip or knee replacement: a meta-analysis of the randomised trials. Lancet 358:9–15

EMA (2008) European Medicines Agency, London, 5 June 2008, Questions and Answers on Heparins, Doc. Ref. EMEA/276814/2008

EMA (2013a) EMA/CHMP/BWP/429241/2013, Committee for Medicinal Products for Human Use (CHMP).Guideline on the use of starting materials and intermediates collected from different sources in the manufacturing of non-recombinant biological medicinal products. http://www.ema.europa.eu/ema/index.jsp?curl=pages/includes/document/document_detail.jsp?webContentId=WC500145739&mid=WC0b01ac058009a3dc. Accessed 11 July 2014

EMA (2013b) EMEA/CHMP/BMWP/118264/2007 Rev. 1, Committee for Medicinal products for Human (CHMP). Guideline on non-clinical and clinical development of similar biological medicinal products containing low molecular-weight-heparins Draft. http://www.ema.europa.eu/docs/en_GB/document_library/Scientific_guideline/2013/01/WC500138309.pdf. Accessed 12 Sept 2014

Ernst S, Rhomberg AJ, Biemann K, Sasisekharan R (1998) Direct evidence for a predominantly exolytic processive mechanism for depolymerization of heparin-like glycosaminoglycans by heparinase I. Proc Natl Acad Sci U S A 95:4182–4187

Falck-Ytter Y, Francis CW, Johanson NA et al (2012) Prevention of VTE in orthopedic surgery patients: antithrombotic therapy and prevention of thrombosis, 9th ed: American college of chest physicians evidence-based clinical practice guidelines. Chest 141(Suppl 2):e278S–e325S

Fan B, Xing Y, Zheng Y et al (2014) pH-responsive thiolated chitosan nanoparticles for oral low-molecular weight heparin delivery: in vitro and in vivo evaluation. Drug Deliv 28:1–10

FDA (2013) Guidance for Industry. U.S. Department of Health and Human Services, Food and Drug Administration, Center for Drug Evaluation and Research (CDER), Center for Veterinary Medicine (CVM), Center for Devices and Radiological Health (CDRH). Heparin for Drug and Medical Device Use: Monitoring Crude Heparin for Quality. www.fda.gov/downloads/drugs/guidancecomplianceregulatoryinformation/guidances/ucm291390.pdf. Accessed 11 July 2014

FDA (2014) Guidance for Industry. U.S. Department of Health and Human Services, Food and Drug Administration, Center for Drug Evaluation and Research (CDER) Immunogenicity-Related Considerations for the Approval of Low Molecular Weight Heparin for NDAs and ANDAs. DRAFT GUIDANCE. http://www.fda.gov/downloads/Drugs/GuidanceCompliance-RegulatoryInformation/Guidances/UCM392194.pdf. Accessed 11 July 2014

Fisher WD, Agnelli G, George DJ et al (2013) Extended venous thromboembolism prophylaxis in patients undergoing hip fracture surgery—the SAVE-HIP3 study. Bone Joint J 95-B:459–466

FRISC study group (1996) Low-molecular-weight heparin during instability in coronary artery disease, Fragmin during Instability in Coronary Artery Disease (FRISC) study group. Lancet 347:561–568

Garcia DA, Baglin TP, Weitz JI et al (2012) Parenteral anticoagulants: antithrombotic therapy and prevention of thrombosis, 9th ed: American college of chest physicians evidence-based clinical practice guidelines. Chest 141(Suppl 2):e24S–e43S

Gasimli L, Glass CA, Datta P et al (2014) Bioengineering murine mastocytoma cells to produce anticoagulant heparin. Glycobiology 24:272–280

Gómez-Outes A, Suárez-Gea ML, Lecumberri R et al (2011) New parenteral anticoagulants in development. Ther Adv Cardiovasc Dis 5:33–59

Gómez-Outes A, Suárez-Gea ML, Calvo-Rojas G et al (2012) Discovery of anticoagulant drugs: a historical perspective. Curr Drug Discov Technol 9:83–104

Gray E, Mulloy B (2009) Biosimilar low molecular weight heparin products. J Thromb Haemost 7:1218–1221

Gray E, Mulloy B, Barrowcliffe TW (2008) Heparin and low-molecular-weight heparin. Thromb Haemost 99:807–818

Gray E, Hogwood J, Mulloy B (2012) The anticoagulant and antithrombotic mechanisms of heparin. Handb Exp Pharmacol 207:43–61 (Heparin—A century of Progress, R. Lever et al. (eds))

Guerrini M, Beccati D, Shriver Z et al (2008) Oversulfated chondroitin sulfate is a contaminant in heparin associated with adverse clinical events. Nat Biotechnol 26:669–675
Guerrini M, Elli S, Gaudesi D et al (2010) Effects on molecular conformation and anticoagulant activities of 1,6-anhydrosugars at the reducing terminal of antithrombin-binding octasaccharides isolated from low-molecular-weight heparin enoxaparin. J Med Chem 53:8030–8040
Harenberg J, Devries JX, Waibel S, Zimmermann R (1983) High-performance size exclusion liquid-chromatography of heparins. Thromb Haemost 50:103
Higashi K, Hosoyama S, Ohno A et al (2012) Photochemical preparation of a novel low molecular weight heparin. Carbohydr Polym 67:1737–1743
Hirsh J, Levine MN (1992) Low molecular weight heparin. Blood 79:1–17
Houiste C, Auguste C, Makrez C et al (2009) Quantitative PCR and disaccharide profiling to characterize the animal origin of low-molecular-weight heparins. Clin Appl Thromb Hemost 15:50–58
Huang Q, Xu T, Wang GY et al (2012) Species-specific identification of ruminant components contaminating industrial crude porcine heparin using real-time fluorescent qualitative and quantitative PCR. Anal Bioanal Chem 402:1625–1634
Hull RD, Schellong SM, Tapson VF et al (2010) Extended-duration venous thromboembolism prophylaxis in acutely ill medical patients with recently reduced mobility: a randomized trial. Ann Intern Med 153:8–18
Iverius PH (1971) Coupling of glycosaminoglycans to agarose beads (sepharose 4B) Biochem J 124:677–683
Jandik KA, Gu K, Linhardt RJ (1994) Action pattern of polysaccharide lyases on glycosaminoglycans. Glycobiology 4:289–296
Johnson EA, Kirkwood TBL, Stirling Y et al (1976) Four heparin preparations: anti-Xa potentiating effect of heparin after subcutaneous injection. Thromb Haemost 35:586–591
Jones CJ, Beni S, Limtiaco JF et al (2011) Heparin characterization: challenges and solutions. Annu Rev Anal Chem (Palo Alto Calif) 4:439–465
Jongen P, de Kaste D (2011) Heparins and changing regulatory requirements in the EU. Pharmaeuropa 23:1
Kane TA, White CL, DeAngelis PL (2006) Functional characterization of PmHS1, a Pasteurella multocida heparosan synthase. J Biol Chem 281:33192–33197
Kearon C, Akl EA, Comerota AJ et al (2012) Antithrombotic therapy for VTE disease: antithrombotic therapy and prevention of thrombosis, 9th ed: American college of chest physicians evidence-based clinical practice guidelines. Chest 141(Suppl 2):e419S–e494S
Kellenbach E, Sanders K, Michiels PJ, Girard FC (2011) (1)H NMR signal at 2.10 ppm in the spectrum of KMnO(4)-bleached heparin sodium: identification of the chemical origin using an NMR-only approach. Anal Bioanal Chem 399:621–628
Kishimoto TK, Viswanathan K, Ganguly T et al (2008) Contaminated heparin associated with adverse clinical events and activation of the contact system. N Engl J Med 358:2457–2467
Knobloch JE, Shaklee PN (1997) Absolute molecular weight distribution of low-molecular-weight heparins by size-exclusion chromatography with multiangle laser light scattering detection. Anal Biochem 245:231–241
Komatsu H, Yoshii K, Ishimitsu S et al (1993) Molecular mass determination of low-molecular-mass heparins. Application of wide collection angle measurements of light scattering using a high-performance gel permeation chromatographic system equipped with a low-angle laser light-scattering photometer. J Chromatogr 644:17–24
Kristensen HI, Tromborg EM, Nielsen JR et al (1991) Development and validation of a size exclusion chromatography method for determination of molecular masses and molecular mass distribution in low molecular weight heparin. Thromb Res 64:131–141
Kuberan B, Beeler DL, Lech M et al (2003a) Chemoenzymatic synthesis of classical and non-classical anticoagulant heparan sulfate polysaccharides. J Biol Chem 278:52613–52621
Kuberan B, Lech MZ, Beeler DL et al (2003b) Enzymatic synthesis of antithrombin III-binding heparan sulfate pentasaccharide. Nat Biotechnol 21:1343–1346

Lassen MR, Dahl OE, Mismetti P (2009) AVE5026, a new hemisynthetic ultra-low-molecular-weight heparin for the prevention of venous thromboembolism in patients after total knee replacement surgery–TREK: a dose-ranging study. J Thromb Haemost 7:566–572
Laurencin CT, Nair L (2008) The FDA and safety—beyond the heparin crisis. Nat Biotechnol 26:621–623
Lawson F, Turpie AG, SAVE-ONCO Investigators (2012) Semuloparin for thromboprophylaxis in patients receiving chemotherapy for cancer. N Engl J Med 366:601–609
Lee S, Raw A, Yu L et al (2013) Scientific considerations in the review and approval of generic enoxaparin in the United States. Nat Biotechnol 31:220–6
Levieux A, Rivera V, Levieux D (2002) Immunochemical control of the species origin of porcine crude heparin and detection of ovine and caprine materials. J Pharm Biomed Anal 27:305–313
Li L, Zhang F, Zaia J, Linhardt RJ (2012) Top-down approach for the direct characterization of low molecular weight heparins using LC-FT-MS. Anal Chem 84:8822–8829
Lim W, Cook DJ, Crowther MA (2004) Safety and efficacy of low molecular weight heparins for hemodialysis in patients with end-stage renal failure: a meta-analysis of randomized trials. J Am Soc Nephrol 15:3192–3206
Lindahl U, Thunberg L, Backstrom G et al (1984) Extension and structural variability of the antithrombin-binding sequence in heparin. J Biol Chem 259:12368–12376
Lindahl U, Li JP, Kusche-Gullberg M et al (2005) Generation of "neoheparin" from E. coli K5 capsular polysaccharide. J Med Chem 48:349–52
Linhardt RJ, Dordick PL, Deangelis PL, Liu J (2007) Enzymatic synthesis of glycosaminoglycan heparin. Semin Thromb Hemost 33:453–465.
Liu H, Zhang Z, Linhardt RJ (2009) Lessons learned from the contamination of heparin. Nat Prod Rep 26:313–321
Liu Z, Ji S, Sheng J, Wang F (2014) Pharmacological effects and clinical applications of ultra low molecular weight heparins. Drug Discov Ther 8(1):1–10
Lord MS, Whitelock JM (2014) Bioengineered heparin. Is there a future for this form of the successful therapeutic? Bioengineered 5(4):1–5 (July/August 2014 (in press))
Ludwig RJ (2009) Therapeutic use of heparin beyond anticoagulation. Curr Drug Discov Technol 6:281–289
Martínez-González J, Rodríguez C (2010) New challenges for a second-generation low-molecular-weight heparin: focus on bemiparin. Expert Rev Cardiovasc Ther 8:625–634
McEwen I, Mulloy B, Hellwig E et al (2008) Determination of oversulfated chondroitin sulfate and dermatan sulfate in unfractionated heparin by ^{1}H NMR. Collaborative study for quantification and analytical determination of LoD. Pharmaeuropa Bio 2008(1):31–39
Mintz CS, Liu JL (2013) China's heparin revisited: what went wrong and has anything changed? J Commerc Biotechnol 19(1):33–39
Mourier AJ, Viskov C (2004) Chromatographic analysis and sequencing approach of heparin oligosaccharides using cetylmethylammonium dynamically coated stationary phases. Anal Biochem 332:299–313
Mourier PA, Guichard OY, Herman F, Viskov C (2011) Heparin sodium compliance to the new proposed USP monograph: elucidation of a minor structural modification responsible for a process dependent 2.10 ppm NMR signal. J Pharm Biomed Anal 54:337–344
Mourier PA, Guichard OY, Herman F, Viskov C (2012) Heparin sodium compliance to USP monograph: structural elucidation of an atypical 2.18 ppm NMR signal. J Pharm Biomed Anal 67–68:169–174
Mulloy B (2012) Structure and physicochemical characterisation of heparin. Handb Exp Pharmacol 207:77–98
Mulloy B, Gee C, Wheeler SF et al (1997) Molecular weight measurements of low molecular weight heparins by gel permeation chromatography. Thromb Haemost 77:668–674
Mulloy B, Gray E, Barrowcliffe TW (2000) Characterization of unfractionated heparin: comparison of materials from the last 50 years. Thromb Haemost 84:1052–1056

Mulloy B, Heath A, Behr-Gross ME (2007) Establishment of replacement batches for heparin low-molecular-mass for calibration CRS, and the International Standard Low Molecular Weight Heparin for Calibration. Pharmeuropa Bio 2007:29–48

Mulloy B, Heath A, Shriver Z et al (2014) USP compendial methods for analysis of heparin: chromatographic determination of molecular weight distributions for heparin sodium. Anal Bioanal Chem 406:4815–4823

Murphy SA, Gibson CM, Morrow DA et al (2007) Efficacy and safety of the low-molecular weight heparin enoxaparin compared with unfractionated heparin across the acute coronary syndrome spectrum: a meta-analysis. Eur Heart J 28:2077–2086

Navarese EP, De Luca G, Castriota F et al (2011) Low-molecular-weight heparins vs unfractionated heparin in the setting of percutaneous coronary intervention for ST-elevation myocardial infarction: a meta-analysis. J Thromb Haemost 9:1902–1915

NC-IUBMB (2014) Nomenclature committee of the international union of biochemistry and molecular biology. Enzyme Nomenclature. Recommendations. http://www.chem.qmul.ac.uk/iubmb/enzyme/EC4/2/2/. Accessed 12 Sept 2014

Nielsen JI (1992) A convenient method for molecular mass determination of heparin. Thromb Haemost 68:478–480

Paty I, Trellu M, Destors JM et al (2010) Reversibility of the anti-FXa activity of idrabiotaparinux (biotinylated idraparinux) by intravenous avidin infusion. J Thromb Haemost 8:722–729

Peters SM, Jones YL, Perrella F et al (2014) Development of a multiplex real-time PCR assay for the detection of ruminant DNA in raw materials used for monitoring crude heparin for quality. Draft. http://www.fda.gov/AnimalVeterinary/ScienceResearch/ToolsResources/ucm350289.htm. Accessed 11 July 2014

Petitou M, van Boeckel CAA (2004) A synthetic antithrombin III binding pentasaccharide is now a drug! What comes next? Angew Chem Int Ed Engl 43:3118–3133

Petitou M, Jacquinet JC, Choay J et al (1989) Process for the organic synthesis of oligosaccharides and derivatives thereof. US Patent 4,818,816

Petitou M, Herault JP, Bernat A et al (1999) Synthesis of thrombin-inhibiting heparin mimetics without side effects. Nature 398:417–422

Rico S, Antonijoan RM, Gich I et al (2011) Safety assessment and pharmacodynamics of a novel ultra low molecular weight heparin (RO-14) in healthy volunteers-a first-time-in-human single ascending dose study. Thromb Res 127:292–298

Rivera V, Levieux A, Levieux D (2002) Immunochemical characterisation of species-specific antigens in bovine crude heparin. J Pharm Biomed Anal 29:431–441

Rosenberg RD, Damus PS (1973) The purification and mechanism of action of human antithrombin-heparin cofactor. J Biol Chem 248:6490–6505

Savi P, Herault JP, Duchaussoy P et al (2008) Reversible biotinylated oligosaccharides: a new approach for a better management of anticoagulant therapy. J Thromb Haemost 6:1697–1706

Schroeder M, Hogwood J, Gray E et al (2011) Protamine neutralisation of low molecular weight heparins and their oligosaccharide components. Anal Bioanal Chem 399:763–771

Senior M (2013) Biosimilars battle rages on, Amgen fights both sides. Nat Biotechnol 31:269–270

Shastri MD, Peterson GM, Stewart N et al (2014) Non-anticoagulant derivatives of heparin for the management of asthma: distant dream or close reality? Expert Opin Investig Drugs 23:357–373

Shriver Z, Sundaram M, Venkataraman G et al (2000) Cleavage of the antithrombin III binding site in heparin by heparinases and its implication in the generation of low molecular weight heparin. Proc Nat Acad Sci U S A 97:10365–10370

Thunberg L, Bäckström G, Lindahl U (1982) Further characterization of the antithrombin-binding sequence in heparin. Carbohydr Res 100:393–410

Tumolo T, Angnes L, Baptista MS (2004) Determination of the refractive index increment (dn/dc) of molecule and macromolecule solutions by surface plasmon resonance. Anal Biochem 333:273–279

van Dedem GW, Nielsen JI (1991) Determination of the molecular mass of low molecular mass (LMM) heparin. Pharmeuropa 3:202–218

van Gogh Investigators, Buller HR, Cohen AT et al (2007) Idraparinux versus standard therapy for venous thromboembolic disease. N Engl J Med 357:1094–104

Vignoli A, Marchetti M, Russo L et al (2011) LMWH bemiparin and ULMWH RO-14 reduce the endothelial angiogenic features elicited by leukemia, lung cancer, or breast cancer cells. Cancer Invest 29:153–161

Viskov C, Bouley E, Hubert P et al (2009a) Isolation and characterization of contaminants in recalled unfractionated heparin and low-molecular-weight heparin. Clin Appl Thromb Hemost 15:395–401

Viskov C, Just M, Laux V et al (2009b) Description of the chemical and pharmacological characteristics of a new hemisynthetic ultra-low-molecular-weight heparin, AVE5026. J Thromb Haemost 7:1143–1151

WHO (2014) WHO International biological reference preparations. http://www.who.int/biologicals/reference_preparations/en/. Accessed 12 Sept 2014

Xu Y, Masuko S, Takieddin M et al (2011) Chemoenzymatic synthesis of structurally homogeneous ultra-low molecular weight heparins. Science 334:498–501

Xu Y, Cai C, Chandarajoti K et al (2014) Homogeneous low-molecular-weight heparins with reversible anticoagulant activity. Nat Chem Biol 10:248–250

Zhang Z, Weïwer M, Li B et al (2008) Oversulfated chondroitin sulfate: Impact of a heparin impurity, associated with adverse clinical events, on low-molecular-weight heparin preparation. J Med Chem 51:5498–5501

Nanoparticle Albumin-Bound Anticancer Agents

Neil Desai

Contents

Introduction 329
Properties and Manufacturing of *nab*-Paclitaxel 330
Nanoparticle Properties 330
Manufacturing 331
Analytical Methods 332
Pharmacology of *nab*-Paclitaxel 333
Mechanism of Action 333
Pharmacokinetics 335
Efficacy and Safety 337
Regulatory Status 338
Regulatory Status of nab Technology-Based Drugs 338
Considerations for Evaluating Innovator Complex Drugs 339
Considerations for Evaluating Generic Complex Drugs 340
Regulatory Guidance 341
Prospects and Future Directions 342
References 343

Abstract Albumin is a key plasmic carrier of hydrophobic molecules and is highly accumulated in tumors. Nanoparticle albumin bound (nab) technology is a nanoparticle drug delivery platform that utilizes the unique transport and binding properties of albumin to achieve enhanced tumor penetration and accumulation of albumin-bound hydrophobic drugs while eliminating the need for toxic solvents coadministered with poorly soluble drugs. The first product in the nab drug family and the first protein nanotechnology-based chemotherapeutic approved by the US Food and Drug Administration and the EMA is nab-paclitaxel (Abraxane®, ABI-007; manufactured by Celgene Corporation, Summit, NJ). It is a Cremophor-free, albumin-bound nanoparticle formulation of paclitaxel with a mean particle size of approximately 130 nm. A proprietary process combines paclitaxel with albumin

N. Desai (✉)
Strategic Platforms, Celgene Corporation (Abraxis BioScience), 11755 Wilshire Blvd., Suite 2300, Los Angeles, CA 90025, USA
e-Mail: ndesai@celgene.com

D. J. A. Crommelin, J. S. B. de Vlieger (eds.), *Non-Biological Complex Drugs,* AAPS Advances in the Pharmaceutical Sciences Series 20, DOI 10.1007/978-3-319-16241-6_10

to create a colloidal suspension of nanoparticles. Paclitaxel and albumin are not covalently linked but rather associated through hydrophobic interactions. The particles of paclitaxel are in a noncrystalline, amorphous, readily bioavailable state, allowing for rapid drug release from the particles following intravenous administration. Nanoparticles of nab-paclitaxel are complex three dimensional constructs that require careful design and engineering, detailed orthogonal analysis methods, and a reproducible scale-up and manufacturing process to achieve a consistent product with the intended physicochemical characteristics, biological behavior, and pharmacological profiles. Due to its complexity, the safety and efficacy may be influenced by minor variations in the physicochemical properties or the manufacturing process and needs to be carefully examined in preclinical and clinical studies.

Preclinical and clinical studies have demonstrated that nanoparticlebased nab-paclitaxel displays distinct pharmacokinetics (PK) and biodistribution profiles compared with conventional Cremophor-paclitaxel. *nab*-Paclitaxel exhibits a linear PK profile with faster clearance and increased volume of distribution, whereas Cremophor-paclitaxel forms micelles leading to prolonged exposure to the systemic circulation, slower tissue distribution, and increased drug toxicity. In preclinical and clinical studies, nab-paclitaxel demonstrated an increased antitumor efficacy and an improved safety profile compared with Cremophor-paclitaxel. Based on significant clinical benefit in pivotal trials, nab-paclitaxel has been approved for use in the treatment of patients with metastatic breast cancer, locally advanced or metastatic non-small cell lung cancer (NSCLC), and for first-line treatment of metastatic adenocarcinoma of the pancreas. Anticancer agents based on nab technology demonstrate broad applications and could target multiple types of malignancies through exploitation of the natural properties of albumin and tumor biology.

Keywords Nanoparticle · Albumin · Paclitaxel · Chemotherapy · Bioequivalence · Drug distribution

Abbreviations

CI	Confidence interval
CrEL	Cremophor EL®
EPR	Enhanced permeability and retention
HR	Hazard ratio
MTD	Maximum tolerated dose
nab	Nanoparticle albumin bound
NSCLC	Non-small cell lung cancer
OS	Overall survival
PFS	Progression-free survival
PK	Pharmacokinetics
TEM	Transmission electron microscopy
Vdss	Volume of distribution steady state

Introduction

Nanoparticle albumin-bound (*nab*™) technology is a proprietary nanotechnology-based drug delivery platform that utilizes the endogenous properties and pathways of albumin to achieve solvent-free and efficient delivery of hydrophobic molecules to target sites. The conventional formulation of hydrophobic therapeutic agents requires the use of toxic solvents and surfactants such as Cremophor EL and Tween, which are associated with their own toxicities and can hinder the distribution and delivery of the active drug ingredient by micellar sequestration. In contrast, *nab* technology enables hydrophobic molecules to associate with albumin through non-covalent hydrophobic interaction to create a colloidal suspension of nanoparticles with a size of 50–150 nm (Fig. 1). Currently, *nab* technology has been successfully applied to a number of small molecule hydrophobic compounds, of which *nab*-paclitaxel has received market approval.

As a protein-based nanotechnology platform, *nab* technology occupies a unique niche between biologics and non biological complex drugs (NBCDs) and shares many similarities with NBCDs. Albumin is the most abundant plasma protein and an important natural transporter of a vast array of biological and chemical molecules, including hormones, bilirubin, metal ions, fatty acids, and hydrophobic drug molecules. Albumin is biologically compatible and has multiple specific and non-specific binding sites for a broad range of molecules, which are important characteristics required for polymers and other carriers used in the formulation of NBCDs.

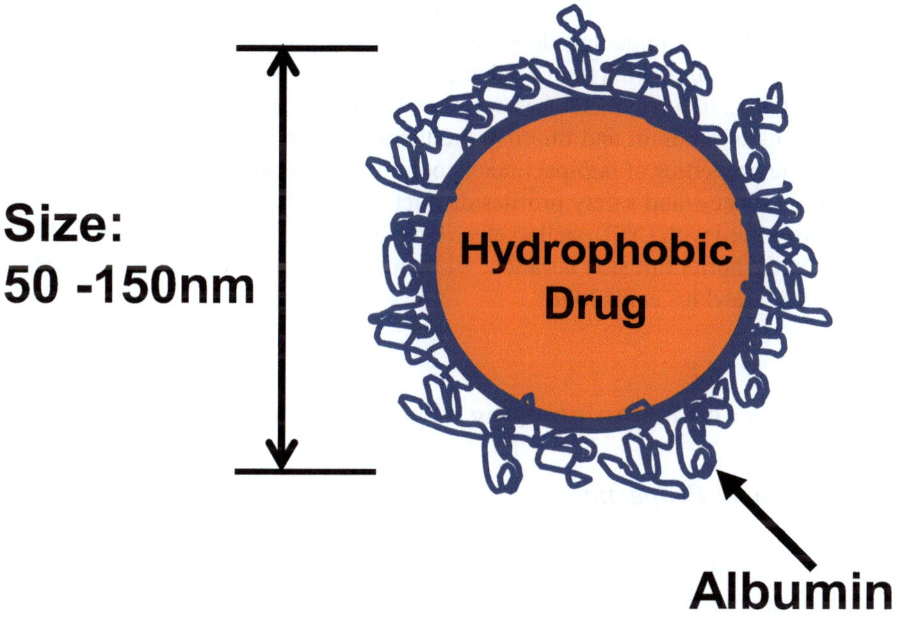

Fig. 1 Schematic representation of a nanoparticle prepared by *nab*-technology

Similar to NBCDs, *nab* nanoparticles are complex three-dimensional constructs of multiple components with a specific spatial arrangement. The pharmacokinetic and pharmacodynamic profiles of these drugs are a result of the complex combination of physicochemical properties, interaction between the components, integrity and stability of the complex constructs, and interaction of each of the components and their combinations within the biological environment. Therefore, unlike simple small molecule drug products, *nab* technology-based drugs demand many similar considerations as NBCDs and biologics during all stages of development, manufacturing, regulatory approval, and pharmacovigilance of generics and biosimilars.

The first *nab* technology-based product and the first protein-based nanoparticle chemotherapeutic approved for the market by the US Food and Drug Administration and European Medicines Agency is *nab*-paclitaxel (ABRAXANE®, Celgene Corporation, Summit, NJ), a Cremophor-free albumin stabilized particle form of paclitaxel. Paclitaxel is a potent antineoplastic agent with a broad spectrum of activity against solid tumors including breast, ovarian, lung, prostate, and other cancers. Paclitaxel stabilizes microtubules and interferes with microtubule reorganization, leading to altered mitosis and cell death (Schiff and Horwitz 1980; Srivastava et al. 1998; Verweij et al. 1994). Due to its poor aqueous solubility, the conventional formulation of paclitaxel consists of the polyoxyethylated castor oil vehicle Cremophor EL® (CrEL) and dehydrated ethanol USP (1:1, v/v). CrEL is associated with significant toxicities in patients including peripheral neuropathy and severe and potentially fatal hypersensitivity reactions that necessitate premedication with corticosteroids and antihistamines (Weiss et al. 1990; Irizarry et al. 2009; Mielke et al. 2006). In addition, CrEL sequesters paclitaxel in circulation by forming micelles with highly hydrophobic interiors and prolongs the exposure to the systemic circulation, thereby impeding drug delivery to target tissues and increasing side effects such as neutropenia (van Zuylen et al. 2001). To overcome these limitations posed by CrEL, *nab*-paclitaxel was developed to leverage the endogenous transport mechanisms of albumin and improve the therapeutic index over CrEL-paclitaxel. The unique properties of *nab*-paclitaxel confer it with pharmacokinetic, pharmacodynamic, efficacy, and safety profiles distinct from conventional CrEL-paclitaxel. When compared with CrEL paclitaxel, *nab*-paclitaxel can be administered at a higher dose with shorter infusion duration, without the need for premedication (Ibrahim et al. 2002; Hawkins et al. 2008).

Properties and Manufacturing of *nab*-Paclitaxel

Nanoparticle Properties

Nanoparticles of *nab*-paclitaxel have a narrow size distribution with a mean particle size of approximately 130 nm as determined by dynamic laser light scattering (DLS) (Gradishar 2006). The nanoparticles in different stages of development were characterized by transmission electron microscopy (TEM) and Cryo-TEM (Fig. 2).

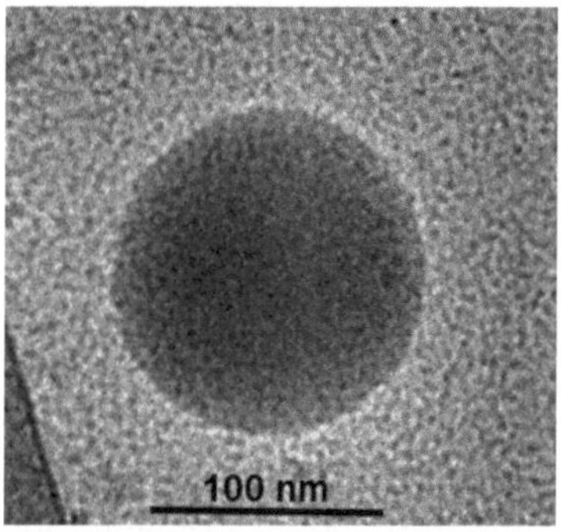

Fig. 2 Cryo-transmission electron micrograph of an early albumin-bound paclitaxel nanoparticle

Albumin has more than six specific and non-specific binding sites for paclitaxel with different affinities and with positive cooperativity (Paal et al. 2001). The nanoparticles of *nab*-paclitaxel utilize the affinity of paclitaxel to albumin, with paclitaxel non-covalently bound to albumin via hydrophobic interactions. A layer of albumin molecules crosslinked to a certain level forms the nanoparticle surface. The highly negative zeta potential of −31 mV and steric repulsion of this albumin surface prevent agglomeration and stabilize nanoparticles in aqueous suspension (Desai 2012a, 2012b, 2013). Reconstituted to 5 mg paclitaxel/mL concentration with 0.9 % (w/v) saline solution, *nab*-paclitaxel nanoparticles remain stable at room temperature for several days. X-ray powder diffraction revealed that paclitaxel within the nanoparticles is amorphous and non-crystalline (Desai 2012a, 2012b, 2013), allowing the drug to be readily available in circulation for rapid drug release and tissue distribution without the time lag and free energy needed to dissolve crystalline paclitaxel as in the case of nanocrystals (Merisko-Liversidge et al. 1996).

Manufacturing

Similar to NBCDs, *nab*-paclitaxel is not a homo-molecular structure. Rather, it has a complex three-dimensional multicomponent structure, with the composition highly dependent on the manufacturing process. A full understanding of the components and their interactions is essential to defining the key characteristics of the product (Ehmann et al. 2013). The *nab* technology uses the affinity of paclitaxel and albumin to assemble these molecules into *nab*-paclitaxel nanoparticles in a bottom-up approach. The manufacturing of *nab*-paclitaxel is a complex multistep process, and *nab*-paclitaxel underwent extensive preliminary testing of a wide range of conditions to ensure scalability and reproducibility (Desai 2012b). Due to the complexity of the product and the process, subtle changes in the manufacturing process may result in substantial changes in pharmacology and safety of the product (Crommelin and Florence 2013). In-process testing for important nanoparticle parameters is informative and vital for a well controlled manufacturing process (Hu et al. 2004; Langer et al. 2003). The manufacturing plan therefore needs to define acceptable limits for key nanoparticle attributes that can impact the product both from a physicochemical and biological perspective and identify process conditions critical to achieve these key attributes and functions (Desai 2012b).

The significant challenges for scale-up of *nab*-paclitaxel are clearly demonstrated by several recent unsuccessful attempts in the market to copy *nab*-paclitaxel. These attempted copy formulations, though claimed by the manufacturers to be copies of approved *nab*-paclitaxel, fail to reproduce the specific size distribution, stability, potency, or physicochemical characteristics of *nab*-paclitaxel, potentially leading to undesirable and unsafe effects (Desai 2012b). For example, one claimed copy had high endotoxin and residual solvent levels greatly exceeding safety limits allowed by regulatory authorities. Another attempted formulation had poor reproducibility in the manufacturing process. There were substantial inter-batch variations in particle size and a wide size distribution with a large portion of particles over 200 nm, resulting in significant drug loss following filtration through a 220 nm sterile filter. Once reconstituted, the nanoparticles also displayed poor stability under specific test conditions and formed large precipitates and aggregates of several micrometers in size within 24 h, whereas *nab*-paclitaxel nanoparticles remained stable under the same condition (Desai 2012b). In these cases, variations in composition and manufacturing process caused fundamental differences in the behavior and safety of the drug products.

Analytical Methods

The failed examples above also illustrate the need to develop orthogonal analytical methods to ensure the properties and consistency of complex drugs such as *nab*-paclitaxel (Feng 2006). Similar to NBCDs, *nab*-paclitaxel is a complex drug product, not simply the sum or mixture of all its individual components. These complex drugs consist of different, yet closely related structures that cannot be fully quantified and characterized by physicochemical analytical tools alone. Therefore, a comprehensive analytical approach is required to measure not only the characteristics of individual components in the drug composition, but also test the properties, structure, and the corresponding functions of the final drug product as a whole (Eifler and Thaxton 2011).

In addition to the standard analytical tests to quantify active and inactive ingredients as well as impurities, more sophisticated techniques are required to fully characterize *nab*-paclitaxel (Hawkins et al. 2008; Desai 2012a, b). The size and size distribution affect the sterile filtration, stability, dissolution, drug release, distribution, and clearance of nanoparticles. Various methods, including dynamic light scattering and transmission electron microscopy (TEM), are used to define the size of *nab*-paclitaxel nanoparticles. The surface charge stabilizes the particles from aggregation, and the zeta potential is measured by dynamic laser light scattering (Desai 2012b). The physical state of paclitaxel encapsulated in the nanoparticles, which is critical for nanoparticle stability, drug release, and drug distribution, is assessed by X-ray diffraction. The overall structure and morphology are analyzed by cryo-TEM. Nanoparticle stability on the shelf and solubility are assessed, and the stability of the reconstituted suspension is tested in different biological media and under various conditions. It is also important to analyze particle dissolution kinetics, which could relate to drug release, distribution, and PK in vivo (Desai 2012b).

Another key aspect of *nab*-paclitaxel characterization is its albumin content. As an integral component of *nab*-paclitaxel nanoparticles, albumin is not merely an inactive excipient, but confers special and unique characteristics that impact in vivo function and clinical performance of the drug, including particle stability, solubility, dissolution, drug release, transport, clearance, distribution, targeting, and tumor accumulation (Desai 2012b). Albumin is a biological molecule, therefore it will share similar analytical, pharmacological, and safety considerations with biologic drugs. The composition of nanoparticles with respect to albumin to paclitaxel ratio and oligomeric albumin content is dependent on the manufacturing process and is critical to particle stability, dissolution rate, biodistribution, safety and immunogenicity of *nab*-paclitaxel. Similar to considerations of a biologic or biosimilar drug, the albumin source and nanoparticle manufacturing steps may lead to molecular changes of albumin in chemical structure, conformation, denaturation, crosslinking, coagulation, and degradation. These modifications have the potential to cause immunogenicity and other safety issues, and alter nanoparticle properties and pharmacology (Zolnik et al. 2010). Therefore, the oligomerization and other characteristics of albumin need to be extensively tested and carefully controlled.

The safety of nanoparticles as a whole also requires monitoring. Interaction of paclitaxel and albumin can possibly result in conformation changes that result in immunological issues (Trynda-Lemiesz 2004). Unstable nanoparticles may form large aggregates in the micrometer size scale, which can be entrapped in the capillary bed of the lungs and pose a serious danger to patients. The complex nanoparticle manufacturing process presents many opportunities for endotoxin contamination, which is also a source for immune response. Other types of nanoparticles have also been associated with other hematologic safety concerns such as hemolysis and thrombogenicity, through nanoparticle-specific antibody or interactions of nanoparticles with erythrocyte and blood coagulation components (Dobrovolskaia and McNeil 2007; Bosi et al. 2004; Greish et al. 2011).

In summary, multiple orthogonal analysis methods are essential for appropriate in-process quality controls and tests for final products to ensure that *nab*-paclitaxel nanoparticles have all the desired properties for the intended therapeutic purpose and safety. Deviations from key nanoparticle parameters and processes could have serious negative impacts on safety and efficacy of *nab*-paclitaxel.

Pharmacology of *nab*-Paclitaxel

Mechanism of Action

By leveraging the natural properties and transport pathways of albumin, *nab*-paclitaxel can achieve enhanced drug delivery and accumulation in tumors. Upon infusion into the circulation, *nab*-paclitaxel nanoparticles undergo a dynamic dissolution process into smaller nanoparticles and eventually to albumin-bound paclitaxel complexes. The blood vessels of proliferating tumors are leaky and highly permeable to nanoparticles and macromolecules due to structural defects with fenestrations ranging between 0.2–1.2 μm, allowing nanoparticles with sizes below 200 nm to extravasate into tumors (Yuan et al. 1995; Hobbs et al. 1998; Haley and Frenkel 2008). The lack of proper lymphatic drainage in tumors reduced the clearance of albumin and other molecules with molecular weight greater than 40 kDa (Maeda et al. 2001). Theoretically, nanoparticles and macromolecules such as *nab*-paclitaxel nanoparticles and albumin-bound paclitaxel could potentially take advantage of the combined enhanced permeability and retention effect (EPR) to accumulate in tumors. However, recent studies raise serious questions into the clinical significance of EPR for the efficacy and safety of nanoparticles, as drug delivery to solid tumors in patients has been strongly hampered by major biological barriers including heterogeneous blood supply, elevated interstitial fluid pressure (IFP), and large transport distances in the tumor interstitum (Crommelin and Florence 2013; Nichols and Bae 2014). Importantly, albumin is transported across blood vessel endothelium through a receptor-mediated, active transcytosis mechanism (Fig. 3). The 60-kDa glycoprotein gp60 (albondin) is an albumin receptor located on endothelial cell surface and binds to native albumin with a high affinity in the nanomolar range (Schnitzer 1992). Albumin binding induces gp60 clustering and association with caveolar-scaffolding protein caveolin-1, leading to the activation of tyrosine kinase Src and the formation of caveolae (Tiruppathi et al. 1997). The plasmalemmal vesicles carrying both gp60-bound and fluid phase albumin migrate from apical to basal membrane, and release their contents by exocytosis into the subendothelial space. As shown by in vitro drug uptake and permeability assays, *nab*-paclitaxel formulation increased the endothelial binding of paclitaxel by 9.9 fold compared with CrEL-paclitaxel ($P<0.0001$) and the transport of paclitaxel across microvessel endothelial cell monolayers by 4.2 fold ($P<0.0001$) (Desai et al. 2006). On the other hand, the presence of clinically relevant levels of CrEL (up to 0.3 %) significantly decreased paclitaxel binding to albumin and the transport of paclitaxel across endothelial cells (Desai et al. 2006). These results clearly demonstrate that *nab*-paclitaxel, but not CrEL-paclitaxel, can utilize and leverage the active albumin transport mechanism for efficient distribution from the circulation.

Further, albumin is highly accumulated in tumors, as tumor cells take up albumin through endocytosis and macropinocytosis and catabolize albumin by lysosomal degradation (Stehle et al. 1997; Commisso et al. 2013). Albumin serves as a major energy and nitrogen source and provides amino acids such as glutamine to meet the great demand by fast proliferating tumor cells (Stehle et al. 1997; Commisso et al. 2013). The natural accumulation of albumin in solid tumors facilitates the delivery of active drug by *nab*-paclitaxel. Preclinical results showed that *nab*-paclitaxel achieved 33 % higher intratumoral paclitaxel concentration than equal dose of CrEL-paclitaxel in mice bearing human breast tumor xenografts (Desai et al. 2006). In pediatric tumor models of rhabdomyosarcoma and neuroblastoma, a 4–7 fold higher tumor/plasma paclitaxel drug ratio was observed for *nab*-paclitaxel compared with DMSO-paclitaxel (Zhang et al. 2013).

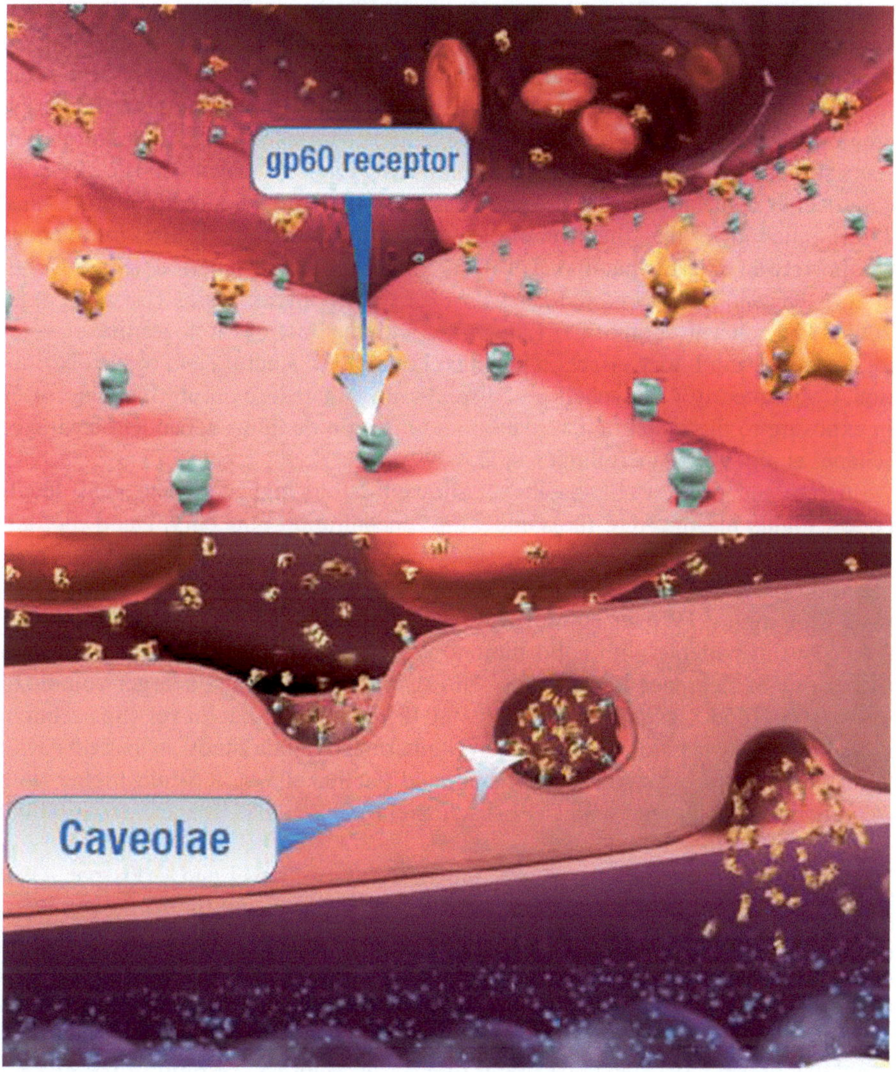

Fig. 3 Mechanisms for the transport of *nab*-paclitaxel into tumors. The transcytosis of albumin-bound paclitaxel complexes across the endothelial barrier is facilitated by the binding to the gp60 receptor and caveolar transport

Pharmacokinetics

Due to its unique albumin-based nanoparticle formulation, *nab*-paclitaxel displays distinct pharmacokinetic (PK) and biodistribution profiles compared with conventional CrEL-paclitaxel. In both preclinical and clinical studies, *nab*-paclitaxel exhibits a linear PK with rapid tissue distribution and increased distribution volume. Clinically, the systemic drug exposure for intravenous *nab*-paclitaxel was approximately dose proportional from 80 to 300 mg/m^2 and was independent of the infusion duration (Ibrahim et al. 2002; Chen et al. 2014). On the other hand, the cremophor EL/ethanol vehicle in CrEL-paclitaxel forms micelles that entrap and sequester paclitaxel in the vascular compartment (van Zuylen et al. 2001). The critical micellar concentration (CMC) of CrEL is 0.009 % in aqueous solution (Kessel 1992), much lower than the peak plasma CrEL level of 0.3–0.5 % after intravenous administration of CrEL-paclitaxel (100–175 mg/m^2, over a 3-h period) (Sparreboom et al. 1998) and the plasma CrEL level of 0.1 % 24 h after infusion (Brouwer et al. 2000). The concentration of CrEL increases with the dose of CrEL-paclitaxel, leading to greater inhibitory effect on paclitaxel binding to albumin and tissue distribution. Consequently, CrEL-paclitaxel displays more than dose proportional increases in exposure to the systemic circulation and infusion duration-dependent clearance (Gianni et al. 1995; van Tellingen et al. 1999).

In rats and mice, CrEL-paclitaxel showed 3-fold higher plasma peak levels (C_{max}), higher plasma AUC (area under the concentration-time curve), and approximately a 7- to 10-fold lower steady state volume of distribution (Vdss) compared with *nab*-paclitaxel (Sparreboom et al. 2005). In a PK study comparing *nab*-paclitaxel (260 mg/m^2 IV over 30 min, q3w) and CrEL-paclitaxel (175 mg/m^2 IV over 3 h, q3w) in patients with solid tumors, *nab*-paclitaxel displayed a significantly higher rate of clearance (21.13 vs. 14.76 L/h/m^2, P=0.048) and a larger volume of distribution (663.8 vs. 433.4 L/m^2, P=0.040) than CrEL-paclitaxel (Sparreboom et al. 2005). In a randomized crossover pharmacokinetic study in patients with solid tumors, the mean fraction of unbound paclitaxel was 2.6-fold higher with *nab*-paclitaxel compared with CrEL-paclitaxel (0.063 ± 0.021 versus 0.024 ± 0.009; P < 0.001) (Gardner et al. 2008).

In addition to distinct plasma PK profiles, *nab*-paclitaxel and CrEL-paclitaxel also show significant differences in distribution to tumor and tissues. Intravenous *nab*-paclitaxel achieved a 33 % higher intratumoral paclitaxel concentration at equal dose than CrEL-paclitaxel in MX-1 human breast tumor xenografts (Desai et al. 2006). In xenograft-bearing mice, radiolabled paclitaxel from *nab*-paclitaxel distributed favorably into tumors versus normal tissues at the early time points, with a *nab*-paclitaxel:CrEL-paclitaxel ratio of 1.25 for tumor vs. a ratio of 0.4–0.8 for normal tissue of different organs at 1 h post dose (Hawkins et al. 2003). A population PK study has demonstrated that the faster tissue distribution by *nab*-paclitaxel causes a shorter duration of high drug concentrations above 720 ng/mL in plasma, reducing the risk of the dose-limiting toxicity neutropenia (Chen et al. 2014). The increased paclitaxel delivery into tumors and reduced drug exposure to the circulation and normal tissues by *nab*-paclitaxel at least in part contributes to its enhanced antitumor efficacy and improved tolerability compared to CrEL-paclitaxel.

Efficacy and Safety

Numerous preclinical and clinical studies demonstrate the improved therapeutic index of *nab*-paclitaxel. In mice, the maximum tolerated dose (MTD) for *nab*-paclitaxel (30 mg/kg, qdx5) was substantially higher than for CrEL-paclitaxel (13.4 mg/kg, qdx5) (Desai et al. 2006). In nude mice bearing various human tumor xenografts (lung, breast, ovarian, prostate, and colon) receiving both agents at MTD, *nab*-paclitaxel resulted in more complete regressions, longer time to recurrence and tumor doubling, and prolonged survival (Desai et al. 2006). The antitumor activity of *nab*-paclitaxel was also better or equal compared with polysorbate-based docetaxel at its MTD in various breast, lung, prostate, and colon tumor xenograft models (Desai et al. 2008). In preclinical studies with pancreatic cancer models, the combination treatment of *nab*-paclitaxel and gemcitabine displayed strong antitumor activity and increased intratumoral gemcitabine levels, potentially due to the ability of *nab*-paclitaxel either to disrupt tumor stroma (Alvarez et al. 2013) or to decrease the protein levels of cytidine deaminase, the primary gemcitabine metabolizing enzyme (Frese et al. 2012). Furthermore, *nab*-paclitaxel was well tolerated and effective in suppressing growth of a wide range of pediatric tumor xenografts, including neuroblastoma, rhabdomyosarcoma, osteosarcoma, and Ewing sarcoma (Zhang et al. 2013; Wagner et al. 2014).

In a Phase 1 clinical study in patients with solid tumors, the relatively lower toxicities of *nab*-paclitaxel allowed the administration of a 70% higher dose than CrEL-paclitaxel (300 versus 175 mg/m^2, q3w) over a shorter infusion time (30 min versus 3 h), without the need for corticosteroid premedication (Ibrahim et al. 2002).

In a randomized Phase 3 study in 460 patients with metastatic breast cancer (Gradishar et al. 2005), compared with CrEL-paclitaxel at 175 mg/m^2 q3w, *nab*-paclitaxel administered at 260 mg/m^2 q3w had statistically significantly higher response rates (33 versus 19%, $P=0.001$), longer time to tumor progression (5.3 versus 3.9 months, $P=0.006$), and increased survival in the subset of patients receiving second-line or greater treatment (12.9 versus 10.7 months, $P=0.024$). The incidence of grade 4 neutropenia was significantly lower with *nab*-paclitaxel than with CrEL-paclitaxel (9 versus 22%, $P=0.001$). No severe hypersensitivity reactions occurred with *nab*-paclitaxel despite the lack of premedication. Grade 3 neuropathy was higher for *nab*-paclitaxel (10 vs. 2%, $P=0.001$) due to the approximately 50% higher dosage, but was easily manageable and improved quickly (median: 22 days).

In a randomized phase 3 study in patients with advanced non-small cell lung cancer (NSCLC) (Socinski et al. 2010), patients received either the combination of *nab*-paclitaxel (100 mg/m^2, qw)/carboplatin (AUC 6, q3w) ($n=521$) or CrEL (200 mg/m^2, q3w)/carboplatin ($n=531$). The *nab*-paclitaxel arm demonstrated a significantly higher overall response rate than the CrEL-paclitaxel arm (33 vs. 25%; response rate ratio, 1.313; 95% confidence interval [CI], 1.082–1.593; $P=0.005$) with a favorable trend in progression-free survival (PFS; median, 6.3 vs. 5.8 months; hazard ratio [HR], 0.902; 95% CI, 0.767–1.060; $P=0.214$) and overall survival (OS; median, 12.1 vs. 11.2 months; HR, 0.922; 95% CI, 0.797–1.066;

$P = 0.271$). Significantly less grade 3 and 4 neuropathy, neutropenia, arthralgia, and myalgia were reported in the *nab*-paclitaxel arm, and less thrombocytopenia and anemia in the CrEL-paclitaxel arm (Socinski et al. 2012).

Furthermore, *nab*-paclitaxel is highly active in the treatment of metastatic pancreatic cancer, whereas solvent-based taxanes failed to demonstrate clinically meaningful activity and adequate safety in multiple Phase 2 studies (Whitehead et al. 1997; Androulakis et al. 1999; Jacobs et al. 1999). In a randomized Phase 3 study in 861 patients with metastatic pancreatic cancer, combination of *nab*-paclitaxel (125 mg/m^2 weekly, 3 out of 4 weeks) and gemcitabine (1000 mg/m^2 weekly, 3 out of 4 weeks) demonstrated significantly longer overall survival and improved clinical outcomes compared with the standard of care treatment of gemcitabine alone (Von Hoff et al. 2013). The median OS was 8.5 months for the *nab*-paclitaxel/gemcitabine arm vs. 6.7 months for the gemcitabine arm (HR: 0.72; 95 % CI, 0.62–0.83; $P < 0.001$). The median PFS was 5.5 months in the *nab*-paclitaxel/gemcitabine arm vs. 3.7 months in the gemcitabine arm (HR: 0.69; 95 % CI, 0.58–0.82; $P < 0.001$); the response rate was 23 vs. 7 % in the two groups ($P < 0.001$). The survival rate was higher with the *nab*-paclitaxel/gemcitabine arm (35 vs. 22 % at 1 year, and 9 vs. 4 % at 2 years). The most common adverse events of Grade 3 or higher were neutropenia (38 % in the *nab*-paclitaxel/gemcitabine arm vs. 27 % in the gemcitabine arm), fatigue (17 vs. 7 %), and neuropathy (17 vs. 1 %). Febrile neutropenia occurred in 3 vs. 1 % of the patients in the two arms. In the *nab*-paclitaxel/gemcitabine arm, neuropathy of Grade 3 improved to Grade ≤ 1 in 29 days (median).

Regulatory Status

Regulatory Status of nab Technology-Based Drugs

Because of the clinical efficacy and safety demonstrated in these pivotal clinical trials, *nab*-paclitaxel has been approved in the US for the treatment of patients with metastatic breast cancer, locally advanced or metastatic NSCLC, and metastatic adenocarcinoma of the pancreas. It is also approved in the European Union (via the centralized procedure) and other countries for metastatic breast cancer and metastatic pancreatic cancer, and in Japan for the treatment of gastric cancer and NSCLC.

In addition to *nab*-paclitaxel, several other drugs based on the *nab* technology platform are under preclinical and clinical development for oncology and vascular disease indications. *nab*-Rapamycin (ABI-009) is an albumin-bound injectable form of rapamycin. The mammalian target of rapamycin, mTOR, is a key regulator of cell proliferation and an important target in cancer and proliferative vascular diseases (Dancey 2010; Goncharova 2013). In a phase 1 study in 26 patients with advanced solid tumors, *nab*-rapamycin was well tolerated with MTD established at 100 mg/m^2 weekly and showed evidence of responses and stable disease in heavily pretreated patients with various solid tumors including renal cell carcinoma and bladder cancer, both of which are known for mTOR overexpression (Gonzalez-Angulo et al. 2013).

Considerations for Evaluating Innovator Complex Drugs

As described above, *nab* technology-based drugs such as *nab*-paclitaxel are highly complex products that are intentionally designed with an arrangement of nanostructures to achieve desired pharmaceutical functions. The nanoparticles impart unique physical and chemical attributes that lead to improved distribution, efficacy, and safety over conventional solvent-based counterparts. The structure and morphology of these albumin nanoparticles are critical for their performance. These drugs utilize the natural transport pathways and tumor accumulation properties of albumin. Albumin is an integral biologic component of the nanoparticles and plays a critical role in the overall efficacy and toxicities of these drugs. The specific modifications of albumin during manufacture need to be fully characterized and carefully maintained to control particle functional attributes and to ensure consistency in drug performance and safety.

In several key aspects, *nab*-based drugs share similar characteristics with NBCDs. The drugs consist of multiple closely related structures, which are highly dependent on the particular manufacturing process. Further, the entire complex contributes to the activity of the pharmaceutical ingredient, and the properties cannot be fully characterized by physicochemical analysis alone. Because of the sophisticated nature and manufacturing process of these complex drugs, they may require additional levels of development effort and regulatory vigilance.

For innovator complex drugs, it is important to identify the key characteristics of the product that are essential for its activity and safety, and ensure those critical characteristics reproduced within acceptable pharmaceutical limits in the manufacturing process. Additional "structure-function" tests may be required to verify the biological functions such as transcytosis, tissue distribution, drug release, and accumulation in target sites.

There are other practical considerations to evaluate the PK of innovator complex drugs like *nab*-based drugs and NBCDs. Small molecule drug products rely more on passive diffusion for tissue distribution, with tissue levels at intended therapeutic sites typically correlating with blood levels; being only a small fraction of the total administered dose. On the other hand, subtle compositional and physicochemical differences in complex drugs can affect their biodistribution and PK, and complex drugs with targeting capability can significantly alter the distribution of the active pharmaceutical ingredients, resulting in major changes in drug efficacy. For *nab*-paclitaxel, the targeting to tumor sites by endogenous albumin pathways confers increased efficacy in breast cancer and NSCLC, and in pancreatic cancer where conventional taxanes were not effective. The PK analysis of *nab*-paclitaxel is further complicated by the presence of dissolving nanoparticles, albumin-bound paclitaxel, and free paclitaxel. Due to the rapid tissue distribution by *nab*-based drugs, there is no direct correlation between plasma concentration and clinical outcome.

There are also unique challenges to evaluate the safety of innovator complex drugs. Because of the complexity and diversity of such drugs, each drug requires case by case review. The standard battery of formal preclinical toxicology studies can identify many potential adverse effects of complex drugs in patients. Additional in vitro assays should be conducted to test the possible interaction of complex drugs with blood components and the immune system (Dobrovolskaia 2008). However, animal models are generally poor predictors of immunological responses in human. In clinical studies, special attention needs to be paid to immune reactions, in response to both individual drug components (non-biological and biological) and the complex drug as a whole.

Considerations for Evaluating Generic Complex Drugs

In recent years, there have been intense public debates on the evaluation of generic versions of complex drugs as their earliest examples are coming off patent protection. For simple, small molecule drugs, a generic version needs to contain the identical active pharmaceutical ingredient and exhibit bioequivalence with the analysis of plasma PK as indicator for efficacy and safety. In contrast, the properties and performance of complex drugs are highly sensitive to variations in the manufacturing process. Full in vitro characterization of composition and structural features is extremely difficult if not impossible, and there is no direct correlation between plasma concentration and clinical activities. Therefore, the process for evaluating generic small molecule drugs may not be appropriate for complex drugs.

For complex drugs such as *nab*-based drugs and NBCDs, a putative generic version requires more careful evaluation on several aspects. First, the generic version must demonstrate sameness or equivalence of critical physicochemical and functional attributes through comprehensive in vitro and in vivo testing. This task is impossible without a full understanding of the nature and structure-function relationship of the drug. Second, bioequivalence cannot be assessed with PK alone. To build on traditional PK studies focusing on evaluating C_{max} and AUC is insufficient to demonstrate comparability in efficacy and safety for complex drugs, without taking into account the drug complexity, variations in drug distribution, and potential targeting mechanisms. This may lead to generic copies with similar plasma PK profiles that however, are not interchangeable or substitutable with the originator. Finally, the safety of a generic complex drug, particularly related to immune responses, cannot be adequately assessed in animal models. Taken together, a totality of evidence approach needs to be adopted for evaluating complex drugs. The limitations of preclinical and traditional PK studies may eventually warrant a full-fledged clinical study program to conclusively demonstrate the efficacy and safety of a generic complex drug.

A recent detailed analysis and simulation study of *nab*-paclitaxel PK has shown the dynamic complexity of *nab*-paclitaxel and the micellar paclitaxel formulations in relation to their kinetics of in vivo breakdown and distribution of paclitaxel into tissues (Li et al. 2014). The analysis suggests that the main fraction of paclitaxel is rapidly distributed over the peripheral compartment (tissues) and in the same timeframe as the breakdown or release of the nanoparticle or micellar based paclitaxel occurs. Because of the ultra-fast distribution and decomposition of paclitaxel-carrier complexes, the sensitivity analyses demonstrated that changes in distribution and decomposition significantly affect the extravascular tissue distribution of paclitaxel (as paclitaxel-carrier complexes, free paclitaxel, and protein bound paclitaxel) but do not significantly affect the plasma concentration of total paclitaxel. Extrapolating beyond *nab*-paclitaxel and CrEL-paclitaxel to different paclitaxel delivery vehicles, these would be expected to deliver a distinct amount of paclitaxel to different tissues and organs, resulting in distinct efficacy and safety profiles. On the other hand, the fast extravascular distribution of the paclitaxel-carrier complexes and rapid decomposition in blood renders total plasma paclitaxel exposure insensitive to the changes. Thus, measuring the total plasma concentration of paclitaxel following IV administration of paclitaxel in a certain formulation does not offer useful insight into the tissue drug distribution and resulting pharmacology, efficacy and safety (Li et al. 2014). Therefore traditional PK bioequivalence studies may not be appropriate to establish therapeutic equivalence between complex formulations, and clinical safety and efficacy studies may be required for this purpose.

Regulatory Guidance

In the near future, there will be a strong surge of new complex drugs entering the market, either as innovator drugs or generic copies of existing drugs. There is an urgent need for regulatory authorities around the world to establish proper regulatory guidance for this field. While some general guidelines have been issued by regulatory agencies for biosimilars, there is a general lack of comprehensive regulatory guidance for nanoparticle drugs and NBCDs.

Currently, the FDA, EMA, and other regulatory agencies examine each new complex drug on a product-by-product basis. The FDA recognizes that each case may be different for nanomedicines, which requires thorough understanding and testing of critical physicochemical properties and structure (FDA 2010). The FDA draft guidance for Doxil indicated that this nanosimilar drug product would need to have the same drug product composition and equivalent liposome characteristics. Moreover, the manufacturing process is critical in addition to other attributes of pharmacokinetic bioequivalence (FDA 2002, 2010). The European Medicines Agency (EMA) has also issued several relevant reflection papers on intravenous liposomal products, nanosized colloidal iron-based preparations, block copolymer micelles, and nanomedicine surface coatings (Ehmann et al. 2013). Similar to the FDA, the EMA recognizes that “follow-on” nanomedicines (nanosimilars) may have safety profiles different from ‘traditional’ medicines and need a case-by-case evaluation. The surface properties such as surface ligand orientation can impact PK, biodistribution, stability, and intracellular fate of nanomedicines and are critical for safety and efficacy. Further, the specific physicochemical properties of nanomedicines are dependent on the manufacturing process, hence deriving the concept: the process is the product.

Specific for albumin-bound paclitaxel formulations, FDA issued a draft guidance in September 2012 (FDA 2012). In the draft guidance, FDA recommended two specific bioequivalence studies: single-dose two-way crossover in vivo bioequivalence study with PK endpoints (AUC and C_{max}) for unbound and total paclitaxel; and in vitro particle size distribution profile measuring population bioequivalence based on D_{50} and span $(D_{90}-D_{10})/D_{50}$ or polydispersity index. In addition, the FDA recommended in vitro characterization tests to demonstrate sameness between the test and reference products in terms of particle morphology, particle size, surface potential, paclitaxel crystallinity, fraction of free and bound paclitaxel or albumin in reconstituted suspension, nature of bond between paclitaxel and albumin, and in vitro release kinetics. The FDA draft guidance also recognized the importance and critical role of albumin and recommended characterization of the oligomeric status of albumin in both the albumin excipient and the final drug product. Finally, it encouraged drug applicants to explore methods to characterize in vitro release.

The FDA draft guidance represents an initial step to evaluate innovator *nab*-based drugs and their purported generic copies. Generics of *nab*-paclitaxel must be treated with increased regulatory vigilance and demonstrate equivalence of critical physicochemical properties essential to product performance. With increasing knowledge of the complexity associated with *nab*-based drugs, additional tests need to be conducted to provide a more comprehensive evaluation, including but not limited to particle solubility, dissolution kinetics, endothelial transport, tissue distribution, intratumor drug penetration and accumulation. A "totality-of-the-evidence" approach should involve considerations of structural and functional characterization, nonclinical evaluation, human PK and PD data, clinical immunogenicity data and clinical safety and efficacy data, to ensure sameness and pharmaceutical interchangeability of a generic *nab*-paclitaxel product.

Prospects and Future Directions

The *nab* technology based platform represents a major breakthrough in the delivery of hydrophobic drugs. The natural properties of albumin are harnessed to enable efficient and safe delivery of drugs to tumor and/or other disease sites, while minimizing systemic exposure and side effects to/in normal tissues.

The knowledge gained and lessons learned from the development and regulatory approval of *nab*-based drugs are highly relevant to other complex drugs such as NBCDs. The challenges facing these multicomponent and multifunctional drugs are similar to those of biologic and biosimilar drug products. More innovative analytical methods are constantly being developed and applied to achieve a better understanding of the drug under development and evaluation. Therapeutic equivalence of a generic version of an innovator's complex drug cannot be assessed by using the same drug product ingredients and comparing plasma PK data with a novel drug. There is an urgent need for science based guidance from regulatory agencies to develop a comprehensive list of assays and a streamlined approval process for these complex drug products.

Acknowledgements Dr Shihe Hou's expert editorial and writing assistance of this manuscript is greatly appreciated.

References

Alvarez R, Musteanu M, Garcia-Garcia E, Lopez-Casas PP, Megias D, Guerra C, Munoz M, Quijano Y, Cubillo A, Rodriguez-Pascual J, Plaza C, de Vicente E, Prados S, Tabernero S, Barbacid M, Lopez-Rios F, Hidalgo M (2013) Stromal disrupting effects of *nab*-paclitaxel in pancreatic cancer. Br J Cancer 109(4):926–933. doi:10.1038/bjc.2013.415

Androulakis N, Kourousis C, Dimopoulos MA, Samelis G, Kakolyris S, Tsavaris N, Genatas K, Aravantinos G, Papadimitriou C, Karabekios S, Stathopoulos GP, Georgoulias V (1999) Treatment of pancreatic cancer with docetaxel and granulocyte colony-stimulating factor: a multicenter phase II study. J Clin Oncol 17(6):1779–1785

Bosi S, Feruglio L, Da Ros T, Spalluto G, Gregoretti B, Terdoslavich M, Decorti G, Passamonti S, Moro S, Prato M (2004) Hemolytic effects of water-soluble fullerene derivatives. J Med Chem 47(27):6711–6715. doi:10.1021/jm0497489

Brouwer E, Verweij J, De Bruijn P, Loos WJ, Pillay M, Buijs D, Sparreboom A (2000) Measurement of fraction unbound paclitaxel in human plasma. Drug Metab Dispos 28(10):1141–1145

Chen N, Li Y, Ye Y, Palmisano M, Chopra R, Zhou S (2014) Pharmacokinetics and pharmacodynamics of *nab*-paclitaxel in patients with solid tumors: disposition kinetics and pharmacology distinct from solvent-based paclitaxel. J Clin Pharmacol. doi:10.1002/jcph.304

Commisso C, Davidson SM, Soydaner-Azeloglu RG, Parker SJ, Kamphorst JJ, Hackett S, Grabocka E, Nofal M, Drebin JA, Thompson CB, Rabinowitz JD, Metallo CM, Vander Heiden MG, Bar-Sagi D (2013) Macropinocytosis of protein is an amino acid supply route in Ras-transformed cells. Nature 497(7451):633–637. doi:nature12138 [pii] 10.1038/nature12138

Crommelin DJ, Florence AT (2013) Towards more effective advanced drug delivery systems. Int J Pharm 454(1):496–511. doi:10.1016/j.ijpharm.2013.02.020

Dancey J (2010) mTOR signaling and drug development in cancer. Nat Rev Clin Oncol 7(4):209–219. doi:10.1038/nrclinonc.2010.21

Desai N (2012a) Albumin drug nanoparticles. In: Kratz F, Senter P, Steinhagen H (eds) Drug delivery in oncology: from basic research to cancer therapy, vol 1, 1 edn. Wiley-VCH Verlag GmbH & Co. KGaA, Weinheim, Germany, pp 1133–1161. doi:DOI: 10.1002/9783527634057

Desai N (2012b) Challenges in development of nanoparticle-based therapeutics. AAPS J 14(2):282–295. doi:10.1208/s12248-012-9339-4

Desai N (2013) Integration of *nab*-technology into clinical drug development. In: Bischoff J (ed) Nanotechnologie beim Mammakarzinom- Grundlagen und aktuelle Perspektiven. UNI-MED Verlag AG, Bremen, Germany, pp 22–31

Desai N, Trieu V, Yao Z, Louie L, Ci S, Yang A, Tao C, De T, Beals B, Dykes D, Noker P, Yao R, Labao E, Hawkins M, Soon-Shiong P (2006) Increased antitumor activity, intratumor paclitaxel concentrations, and endothelial cell transport of cremophor-free, albumin-bound paclitaxel, ABI-007, compared with cremophor-based paclitaxel. Clin Cancer Res 12(4):1317–1324

Desai NP, Trieu V, Hwang LY, Wu R, Soon-Shiong P, Gradishar WJ (2008) Improved effectiveness of nanoparticle albumin-bound (*nab*) paclitaxel versus polysorbate-based docetaxel in multiple xenografts as a function of HER2 and SPARC status. Anticancer Drugs 19(9):899–909. doi:10.1097/CAD.0b013e32830f904600001813-200810000-00007 [pii]

Dobrovolskaia MA, McNeil SE (2007) Immunological properties of engineered nanomaterials. Nat Nanotechnol 2(8):469–478. doi:nnano.2007.223 [pii] 10.1038/nnano.2007.223

Dobrovolskaia MA, Aggarwal P, Hall JB, McNeil SE (2008) Preclinical studies to understand nanoparticle interaction with the immune system and its potential effects on nanoparticle biodistribution. Mol Pharm 5(4):487–495

Ehmann F, Sakai-Kato K, Duncan R, Hernan Perez de la Ossa D, Pita R, Vidal JM, Kohli A, Tothfalusi L, Sanh A, Tinton S, Robert JL, Silva Lima B, Amati MP (2013) Next-generation nanomedicines and nanosimilars: EU regulators' initiatives relating to the development and evaluation of nanomedicines. Nanomedicine (London, England) 8(5):849–856. doi:10.2217/nnm.13.68

Eifler AC, Thaxton CS (2011) Nanoparticle therapeutics: FDA approval, clinical trials, regulatory pathways, and case study. Methods Mol Biol 726:325–338. doi:10.1007/978-1-61779-052-221

FDA (2002) Guidance for Industry: Liposome Drug Products. http://www.fda.gov/downloads/drugs/guidancecomplianceregulatoryinformation/guidances/ucm070570.pdf
FDA (2010) Draft Guidance on Doxorubicin Hydrochloride. http://www.fda.gov/downloads/Drugs/.../Guidances/UCM199635.pdf
FDA (2012) Draft Guidance on Paclitaxel. http://www.fda.gov/downloads/drugs/guidancecomplianceregulatoryinformation/guidances/ucm320015.pdf
Feng SS (2006) New-concept chemotherapy by nanoparticles of biodegradable polymers: where are we now? Nanomed 1(3):297–309. doi:10.2217/17435889.1.3.297
Frese KK, Neesse A, Cook N, Bapiro TE, Lolkema MP, Jodrell DI, Tuveson DA (2012) *Nab*-Paclitaxel potentiates gemcitabine activity by reducing cytidine deaminase levels in a mouse model of pancreatic cancer. Cancer Discov 2(3):260–269. doi:2159-8290.CD-11-0242 [pii] 10.1158/2159-8290.CD-11-0242
Gardner ER, Dahut WL, Scripture CD, Jones J, Aragon-Ching JB, Desai N, Hawkins MJ, Sparreboom A, Figg WD (2008) Randomized crossover pharmacokinetic study of solvent-based paclitaxel and *nab*-paclitaxel. Clin Cancer Res 14(13):4200–4205. doi:10.1158/1078-0432.ccr-07-4592
Gianni L, Kearns CM, Giani A, Capri G, Vigano L, Locatelli A, Bonadonna G, Egorin MJ (1995) Nonlinear pharmacokinetics and metabolism of paclitaxel and its pharmacokinetic/pharmacodynamic relationships in humans. J Clin Oncol 13(1):180–190
Goncharova EA (2013) mTOR and vascular remodeling in lung diseases: current challenges and therapeutic prospects. FASEB J 27 (5):1796–1807. doi:10.1096/fj.12-222224
Gonzalez-Angulo AM, Meric-Bernstam F, Chawla S, Falchook G, Hong D, Akcakanat A, Chen H, Naing A, Fu S, Wheler J, Moulder S, Helgason T, Li S, Elias I, Desai N, Kurzrock R (2013) Weekly *nab*-rapamycin in patients with advanced nonhematologic malignancies: final results of a phase I trial. Clin Cancer Res 19(19):5474–5484. doi:10.1158/1078-0432.CCR-12-3110
Gradishar WJ (2006) Albumin-bound paclitaxel: a next-generation taxane. Expert Opin Pharmacother 7(8):1041–1053
Gradishar WJ, Tjulandin S, Davidson N, Shaw H, Desai N, Bhar P, Hawkins M, O'Shaughnessy J (2005) Phase III trial of nanoparticle albumin-bound paclitaxel compared with polyethylated castor oil-based paclitaxel in women with breast cancer. J Clin Oncol 23(31):7794–7803. doi:10.1200/jco.2005.04.937
Greish K, Thiagarajan G, Herd H, Price R, Bauer H, Hubbard D, Burckle A, Sadekar S, Yu T, Anwar A, Ray A, Ghandehari H (2011) Size and surface charge significantly influence the toxicity of silica and dendritic nanoparticles. Nanotoxicology. doi:10.3109/17435390.2011.604442
Haley B, Frenkel E (2008) Nanoparticles for drug delivery in cancer treatment. Urol Oncol 26(1):57–64
Hawkins MJ, Desai N, Soon-Shiong P (2003) Rationale, Preclinical Support, and Clinical Proof-of-Concept for Formulating Waterinsoluble Therapeutics as Albumin-stabilized Nanoparticles: Experience with Paclitaxel. [abstr 442]. American Association for Cancer Research (AACR) Annual Meeting, Anaheim, CA, 2003. pp Control/Tracking Number: 03-AB-442-AACR
Hawkins MJ, Soon-Shiong P, Desai N (2008) Protein nanoparticles as drug carriers in clinical medicine. Adv Drug Deliv Rev 60(8):876–885. doi:S0169-409X(08)00042-2 [pii] 10.1016/j.addr.2007.08.044
Hobbs SK, Monsky WL, Yuan F, Roberts WG, Griffith L, Torchilin VP, Jain RK (1998) Regulation of transport pathways in tumor vessels: role of tumor type and microenvironment. Proc Natl Acad Sci U S A 95(8):4607–4612
Hu J, Johnston KP, Williams RO 3rd (2004) Nanoparticle engineering processes for enhancing the dissolution rates of poorly water soluble drugs. Drug Dev Ind Pharm 30(3):233–245
Ibrahim NK, Desai N, Legha S, Soon-Shiong P, Theriault RL, Rivera E, Esmaeli B, Ring SE, Bedikian A, Hortobagyi GN, Ellerhorst JA (2002) Phase I and pharmacokinetic study of ABI-007, a cremophor-free, protein-stabilized, nanoparticle formulation of paclitaxel. Clin Cancer Res 8(5):1038–1044

Irizarry L, Luu T, McKoy J, Samaras A, Fisher M, Carias E, Raisch D, Calhoun E, Bennett C (2009) Cremophor EL-containing paclitaxel-induced anaphylaxis: a call to action. Community Oncol 6(3):132–134

Jacobs AD, Otero H, Picozzi V (1999) Gemcitabine (G) and Taxotere® (T) in patients with unresectable pancreatic carcinoma. Proc Am Soc Clin Oncol 18:1103A

Kessel D (1992) Properties of cremophor EL micelles probed by fluorescence. Photochem Photobiol 56(4):447–451

Langer K, Balthasar S, Vogel V, Dinauer N, von Briesen H, Schubert D (2003) Optimization of the preparation process for human serum albumin (HSA) nanoparticles. Int J Pharm 257(1–2):169–180

Li Y, Chen N, Palmisano M, Zhou S (2014) Pharmacology of paclitaxel sensitive to its delivery vehicles drives distinct clinical outcome of paclitaxel formulations. J Pharmacol Exp Ther (Submitted)

Maeda H, Sawa T, Konno T (2001) Mechanism of tumor-targeted delivery of macromolecular drugs, including the EPR effect in solid tumor and clinical overview of the prototype polymeric drug SMANCS. J Control Release 74(1–3):47–61

Merisko-Liversidge E, Sarpotdar P, Bruno J, Hajj S, Wei L, Peltier N, Rake J, Shaw JM, Pugh S, Polin L, Jones J, Corbett T, Cooper E, Liversidge GG (1996) Formulation and antitumor activity evaluation of nanocrystalline suspensions of poorly soluble anticancer drugs. Pharm Res 13(2):272–278

Mielke S, Sparreboom A, Mross K (2006) Peripheral neuropathy: a persisting challenge in paclitaxel-based regimes. Eur J Cancer 42(1):24–30

Nichols JW, Bae YH (2014) EPR: evidence and fallacy. J Control Release 190:451–464. doi:10.1016/j.jconrel.2014.03.057

Paal K, Muller J, Hegedus L (2001) High affinity binding of paclitaxel to human serum albumin. Eur J Biochem 268(7):2187–2191

Schiff PB, Horwitz SB (1980) Taxol stabilizes microtubules in mouse fibroblast cells. Proc Natl Acad Sci U S A 77(3):1561–1565

Schnitzer JE (1992) gp60 is an albumin-binding glycoprotein expressed by continuous endothelium involved in albumin transcytosis. Am J Physiol 262(1 Pt 2):H246–254

Socinski MA, Vinnichenko I, Okamoto I, Hon JK (2010) Hirsh V Results of a Randomized, Phase 3 Trial of *nab*-Paclitaxel (*nab*-P) and Carboplatin (C) Compared With Cremophor-based Paclitaxel (P) and Carboplatin as First-line Therapy in Advanced Non-small Cell Lung Cancer (NSCLC). Proceedings of the 46th American Society of Clinical Oncology Annual Meeting (ASCO), Chicago, IL, June 4–8 2010. p abstr LBA7511

Socinski MA, Bondarenko I, Karaseva NA, Makhson AM, Vynnychenko I, Okamoto I, Hon JK, Hirsh V, Bhar P, Zhang H, Iglesias JL, Renschler MF (2012) Weekly *nab*-paclitaxel in combination with carboplatin versus solvent-based paclitaxel plus carboplatin as first-line therapy in patients with advanced non-small-cell lung cancer: final results of a phase III trial. J Clin Oncol 30(17):2055–2062. doi:10.1200/JCO.2011.39.5848

Sparreboom A, Verweij J, van der Burg ME, Loos WJ, Brouwer E, Vigano L, Locatelli A, de Vos AI, Nooter K, Stoter G, Gianni L (1998) Disposition of Cremophor EL in humans limits the potential for modulation of the multidrug resistance phenotype in vivo. Clin Cancer Res 4(8):1937–1942

Sparreboom A, Scripture CD, Trieu V, Williams PJ, De T, Yang A, Beals B, Figg WD, Hawkins M, Desai N (2005) Comparative preclinical and clinical pharmacokinetics of a cremophor-free, nanoparticle albumin-bound paclitaxel (ABI-007) and paclitaxel formulated in Cremophor (Taxol). Clin Cancer Res 11(11):4136–4143

Srivastava RK, Srivastava AR, Korsmeyer SJ, Nesterova M, Cho-Chung YS, Longo DL (1998) Involvement of microtubules in the regulation of Bcl2 phosphorylation and apoptosis through cyclic AMP-dependent protein kinase. Mol Cell Biol 18(6):3509–3517

Stehle G, Sinn H, Wunder A, Schrenk HH, Stewart JC, Hartung G, Maier-Borst W, Heene DL (1997) Plasma protein (albumin) catabolism by the tumor itself–implications for tumor metabolism and the genesis of cachexia. Crit Rev Oncol Hematol 26(2):77–100

Tiruppathi C, Song W, Bergenfeldt M, Sass P, Malik AB (1997) Gp60 activation mediates albumin transcytosis in endothelial cells by tyrosine kinase-dependent pathway. J Biol Chem 272(41):25968–25975

Trynda-Lemiesz L (2004) Paclitaxel-HSA interaction. Binding sites on HSA molecule. Bioorg Med Chem 12(12):3269–3275. doi:10.1016/j.bmc.2004.03.073S0968089604002767 [pii]

van Tellingen O, Huizing MT, Panday VR, Schellens JH, Nooijen WJ, Beijnen JH (1999) Cremophor EL causes (pseudo-) non-linear pharmacokinetics of paclitaxel in patients. Br J Cancer 81(2):330–335

van Zuylen L, Karlsson MO, Verweij J, Brouwer E, de Bruijn P, Nooter K, Stoter G, Sparreboom A (2001) Pharmacokinetic modeling of paclitaxel encapsulation in Cremophor EL micelles. Cancer Chemother Pharmacol 47(4):309–318

Verweij J, Clavel M, Chevalier B (1994) Paclitaxel (Taxol) and docetaxel (Taxotere): not simply two of a kind. Ann Oncol 5(6):495–505

Von Hoff DD, Ervin T, Arena FP, Chiorean EG, Infante J, Moore M, Seay T, Tjulandin SA, Ma WW, Saleh MN, Harris M, Reni M, Dowden S, Laheru D, Bahary N, Ramanathan RK, Tabernero J, Hidalgo M, Goldstein D, Van Cutsem E, Wei X, Iglesias J, Renschler MF (2013) Increased survival in pancreatic cancer with *nab*-paclitaxel plus gemcitabine. N Engl J Med 369(18):1691–1703. doi:10.1056/NEJMoa1304369

Wagner LM, Yin H, Eaves D, Currier M, Cripe TP (2014) Preclinical evaluation of nanoparticle albumin-bound paclitaxel for treatment of pediatric bone sarcoma. Pediatric Blood Cancer. doi:10.1002/pbc.25062

Weiss RB, Donehower RC, Wiernik PH, Ohnuma T, Gralla RJ, Trump DL, Baker JR Jr, Van Echo DA, Von Hoff DD, Leyland-Jones B (1990) Hypersensitivity reactions from taxol. J Clin Oncol 8(7):1263–1268

Whitehead RP, Jacobson J, Brown TD, Taylor SA, Weiss GR, Macdonald JS (1997) Phase II trial of paclitaxel and granulocyte colony-stimulating factor in patients with pancreatic carcinoma: a Southwest Oncology Group study. J Clin Oncol 15(6):2414–2419

Yuan F, Dellian M, Fukumura D, Leunig M, Berk DA, Torchilin VP, Jain RK (1995) Vascular permeability in a human tumor xenograft: molecular size dependence and cutoff size. Cancer Res 55(17):3752–3756

Zhang L, Marrano P, Kumar S, Leadley M, Elias E, Thorner P, Baruchel S (2013) *Nab*-Paclitaxel is an active drug in preclinical model of pediatric solid tumors. Clin Cancer Res 19(21):5972–5983. doi:10.1158/1078-0432.CCR-13-1485

Zolnik BS, Gonzalez-Fernandez A, Sadrieh N, Dobrovolskaia MA (2010) Nanoparticles and the immune system. Endocrinology 151(2):458–465. doi:en.2009-1082 [pii] 10.1210/en.2009–1082

Part IV
Regulatory landscape and outlook

The EU Regulatory Landscape of Non-Biological Complex Drugs

Ruben Pita

Contents

Introduction 352
EU Pharmaceutical Legislation and NBCDs 353
Legislating NBCDs 355
Legal Basis of Submission to Obtain a Marketing Authorisation 360
Post-Authorisation Considerations 365
Medical Devices, Borderline and Combination Products 366
Scientific Guidance 367
Lessons from ATMPs 371
Conclusion 373
References 374

Abstract Non-biological complex drugs (NBCDs) present today and in the future therapeutic opportunities to, inter alia, target the delivery of active ingredients, reduce toxicity and improve efficacy. It has been widely debated whether the EU regulatory system is ready to embrace this group of medicinal products. This chapter introduces the EU pharmaceutical legislation and explains how NBCDs are already integrated into the existing legislative framework. Supported by a recent example of legislative measures constructed purposefully to account for a new class of medicinal products (Advanced Therapy Medicinal Products), we explain what is currently done by EU competent authorities to foster the development of NBCDs and what are the options and challenges faced by the regulatory system to evolve in parallel with the progress made with this innovative and promising class of medicinal products.

R. Pita (✉)
European Medicine Agency, EMA, London, UK
e-mail: Ruben.Pita@ema.europa.eu

D. J. A. Crommelin, J. S. B. de Vlieger (eds.), *Non-Biological Complex Drugs*, AAPS Advances in the Pharmaceutical Sciences Series 20, DOI 10.1007/978-3-319-16241-6_11

Keywords EU legislation · Marketing authorisation · Regulatory guidance · European medicines agency

Note: Text in italics is transcription from EU legislation

Abbreviations

API	Active pharmaceutical ingredient
ATMP	Advanced therapy medicinal products
CHMP	Committee for medicinal products for human use
COMP	Committee for orphan medicinal products
ECHA	European chemicals agency
EEA	European environment agency economic area
EMA	European medicines agency
EFSA	European food safety authority
EU	European Union, formerly European community
FDA	US Food and Drug Agency
HTA	Health technology assessment
ICH	International Conference on Harmonisation of technical requirements for registration of pharmaceuticals for human use
INN	International nonproprietary name
ITF	Innovation Task Force
LVEF	Left ventricular ejection fraction
MAA	Marketing authorisation application
MAH	Marketing authorisation holder
MUGA	Multiple gated acquisition
NBCD	Non-biological complex drugs
ODD	Orphan Drug Designation
PhV, PSE	Pharmacovigilance, product safety evaluation
PIP	Paediatric investigation plan
SA	Scientific advice
SAWP	Scientific advice working party
SME	Small and medium sized enterprises
SME	Small and medium enterprises
SmPC	Summary of product characteristics

Introduction

Nanomedicines constitute the main group of non-biological complex drugs (NBCDs). Nanotechnology was identified by the European Commission[1] as one of the six Key Enabling Technologies with an expected market value of 2 trillion €

[1] Communication from the European Commission from 03 October 2012 on Second Regulatory Review on Nanomaterials, http://ec.europa.eu/research/industrial_technologies/pdf/policy/communication-from-the-commission-second-regulatory-review-on-nanomaterials_en.pdf.

by 2015, representing an opportunity for investment and growth in Europe. The potential uses for nanotechnology for medical use are extensive: targeted drug delivery systems, modified release formulations, carriers, diagnostics/imaging, and *theranostics*[2]. The innovation brought by this class of products also brings new challenges resulting from gaps in the current scientific understanding about, *inter alia*, their manufacture, physical/chemical/biological characterisation, toxicology, pharmacokinetics, pharmacodynamics and environmental impact.

This chapter aims to provide an overview of European pharmaceutical legislation, the main aspects of marketing authorisation procedures and how NBCDs fit in the current legislative framework. The regulatory challenges of NBCDs will be discussed as well as the need to revise existing EU legislation and scientific guidance to accommodate advances in this field while guaranteeing an adequate level of public health protection.

EU Pharmaceutical Legislation and NBCDs

European pharmaceutical legislation was fundamentally shaped by the thalidomide catastrophe in the mid-twentieth century (Feick 2005). With the first European legal measure in the pharmaceutical sector—Council Directive 65/65/EC of 26 January 1965—came the imperative requirement of having a marketing authorisation issued by a competent authority before a medicinal product could be placed in the market of a Member State. The pharmaceutical legislation of the European Union has since changed significantly but the need for a marketing authorisation remains unaltered[3]. This requirement aims ultimately to safeguard public health. This fundamental objective must nonetheless be achieved without hindering the development of the pharmaceutical industry or trade in medicinal products in the Union, which implies the harmonisation of national provisions and promotion of a uniform application of technical standards regarding quality, safety and efficacy.

It is under these principles that a complex environment of regulatory norms and scientific guidance has evolved during the last 50 years defining legal responsibilities and the level of protection of marketing authorisation applicants and holders, whilst publishing technical guidance to increase regulatory certainty for stakeholders in both private and public sectors.

The body of European Union legislation in the pharmaceutical sector for human medicinal products is compiled in *The rules governing medicinal products in the European Union*: Volume 1—EU pharmaceutical legislation for medicinal products for human use. The legislation is then complemented by a series of guidelines:

[2] Medicinal product with combined diagnostic and therapeutic functions.

[3] Article 6 of Directive 2001/83/EC of the European Parliament and the Council on the [Union] Code Relating to Medicinal Products for Human Use

- Volume 2—Notice to applicants and regulatory guidelines for medicinal products for human use
- Volume 3—Scientific guidelines for medicinal products for human use
- Volume 4—Guidelines for good manufacturing practices for medicinal products for human and veterinary use
- Volume 6—Notice to applicants and regulatory guidelines for medicinal products for veterinary use
- Volume 7—Scientific guidelines for medicinal products for veterinary use
- Volume 8—Maximum residue limits
- Volume 9—Pharmacovigilance guidelines
- Volume 10—Guidelines for clinical trial

There are four marketing authorisation procedures in the EU, three leading to a national authorisation: the purely national procedure, the mutual recognition procedure and the decentralised procedure; and the EU-wide procedure: the centralised procedure.

The centralised procedure is compulsory for:

- Medicinal products developed by recombinant DNA, controlled expression of genes coding for biotechnologically active proteins in prokaryotes and eukaryotes including transformed mammalian cells, hybridoma and monoclonal antibody methods;
- Medicinal products for human use containing a new active substance for the treatment of acquired immune deficiency syndrome, cancer, neurodegenerative disorder, diabetes, autoimmune diseases and other immune dysfunctions, viral diseases;
- Designated orphan medicinal products.

The optional scope of the centralised produce is open for applicants showing that their medicinal product:

- Contains a new active substance not authorised in the union;
- Constitutes a significant therapeutic, scientific or technical innovation or that the granting of the authorisation is in the interest of patient health at Union level.

National procedures are managed by the National Competent Authorities of each Member State and the centralised procedure is managed by the European Medicines Agency (EMA).

The EMA was established in the 1995 in accordance with Council Regulation No. 2309/93 (currently Council Regulation No. 726/20014) and is responsible for:

- The scientific evaluation of applications for EU marketing authorisations for human and veterinary medicines in the centralised procedure;
- Coordinating the EU's safety-monitoring or 'pharmacovigilance' system for medicines;
- Referral procedures;

- Coordinating inspections;
- Implementing the EU telematics programme (e.g. EudraPharm, EudraGMDP);
- Stimulating innovation and research in the pharmaceutical sector.

The assessment of medicines by the EMA benefits from a peer-review system composed of a network of over 4500 experts appointed by EU Member States[4]. These experts serve as members of the Agency's scientific committees, working parties or scientific-assessment teams, and are responsible for addressing questions raised during the assessment of the benefit-risk ratio of a medicinal product, or by identifying and mitigating gaps resulting from the emergence of innovative technologies, e.g. by drafting guidance on nanomedicines.

Legislating NBCDs

NBCDs are defined scientifically as not being a biological medicine where the active substance is not a homo-molecular structure, but consists of different closely related and often nanoparticulate structures that cannot be isolated and fully quantitated, characterised and/or described by physicochemical analytical means, where the structural elements that might impact the therapeutical performance are unknown (Schellekens et al. 2014).

In order to understand how NBCDs are regulated it is first necessary to understand how the legislation categorises the different classes of medicinal products.

In the EU pharmaceutical legislation a medicinal product is defined as:

a. *Any substance or combination of substances presented as having properties for treating or preventing disease in human beings; or*
b. *Any substance or combination of substances which may be used in or administered to human beings either with a view to restoring, correcting or modifying physiological functions by exerting a pharmacological, immunological or metabolic action, or to making a medical diagnosis.*

In its turn, substance is defined as:

Any matter irrespective of origin which may be:

- human, e.g. human blood and human blood products;
- animal, e.g. micro-organisms, whole animals, parts of organs, animal secretions, toxins, extracts, blood products;
- vegetable, e.g. micro-organisms, plants, parts of plants, vegetable secretions, extracts;
- chemical, e.g. elements, naturally occurring chemical materials and chemical products obtained by chemical change or synthesis.

---

4 EMA European expert list, http://www.ema.europa.eu/ema/index.jsp?curl=pages/about_us/landing/experts.jsp&mid=WC0b01ac058043244a.

Directive 2001/83/EC goes further to describe specific classes of medicinal products:

- Chemical:
  Chemical medicinal products do not have a standalone definition, but there is instead a definition of what is a chemical substance, as described above.
- Biological:
  A biological medicinal product is a product, the active substance of which is a biological substance. A biological substance is a substance that is produced by or extracted from a biological source and that needs for its characterisation and the determination of its quality a combination of physico-chemical-biological testing, together with the production process and its control.
- Herbal:
  (...) Any medicinal product, exclusively containing as active ingredients one or more herbal substances or one or more herbal preparations, or one or more such herbal substances in combination with one or more such herbal preparations.
  A herbal substance is defined as all mainly whole, fragmented or cut plants, plant parts, algae, fungi, lichen in an unprocessed, usually dried, form, but sometimes fresh (...)

The three main groups above contain subgroups of particular medicinal products such as plasma-derived medicinal products, vaccines, radiopharmaceuticals and precursors, homeopathic medicinal products, and advanced therapy medicinal products as defined in Annex I of Directive 2001/83/EC to which specific requirements apply. These categories of medicinal products are mainly defined by the origin of their active substance(s). The origin of an active substance in a medicinal product is therefore a major driver to establish the legal requirements applicable to the contents of an application for its marketing authorisation. It is, *inter alia*, the subsequent pharmaceutical complexity inherent to the active substance, delivery system of the pharmaceutical formulation, pharmacodynamics, pharmacokinetics, indication and target population that will define the additional volume of quality, non-clinical and clinical data that will be necessary to establish its benefit-risk ratio.

A legal definition for a nanomedicine is yet to be devised. In October 2011 the European Commission published a Recommendation on the definition of nanomaterial[5]:

> as "a natural, incidental or manufactured material containing particles, in an unbound state or as an aggregate or as an agglomerate and where, for 50 % or more of the particles in the number size distribution, one or more external dimensions is in the size range 1 nm–100 nm. In specific cases and where warranted by concerns for the environment, health, safety or competitiveness the number size distribution threshold of 50 % may be replaced by a threshold between 1 and 50 %. [...]

The EU Scientific Committee on Emerging and Newly Identified Health Risks from the European Commission has nonetheless acknowledged that the 100 nm threshold

[5] European Commission Recommendation of 18 October 2011 on the definition of nanomaterial, http://ec.europa.eu/research/industrial_technologies/pdf/policy/commission-recommendation-on-the-definition-of-nanomater-18102011_en.pdf.

is not scientifically justified and noted that the *special circumstances prevailing in the pharmaceutical sector*, stating that this definition *should not prejudice the use of the term 'nano' when defining certain pharmaceuticals and medical devices*[6].

During the many years of pharmaceutical legislation, several innovator and follow-on chemical medicinal products submitted to Competent Authorities in the context of a marketing authorisation application, variation application or other regulatory procedure (e.g. orphan drug designation) have challenged the standard approach applied to the review of quality, safety and efficacy data of medicines containing small, well-characterised molecules. Examples are given of the complexity of already approved chemical medicinal products:

- The characterisation of medicinal products with multi-component active substances, when each component has a different pharmacokinetic and pharmacodynamic profile, e.g. medicines containing polymyxins (Li et al. 2006), teicoplanin[7];
- Establishing therapeutic equivalence between reference and generic/hybrid modified-release oral formulations of narrow therapeutic index drugs with different indications, e.g. prolonged release tablets of tacrolimus[8];
- Establishing therapeutic equivalence between reference and generic/hybrid medicinal products of different routes of administration and pharmaceutical forms, e.g. immediate release tablets and prolonged release transdermal patch containing granisetron[9];
- Establishing therapeutic equivalence between reference and generic/hybrid medicinal products of non-oral and non-parenteral medicinal products, e.g transdermal patches[10] and inhalation formulations[11].

---

[6] EU Scientific Committee on Emerging and Newly Identified Health Risks. *Scientific Basis for the Definition of the Term "Nanomaterial". European Commission, Brussels, Belgium, 2010.*

[7] Assessment Report of Teicoplanin, Procedure no: EMEA/H/A-5(3)/1315. http://www.ema.europa.eu/docs/en_GB/document_library/Report/2013/04/WC500142229.pdf.

[8] Envarsus® Summary of CHMP positive webpage http://www.ema.europa.eu/docs/en_GB/document_library/EPAR_-_Public_assessment_report/human/002655/WC500170414.pdf.

[9] Sancuso® European Public Assessment Report, http://www.ema.europa.eu/docs/en_GB/document_library/EPAR_-_Public_assessment_report/human/002296/WC500127130.pdf.

[10] EMA/CHMP/QWP/202350/2010, Concept paper on the revision of the note for guidance on quality of modified release oral dosage forms and transdermal dosage forms: Sect. I (quality). http://www.ema.europa.eu/ema/pages/includes/document/open_document.jsp?webContentId=WC500095366 and EMA/CHMP/EWP/1303/2010 Concept paper on the need for revision of the note for guidance on modified release oral and transdermal dosage forms: Sect. II (pharmacokinetic and clinical evaluation) http://www.ema.europa.eu/ema/pages/includes/document/open_document.jsp?webContentId=WC500091662.

[11] http://www.ema.europa.eu/docs/en_GB/document_library/Scientific_guideline/2009/09/WC500003508.pdf and http://www.ema.europa.eu/docs/en_GB/document_library/Scientific_guideline/2009/09/WC500003568.pdf.

- Non-standard delivery systems, e.g. active substance released from enucleate autologous cells[12];
- Conjugation of biological and chemical components into one active substance, e.g. antibody conjugated to a chemical cytotoxic[13].

In addition to the above, a non-exhaustive list is provided in the table of chemical nanotechnology-based medicinal products in the EU[14].

Trade name/ API—INN	Platform/technology	Indication	MAH	Approval
Liposomes				
Caelyx® doxorubicin hydrochloride	API in sterically stabilised (Stealth®) pegylated liposomes, to increase blood circulation (long-acting) and reduce cardiotoxicity	Multiple myeloma, ovarian neoplasms, breast neoplasms, Kaposi sarcoma	Janssen-Cilag International N.V	21/06/1996
Myocet® doxorubicin	Liposome-encapsulated doxorubicin-citrate complex to reduce cardiac toxicity and to increase tumor tissue distribution	Breast neoplasms	Cephalon Europe	13/07/2000
Visudyne® verteporfin	Liposomal formulation of semisynthetic mixture of porphyrins	Degenerative myopia, age-related macular degeneration	Novartis Europharm Ltd	27/07/2000
DepoCyte® cytarabine	Multivesicular liposomes with unique structure of multiple non-concentric aqueous chambers (DepoFoam®)	Meningeal neoplasms	Pacira Limited	11/07/2001
Mepact® mifamurtide	Fully synthetic analogue of a component of *Mycobacterium sp.* cell wall encapsulated in multilamellar liposomes to facilitate activation of macrophages	High-grade resectable non-metastatic osteosarcoma	IDM PHARMA SAS	06/03/2009
Nanoparticles				
Rapamune® sirolimus	API particles in nanocrystal colloidal nanodispersion stabilised with poloxamer to reduce particle size for increased stability and bioavailability	Prophylaxis of organ rejection in renal transplant	Wyeth Europa Ltd	13/03/2001

[12] http://www.ema.europa.eu/docs/en_GB/document_library/Orphan_designation/2013/08/WC500147978.pdf.

[13] http://www.ema.europa.eu/docs/en_GB/document_library/EPAR_-_Public_assessment_report/human/002455/WC500135054.pdf.

[14] Adapted from Vamvakas et al. 2011.

Trade name/ API—INN	Platform/technology	Indication	MAH	Approval
Emend® aprepitant	Colloidal dispersion of nanocrystals to increase bioavailability (wet milling method)	Nausea and vomiting	Merck Sharp & Dohme Ltd	11/11/2003
Abraxane® paclitaxel	Solvent-free colloidal suspension of albumin-bound spherical nanoparticles to increase water solubility	Metastatic breast cancer	Abraxis BioSciences Ltd	11/01/2008
Polymer-conjugates				
Macugen® pegaptanib	Pegylated modified oligonucleotide	Wet macular degeneration	Pfizer Limited	31/01/2006
Gas dispersions				
SonoVue® sulphur hexafluoride	Sulphur hexafluoride gas as 'microbubbles' dispersion	Contrast agent for echocardiography and ultrasonography	Bracco International BV	26/03/2001

The regulatory and scientific aspects of the examples above were assessed under the existing legislation and adopted guidance.

Scientific guidance is available from different competent authorities/organisations: EMA, European Commission, International Conference on Harmonisation, United States Food and Drug Administration, World Health Organisation, Organisation for Economic Co-operation and Development, International Organization for Standardisation, European Directorate for the Quality of Medicines and Health-Care, etc. This body of guidance not only reflects the precise requirements under the scope of a particular document but also mirrors the underlying regulatory thinking on the critical points to be addressed and data to be generated when demonstrating quality, safety and efficacy. These regulatory principles can be extended beyond the defined scope of a particular note for guidance and be borrowed, as and where applicable, to the assessment of the benefit/risk of any product. This all-encompassing guidance principle is very important and can be observed, for example, in the cross-referencing across guidance for chemical and biological products or bioequivalence and efficacy. Applicants of NBCDs should make use of the thinking already available on published guidance to infer to which extent it can be adapted and applied to their products.

> The existing framework does accommodate NBCDs in so far as NBCDs are non-standard chemical products for which special considerations are necessary when defining their quality, safety, efficacy (including therapeutic equivalence); and that there are already approved medicinal products that fall under this category.

Another relevant question often asked by stakeholders is whether the existing framework should be revised to accommodate the particularities of NBCDs regarding quality, safety, efficacy, environmental risk assessment and pharmacovigilance. The different regulatory dimensions to this question will be analysed in the next sections of this chapter.

Legal Basis of Submission to Obtain a Marketing Authorisation

An important consideration to take into account when determining the necessary level of quality, safety and efficacy data to substantiate a marketing authorisation application of a medicinal product is the legal basis under which the submission is made.

The possible legal bases for submission, as detailed in Directive 2001/83/EC, are:

- Article 8(3)—stand alone application comprised of a full set of pharmaceutical, preclinical and clinical data, or mixed applications comprised of a full set of pharmaceutical data with the option of a combination of reports of limited non-clinical and/or clinical data carried out by the applicant and of bibliographical references.
- Article 10:
  - Paragraph 1, so called *generic medicinal product*;
  - Paragraph 3, so called *hybrid medicinal product*;
  - Paragraph 4, so called *similar biological medicinal product* or *biosimilar*;
- Article 10a—well-established use application, where it is possible to replace pre-clinical and clinical data by detailed references to published scientific literature if it can be demonstrated that the active substance(s) has been in well established medical use within the EU for at least 10 years, with recognised efficacy and acceptable safety;
- Article 10b—fixed-combination application, where active substances used in the composition of medicinal products but not previously used in combination in the same medicinal product for therapeutical purposes;
- Article 10c—informed consent applications, where the holder of an authorised medicinal product allows the quality, non-clinical and clinical data submitted for its authorisation to be used for future applications of other medicinal products with the same qualitative and quantitative composition of active substance(s) and same pharmaceutical form(s).

For Article 10 applications (applications for a generic, hybrid or biosimilar) the applicant is not required to provide results of pre-clinical tests and clinical trials if equivalence is demonstrated to the reference medicinal; the following definitions are critical:

- *"reference medicinal product" shall mean a medicinal product authorised under Article 6, in accordance with the provisions of Article 8*, i.e. a medicinal product authorised in one or more Member States of the EU through a national or centralised procedure as a stand alone application with a complete set of quality, non-clinical and clinical data;
- *"generic medicinal product" shall mean a medicinal product which has the same qualitative and quantitative composition in active substances and the same pharmaceutical form as the reference medicinal product, and whose bioequiva-*

lence with the reference medicinal product has been demonstrated by appropriate bioavailability studies. The different salts, esters, ethers, isomers, mixtures of isomers, complexes or derivatives of an active substance shall be considered to be the same active substance, unless they differ significantly in properties with regard to safety and/or efficacy. In such cases, additional information providing proof of the safety and/or efficacy of the various salts, esters, or derivatives of an authorised active substance must be supplied by the applicant. The various immediate-release oral pharmaceutical forms shall be considered to be one and the same pharmaceutical form. Bioavailability studies need not be required of the applicant if he can demonstrate that the generic medicinal product meets the relevant criteria as defined in the appropriate detailed guidelines;

- *In cases where the medicinal product does not fall within the definition of a generic medicinal product as provided in paragraph [above] or where the bioequivalence cannot be demonstrated through bioavailability studies or in case of changes in the active substance(s), therapeutic indications, strength, pharmaceutical form or route of administration, vis-à-vis the reference medicinal product, the results of the appropriate pre-clinical tests or clinical trials shall be provided*—in this case the medicinal product will be considered as a hybrid medicinal product.

In the context of NBCDs it is important to make the following NBCDs-specific reflections with regards the possible legal bases for submission.

The EU legislation provides for different levels of data exclusivity, market protection and market exclusivity through the above-mentioned different legal bases and regulatory procedures[15]. The standard level of protection of 8 years data exclusivity and 10 years market protection is granted under Article 8(3) submissions for active substances not previously authorised as a medicinal product in the EU provided they do not fall under the same global marketing authorisation[16]. A medicinal product will fall under the same global marketing authorisation if its active substance(s) has previously been authorised as a medicinal product to the same marketing authorisation holder. Here it is important to introduce the concept of new active substance[17] and known active substance, as this will confirm whether or not the potential marketing authorisation will be part of the same global marketing authorisation and, consequently, be eligible or not for the data exclusivity and data protection accorded by the Directive.

Annex I of the Notice to Applicants, Volume 2A, Chap. 1 states the following:

A new chemical, biological or radiopharmaceutical active substance includes:

[15] Please see the recommend references for more information.

[16] Consult Notice to Applicants Volume 2A, Chap. 1, Sect. 2.3 for the Notion of "global marketing authorisation".

[17] Reflection paper on considerations given to designation of a single stereo isomeric form (enantiomer), a complex, a derivative, or a different salt or ester as new active substance in relation to the relevant reference active substance, http://www.ema.europa.eu/docs/en_GB/document_library/Scientific_guideline/2012/11/WC500134993.pdf.

- a chemical, biological or radiopharmaceutical substance not previously authorised as a medicinal product in the European Union;
- an isomer, mixture of isomers, a complex or derivative or salt of a chemical substance previously authorised as a medicinal product in the European Union but differing significantly in properties with regard to safety and efficacy from that chemical substance previously authorised;
- a biological substance previously authorised as a medicinal product in the European Union, but differing in molecular structure, nature of the source material or manufacturing process;
- a radiopharmaceutical substance which is a radionuclide, or a ligand not previously authorised as a medicinal product in the European Union, or the coupling mechanism to link the molecule and the radionuclide has not been authorised previously in the European Union.

NBCDs due to their complexity, when compared to *conventional* chemical medicinal products, may be required to provide additional data to demonstrate that the new NBCDs qualifies for the new active substance status if the application is being submitted under Article 8(3). The deliberation on the new active substance status of a chemical substance is made primarily on the significance of the differences observed in its molecular structure when compared to active substances present in authorised medicinal products, and whether such differences have any demonstrated, or biologically plausible, impact to the safety and efficacy of the product[18]. NBCDs may require a more challenging battery of data than conventional chemical substances to justify a claim of new active substance, e.g.:

- Comparison between substance X and the same substance X manufactured with nanotechnology[19];
- Substance X manufactured with nanotechnology and the same substance X with the same molecular structure manufactured with differences in the nanotechnology production process (e.g. different size distribution, shape);
- Differences between two active substances composed by groups of closely related non-homo-molecular structures with claims of differences in safety and efficacy;
- Differences in safety and efficacy accorded by different functional excipients coupled to the same active substance (different nanocarriers).

The new active substance status is critical in several aspects, e.g. determining whether marketing protection should be granted to an applicant with a marketing authorisation granted for a potentially equivalent medicinal product, substantiate claims of therapeutic superiority against existing options, defining pharmacovigilance obligations and health technology assessment considerations.

Regardless of the new active substance issue, it is important to highlight how differences in safety and efficacy accorded to a product from the use of nanotechnology when compared to its non-nanotechnology equivalent are often included in the Summary of Product Characteristics (SmPC):

---

[18] See example of Aubagio®, EPAR Sect. 2.9: http://www.ema.europa.eu/docs/en_GB/document_library/EPAR_-_Public_assessment_report/human/002514/WC500148684.pdf.

[19] Expert discussion on Bawa 2008.

- Myocet® SmPC: (…) *Analyses of cardiotoxicity in clinical trials have shown a statistically significant reduction in cardiac events in patients treated with Myocet compared to patients treated with conventional doxorubicin at the same dose in mg. A meta-analysis showed a statistically significant lower rate of both clinical heart failure ($RR = 0.20$, $p = 0.02$) and clinical and subclinical heart failure combined ($RR = 0.38$, $p < 0.0001$) in patients treated with Myocet versus conventional doxorubicin*[20] (…);
- AmBisome® SmPC: (…) *has been shown to be substantially less toxic than conventional amphotericin B, particularly with respect to nephrotoxicity* (…)[21];
- Caelyx® SmPC: (…) *Of the 418 patients treated with Caelyx 50 mg/m²/cycle, and having a baseline measurement of left ventricular ejection fraction (LVEF) and at least one follow-up measurement assessed by MUGA scan, 88 patients had a cumulative anthracycline dose of >400 mg/m² an exposure level associated with an increased risk of cardiovascular toxicity with conventional doxorubicin* (…)[22];
- Abraxane® SmPC: (…) *Results for overall response rate, median time to disease progression, and progression-free survival as assessed by the investigator of Abraxane versus Solvent-based paclitaxel* (…)[23].

In contrast to the objective of an applicant submitting an application under Article 8(3), if the submission is made under Article 10 data may be required to justify that differences in safety and/or efficacy between the proposed active substance and the active substance of the reference product do not amount to the two substances being considered significantly different. In that regards, the same chapter of Notice to Applicants also states that:

> The different salts, esters, ethers, isomers, mixtures of isomers, complexes or derivatives of an active substance must be considered to be the same active substance, unless they differ significantly in properties with regard to safety and/or efficacy. In such cases, additional information providing proof of the safety and/or efficacy of the various salts, esters, ethers, isomers or mixtures thereof or derivatives of an authorised active substance must be supplied by the applicant. If additional information concerning changes to the nature of the active substance cannot establish the absence of a significant difference with regard to safety or efficacy then it would be necessary to submit the results of appropriate pre-clinical tests and clinical trials in accordance with the requirements of Article 10(3) (…). To the extent that the active substance may be considered as a new active substance (…), the applicant may consider the submission of an application in accordance with Article 8(3) of Directive 2001/83/EC.

[20] Myocet® product information, last consulted on 22 July 2014, http://www.ema.europa.eu/ema/index.jsp?curl=pages/medicines/human/medicines/000297/human_med_000916.jsp&mid=WC0b01ac058001d124.

[21] AmBisome® product information, last consulted 22 July 2014, http://www.mhra.gov.uk/home/groups/spcpil/documents/spcpil/con1399707080555.pdf.

[22] Caelyx® product information, last consulted on 22 July 2014, http://www.ema.europa.eu/ema/index.jsp?curl=pages/medicines/human/medicines/000089/human_med_000683.jsp&mid=WC0b01ac058001d124.

[23] http://www.ema.europa.eu/docs/en_GB/document_library/EPAR_-_Product_Information/human/000778/WC500020435.pdf.

The main difference to the bioequivalence study model applied to conventional chemical medicinal products for NBCDs is that measuring the free active substance in plasma alone may not be meaningful; due to the nanoproperties of the active substance or its nanocarrier it may be necessary to go beyond the standard bioequivalence study and perform biodistribution, safety or efficacy studies. The published guidance on the comparability of NBCDs follows a stepwise approach, similarly to that adopted for biosimilars. The baseline level of uncertainty on the differences between reference and test NBCDs are initially addressed at CMC level: identifying the critical quality attributes of the reference product and their batch variability, then assessing the impact on individual and cumulative differences of the critical quality attributes to the test product, justifying an adequate control of manufacturing process parameters to guarantee clinical performance throughout the lifecycle of the product. The differences identified at CMC level will determine the extent of non-clinical and clinical data necessary to supplement the comparability data needed to support the therapeutic equivalence between the two products.

The limitations inherent to the available analytical methods and *in vitro* and *in vivo* models, associated with the idiosyncrasies of NBCDs leading to a regulatory case-by-case approach are recognised by all parties, regulators included. A conclusion regarding an Article 10 application may be: if an authorised NBCD and a NBCD under review are considered significantly different with regards safety and/or efficacy, then the latter cannot be considered as a generic or hybrid version of the former.

As reflected by Schellekens[24] and Borchard (Borchard et al. 2012), NBCDs are made of structures that cannot be isolated and fully quantitated, characterised, and/or described by analytical means, to which it is unknown which therapeutic elements might impact therapeutic performance and with clinical performance highly dependent on the production process. Follow-on products ("generics") of nanomedicines have been called *similar nanomedicines* or *nanosimilars* (Ehmann et al. 2013), names that adequately translate the new challenges presented by these products due to a complexity closer to biosimilars than their chemical counterparts.

Although by definition article 10(4)—biosimilar applications—is not applicable to NBCDs it is relevant to analyse the legislative provisions for these applications. Similarly to chemical medicinal products, biosimilar applications are first expected to comply with the definition for generic medicinal products. It is only failing this definition, *owing to, in particular, differences relating to raw materials or differences in manufacturing processes of the biological medicinal product and the reference biological medicinal product, the results of appropriate pre-clinical tests or clinical trials relating to these conditions must be provided.* Reference is then made to the relevant criteria set out in Annex I in the Directive. In this annex general requirements and reference to the relevant scientific guidelines are described. As for biosimilars, the Directive also directs the requirements for chemical generic and hybrid applications to the *appropriate detailed guidelines (Article 10.2(b))*. This is to say that regarding legal provisions to dossier contents there is no significant difference to how simple or complex chemical products and biosimilars are addressed

[24] Ibid. 6.

as for both general requirements and a reference to the applicable guidance is mentioned in the Directive.

Another important aspect of biosimilars is the innovation being brought by the revised Guideline on Similar Biological Medicinal Products (currently under revision[25]) to the origin of batches used in the comparability studies; the following is stated in that regard:

> (…), with the aim of facilitating the global development of biosimilars and to avoid unnecessary repetition of clinical trials, it may be possible for an Applicant to compare the biosimilar in certain clinical studies and in vivo non-clinical studies (where needed) with a non-EEA authorised comparator (i.e. a non-EEA authorised version of the reference medicinal product) which will need to be authorised by a regulatory authority with similar scientific and regulatory standards as EMA (i.e. ICH countries). In addition, it will be the Applicant's responsibility to establish that the comparator authorised outside the EEA is representative of the reference product authorised in the EEA.

As some *follow-on* NBCDs may also require extensive comparability data including clinical and non-clinical studies, as per the relevant scientific guidelines, the reasoning behind this new approach[26] for biosimilars, i.e. *facilitating the global development of biosimilars and to avoid unnecessary repetition of clinical trials,* could be envisaged for NBCDs allowing for non-EEA reference medicinal products.

Fixed dose combination applications may also present additional complexities as a result of the use of nanotechnology resulting from clinically significant differences of pharmacokinetics and pharmacodynamics of, for example, delivery systems presenting target cells with specific combined concentrations of active substances not possible through traditional intravenous administration.

Post-Authorisation Considerations

Independently of the authorisation procedure or legal basis of submission, once in the market the product has to comply with Variation Regulation EC/1234/2008[27]. This regulation defines what are minor (Type IA and Type IB) and major variations (Type II), and what are the changes to the active substance, strength pharmaceutical form and route of administration that would lead to an extension of the existing marketing authorisation.

Due to all reasons aforementioned, NBCDs may represent an increased challenge when compared to their conventional chemical counterparts with regard to the categorisation of changes to the terms of their marketing authorisations (minor

[25] Draft of the Guideline on Similar Biological Medicinal Products consulted on 15 July 2014, http://www.ema.europa.eu/docs/en_GB/document_library/Scientific_guideline/2013/05/WC500142978.pdf.

[26] EMA News item from 28/09/2012 on European Medicines Agency to accept biosimilar reference medicines sourced outside European Economic Area, http://www.ema.europa.eu/ema/index.jsp?curl=pages/news_and_events/news/2012/09/news_detail_001615.jsp&mid=WC0b01ac058004d5c1.

[27] http://ec.europa.eu/health/files/eudralex/vol-1/reg_2008_1234/reg_2008_1234_en.pdf.

vs. major variations) and face an increased complexity in submission for extension of applications due to potential changes to the safety and efficacy of the product. In this regard, biological medicinal products[28] more often than conventional chemical products resort to post-approval change management protocols[29]. Post-approval change management protocols are reviewed (1) as part of the initial marketing authorisation application, (2) extension to the marketing authorisation or (3) as a stand-alone procedure, in anticipation of potential changes during the product lifecycle such as changes in raw materials, manufacturing process or manufacturing site and which data would be necessary to support such changes, e.g. comparability studies between an existing product and a product subject to the proposed changes. Applicants and Marketing Authorisation holders of NBCDs would greatly benefit from this post approval management tool to be made available by the regulation.

The new pharmacovigilance legislation (Directive 2010/84/EU and Regulation (EU) No 1235/2010, amended in 2012) was recently adopted with the goal of promoting and protecting public health by strengthening the existing Europe-wide system for monitoring the safety and benefit-risk balance of medicines. From the many improvements brought by the new legislation it is worth highlighting one of particular interest to NBCDs: the strengthened legal basis for requesting post-authorisation safety and efficacy studies from the pharmaceutical industry. Post-authorisation safety and efficacy studies aim to support the decision-making on the safety and benefit-risk profile of a medicine, to enhance its safe and effective use. These studies can be particularly relevant when deciding on the entry to the market of innovative medicines challenging the existing models of safety and efficacy assessment.

Medical Devices, Borderline and Combination Products

Finally, it is likely that novel applications of nanotechnology will cross the regulatory boundaries between medicinal products and medical devices, challenging current criteria for classification and evaluation[30,31].

According to Article 1(2)a of Directive 93/42/EEC, as amended, a medical device is defined as:

> (a) 'medical device' means any instrument, apparatus, appliance, material or other article, whether used alone or in combination, including the software necessary for its proper application intended by the manufacturer to be used for human beings for the purpose of:

---

[28] Note for guidance on biotechnological/ biological products subject to changes in their manufacturing process, http://www.ema.europa.eu/ema/pages/includes/document/open_document.jsp?webContentId=WC500002805.

[29] http://www.ema.europa.eu/docs/en_GB/document_library/Scientific_guideline/2012/04/WC500125400.pdf.

[30] Reflection Paper on the Nanotechnology-based medicinal products for human use, http://www.ema.europa.eu/docs/en_GB/document_library/Regulatory_and_procedural_guideline/2010/01/WC500069728.pdf.

[31] European Commission, Medical Devices Borderline and classification issues webpage http://ec.europa.eu/health/medical-devices/documents/borderline/index_en.htm.

- diagnosis, prevention, monitoring, treatment or alleviation of disease,
- diagnosis, monitoring, treatment, alleviation of or compensation for an injury or handicap,
- investigation, replacement or modification of the anatomy or of a physiological process,
- control of conception,

and which does not achieve its principal intended action in or on the human body by pharmacological, immunological or metabolic means, but which may be assisted in its function by such means.

Under the current legislation, the mechanism of action is key to decide whether a product should be regulated as a medicinal product or a medical device (Directive 2001/83/EC, Article 1.2(b)), e.g. a nanoproduct with a primary physical action would fall under the medical devices regulation. The emerging complexity of nanomedicines may lead to borderline cases where the boundary between medical device and medicinal product may be more difficult to determine, e.g. devices for administering medicinal products where the device and the medicinal product form a single integral product designed to be used exclusively in the given combination and which are not re-usable or refillable[32], or where there is a combination of multiple modes of action with which neither of the mode is secondary to the other (Lebourgeois 2008). The differences in regulation between the medicinal products and medical devices are extensive[33], (Parvizi and Woods 2014; Chowdhury 2010). Developers of novel medical products should at an early stage of development understand which requirements would be applicable to their technologies, and should seek advice of the relevant national competent authority or of EMA's Innovation Task Force[34]

It is clear: innovations challenge the regulatory framework again and again, But, as stated by Dorbeck-Jung and Chowdhury (Dorbeck-Jung and Chowdhury 2011), pharmaceutical legislation *has proved to be robust in terms of well-established rules that have been kept up to date and relevant by continuous and careful adaptation to new product development and new insights into product safety, quality and efficacy* allowing for the authorisation of nanomedicines. That attitude of continuous improvement should be sustained.

Scientific Guidance

Scientific and regulatory guidance may be available in several formats, such as guidelines, public statements, reflection papers, questions and answers documents, recommendations; these documents are often referred to as *soft law*, contrarily to *hard law* which constitutes legally binding documents, such as regulations, directives and decisions adopted by a legal act as detailed in the provisions of the Treaty on the Functioning of the European Union.

---

[32] Guidance on Legislation—Borderlines Between medical devices and medicinal products, June 2013—MHRA, consulted on 14 July 2014, http://www.mhra.gov.uk/home/groups/dts-bs/documents/publication/con286964.pdf.

[33] European Commission Medical Devices Regulatory Framework webpage http://ec.europa.eu/health/medical-devices/regulatory-framework/index_en.htm

[34] ITF webpage, http://www.ema.europa.eu/ema/index.jsp?curl=pages/regulation/general/general_content_000334.jsp&mid=WC0b01ac05800ba1d9.

A guideline is a EU document with an explicit legal basis referred to in legislation as intended to fulfil a legal obligation laid down in the Union pharmaceutical legislation[35]. Guidelines aim to provide a basis for practical harmonisation of the manner in which Member States and EMA interpret and apply the detailed requirements for the demonstration of quality, safety and efficacy by providing advice to applicants or marketing authorisation holders, competent authorities and/or other interested parties on the best or most appropriate way to fulfil an obligation laid down in EU pharmaceutical legislation[36].

A reflection paper is developed to communicate the current status of discussions or to invite comment on a selected area of medicinal product development or a specific topic. It can provide a framework for discussion or clarification particularly in areas where scientific knowledge is fast evolving or experience is limited. A reflection paper does not provide scientific, technical or regulatory guidance, but may contribute to future development of such guidelines, or related documents[37]. Reflection papers have thus been chosen as the best communication tool to convey the current regulatory thinking on medicinal products using innovative technologies, such as NBCDs.

Following on from the principle on the overarching nature of scientific guidance, the documents listed below although not always targeted to NBCDs may be relevant during their development. All documents are available online on the EMA website.

Quality
International conference on harmonisation of technical requirements for registration of pharmaceuticals for human use guideline ich q11 on development and manufacture of drug substances (chemical entities and biotechnological/biological entities)
Process validation, including annex ii on process validation of non-standard processes manufacture of the finished dosage form
Setting specifications for related impurities in antibiotics
Impurities: guideline for residual solvents
Specifications: test procedures and acceptance criteria for new drug substances and new drug products: chemical substances
Specifications and control tests on the finished product
Similar biological medicinal products containing biotechnology-derived proteins as active substance: quality issues
Note for guidance on biotechnological/biological products subject to changes in their manufacturing process
Similar biological medicinal product
Excipients in the dossier for application for marketing authorisation of a medicinal Product
Pharmaceutical development
Pharmaceutical development of intravenous medicinal products containing active substances solubilised in micellar systems (non-polymeric surfactants)
Requirements to the chemical and pharmaceutical quality documentation concerning investigational medicinal products in clinical trials
Questions and answers on post approval change management protocols

[35] EMA Guidelines as published in EudraLex Volume 3, http://ec.europa.eu/health/documents/eudralex/vol-3/index_en.htm.

[36] Procedure for European Union guidelines and related documents within the pharmaceutical legislative framework, http://www.ema.europa.eu/docs/en_GB/document_library/Scientific_guideline/2009/10/WC500004011.pdf.

[37] Ibid. 45.

Safety
Safety pharmacology studies for human pharmaceuticals
Pharmacokinetics: Guidance for repeated dose tissue distribution studies
Duration of chronic toxicity testing in animals (rodent and non-rodent toxicity testing)
Limits of genotoxic impurities
Clinical
Use of pharmacogenetic methodologies in the pharmacokinetic evaluation of medicinal products
Investigation of bioequivalence
Reflection paper on considerations given to designation of a single stereo isomeric form (enantiomer), a complex, a derivative, or a different salt or ester as new active substance in relation to the relevant reference active substance
Clinical evaluation of diagnostic agents
Fixed combination medicinal products
Population exposure: The extent of population exposure to assess clinical safety
General considerations for clinical trials
Nanomedicines
Data requirements for intravenous iron-based nano-colloidal products developed with reference to an innovator medicinal product
Surface coatings: general issues for consideration regarding parenteral administration of coated nanomedicine products
Data requirements for intravenous liposomal products developed with reference to an innovator liposomal product
Development of block-copolymer-micelle medicinal products—Joint EMA and Ministry of Health, Labour and Welfare—Japan
Non-clinical studies for generic nanoparticle iron medicinal product applications

The publication of new and revised guidance is subject to identified demand, scientific agreement and regulatory experience in the area. There is no question regarding demand for new or revised guidance with the increasing number of nanomedicines/complex medicines under development and being submitted for scientific advice, orphan designation or marketing authorisation[38]. However it is also recognised that there is a wide range of nanotechnologies being developed, where existing scientific insights cannot be extrapolated from one platform/family (liposomes, iron-sugar complexes, glatiramoids, polymeric micelles) to another, as well as to modifications within the same technological platform. This makes it very necessary and challenging to develop new guidance on pharmaceutical, non-clinical and clinical studies. The need to revise or draft new guidance adapted to (families of) nanoproducts/complex medicines will need to be adequately assessed to avoid creating an unnecessary burden on future applicants but also to guarantee that the requested models are sufficiently predictive of the clinical performance of the product.

While it has been possible to publish draft product-specific guidance on demonstration of bioequivalence for classical chemical products[39] or for intravenous iron-based nano-colloidal products developed with reference to an innovator medicinal

[38] Ibid. 17 and 29.

[39] Development of product-specific guidance on demonstration of bioequivalence, http://www.ema.europa.eu/docs/en_GB/document_library/Scientific_guideline/2013/07/WC500147001.pdf.

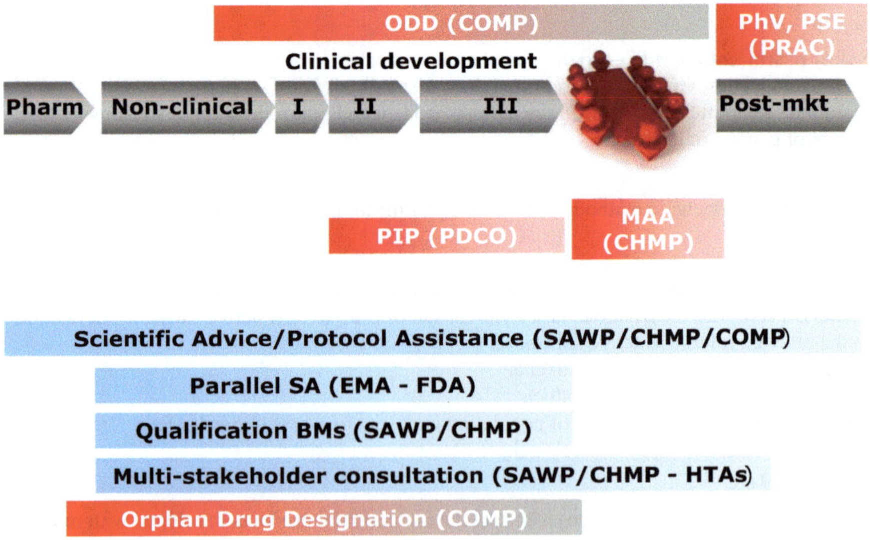

Fig. 1 EMA Involvement at different stages of drug development. *BM* Biomarkers; *CHMP* Committee for Medicinal Products for Human Use; *COMP* Committee for Orphan Medicinal Products; *HTA* Health Technology Assessment [bodies]; *ITF* Innovation Task Force; *MAA* Marketing Authorisation Application; *ODD* Orphan Drug Designation; *Pharm* Pharmaceutical Development; *PhV, PSE* pharmacovigilance, product safety evaluation; *PIP* Paediatric Investigation Plan; *SA* Scientific Advice; *SAWP* Scientific Advice Working Party

product, at present regulators may not be at the stage to start drafting product-specific guidance for NBCDs due to the levels of complexity already discussed.

Efforts are also being made to harmonise requirements across different legislative requirements for different nano-products (through EMA, ECHA, EFSA)[40] and across the regions, e.g. the International Regulators Subgroup on Nanomedicines involving the EU (EMA), USA (US FDA), Japan (Ministry of Health, Labour and Welfare) and Canada (Health Canada).

Recognising that it is not possible to make product-specific guidance available at every occasion, regulators acknowledge the need for applicants to receive product-specific advice at the different stages of development of a medicinal product. Competent Authorities have prepared different platforms of communication with stakeholders to facilitate the development and availability of high-quality, effective and acceptably safe medicines, for the benefit of patients. Examples of those platforms at the EMA include (see Fig. 1[41]):

- Scientific advice and protocol assistance[42];

[40] Ibid. 1.

[41] Adapted from Vamvakas et al. 2011.

[42] European Medicines Agency guidance for companies requesting scientific advice and protocol assistance, http://www.ema.europa.eu/ema/index.jsp?curl=pages/regulation/general/general_content_000049.jsp&mid=WC0b01ac05800229b9#.

- Qualification of innovative development methods for a specific intended use in the context of research and development into pharmaceuticals[43];
- Innovation Task Force (ITF)[44];
- CHMP Expert Group on Nanomedicines;
- Micro-, small- and medium-sized-enterprise (SME) office[45];
- 2010 Nanomedicines Workshop[46].

Stakeholders are strongly encouraged to contact the EMA through the platforms described in order to obtain at early stages of development the required scientific, regulatory and procedural guidance.

Lessons from ATMPs

At the end of 2008, Regulation 1394/2007 on Advanced Therapy Medicinal Products[47] (ATMP) was adopted to take into account developments of advanced therapies such as gene therapy, somatic cell therapy and tissue engineering. This new class of products is included under the umbrella of biological medicinal products.

It is relevant to analyse the ATMP regulation. Not to establish any scientific parallelism with NBCDs but to identify which legal mechanisms were triggered on the arrival of a technological advance.

Recital 9 of the ATMP regulation, dictates that similarly to all other modern biotechnological medicinal products, the centralised procedure should also be made mandatory for ATMPs in order to *overcome the scarcity of expertise in the [Union], ensure high level of scientific evaluation (...), preserve confidence of patients and medical professions in the evaluation and facilitate [Union] market access for these innovative technologies*. Should the same arguments justify including NBCDs in the mandatory scope of the centralised procedure?

Chemical medicinal products are made mandatory to the centralised procedure not because of their technological platform or complexity but because of the therapeutic area of their indications or orphan designation. Applicants have the opportunity, if desired, to justify their access to the centralised procedure on the basis of the optional scope of new active substance or significant therapeutic, scientific or technical innovation. Even if the application is submitted through a national proce-

[43] Guidance to Applicants, http://www.ema.europa.eu/docs/en_GB/document_library/Regulatory_and_procedural_guideline/2009/10/WC500004201.pdf.

[44] ITF webpage, http://www.ema.europa.eu/ema/index.jsp?curl=pages/regulation/general/general_content_000334.jsp&mid=WC0b01ac05800ba1d9.

[45] SME office webpage, http://www.ema.europa.eu/ema/index.jsp?curl=pages/regulation/general/general_content_000059.jsp&mid=WC0b01ac05800240cc.

[46] 2010 Nanomedicines Workshop webpage, http://www.ema.europa.eu/ema/index.jsp?curl=pages/news_and_events/events/2009/12/event_detail_000095.jsp&mid=WC0b01ac058004d5c3.

[47] Recital 1 and 2 of Regulation 1394/2007 on Advanced Therapy Medicinal Products.

dure, the Member State(s) always have the possibility to request an opinion from EMA's scientific groups.

In recital 19 of the ATMP regulation, reference is made to the need to adapt the summary of product characteristics, labelling and package leaflet to the specificities of ATMPs. Again, should such provision also be considered for NBCDs? In addition, article 14 makes special reference to the requirement, where justified, of having a risk management system adopted as part of the marketing authorisation to identify, characterise, prevent or minimise risks related to ATMPs.

It can be argued that the same could be achieved for NBCDs without the need for a revised regulation but by drafting new standards in the form of scientific guidance since the existing legislation already allows the needed flexibility both in terms of product information and post authorisation safety or efficacy studies or risk management systems.

Articles 16 and 19 provide incentives for the development and submission of ATMPs marketing authorisation applications by reducing the fee required for scientific advice and regulatory submissions.

Here again, the EMA has already created the ITF to assist companies in the development of emerging therapies and technologies. ITF briefing meetings are free of charge and are intended to facilitate informal exchange of information and the provision of guidance early in the development process. Meetings with EMA's ITF are also available for NBCDs.

On the 1st of April 2014 the European Commission published a report on the application of the ATMP regulation[48]. The report reflects on several elements that exemplify the benefits of implementing legislation for a class of specific medicinal products and the lessons learned from regulating such complex and diverse products. The following lessons learned from the application of the ATMP regulation are very pertinent if the time comes to discuss a specific legal framework for NBCDs:

- Empowering the European Commission to adopt specific requirements on the content of marketing authorisation applications, good manufacturing practice, good clinical practice and traceability of ATMPs;
- The ATMPs introduced to the EU market prior and post implementation of the ATMP regulation are subject to market surveillance;
- The lack of harmonisation of the conditions required by Member States for the application of the derogatory provisions in the regulation resulted in competitive market disadvantages;
- Understanding the importance of a clear definition for the different classes of ATMPs and recognising that those definitions require continuous review due to rapid scientific progress;

[48] REPORT FROM THE COMMISSION TO THE EUROPEAN PARLIAMENT AND THE COUNCIL in accordance with Article 25 of Regulation (EC) No 1394/2007 of the European Parliament and of the Council on advanced therapy medicinal products and amending Directive 2001/83/EC and Regulation (EC) No 726/2004, http://ec.europa.eu/health/files/advtherapies/2014_atmp/atmp_en.pdf.

- Avoiding difficulty in determining which regulatory regime would be applicable (e.g. medicines, medical devices or cosmetics) given the complex nature and mode of action of the products;
- Recognising the need to introduce flexibility to the requirements for ATMP's marketing authorisation submissions, particularly regarding quality;
- Managing the complexity of regulatory procedures resulting from the addition of an additional scientific committee.

Conclusion

Different issues related to NBCDs have been discussed in this chapter, such as the lack of a legal definition for NBCDs and the resulting uncertainty as to which requirements would be applicable (medicinal products, medical devices, combination products); and the need for more NBCD specific guidance on quality, safety, efficacy, pharmacovigilance and environmental analysis.

To date NBCDs have been legally addressed as conventional non-biological medicines with the necessary adjustments to the content of their marketing authorisation dossiers but without specific regulatory provisions. It cannot be denied that many NBCDs have been authorised and proven to be clinically successful under this framework, corroborating the ability of the existing framework to address the specificities of NBCDs. Stakeholders question the suitability of the existing regulatory platform to deal with NBCDs currently under development and follow-ons of existing NBCDs. A discussion on the necessity to revise scientific guidance and the legislative framework cannot be ignored. In order to understand the regulatory requirements of NBCDs, it is prudent to wait until more scientific data is gathered and more regulatory experience is gained (e.g. through scientific advice or marketing authorisation applications) so it can be concluded if the right way forward is to continue adapting the existing framework to new technologies by revising existing guidance or if new specific legislation should be considered to cater for the specificity of NBCDs, similar to what was done for ATMPs. Either way, in their mission to increase the certainty and clarity of NBCDs regulation, competent authorities need to strike the right balance between protecting public health, promoting access of new more effective medicines and not hindering the development of the pharmaceutical industry or trade of medicinal products in the EU.

Conflict of Interest European Medicines Agency (EMA), London, UK

The views expressed in this chapter are the personal views of the author and may not be understood or quoted as being made on behalf of or reflecting the position of the European Medicines Agency or one of its committees or working parties.

References

Bawa R (2008) Nanoparticle-based therapeutics in humans: a survey. Nanotech. L. Bus 5(2):135–155

Borchard G, Flühmann B, Mühlebach S (2012) Nanoparticle iron medicinal products–Requirements for approval of intended copies of non-biological complex drugs (NBCD) and the importance of clinical comparative studies. Regul Toxicol Pharmacol 64(2):324–328

Chowdhury N (2010) Regulation of nanomedicines in the EU: distilling lessons from the pediatric and the advanced therapy medicinal products approaches. Nanomedicine 5(1):135–142

Dorbeck-Jung BR, Chowdhury N (2011) Is the european medical products authorisation regulation equipped to cope with the challenges of nanomedicines? Law Policy 33(2):276–303

Ehmann F, Sakai-Kato K, Duncan R, Pérez de la Ossa DH, Pita R, Vidal J-M, Kohli A, Tothfalusi L, Sanh A, Tinton S (2013) Next-generation nanomedicines and nanosimilars: EU regulators' initiatives relating to the development and evaluation of nanomedicines. Nanomedicine 8(5):849–856

Feick J (2005) Learning and interest accommodation in policy and institutional change: EC risk regulation in the pharmaceuticals sector. Centre for analysis of risk and regulation, London School of Economics and Political Science

Lebourgeois C (2008) Device and drug combination products: a new regulatory, reimbursement and marketing challenge lifescience online. www.lifescience-online.com/articles.html?portalPage=Lifescience+Today.Articles&a=1014. Accessed 17 August 2009

Li J, Nation RL, Turnidge JD, Milne RW, Coulthard K, Rayner CR, Paterson DL (2006) Colistin: the re-emerging antibiotic for multidrug-resistant Gram-negative bacterial infections. Lancet Infect Dis 6(9):589–601

Parvizi N, Woods K (2014) Regulation of medicines and medical devices: contrasts and similarities. Clin Med 14(1):6–12

Schellekens H, Stegemann S, Weinstein V, de Vlieger JS, Flühmann B, Mühlebach S, Gaspar R, Shah VP, Crommelin DJ (2014) How to regulate nonbiological complex drugs (NBCD) and their follow-on versions: points to consider. AAPS J 16(1):15–21

Vamvakas S, Martinalbo J, Pita R, Isaac M (2011) On the edge of new technologies (advanced therapies, nanomedicines). Drug Discov Today: Technol 8(1). doi:10.1016/j.ddtec.2011.04.001

Epilogue: What Did We Learn? What Can We Expect in the Future? Concluding Remarks and Outstanding Issues

Daan J. A. Crommelin and Jon S. B. de Vlieger

Contents

Similarity Approach for Follow on Versions: No Reinvention of the Wheel 376
Terminology 377
NBCDs: Other Product Families? 378
Fact Finding: Performance of Follow on and Innovator's NBCD Products 379
Global Bias 380
Analytical Challenges 380
Educational Activities 381
References 381

Abstract At the end of this book, it is the place to draw a number of key conclusions. Information is provided and opinions are expressed in the 11 chapters describing the basic concepts of NBCDs, different NBCD and closely related product families, and analytical toolboxes for NBCDs. Going through the contents one can come up with a number of observations and conclusions. In this epilogue: 'What did we learn? What can we expect in the future? Concluding remarks and outstanding issues' we will discuss some of these observations and provide (a beginning of) some answers.

Keywords Similarity approach · Interchangeability · Substitution · Therapeutic equivalence · Performance of follow-on versions · NBCD working group · Science based discussions

D. J. A. Crommelin (✉)
Department of Pharmaceutics, Utrecht University, Utrecht, The Netherlands
e-mail: d.j.a.crommelin@uu.nl

J. S. B. de Vlieger
NBCD Working Group, Top Institute Pharma, Leiden, The Netherlands

D. J. A. Crommelin, J. S. B. de Vlieger (eds.), *Non-Biological Complex Drugs,* AAPS Advances in the Pharmaceutical Sciences Series 20, DOI 10.1007/978-3-319-16241-6_12

Similarity Approach for Follow on Versions: No Reinvention of the Wheel

One aspect of NBCDs received ample attention in the book: the development of follow on versions of NBCDs. In the Introduction we proposed to have a good look at the experience obtained with the approval process for biosimilars and to develop a 'similar approach' for NBCDs (cf. Fig. 1 as used in the Introduction), but now without a question mark). It would be worthwhile to see how much of the biosimilar experience obtained in the last decade would be applicable to the NBCD paradigm. Indeed, at several places in the book the authors concur with the idea to use the biosimilar approach as a guiding principle for the introduction of follow on products whilst applying a case-by-case or product/class specific approach when necessary. Why reinvent the wheel? The 'totality of evidence approach/stepwise (physico-chemical, non-clinical and clinical data collection) approach' as adopted by the FDA and EMA to build a successful dossier looks a very logical pathway.

Taking this parallel with biosimilars a step further would imply that the only pathway for approval in Europe has to be the 'centralized procedure', closing the door for the 'mutual recognition' pathway, which was even-recently used to introduce a follow on version of iron sucrose complexes in Sweden (Rechon Life Sciences 2014). Moreover, this would imply that in the EU or the USA the rules for interchangeability for biosimilars and NBCDs would be the same. That means that in the EU countries the decision on interchangeability would be taken by the national competent authorities.

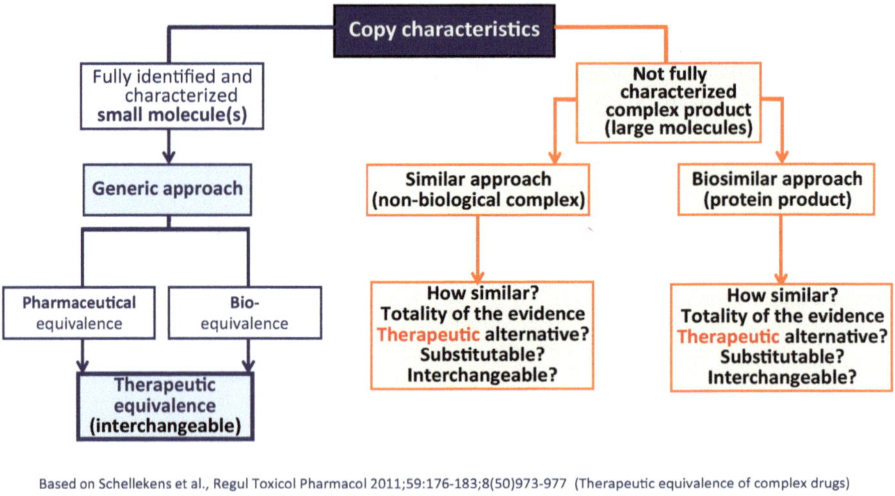

Fig. 1 Similarity approach for complex drugs

Terminology

When discussing biosimilars and NBCDs there is a constant struggle with the exact meaning of the terms used. FDA, EMA and WHO have issued their definitions for a number of terms, such as biologicals, but these definitions differ and e.g. include/exclude different groups of medicinal products. Some time ago members of the NBCD steering committee (see Introduction) published a paper (Crommelin et al. 2014) where the terminology pertaining to biologicals and NBCDs used in different parts of the world was discussed and where a proposal for standardization was made for the terms: biological product, non-biological complex drug product, therapeutic equivalence, interchangeability, substitution, switchability, traceability and extrapolation (see below).

Biological product
A biological product is a product derived from living material (such as cells or tissues) used to treat or cure disease. Biological products include a wide range of products such as vaccines, blood and blood components, allergenics, somatic cells, gene therapeutics, tissues, and recombinant therapeutic proteins. Biological products can be composed of sugars, proteins, or nucleic acids or complex combinations of these substances, or may be living entities such as cells and tissues. Biological products are isolated from a variety of natural sources—human, animal, or microorganism based—and may be produced by biotechnology methods.

Biosimilar product/similar biological medicinal products
A similar biological medicinal product (also known as biosimilar) is a biological product authorized by an abbreviated regulatory pathway requiring similarity to an already licensed biological product (the reference product) in physico-chemical, in vitro and in vivo biological characteristics, and clinical data showing similarity in efficacy, safety and immunogenicity.

Non-Biological Complex Drugs (NBCD) products
A medicinal product, not being a biological medicine, where the active substance is not a homo-molecular structure, but consists of different (closely related) structures that can't be isolated and fully quantitated, characterized and described by (physico-)chemical analytical means. Their manufacturing process is complex and needs to be carefully controlled to assure reproducibility. Examples of NBCD are, amongst others, liposomes, iron –carbohydrate ('iron-sugar') drugs and glatiramoids.

Therapeutic equivalent
Two medicinal products are therapeutically equivalent if they are pharmaceutically equivalent or pharmaceutical alternatives and if their bioavailability after administration in the same molar dose are similar to such a degree that their effects, with respect to both efficacy and safety, will be essentially the same. This is considered demonstrated if the 90 % confidence intervals (90 % CI) of the ratios for log AUC0-t and Cmax between the two preparations lie in the range 80.00–125.00 %. Pharmaceutical equivalence implies the same amount of the same active substance(s), in the same dosage form, for the same route of administration and meeting the same or comparable standards.
Drug products classified as therapeutically equivalent can be interchanged with the full expectation that the substituted product will produce the same clinical effect and safety profile as the prescribed product.

Interchangeability
Therapeutic equivalence of two different products enables the products to be interchanged. Interchangeability can be at the population level meaning both products can be used for

treatment for the same condition in the same population. Interchangeability at the individual level means that in an individual patient the products can be alternated or switched. Interchangeability at the individual level is a condition for substitution.

Substitution
A policy to allow replacement at the individual level of a medicinal product for a similar/bioequivalent product without the prescriber's consent.

Switchability
Changing the product (e.g. from reference product to biosimilar or vice versa) in a patient during the course of treatment.
Traceability
The ability to trace each individual unit of a medicinal product from the source to its final destination, and vice versa.

Extrapolation
The possibility to use the clinical data showing safety and efficacy in one indication (reference indication) to claim safety and efficacy in other indications. Extrapolation concerns the extrapolation of four different aspects: efficacy, safety, immunogenicity and interchangeability and may concern the indication, population or both.
(from Crommelin et al. 2014)

The global community of pharmaceutical scientists and health care professionals is still far away from using this/such a standardized terminology when discussing the development and use of complex drugs. Again and again this leads to confusion, unnecessary extra work and loss of time and money. As we state in the conclusion of the Crommelin et al. paper (2014): 'The pharmaceutical "rules of engagement" are becoming more and more global in character. Common, accepted terminology is a first requirement for global harmonization of regulatory rules and actions. It is critically important for authorities, health care professionals, scientific experts, and patients to have one unified terminology to guarantee a consistent quality and use of follow on versions of complex innovator products'.

NBCDs: Other Product Families?

In Chapters 2–6 a number of NBCD families were discussed in depth and the challenges for the different stakeholders, i.e. industry scientists, policy makers, regulatory scientists and users were outlined. But, is this an exhaustive list? The answer is no. There are other groups of complex drugs that fall under the definition of NBCDs (cf. Introduction) and could join the activities of the NBCD working group. Examples are: (1) oral bioactive polymers such as sevelamers, other complex phosphate binders; (2) parenteral lipid emulsions and (3) dry powder inhaler drugs.

Re 1) Sevelamer hydrochloride and sevelamer carbonate are cross-linked polymers that are taken orally to treat hyperphosphatemia in dialysis and chronic kidney disease patients. Sevelamer binds phosphate liberated during the digestive process, sequestering it in the polymer before it can be absorbed into the body, resulting in lower serum phosphorus concentrations. This mode of action is crucial in maintaining a clinically acceptable level of serum phosphorus in hyperphosphataemic patients. Due to the complex nature of sevelamer, however, the manner of demonstrating qualitative and quantitative equivalence as well as bioequivalence are different for sevelamer than for a conventional small molecule. Sevelamer is manufactured by chemically cross-linking a partially neutralized solution of poly(allylamine hydrochloride) with epichlorohydrin and subsequent washing of the resultant gel to remove impurities, followed by drying, particle size reduction and packaging. Differences in processing conditions are potentially of great importance, as they affect the physical and performance characteristics of the resultant hydrogel (report on file and Hudson et al. 2012).

Re 2) Parenteral lipid emulsions have been used for parenteral nutrition in the clinic for years. Intralipid is the prime example of a successful, innovator product in this category. But the search for safer products drives innovation and new and safer emulsions, e.g. based on fish oil rich in ω-3 polyunsaturated fatty acids, were developed (Hippalgaonkar et al. 2010). Moreover, parenteral lipid emulsions have been used as carriers of poorly water soluble drugs. Propofol is one of the best known examples. Looking at the definition of NBCDs, parenteral lipid emulsions as such and as carrier of drugs would fit very well in the NBCD group: they are non-biological complex drugs.

Re 3) Dry Powder Inhaler (DPI) products consist of a medical device to administer a bioactive to different sites in the lung. The construction of the device, selection of excipients, the design of the formulation and manufacturing of the drug product play a critical role in the clinical performance of these complex drug delivery systems. The regulatory guidance documents at the EMA and FDA site were developed over time. In 2009 the EMA published a document describing the requirements for generic versions of inhalers, including dry powder inhalers). In general, pharmacodynamic and/or clinical studies are requested to establish therapeutic equivalence (EMA 2009; Stegeman et al. 2013). Considering the above, DPIs fall under the definition of the NBCDs.

Fact Finding: Performance of Follow on and Innovator's NBCD Products

Do follow on versions of NBCDs behave differently from the innovator's product? A simple question with a not so simple answer. Which information sources provide valuable and which ones immaterial data? We propose only to use peer-reviewed/regulatory body literature sources. If that guiding principle is followed, it becomes clear that relevant factual material is scarce. These comparisons of a follow on/innovator product in publications can be at the level of physico-chemical characterization/in vitro/preclinical tests (e.g., Toblli et al. 2009, 2012, peer-reviewed publications mentioned in the Citizens Petition 2014), PK (CHMP assessment report Doxorubicin SUN (2011), or clinical comparisons (Rottembourg et al. 2011; Lee et al. 2013; Aguera et al. 2014; Citizens Petition 2014); Tzanno-Martins et al. 2014).

For biosimilars the term bioquestionables has been coined for follow on versions of innovator-biologicals that were not scrutinized by rigorous, well described, publicly accessible dedicated protocols as those described by the EMA or FDA and a number of other regulatory offices (e.g. Health Canada, Australian Therapeutic Goods Administration). Mutatis mutandis, for NBCD products such as the glatiramoids, doxorubicin or amphotericin liposomes and iron sucrose complexes the term '*NBCD*-questionables' may be used. In this book NBCD-questionables for GA, Doxil/Ambisome and the iron sucrose innovator product were identified/discussed. It is highly desirable that these products are tested and evaluated on the basis of the totality of evidence approach and that the results are published in peer-reviewed, publicly accessible journals; not only the positive but also the negative results.

Global Bias

The authors contributing to this book were selected from the USA and Europe. That doesn't mean that other parts of the world were deliberately excluded. Not at all. But, as a consequence, relevant websites and literature in languages other than the English, German and French language are difficult to be found and, even more difficult, to understand.

However, the issues and challenges discussed in this book are not restricted to the above language areas. It's clear that other regulatory systems, such as those used in Japan, other Asian countries, Latin America and China are facing the same challenges and also need and may have found ways to ensure the safety and efficacy of NBCD follow on products for patients.

We realize that we may have missed this information, introducing a certain bias, and hope that our call to publish in the English language will be heard and lead to peer-reviewed (translated) publications in this lingua franca of modern science to widen our audience and share the experience obtained with NBCDs.

Analytical Challenges

The lack of full characterization of the physico-chemical characteristics is a critical attribute of the NBCDs. Extrapolation of the advances in our analytical toolbox over the last decennia to the future is not possible. But, impressive advances have been made during these years, both on the 'hardware side', e.g. with (mass) spectroscopy, microscopy (AFM, cryoTEM), chromatography, light scattering, PAT, and on the formulation design side, e.g. the introduction of the QbD paradigm.

The use of orthogonal techniques has been advocated to provide more detailed physicochemical insights. For some NBCDs full characterization may be more close by (liposomes) than for other NBCD families (glatiramoids and iron carbohydrate complexes). Time will tell if and when liposomes will be taken off the NBCD list, although challenges will still exist to translate in vitro changes to in vivo performance.

A point that is brought up a number of times in the book (cf. Chapters "Liposomes: The Science and the Regulatory Landscape" and "NBCD Pharmacokinetics and Bioanalytical Methods to Measure Drug Release") are the challenges to develop validated protocols to follow the fate of NBCDs upon administration: i.e. (1) the concentration of free drug, and (2) the concentration of blood protein/cells associated, and (3) the carrier associated drug all have to be monitored. The present approaches to separate these fractions are—to say the least- cumbersome and center of criticism. Concerted actions by the scientific community should be encouraged. The NCL (Nanotechnology Characterization Laboratory, National Cancer Institute, NIH) is an institute where protocols for such detailed drug monitoring for nanoparticulate carriers are being developed. This important initiative is worth following—in one or another form—by colleagues in other parts of the world. In Europe, a similar initiative is currently under discussion and proposed by the European Technology Platform on Nanomedicine as part of the Horizon 2020 innovation program.

Educational Activities

This book is the first one on NBCDs. The science base for the development and use of these products and the regulatory positions are discussed at length. In the Introduction the target audience for this book is defined: firstly regulatory scientists in industry and regulatory bodies, secondly skilled professionals working in hospitals and, thirdly, those who work on the development of nanomedicines. But there are more stakeholders, such as health insurance companies, patient organizations and, last but not least, the patients themselves. For these groups of stakeholders, who are not reached through this book, alternative, target group specific, publications explaining the relevant specifics of NBCDs should be prepared. The NBCD working group (see Introduction) is committed to provide a platform for science-based discussion on NBCDs, and to reach out to all stakeholders including regulatory authorities, industry, academia and health care providers to jointly ensure patient benefit and safety.

References

Aguera ML, Martin-Malo A, Álvares de Lara MA et al (2014) The treatment with generic iv iron needs a higher dose of iron and ESA to keep hemoglobin stable. Nefrologia 34(S1):abstr 319

CHMP assessment report Doxorubicin SUN (2011) http://www.ema.europa.eu/docs/en_GB/document_library/Application_withdrawal_assessment_report/human/002049/WC500112957.pdf. Accessed 27 April 2015

Citizens petition (2014) Docket No. FDA 2014-P-0933. http://www.regulations.gov/#!documentDetail;D=FDA-2014-P-0933-0001. Accessed 10 Sept 2014

Crommelin DJA, de Vlieger JSB, Weinstein V, Mühlebach S, Shah VP, Schellekens H (2014) Different pharmaceutical products need similar terminology. AAPS J 16:11–14

EMA (2009) Guideline on the requirements for clinical documentation for orally inhaled products (oip) including the requirements for demonstration of therapeutic equivalence between two inhaled products for use in the treatment of asthma and chronic obstructive pulmonary disease (copd) in adults and for use in the treatment of asthma in children and adolescents. http://www.ema.europa.eu/docs/en_GB/document_library/Scientific_guideline/2009/09/WC500003504.pdf
Hippalgaonkar K, Majumdar S, Kansara V (2010) Injectable lipid emulsions—advancements, opportunities and challenges. AAPS Pharm Sci Tech 11:1526–1540
Hudson SP, Owens E, Hughes H, McLoughlin P (2012) Enhancement and restriction of chain motion in polymer networks. Int J Pharm 430:34–41
Lee ES, Park BR, Kim JS, Choi GY, Lee JJ, Lee IS (2013) Comparison of adverse event profile of intravenous iron sucrose and iron sucrose similar in postpartum and gynecologic operative patients. Curr Med Res Opin 29:141–147
Rechon Life Science (2014) http://www.rechon.com/products/rechon-products/
Rottembourg J, Kadri A, Leonard E, Dansaert A, Lafuma A (2011) Do two intravenous iron sucrose preparations have the same efficacy? Nephrol Dial Transplant 26:3262–3267
Stegemann S, Kopp S, Borchard G, Shah VP, Senel S, Dubey R, Urbanetz N, Cittero M, Schoubben A, Hippchen C, Cade D, Fuglsangm A, Morais J, Borgström L, Farshi P, Seyfang KH, Hermann R, van de Putte A, Klebovich I, Hinca l A (2013) Developing and advancing dry powder inhalation towards enhanced therapeutics. Eur J Pharm Sci 48:181–194
Toblli JE, Cao G, Oliveri L, Angerosa M (2009) Differences between original intravenous iron sucrose and iron sucrose similar preparations. Arzneimittelforschung 59(4):176–190
Toblli JE, Cao G, Oliveri L, Angerosa M (2012) Comparison of oxidative stress and inflammation induced by different intravenous iron sucrose similar preparations in a rat model. Inflamm Allergy Drug Targets 11:66–78
Tzanno-Martins C, Biavo BM, Ferreira-Filho O, Ribeiro-Junior E, João-Luiz MV, Degaspari S, Scavone C, Kawamoto E (2014) Clinical efficacy, safety and anti-inflammatory activity of two sevelamer tablet forms in patients on low-flux hemodialysis. Int J Immunopathol Pharmacol 27:25–35

Index

A
Active pharmaceutical ingredient (API), 98
  and drug delivery vehicles, 198
  biodistribution and PK of, 347
  federal agencies approval, 196
  generic complex drugs, evaluation of, 348
Adverse events, 346
  due to heparin therapy, 305
Albumin, 337
  abraxane, 6
  and paclitaxel, 341
  based nanomedicines, 6
  based nanoparticles, 274
  binding sites of, 339
  Cremophor-free, 338
  endogenous transport mechanisms of, 338
  molecular changes of, 341
  nanoparticle albumin-bound (*nab*™) technology, 337
  nanoparticulate, 266
  role in nab-paclitaxel nanoparticles, 341
Ambisome™, 80, 100, 201
Anticoagulant, 294, 295, 322
  action and relationship of, 298
  activity of, 307, 314, 323

B
Bioanalytical methods, 276, 281
  for estimation of NBCD drug release, 276
Biodisposition and in vivo performance, 156, 160
Bioequivalence, 95, 163, 165, 264, 349, 375
  assessment, 266, 279
Biological activity
  measure of, 307
Biosimilar, 320, 338, 349, 370
  side-effects, 321
Block copolymer, 349

C
Caelyx™, 80
Capillary electrophoresis (CE), 306
Characterization, 312
  for liposome, 102
  physical, 255
  physico-chemical, 4, 87, 91, 385, 386
Colloidal suspension, 205, 250
  of nanoparticles, 337
Critical micelle concentration (CMC), 251, 344

D
Doxil™, 80
Drug distribution, 340, 348, 349
Drug product, 167, 239, 350
  excipients of, 196
  of liposome, 84, 86, 98
Drug release methods, 264, 267
  estimation of NBCD, 276, 277
  measurement of, 267, 272
  process-induced, 275
Drug vehicle, 255
Dynamic light scattering (DLS), 87, 199, 255, 340
  protein of, 207
  technique of, 246

E
Efficacy
  non-clinical, 266
Safety of MPs, 162, 163
Enhanced permeability and retention (EPR), 91, 92, 342
Equilibrium dialysis, 273, 275
EU legislation, 359, 367
European medicines agency (EMA), 2, 196, 266, 321, 349, 360

D. J. A. Crommelin, J. S. B. de Vlieger (eds.), *Non-Biological Complex Drugs,* AAPS Advances in the Pharmaceutical Sciences Series 20, DOI 10.1007/978-3-319-16241-6

F

Food and drug administration (FDA), 2, 90, 196, 266
Formulation, 270, 340, 349
  absence of, 274
  amphotericin-lipid, 100
  definition of, 198
  nab-paclitaxel, 342
  of liposome, 96
  pro-drug, 156

G

Generic, 264, 321
  development of, 96
  evaluation of, 348
Generic/follow-on liposomes, 3
Glatiramoids, 5, 375
Glycosaminoglycan (GAG), 294, 296, 322, 324

H

Heparin, 296, 305, 307
  analysis of, 308
  depolymerization of, 294, 296, 303, 304, 315, 321
  in mammalian cells, 324, 325
  of unfractionated, 303
  pentasaccharide, 300, 322
  pentasaccharide-containing, 313
  synthesis of, 322–324
  synthetic analogs of, 322

I

Innate immune system, 95, 160
Interchangeability, 383
International conference on harmonization (ICH), 97, 164
IV iron carbohydrates, 164

L

Large-Diameter tail (LDT), 205
Liposomes, 79, 97
Liquid-liquid extraction (LLE), 267, 272
Low molecular weight heparins (LMWHs), 294–296, 303, 307, 313, 315
  analysis of, 308
  depolymerization of, 303, 304
  safety of, 306

M

Manufacturing, 86, 302, 340
  of nab-paclitaxel, 338
Marketing authorization, 320, 359, 360
  legal basis of submission, 366
Mechanism of action, 341, 342
Metabolite pharmacokinetics, 276
  analysis of, 279, 280
Modeling and simulation, 277, 278, 281

N

Nano-colloidal, 162
  iron-based, 375
Nanomedicine, 102, 166, 358
Nanoparticle, 206, 337
  properties, 338
NBCD working group, 5, 384, 387
Non-biological complex drugs (NBCDs), 5, 166, 197, 205, 294, 337, 347
  category of, 256
  definition of, 5–7, 80, 384
  legislating, 361
  regulatory position of, 6

P

Paclitaxel, 338
Parenteral dispersions, 198
Performance of follow-on versions, 155
Pharmacokinetics, 315, 343
  of liposomes, 84
Pharmacopoeial methods, 199
Polydispersity, 207, 248, 251, 295, 307
  degree of, 207, 241, 264

R

Regulatory guidance, 349, 350

S

Safety, 345
  of pharmaceutical dosage forms, 197, 198
Similarity approach, 382, 383
Size exclusion chromatography, 268, 270
Solid phase extraction, 271
Stability-indicating methods, 198
Substitution, 384

T

Therapeutic equivalence, 264, 350

U

Ultracentrifugation, 275
Ultrafiltration, 274
Unfractionated heparin, 296

MIX
Papier aus verantwortungsvollen Quellen
Paper from responsible sources
FSC® C105338

If you have any concerns about our products,
you can contact us on
ProductSafety@springernature.com

In case Publisher is established outside the EU,
the EU authorized representative is:
Springer Nature Customer Service Center GmbH
Europaplatz 3, 69115 Heidelberg, Germany

Printed by Libri Plureos GmbH
in Hamburg, Germany